Wheat Gluten

The two 'fathers' of wheat protein chemistry.

Jacopo Bartholomew Beccari (top) was Professor of Chemistry at the University of Bologna when he described the isolation of gluten in 1745.

Thomas Burr Osborne (bottom) worked at the Connecticut Agricultural Experiment Station from 1886 until 1928 publishing studies of seed proteins from 32 plant species including wheat (oil portrait provided by the Connecticut Agricultural Experiment Station)

Wheat Gluten

Edited by

Peter R. Shewry
University of Bristol, UK

Arthur S. Tatham
University of Bristol, UK

RS•C
ROYAL SOCIETY OF CHEMISTRY

The proceedings of the 7th International Workshop Gluten 2000 held at the University of Bristol on 2–6 April 2000.

The front cover shows a molecular model of a β-spiral structure based on the repetitive domain of a high molecular weight subunit of gluten. The figure was kindly supplied by Dr. David Osguthorpe, University of Bath.

Special Publication No. 261

ISBN 0-85404-865-0

A catalogue record for this book is available from the British Library

Published by The Royal Society of Chemistry,
Thomas Graham House, Science Park, Milton Road,
Cambridge CB4 0WF, UK

Printed by MPG Books Ltd, Bodmin, Cornwall, UK

Preface

Over 600 million tonnes of wheat are grown in the world each year, making it the single most important crop. Much of this is consumed by humans, almost exclusively after processing into bread, pasta and noodles (made with durum or bread wheats, respectively) or a range of other foods. Bread, in particular, occurs in a vast range of forms in different cultures. The ability of wheat flour to be processed into these foods is largely determined by the gluten proteins, which confer unique visco-elastic properties to doughs. It is not surprising, therefore, that gluten proteins have been the subject of intensive study for a period exceeding 250 years. This has revealed that gluten proteins have unusual structures and properties, making them of interest for basic studies as well as more applied work on their functional properties. As a consequence the series of Wheat Gluten Workshops, which were initiated in Nantes, France, in 1980, have attracted a wide range of participants from academia, government laboratories and industry.

This volume contains papers based on presentations made at the 7th Workshop, which was held in Bristol, UK, from 2 to 6 April 2000. The topics range from genetics through structural and functional studies to genetic engineering providing a unique snapshot of the current status of research and indicating exciting opportunities for future work.

We would like to take this opportunity to thank fellow members of the organising committee, Professor Peter Frazier (Cambridge, UK) Professor David Schofield (University of Reading, UK) Dr Peter Payne (PBI Cambridge, UK) and Mr Harry Anderson (IACR-Long Ashton) for putting together an excellent scientific programme and Mrs Christine Cooke, Mrs Pat Baldwin (IACR-Long Ashton) for assistance with organisation of the meeting. Finally, we are greatly indebted to Mrs Valerie Topps and Mrs Sue Richens (IACR-Long Ashton) for assistance with preparation of this volume.

P. R. Shewry
A. S. Tatham

Contents

Disulphide Bonds and Redox Reactions

Non-Gluten Components

Genetics and Quality Considerations

THE GENETICS OF WHEAT GLUTEN PROTEINS: AN OVERVIEW

Lafiandra D.[1], Masci S.[1], D'Ovidio R.[1], Margiotta B.[2]

1. Dept. of Agrobiology & Agrochemistry, University of Tuscia, via S.C. De Lellis, 01100 Viterbo, Italy. 2. Germplasm Institute, C.N.R. , via Amendola 165/a, 70126 Bari, Italy.

1 INTRODUCTION

The end use characteristics of flours produced from bread wheat or semolinas obtained from durum wheat are strongly determined by the gluten proteins and their covalent and non-covalent interactions. This complex mixture of proteins has been shown to consist of two types of protein: the monomeric gliadins and the polymeric glutenins; the first group includes proteins subdivided into α-, γ- and ω–types according to their *N*-terminal amino acid sequences, where disulphide bonds, if present, are intramolecular. The glutenin fraction is formed of a mixture of polymers with a wide size distribution, ranging from dimers to polymers with molecular weights into the millions[1-3]. The constituent subunits, termed high- and low-M_r glutenin subunits, are linked through intermolecular disulphide bonds. Low-M_r glutenin subunits have been shown to be very heterogeneous in size and charge and have been subdivided into B, C and D groups according to their biochemical characteristics. On the basis of their *N*-terminal sequences, the B group includes two major classes, termed LMW-s and LMW-m which start with the amino acids serine and methionine, respectively. The C group includes mainly subunits with sequences homologous to γ- and α-gliadins, whereas the D group subunits show *N*-terminal sequences homologous to the ω-gliadins[4-6].

It has been demonstrated that variation in the types and amounts of high- and low-M_r glutenin subunits correlate with dough rheological properties by affecting the molecular weight distribution of the glutenin polymers[7-9], but the detailed structure and composition of the glutenin polymer is still unclear. Different hypothetical models have been proposed[1,10,11]. Kasarda[1], based on the hypothesis of Ewart[10], suggested that glutenin polymer is formed of different subunits, randomly linked through disulphide bonds in a linear fashion, and that polymer size is modulated by the incorporation of chain extender or chain terminator subunits, the former having two or more cysteine residue available for intermolecular disulphide bonds and the latter a single cysteine.

The acquired genetical knowledge of the gluten components together with the identification and production of interesting genetic material has already proved useful in clarifying some of the factors affecting protein composition-quality relationships and in directing breeding strategies.

The availability of different aneuploid stocks, together with the refinement of electrophoretic techniques, has led to the complete assignment of genes encoding gluten proteins. The *Gli-1* and *Gli-2* loci, present on the short arms of the homoeologous group 1 and 6 chromosomes, contain tightly associated clusters of genes corresponding to ω-, γ- and α-gliadins[12].

Six genes have been identified in hexaploid wheat corresponding to high-M_r glutenin subunits, two on each of the long arms of the homoeologous group one chromosomes (*Glu-1* loci). These encode an x-type subunit of higher M_r and a y-type subunit of lower M_r.

These subunit types differ also in structural characteristics such as the number of cysteine residues and the size and composition of the repetitive domain[12,13]. The y-type gene present at the *Glu-A1* locus is always silent in tetraploid and hexaploid cultivated wheats, whereas the x-type gene at the same locus and the y-type gene at the *Glu-B1* locus are expressed only in some cultivars; this leads to variation in the number of subunits from three to five in bread wheat and from two to three in durum wheat.

Low-M_r glutenin subunits are encoded by multigene families located on the homoeologous group 1 chromosomes at the *Glu-3* loci which are tightly linked to the *Gli-1* loci[14]. In bread wheat the low M_r subunit gene family is represented by 30-40 members as estimated by Southern blot analysis[15,16]. The existence of additional loci corresponding to low-M_r glutenin subunits on group 7 chromosomes has recently been reported[17,18]. Genetical analyses carried out in the past few years have indicated the complexity of organisation of gene loci on the short arms of the homoeologous group 1 chromosomes with the presence of minor dispersed loci corresponding to gliadin and low-M_r glutenin subunit proteins (see Shepherd[19] for a more detailed description).

2 HIGH M_r SUBUNITS

The high-M_r glutenin subunits have been shown to be critical components in determining gluten viscoelastic properties, therefore their study has attracted the interest of many different research groups. The availability of different genetic materials (aneuploids, isogenic lines, recombinant inbred lines, biotypes, etc) has been very useful in addressing composition-functionality relationships.

After the extensive studies of Payne and coworkers[12] which demonstrated that breadmaking properties are strongly influenced by allelic variation existing at each of the different *Glu-1* loci, further studies have contributed to unravelling the critical role of these components. Another major breakthrough in establishing the role of high-M_r glutenin subunits, was the detection of null alleles at each of the three *Glu-1* loci and the consequent creation of bread wheat lines with a variable number of subunits from two to five[20] or from zero to five[21]. These lines allowed the role of a number of subunits in influencing gluten viscoelastic properties to be assessed. Subsequently, experiments carried out by Popineau *et al.*[8] and Gupta *et al.*[9] on the same materials demonstrated a direct effect of the removal of the high-M_r glutenin subunits in reducing the size distribution of large glutenin polymers.

After the establishment of the relative importance of certain subunits compared to others, the molecular mechanisms by which certain allelic subunits confer superior dough properties has been a matter of intensive investigations. Qualitative effects can be related to differences in the amounts of subunits produced by the different alleles or resulting from differences in their structure which can in turn affect their ability to form polymers with other high- or low-M_r subunits. Structural differences can also be important in determining non-covalent interactions such as hydrogen bonding between glutamine residues which can stabilise certain conformations of gluten proteins and influence protein-protein interactions. Results of studies carried out over the past few years have produced convincing evidence that the number and arrangement of cysteine residues and the length of the repetitive domain play a major role in this respect. Studies by Gupta *et al.*[22] on recombinant inbred lines or biotypes differing in allelic composition at the *Glu-B1* (17+18 vs 20) or at the *Glu-D1* loci (5+10 vs 2+12) demonstrated that the superiority of the subunit pairs 17+18 or 5+10 was associated with the production of larger amounts of large-sized glutenin polymers. No quantitative differences were found between each pair of allelic combinations tested, the most striking difference being the presence of an extra cysteine residue in subunit 5 compared to subunit 2 and the presence of only two cysteine residues in subunit 20[23,24] compared to the four present in subunit 17. Similar results have been obtained by Margiotta *et al.* (these proceedings), who, by means of isogenic lines, have compared the effects of a novel 1Bx subunit with two cysteine residues versus subunit 7 possessing four cysteine residues. The results of this study have indicated that a lower amount of large glutenin polymers and poor mixograph parameters are associated with the wheat line carrying the novel 1Bx subunit.

The possibility offered by molecular biology to engineer proteins with desired structural characteristics combined with small scale functionality testing have become powerful tools for exploring structure-function relationships of high-M_r glutenin subunits and gluten protein in general[25]. Using these *in vitro* approaches it has been possible to demonstrate that an increase in the length of the repetitive domain results in stronger dough mixing properties. This is probably the result of more extensive hydrogen bonds, formed by the glutamine residues present in the repetitive domains, within and between subunits, which have been suggested to influence dough rheological properties[26,27].

A different way to approach these points is the development of proper genetic material. Near isogenic lines, developed by crossing a donor genotype carrying a given allele of interest to a recipient variety, have been already used in wheat quality studies[20,28]. Although their production requires time and effort, these materials are extremely valuable for directly comparing the effects due to the change of a single subunit and clarifying some of the aspects described. Recently, a new set of isogenic lines, differing in number of high-M_r glutenin subunits (from three up to six) or containing subunits modified in the size of their repetitive domain, have been obtained using the Italian bread wheat cultivar Pegaso as recipient[29]. These will be used to further assess the role of the number of subunits and the size of the repetitive domain.

Wheat flour has many different end uses besides bread and the potential to explore these, by manipulating high-M_r glutenin subunits, was demonstrated by Payne and Seekings[30]. These researchers have produced isogenic lines in the bread wheat cultivar Galahad with single high-M_r glutenin subunits which have proved to be very extensible and suitable for biscuit production. A new set of near isogenic bread wheat lines with single subunits is reported in Fig.1. This material allows the previous observations on Galahad to be extended and will make it possible to obtain lines with unusual high M_r-glutenin subunit composition (e.g. a combination of only x- or y- type subunits) which will be used to provide additional information on glutenin polymer structure. In fact, it has been hypothesised that disulphide linkages between x- or y-type high M_r glutenin subunits are an important feature of glutenin polymers. Lack of recombination between x- and y-type genes present at the same *Glu-1* locus has not allowed us to establish which type of subunit is more effective in affecting dough properties. The genetic material being generated should prove useful in assessing the role and importance of different types of subunit in glutenin polymer structure.

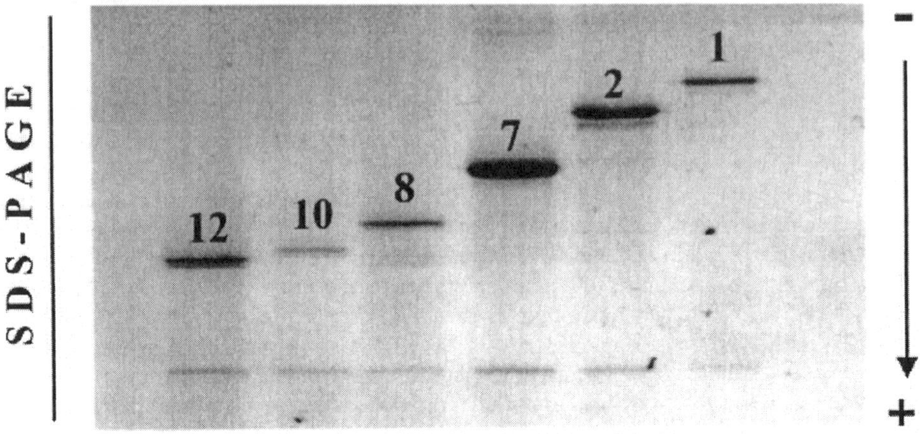

Figure 1 *SDS-PAGE of bread wheat lines with single x- or y-type high-*M_r *glutenin subunit.*

The development of wheat transformation protocols has offered another approach for the manipulation of high-M_r glutenin subunit composition in both durum and bread wheat[31-33]. The results are still far from being directly applicable in wheat breeding, but offer a new way to manipulate gluten composition. An example of a durum wheat line with the 1Dx5 subunit, recently obtained by He *et al.*[33] is shown in Fig.2, lane 7. Crossing can be performed and the transgene can be incorporated in lines with different compositions of high-M_r glutenin subunits resulting in the production of new useful material for quality studies.

In durum wheat the role of high-M_r glutenin subunit has not been firmly established . The limited genetic variation at the *Glu-B1* locus and the constant presence of the null allele at the *Glu-A1* locus has prevented conclusive assessment of the role of high-M_r glutenin subunits in this species. Proper genetic material is being developed and durum wheat lines differing in number and type of high-M_r glutenin subunits are being produced (Fig. 2). Particularly interesting in this respect is a recently developed set of durum wheat lines, in which the D-genome related subunits have been introduced[34-36]. In particular, through chromosome engineering, chromosome segments carrying the pairs of subunits 5+10 or 2+12, encoded by genes present at the *Glu-D1* locus in bread wheat, have been transferred to chromosome 1A of durum wheat, replacing the *null* allele present at the *Glu-A1* locus (see Fig. 2, lanes 6, 9 and 10 from left). Quality studies on these materials have demonstrated that insertion of either pair of subunits results in a large increase in SDS-unextractable polymeric glutenin and a substantial increase in gluten strength[36] (see also Lafiandra *et al.*, these proceedings).

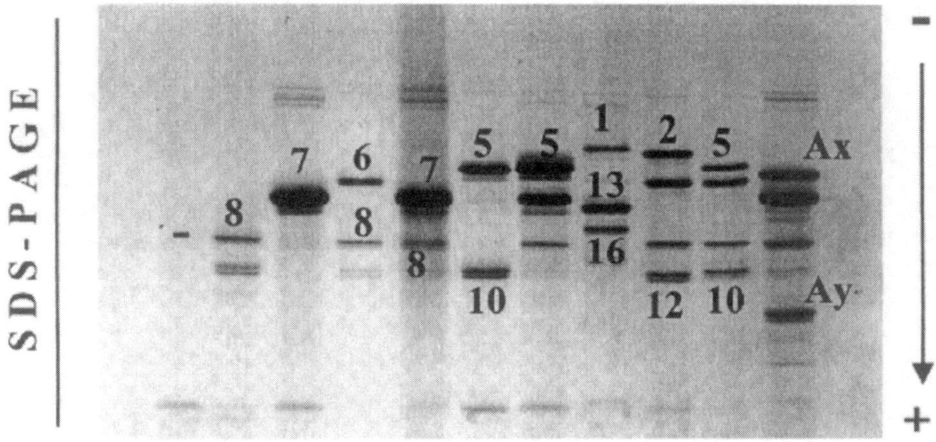

Figure 2 *SDS-PAGE of durum wheat lines differing in high-*M_r *glutenin subunit composition. Durum wheat lines possessing typical subunits (lanes 4, 5 and 8 from left), null lines (lane 1), with single y- (lane 2) or x-type subunits (lane 3) and with D-genome associated subunits are shown (lanes 6, 7, 9, and 10). A durum wheat line with both x- and y-type subunit at the* Glu-A1 *locus is also included (lane 11).*

3 LOW M_r SUBUNITS

Allelic variation in low-M_r glutenin subunits has been described[37-39] and the influence on dough properties both in durum and bread wheat has been reported[22,40,41], but detailed information on this class of gluten components is still scarce.

One of the reasons for limited information on this class of proteins results from their large number and similarity that render the identification and determination of the specific contribution to flour quality of single components very difficult. Only for two group of proteins, namely LMW-1 and LMW-2, has their relative contribution to durum wheat quality been demonstrated[40,41]. A similar situation is present at the DNA level, where a direct correlation between specific *lmw-gs* genes and their corresponding products is limited to the so-called 42 K low-M_r glutenin subunit present in the cultivar Yecora Rojo[42]. Recently, a similar result has been also obtained in durum wheat, where two allelic genes related to polypeptides belonging to the LMW-1 and LMW-2 groups have been characterised[43]. Sequence analysis of the 42K subunit and of the two allelic genes showed that the encoded proteins had similar characteristics and are very likely act as chain extenders[5] of the growing glutenin polymer. In fact, both genes possess eight cysteine codons located at corresponding positions. The first and the seventh cysteines should form inter-molecular disulphide bonds whereas the remaining cysteines should be involved in intra-molecular disulphide bonds (for review see Shewry and Tatham[44]). Moreover, the differences between the proteins encoded by the allelic *lmw-gs* genes, consisting of a few amino acid substitutions and the deletion of two hexapeptide repeats, seem by themselves to be insufficient to explain the different effects on quality observed between the two group of proteins. If the intrinsic structure of the allelic low-M_r glutenin subunits belonging to the LMW-1 and LMW-2 groups cannot completely explain their different contribution to the viscoelastic properties of wheat dough, then the differences in their relative amounts can account for their contrasting performance. In this regard, results on qualitative analyses of LMW-1 and LMW-2 demonstrated that the LMW-2 are present in a significantly greater amounts[45,46]. A deeper understanding of the role of specific subunits in the technological properties of dough could be obtained by extending the characterisation of the *lmw-gs* gene family. In particular, the increasing number of characterised *lmw-gs* genes from a single genotype, such as those reported for the bread wheat cultivar Cheyenne[16] and the durum wheat cultivar Langdon[47], should allow specific genes to be correlated with their encoded products. This information should provide the basis for a deeper analysis of structure/function relationships between different low-M_r glutenin subunits that subsequently should allow more accurate manipulation of gluten properties, perhaps through biotechological approaches.

As described above, the C and D groups of low-M_r glutenin subunits show sequences characteristic of monomeric gliadins. Some of these gliadin-type glutenin subunits have an odd number of cysteine residues, with one cysteine residue that is likely to participate in intermolecular disulphide bonds. It has, therefore, been suggested that they would act as chain terminators and prevent elongation of developing glutenin polymers[1,48]. As a consequence, a negative effect on dough strength, through a tendency to decrease the average molecular weight of the glutenin polymer, is expected. This has been proved to be the case for the 1D-coded D group of low M_r glutenin subunits[49,50]. Although the gliadin-type glutenin subunits make a significant contribution to the glutenin polymers[5], they have not been studied in detail. The isolation of mutants and the development of suitable genetic material will prove useful in elucidating their role in determining gluten functional properties.

Another important parameter which has been suggested to influence gluten viscoelastic properties is the glutenin to gliadin ratio, as it has been suggested that a change in this ratio toward higher values would result in stronger doughs[2]. Identification of null alleles at the *Gli-1* and *Gli-2* loci in both durum and bread wheats is offering the possibility to investigate these aspects more precisely, although some preliminary results are conflicting. In fact, Pogna *et al.*[51] have reported a remarkable negative effect on dough properties associated with the absence of both *Gli-A2* and *Gli-D2* controlled gliadin components, in contrast to the expected results[2]. Additional genetical lines are being generated in order to extend these studies.

4 CONCLUSIONS

Great progress has been made in understanding genetical aspects of gluten proteins in the past decades and in unraveling the molecular basis of gluten functionality. Research has demonstrated that effects exerted by the different high- and low-M_r glutenin subunits are related to differences in amounts and/or types of glutenin subunits produced by the different alleles[22]. This allows predictable manipulation of gluten components and improved breeding, which is resulting in the production of high quality durum and bread wheat cultivars. Production of new genetic stocks will benefit from novel approaches with the aim to further clarify the role of different gluten components and open the possibility of developing novel gluten functionality.

References

1. D.D. Kasarda, 'Wheat is unique', Y. Pomeranz, Ed. Am. Assoc. Cereal Chem., St. Paul, 1989, p. 277.
2. F. MacRitchie and D. Lafiandra, 'Food Proteins and their Applications', S. Damodaran, A. Paraf Eds, 1997, p. 293.
3. C.W. Wrigley, *Nature*, 1996, **381**, 738.
4. H.P. Tao and D.D. Kasarda, *J. Exp. Botany*, 1989, **40**, 1015.
5. E.J.L. Lew, D.D. Kuzmicky and D.D. Kasarda, *Cereal Chem.*, 1992, **69**, 508.
6. S.M. Masci, D. Lafiandra, E. Porceddu, E.J.-L. Lew, H.P. Tao and D.D. Kasarda, *Cereal Chem.*, 1993, **70**, 581.
7. R.B. Gupta, K. Khan and F. MacRitchie, *J. Cereal Sci.*, 1993, **18**, 23.
8. Y. Popineau, M. Cornec, J. Lefebvre and B. Marchylo. *J. Cereal Sci.*, 1994, **19**, 231.
9. R.B. Gupta, Y. Popineau, J. Lefebvre, M. Cornec, G.J. Lawrence and F. MacRitchie, *J. Cereal Sci.*, 1995, **21**, 103.
10. J.A.D. Ewart, *J. Sci. Food Agric.*, 1979, **30**, 482.
11. Graveland, P. Bosveld, W.J. Lichtendorf, J.P. Marseille, J.H.E. Moonen and A. Scheepstra, *J. Cereal Sci.*, 1985, **3**, 1.
12. P.I. Payne, *Ann. Rev. Plant Physiol.*, 1987, **38**, 141.
13. P.R. Shewry, N.G. Halford and A.S. Tatham, *J. Cereal Sci.*, 1992, **15**, 105.
14. N.K. Singh and K. W. Shepherd, *Theor. Appl. Genet.*, 1988, **75**, 628.
15. P. Sabelli and P.R. Shewry, *Theor. Appl. Genet.* 1991, **83**, 209.
16. B.G. Cassidy, J. Dvorak and O.D. Anderson, *Theor. Appl. Genet.* 1998, **96**, 743.
17. G. Sreeramulu and N.K. Singh, *Genome*, 1997, **40**, 41.
18. J. Dubcovsky, M. Echaide, S. Giancola, M. Rousset, M.C. Luo, L.R. Joppa and J. Dvorak, *Theor. Appl. Genet.*, 1997, **95**, 1169.
19. K.W. Shepherd, 'Proc. 6th Int. Gluten Workshop', C.W. Wrigley Ed., 1996, p. 8.
20. P.I. Payne, L.M. Holt, K. Harinder, D.P. McCartney and G.J. Lawrence, 'Proc 3rd Int. Gluten Workshop', R. Làsztity, F Bèkès Eds. 1987, p. 216.
21. G.J. Lawrence, F. MacRitchie and C.W. Wrigley, *J. Cereal Sci.,* 1988, **7**, 109.
22. R. B. Gupta and F. MacRitchie, *J. Cereal Sci.*, 1994, **19**, 19.
23. A.S. Tatham, J.M. Field, J.N. Keen, P.J. Jackson and P.R. Shewry, *J. Cereal Sci.*, 1991, **14**, 11.
24. F. Buonocore, C. Caporale and D. Lafiandra, *J. Cereal Sci.*, 1995, 23, 195.
25. F. Bèkès and P. Gras, *Cereal Foods World*, 1999, **44**, 580.
26. P.S. Belton, *J. Cereal Sci.*, 1999, **29**, 103.
27. D.D. Kasarda, *Cereal Foods World*, 1999, **44**, 566.
28. W.J. Rogers, P.I. Payne, J.A. Seekings and E.J. Sayers, *J. Cereal Sci.*, 1991, **14**, 209.
29. D. Lafiandra, S. Masci, B. Margiotta and E. De Ambrogio, 'Proc. 9th Int. Wheat Genet. Symp.', A.E. Slinkard Ed., University of Saskatchewan, University Extension Press, 1998, vol. 1 p. 261.
30. P.I. Payne and J.A. Seekings, 'Proc. 6th Int. Gluten Workshop', C.W. Wrigley Ed., 1996, p. 14.
31. A.E. Blechl and O.D. Anderson, *Nature Biotechnology*, 1996, **14**, 875.
32. F. Barro, L. Rooke, F. Békés, P.W. Gras, A.S. Tatham, R. Fido, P.A. Lazzeri, P.R. Shewry and P. Barcelò, *Nature Biotechnology*, 1997, **15**, 1295.

33. G.Y. He, L. Rooke, S. Steele, F. Békés, P. Gras, A.S. Tatham, R. Fido, P. Barcelo, P.R. Shewry and P.A. Lazzeri. *Molecular Breeding*, 1999, **5**, 377.
34. Ceoloni, M. Ciaffi, D. Lafiandra and B. Giorgi, 'Proc. 8th Int. Wheat Genet. Symp.', Z.S. Li and Z.Y Xin Eds. 1993, p. 159.
35. F. Vitellozzi, M. Ciaffi, L. Dominici and C. Ceoloni, *Agronomie*, 1997, **17**, 413.
36. 8. K. Ammar, A.J. Lukaszewski and G.M. Banowetz, *Cereal Foods World*, 1997, **42**, 610.
37. R.B. Gupta and K.W. Shepherd, *Theor. Appl. Genet.*, 1990, **80**, 65.
38. E.A. Jackson, M.-H. Morel, T. Sontag-Strohm, G. Branlard, E.V. Metakovsky and R. Redaelli, *J. Genet. Breed.*, 1996, **50**, 321.
39. M.T. Nieto-Taladriz, M. Ruiz, M.C. Martinez, J.F. Vasquez and J.M. Carrillo, *Theor. Appl. Genet.* 1997, **95**, 1155.
40. P.I. Payne, E.A. Jackson and L.M. Holt, *J. Cereal Sci.*, 1984, **2**, 73.
41. N.E. Pogna, D. Lafiandra, P. Feillet and J.C. Autran, *J. Cereal Sci.*, 1988, **7**, 211.
42. S. Masci, R. D'Ovidio, D. Lafiandra and D.D. Kasarda, *Plant Physiol.*, 1998, **118**, 1147.
43. R. D'Ovidio, C. Marchitelli, L. Ercoli Cardelli and E. Porceddu, *Theor. Appl. Genet.* 1999, **98**, 455.
44. P.R. Shewry and A.S. Tatham, *J. Cereal Sci.*, 1997, **25**, 207.
45. J.C. Autran, B. Laignelet and M.H. Morel, *Biochimie,* 1987, **69**, 699.
46. S. Masci, E.J.-L. Lew, D. Lafiandra, E. Porceddu and D.D. Kasarda, *Cereal Chem.*, 1995, **72**, 100.
47. R. D'Ovidio, C. Marchitelli, S. Masci, P. Tosi, M. Simeone, L. Ercoli Cardelli and E. Porceddu, 'Proc. 8th Int. Wheat Genet. Symp.', Z.S. Li and Z.Y Xin Eds. 1993, p. 269.
48. Lafiandra, S. Masci, C. Blumenthal and C.W. Wrigley, *Cereal Foods World*, 1999, **44**, 572.
49. S. Masci, D. Lafiandra, E. Porceddu, E. J.-L. Lew, H.P. Tao and D.D. Kasarda, *Cereal Chem.*, 1993, **70**, 581.
50. S. Masci, T.A. Egorov, C. Ronchi, D.D. Kuzmicky, D.D. Kasarda and D. Lafiandra, *J. Cereal Sci.*, 1999, **29**, 17.
51. N.E. Pogna, A.M. Monari, P. Cacciatori, R. Redaelli and P.K.W. Ng, 'Proc. 8th Int. Wheat Genet. Symp.', Z.S. Li and Z.Y Xin Eds. 1993, p. 265.

IMPROVED QUALITY 1RS WHEATS VIA GENETICS AND BREEDING

R. A. Graybosch

USDA-ARS, University of Nebraska - Lincoln, Lincoln, NE, USA, 68583

1 INTRODUCTION

Wheat lines carrying rye chromosome 1RS, generally in the form of 1AL.1RS or 1BL.1RS wheat/rye chromosomal translocations, have become widespread in many of the world's wheat breeding populations. 1RS confers to wheat distinct advantages, including resistance to a number of diseases and insect pests, and improved yield and adaptation, at least in some environments. 1RS also conditions diminished gluten strength, easily measured by diminished SDS sedimentation volumes or decreased tolerance to dough overmixing. The loss of gluten strength in 1RS wheats is attributed to the decline in HMW glutenin polymers, brought about by the substitution of monomeric rye secalin proteins for polymer-forming wheat glutenin proteins. Two backcrossing schemes were designed to increase glutenin polymer and glutenin strength in 1RS wheats. In most hexaploid wheats, the *Glu-A1y* gene is inactive. However, in some tetraploid and diploid wheats, functional alleles occur at this locus[1]. Chromosome 1A-encoded HMW glutenins from the tetraploid wheat *Tricticum dicoccoides* were backcrossed into 1AL.1RS and 1BL.1RS wheats. This procedure increased the number of genes encoding HMW glutenins from 5 to 6. 1BL.1RS also was backcrossed into a high protein, strong gluten wheat, N86L177. N86L177 carries high protein genes ultimately derived from NapHal, as well as high protein, strong gluten genes derived from Plainsman V. This paper describes the relative successes of these two approaches.

2 MATERIALS AND METHODS

2.1 Experiment I

T. dicoccoides accessions were obtained from the USDA-ARS Small Grains collection housed in Aberdeen, Idaho. Total protein extracts were separated by SDS-PAGE for the analysis of HMW glutenin subunit composition. Two accessions, PI 471075 and PI 467024, found to produce four HMW glutenin subunits, were assumed to carry active *Glu-A1y* genes. Both accessions were first crossed to the experimental wheat Chinese Spring, and thereafter backcrossed to hexaploid wheat cultivars carrying either 1AL.1RS or 1BL.1RS. After each crossing cycle, glutenin protein was extracted from ½ kernels, and analyzed by SDS-PAGE and silver staining[2]. Seed producing the *T. dicoccoides Glu-A1* HMW glutenins were planted, and used for subsequent crossing cycles. After four backcrosses, lines were advanced to homozygosity via self-pollination, and sets of sister lines, with and without *T. dicoccoides* encoded HMW glutenin subunits (designated TDHMW), were derived. Lines lacking TDHMW glutenin carried the *Glu-A1x* subunit 2*. Hence, the experiment contrasted lines with five HMW glutenin subunits with lines carrying six. Lines were grown in Yuma, Arizona, USA in 1998/99. The number of lines or each respective genotype was: 1AL.1RS + 2*, 24; 1AL.1RS + TDHMW, 18; 1BL.1RS + 2*, 40; 1BL.1RS + TDHMW, 35; non-1RS + 2*, 16; non-1RS + TDHMW, 26. Harvested grain was ground in a Udy cyclone mill, and quality assessed via SDS sedimentation tests. Paired t-tests were used to compare means in all possible combinations.

2.2 Experiment II

1BL.1RS was introduced into the high protein strong gluten wheat N86L177, also via backcrossing. The wheat cultivar 'Siouxland' (1BL.1RS) was crossed to N86L177; F_1 progeny were backcrossed to N86L177, and 1RS was confirmed in seed of the $BC1F_1$ generation by extracting ½ kernels with 70% ethanol, and analyzing the extracted proteins via SDS-PAGE and silver staining. The presence of 1RS was inferred by the presence of ω-secalins[2]. After the $BC1F_3$ generation, a few homogeneous 1BL.1RS and 1BL.1BS lines (Siouxland/2*N86L177) were derived. Crossing, however, continued until the BC3 generation. After the $BC3F_3$ (Siouxland/4*N86L177) three classes of 1BL.1RS in the N86L177 background were derived: homogeneous 1BL.1RS; homogeneous 1BL.1BS (non-1RS or wild-type) and heterogenous (+/-). The majority of the tested lines were of BC3 origin. Lines derived from Siouxland/2*N86L177 and from Siouxland/4*N86L177 were grown in 5 Nebraska locations in 1998 and 1999, along with Siouxland, N86L177, and the wheat cultivar 'Arapahoe', the most widely grown wheat in the 1990's in Nebraska. Grain yield was determined in all five environments. Quality attributes (Mixograph analyses and 100 gram bake tests) were determined from 3 environments. Data were analyzed via analysis of variance, and sample means were compared via calculation of l.s.d. (0.05) values

3 RESULTS AND DISCUSSION

3.1 Experiment I

The introgression of *Glu-D1* HMW glutenin subunits from *T. dicoccoides* (TDHMW) did little to improve the SDS sedimentation volumes of lines carrying either 1AL.1RS or 1BL.1RS (Table 1). Introgression of TDHMW did result in a statistically significant increase in SDS sedimentation volumes in non-1RS wheats, relative to both 1RS classes, and to non-1RS wheats producing HMW glutenin subunit 2*. Evidently, while capable of increasing glutenin content in non-1RS wheats, TDHMW cannot compensate for the loss of LMW glutenin subunits observed when 1RS is present.

Table 1. SDS sedimentation volumes of genotypes in Experiment I.

1RS genotype	HMW glutenin genotype	SDS sedimentation vol. (ml)
1AL.1RS	2*	17.8
1AL.1RS	TDHMW	16.9
1BL.1RS	2*	16.0
1BL.1RS	TDHMW	16.4
Non-1RS	2*	17.9
Non-1RS	TDHMW	20.0*

*significantly higher than all other means.

3.2. Experiment II

Lines used in the Experiment II, and mean grain yields from 5 Nebraska environments, are given in Table 2.

Table 2. *Genotypes and mean grain yield of lines in Experiment II.*

Line	Genotype	Grain yield (kg/ha)
Arapahoe	Non-1RS, control	3118
N95L11873*	1BL.1RS	3075
N95L11881*	1BL.1RS	3049
N97L9539**	1BL.1RS	3035
N97L9525**	1BL.1RS +/-	3003
N97L9527**	1BL.1RS +/-	2996
N95L11880*	Non-1RS	2977
N97L9538**	1BL.1RS	2973
N97L9517**	1BL.1RS +/-	2961
N95L11866*	Non-1RS	2887
N97L9537**	Non-1RS	2870
N97L9530**	Non-1RS	2847
N97L9521**	1BL.1RS	2830
N97L9534**	1BL.1RS	2826
N95L11869*	1BL.1RS	2815
N97L9540**	1BL.1RS +/-	2792
N97L9520**	Non-1RS	2767
N97L9518**	Non-1RS	2761
Siouxland	1BL.1RS	2734
N97L9536**	Non-1RS	2711
N97L9504**	1BL.1RS +/-	2698
N97L9545**	Non-1RS	2671
N97L9532**	Non-1RS	2664
N97L9533**	1BL.1RS	2648
N86L177	Non-1RS	2614
N97L9531**	1BL.1RS	2553
l.s.d.(0.05)		285

*Derived from Siouxland/2*N86L177

**Derived from Siouxland/4*N86L177

Two lines, N95L11873 and N95L11881, produced significantly higher grain yields than both parental lines, Siouxland and N86L177. Nine lines (those with grain yields > 2899 kg/ha) were significantly higher than N86L177; of these nine, all but one carried 1BL.1RS, either in homogeneous or heterogeneous condition.

Quality characteristics of lines in Experiment II, sorted in order of decreasing Mixograph tolerance, demonstrated a significant improving effect of the N86L177 genetic background, especially on dough strength (Table 3).

Table 3. *Quality characteristics of lines in Experiment II.*

Line	Genotype	Mixograph time (min)	Mixograph tolerance (0-9)	Bake mix time (min)	Loaf volume (ml)
N97L9545**	Non-1RS	5.8	5.3	8.1	910
N97L9563**	Non-1RS	6.7	5.0	9.5	900
N97L9520**	Non-1RS	4.8	4.7	7.0	910
N97L9518**	Non-1RS	5.1	4.7	7.3	948
N97L9537**	Non-1RS	5.2	4.7	7.7	892
N95L11880*	Non-1RS	6.4	4.6	8.9	895
N97L9530**	Non-1RS	4.9	4.6	7.2	920
N97L9539**	1BL.1RS +/-	5.5	4.6	8.5	912
N86L177	Non-1RS	5.3	4.5	7.3	927
N95L11866*	Non-1RS	5.2	4.4	7.3	947
N97L9504**	1BL.1RS +/-	4.9	4.4	7.3	917
N97L9535**	1BL.1RS +/-	4.7	4.2	6.6	900
N97L9534**	1BL.1RS	4.6	4.2	4.2	895
N97L9531**	1BL.1RS	5.3	4.1	7.5	895
N97L9517**	1BL.1RS +/-	5.1	4.1	7.3	933
N97L9522**	1BL.1RS	4.8	4.1	6.9	925
N97L9527**	1BL.1RS +/-	5.1	4.0	7.2	898
N97L9540**	1BL.1RS +/-	5.9	4.0	7.5	920
N97L9532**	Non-1RS	5.6	3.7	7.6	903
N97L9533**	1BL.1RS	4.5	3.8	6.7	878
N95L11869*	1BL.1RS	4.5	3.4	5.9	886
N97L9538**	1BL.1RS	5.2	3.4	7.3	902
Arapahoe	Non-1RS	4.8	3.4	6.1	838
N95L11881*	1BL.1RS	5.6	3.3	7.6	868
N97L9521**	1BL.1RS	4.6	3.1	6.4	892
N95L11873*	1BL.1RS	4.0	2.6	4.8	880
Siouxland	1BL.1RS	3.8	2.4	4.9	913
l.s.d. (0.05)		0.9	0.9	0.9	38

*Derived from Siouxland/2*N86L177

**Derived from Siouxland/4*N86L177

All but three of the derived lines had significantly greater Mixograph tolerance scores than Siouxland, all but two had significantly greater Mixograph times, and all but one had significantly greater bake mix times. Nearly all of the derived lines did not differ in these three variables from the parental line N86L177. The two highest yielding derived lines, N95L11881 and N95L11873, were closest to the 1BL.1RS parent, Siouxland, in terms of quality characteristics. There also was still somewhat of a negative effect of 1BL.1RS on quality, as, of the top ten lines, in terms of Mixograph tolerance only one carried 1BL.1RS, and it was in the heterogeneous condition. Thus, no matter what the genetic background, the effects of 1RS cannot be completely masked. They can, however, be dramatically alleviated.

References

1. A.A. Levy, G. Galili and M. Feldman, *Heredity,* 1988, **61**, 63-72.
2. R. Graybosch, J.H. Lee, C.J. Peterson, D.R. Porter and O.K. Chung, *Plant Breeding*, **118**, 125-130.

CHARACTERISATION OF A LMW-2 TYPE DURUM WHEAT CULTIVAR WITH POOR TECHNOLOGICAL PROPERTIES

Masci S.[1], Rovelli L.[1], Monari A.M.[1], Pogna N.E.[2], Boggini G.[3] and Lafiandra D.[1]

1. Dipartimento di Agrobiologia e Agrochimica, Università degli Studi della Tuscia, Via S. Camillo de Lellis, 01100 Viterbo, Italy. 2. Istituto per la Cerealicoltura, Via Cassia 176, 00191 Roma, Italy. 3. Istituto per la Cerealicoltura, Via Mulino 3, 20079 S. Angelo Lodigiano (LO), Italy

1 INTRODUCTION

The visco-elastic properties of durum wheat semolina are mainly due to a particular allelic form of low molecular weight glutenin subunits (LMW-GS), named LMW-2. Genes coding for LMW-2 are genetically linked to γ-gliadin 45, and this latter is used as a genetic marker for quality. The good quality of LMW-2, as opposed to that shown by LMW-1 type durum wheats, seems mainly due to the high amount of glutenin subunits, although structural differences cannot be excluded[1,2,3].

Here we have analysed the Italian durum wheat cultivar Demetra, that, although showing the LMW-2 type pattern, presents poor technological properties, compared to other LMW-2 type durum wheat varieties grown in the same conditions.

2 MATERIALS AND METHODS

Cultivar Demetra was developed at Istituto per la Cerealicoltura (Catania, Italy) from the cross between the durum wheat cultivars Messapia and Gioia. Demetra, together with other LMW-2 type durum wheats (indicated directly in Table 1 and in the legends to the figures), were analysed by alveographic measurements, acid polyacrylamide gel electrophoresis (APAGE) according to Khan et al[4], by SDS-PAGE (T=12 and C=1.28) and two-dimensional (A-PAGE vs. SDS-PAGE) electrophoresis, according to Morel[5], with minor modifications. Reversed phase high performance liquid chromatography (RP-HPLC) of glutenin subunits was also perfomed in order to compare LMW-GS patterns. Conditions were as described in Masci et al[6]. Plant material was grown at three different locations (North, Central and South Italy).

3 RESULTS AND DISCUSSION

Flours obtained from cultivar Demetra and from other Italian durum wheat cultivars possessing LMW-2, grown at three different locations, were submitted to alveographic measurements (Table 1).

Although environmental differences are present, cultivar Demetra shows consistently lower values of W, indicative of poor technological properties. In order to understand the bases of such behaviour, we have first used APAGE and SDS-PAGE to determine gliadin and glutenin subunits composition, respectively. All the cultivars show the quality correlated γ-gliadin 45 and LMW-2 (Figure 1), this latter characterised by the presence of the so-called 42K LMW-GS[3,7], not present in LMW-1. Although the amount of HMW-GS is comparable among cultivars, the amount of LMW-GS present in Demetra is notably lower.

Table 1: *Alveographic measurements (W) relative to the durum wheat cultivars analysed*

Cultivar	W: North	W: Centre	W: South
Ares	254	243	349
Baio	219	204	323
Demetra	**74**	**52**	**106**
Flaminio	303	256	410
Nefer	182	191	297
Preco	193	146	246
Appio	225	103	131
Grazia	232	162	311
Svevo	286	249	324

Figure 1: *APAGE (left side) and SDS-PAGE (right side) of different Italian durum wheat cultivars. Ares (1, 11), Appio (2, 10), Baio (3, 12), Demetra (4, 13), Flaminio (14), Flavio (5), Grazia (6, 15), Nefer (7, 16), Preco (8, 17), Svevo (9, 18).*

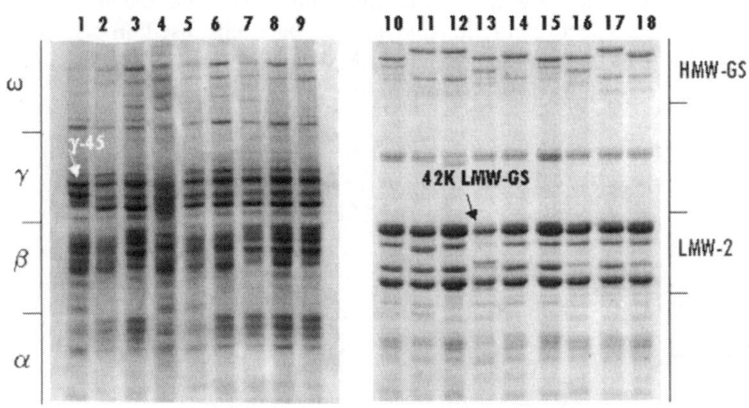

This latter observation was confirmed by the RP-HPLC analysis, that showed that the area corresponding to LMW-2 is lower compared to other LMW-2 type durum wheats (Figure 2). It is also of interest that the area corresponding to C subunits, mainly made up of gliadin-like LMW-GS, is higher in Demetra. If C subunits are mostly chain terminators, this aspect might contribute negatively to the poor technological properties found in Demetra.

Figure 2: *RP-HPLC of glutenin subunits extracted from LMW-2 durum wheat cultivars Demetra and Svevo.*

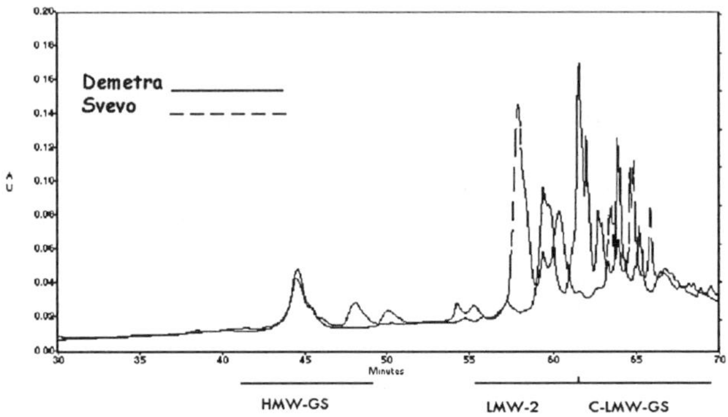

Figure 3 shows the two dimensional pattern of glutenin subunits of cultivar Demetra as compared to Svevo, a high quality durum wheat cultivar, that also shows the LMW-2 pattern. Other LMW-2 cultivars were also analysed by two dimensional electrophoresis (data not shown). This kind of analysis confirmed what was already indicated by the other analyses, namely the presence of a lower amount of LMW-GS in Demetra, due both to a lower number of spots and to a low expression level of the polypeptides present. The lower number of spots might be due to the absence of some 1A coded polypeptides. Electrophoretic analyses of parentals used in the production of Demetra revealed that the combination γ-45/LMW-2 was inherited by Messapia (data not shown), since cultivar Gioia does not contain this allelic form.

Figure 3: *Two-dimensional pattern of cultivar Demetra (A) and cultivar Svevo (B). Arrows indicate a putative 1A-coded protein spot, that is absent in Demetra.*

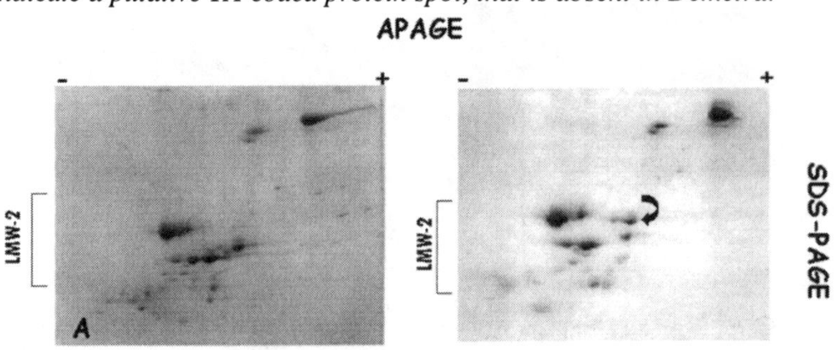

4 CONCLUSIONS

The great majority of durum wheat cultivars grown worldwide typically have the LMW-2 type pattern, together with the associated γ-gliadin 45, because of the positive effect on gluten visco-elasticity exerted by the former[8]. Whether quantitative or structural differences are mainly responsible for such quality differences is still a matter of debate. However, it is likely that the high amount of LMW-GS, usually associated with LMW-2, plays the major role[1,2], since structural differences, although present, do not appear to be sufficient to explain the contrasting effects of LMW-1 and LMW-2[3,9].

The data presented here also provide support for a quantitative effect, since a lower amount of LMW-GS is present in Demetra, as assayed by SDS-PAGE and RP-HPLC. This lower amount is due both to a smaller number of protein subunits present in Demetra (in particular 1A-coded polypeptides might be missing) and to a low level of expression of the polypeptides present. Moreover, a higher amount of C subunits seems to be present in the cultivar Demetra, as assayed by RP-HPLC. Because C subunits are likely to be chain terminators[10], this might contribute to the poor quality characteristics shown by Demetra.

References

1. J.C. Autran, B. Laignelet, M.H. Morel, 1987, *Biochimie*, **69**, 699
2. S. Masci S., E.J.L. Lew, D. Lafiandra, E. Porceddu and D.D. Kasarda, 1995, *Cereal Chem*, **72**, 100
3. S. Masci, R. D'Ovidio, D. Lafiandra and D.D. Kasarda, 1998, *Plant Physiol.*, **118**, 1147
4. K. Khan, A.S. Hamada, J. Patek, *Cereal Chem.*, 1985, **62**, 310
5. M.H. Morel, *Cereal Chem.*, 1994, **71**, 238
6. S. Masci, E.J.L. Lew, D. Lafiandra, E. Porceddu and D.D. Kasarda, *Cereal Chem.*, 1995, **72**, 100
7. S. Masci, R. D'Ovidio. D. Lafiandra and D.D. Kasarda, *Theor. Appl. Genet.*, **3/4**, 396
8. N. Pogna, D. Lafiandra, P. Feillet and J.C. Autran, *J. Cereal Sci.*, 1988, **7**, 211
9. R. D'Ovidio, C. Marchitelli, L. Ercoli Cardelli and E. Porceddu, *Theor. Appl. Genet.*, 1999, **98**, 455
10. H.P. Tao and D.D. Kasarda, *J. Exp. Bot.*, 1989, **40**, 1015

Acknowledgments:

Research supported by the Italian Ministero dell'Università e della Ricerca Scientifica e Tecnologica (M.U.R.S.T.), National Research Project "Studio delle proteine dei cereali e loro relazioni con aspetti tecnologici e nutrizionali". Barilla Alimentare is gratefully acknowledged for having grown plant material and performed alveographic measurements.

EFFECT OF THE *GLU-3* ALLELIC VARIATION ON BREAD WHEAT GLUTEN STRENGTH

M. Rodríguez-Quijano, M.T. Nieto-Taladriz, M. Gómez and J.M.Carrillo.

Unidad de Genética, Escuela Técnica Superior de Ingenieros Agrónomos, Universidad Politécnica, Madrid, Spain.

ABSTRACT

High molecular weight (HMW) and low molecular weight (LMW) glutenins are the main seed storage proteins related to the quality of bread wheat doughs. Extensive variation has been detected at the *Glu-1* (coding for HMW glutenin subunits) and *Glu-3* (coding for LMW glutenin subunits) loci. The relative influence of allelic variation at both loci varies, and interactions between them have been detected. The objective of this research was to determine the effect of the allelic variation at the *Glu-3* loci on gluten strength in an F_2 cross between parents, which had the same allelic composition at the *Glu-1* loci. The parents, 'Blanquillo de Cáceres' (MCB-959) and 'Barbilla de Alcañices' (MCB-1293), differed at the *Glu-A3* and *Glu-B3* loci. Gluten strength was measured by SDS-sedimentation (SDSS) test. Results showed that the allelic variation at the *Glu-A3* locus had no influence on SDSS volume. In contrast, allelic variation at the *Glu-B3* locus had a highly significant effect on gluten strength.

1 INTRODUCTION

Prolamins (gliadins and glutenins) are the main seed storage proteins responsible for the end-use quality of bread wheat. Glutenins are polymeric structures whose subunits are held together by disulphide bonds. When reduced, glutenins are divided into two groups, high molecular weight (HMW) and low molecular weight (LMW) glutenin subunits. HMW glutenin subunits are encoded by genes at the *Glu-1* loci on the long arms of homoeologous group 1 chromosomes. LMW glutenin subunits are encoded by genes at the *Glu-3* loci which is tightly linked to the *Gli-1* loci, coding for gliadins, and located on the short arms of the same chromosomes[1].

Combined studies of HMW and LMW glutenin subunit composition in different wheat cultivars and progenies have revealed that their relative influence on dough properties varies[2, 3]. Further research showed the existence of interaction between alleles at the *Glu-1* and *Glu-3* loci[4-7].

The objective of this research was to determine the effect of the allelic variation at the *Glu-3* loci on gluten strength in an F_2 cross between parents which had the same allelic composition at the *Glu-1* loci.

2 MATERIAL AND METHODS

2.1 Plant Material

The F_2 progeny from the cross between the Spanish bread wheat landraces: 'Blanquillo de Cáceres' (MCB-959) and 'Barbilla de Alcañices' (MCB-1293) were analysed. Plants were sown in a complete randomised trial under normal field conditions. Additional F_2 progeny between 'Blanquillo de Cáceres' and 'Chinese Spring' were used for genetic analysis. Bread wheat cultivars 'Adalid', 'Barbilla', 'Cabezón' and 'Gabo' were used as standards.

2.2 Protein extraction and electrophoresis

The sequential extraction procedure[8] was used to obtain HMW and LMW glutenin subunits. SDS-PAGE was performed as described by Nieto-Taladriz et al.[6]. Gliadins were extracted and separated according to Lafiandra and Kasarda[9]. Nomenclature for HMW glutenin subunits was that of Payne and Lawrence[10]. Gliadin alleles were named according to Metakovsky[11, 12].

2.3 Quality evaluation

Gluten strength was estimated in duplicate by the SDS-sedimentation (SDSS) test[13] on grains from each F_2 plants. Parents were also analysed.

3 RESULTS AND DISCUSSION

The choice of parental varieties for the study of the relationships between LMW glutenin subunits and gluten strength was made so as to limit the variability of HMW glutenin subunits. Thus, 'Blanquillo de Cáceres' and 'Barbilla de Alcañices' had the same alleles at the *Glu-1* loci (Table 1 and Figure 1A), but the SDSS value was higher in 'Barbilla de Alcañices' (75 mm) than in 'Blanquillo de Cáceres' (57 mm).

Table 1. *Allelic composition of the parents at the Glu-1 and Gli-1/Glu-3 loci. Sedimentation volumes (SDSS, in mm) are also included.*

	HMW glutenin subunits			Gliadins/LMW glutenin subunits			
Parents	*Glu-A1*	*Glu-B1*	*Glu-D1*	*Gli-A1/Glu-A3*	*Gli-B1/Glu-B3*	*Gli-D1/Glu-D3*	SDSS
'Blanquillo de Cáceres'	2*	13+16	2+12	r	g	f	57
'Barbilla de Alcañicer'	2*	13+16	2+12	d	h	f	75

Figure 1. (**A**) *SDS-PAGE of HMW and LMW glutenin subunits from bread wheat landraces 'Blanquillo de Cáceres' (lane 7), 'Barbilla de Alcañices' (lane 8) and some F_2 grains from the cross between them (lanes 1-6 and 9-15). HMW glutenin subunits are numbered. Arrowheads and arrows indicate subunits encoded at the Glu-A3 and Glu-B3 loci. The Glu-B3 encoded D glutenin subunit d_4 is also indicated.* (**B**) *A-PAGE of gliadins from the bread wheat cultivars 'Chinese Spring' (lane 1), 'Blanquillo de Cáceres' (lane 2) and 'Barbilla de Alcañices' (lane 3).*

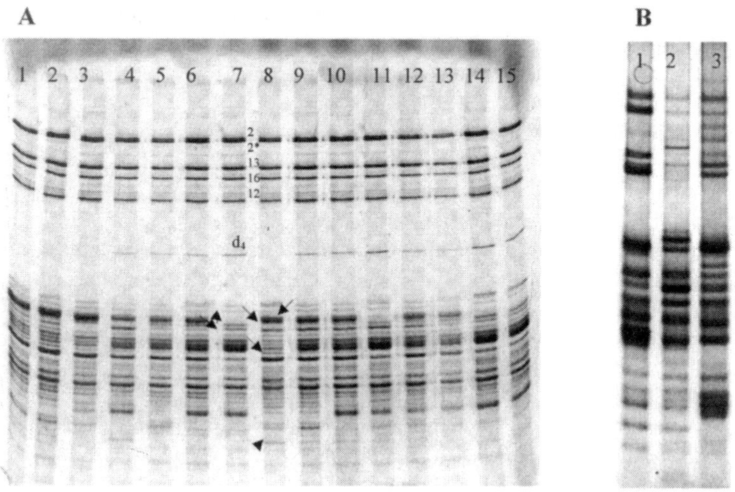

The electrophoretic analysis of the parents showed differences in both LMW glutenin subunits and gliadins (Figure 1A and B). The determination of the *Glu-3* and *Gli-1* alleles present in 'Blanquillo de Cáceres' was done by analysis of the F_2 progeny from its cross with 'Chinese Spring' (data not shown). Thus, analysis of the F_2 progeny from the 'Blanquillo de Cáceres'/'Barbilla de Alcañices' cross, allowed allelic variation at the *Gli-A1/Glu-A3* and *Gli-B1/Glu-B3* loci to be identified (Figure 1A and B). No recombination was detected between *Gli-1* and *Glu-3* loci and the nomenclature proposed for *Gli-1* was adopted for the complex *Gli-1/Glu-3* loci. Table 1 summarises the alleles detected in both parents.

The effect of the allelic variation at each of the loci on gluten strength was determined (Table 2). Heterozygous plants were eliminated from the analysis. Results showed that the allelic variation at the *Gli-A1/Glu-A3* had no significant influence on SDSS volume. Otherwise, the allelic variation at the *Gli-B1/Glu-B3* had a highly significant effect on gluten strength, and the mean SDSS volume of lines with the *Gli-B1/Glu-B3* allele from 'Barbilla de Alcañices' had a significantly higher (P<0.01) volume than the mean of those with the allele from 'Blanquillo de Cáceres' (Figure 2).

Table 2. *Analysis of the variance of gluten SDSS volume of the F_2 plants from the 'Blanquillo de Cáceres'/'Barbilla de Alcañices' cross [mean square (MS) and F value].*

Source	df	MS	F
Glu-A3	1	54.77	0.30 ns
Glu-B3	1	1923.14	7.92 **

ns: not significant, **: significant at the 1% level of probability

Figure 2. *Mean SDSS values for the Glu-A3 and Glu-B3 allelic variants found at the F_2 progeny from the 'Blanquillo de Cáceres'/'Barbilla de Alcañices' cross*

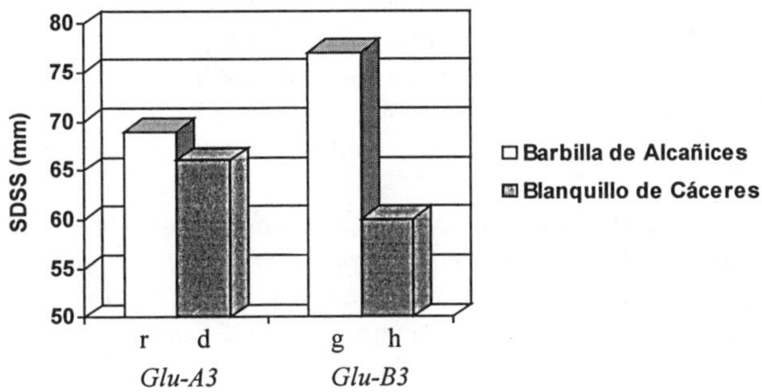

Several studies have been focused on the effects of allelic variation at both the *Glu-1* and *Glu-3* loci on bread wheat quality. Allelic variation at the *Glu-1* loci generally had a greater effect on quality than *Glu-3*[14].

Results obtained here show that, when there is no variation at the *Glu-1* loci, the allelic variation at the *Gli-B1/Glu-B3* locus had a high significant effect. Results agree with those of Gupta *et al.*[4] and Nieto-Taladriz *et al.*[6].

It should be noted that the new *Glu-B3* allele from 'Blanquillo de Cáceres' includes the D glutenin subunit d_4. The same subunit is also present in the *Glu-B3* allele of the bread wheat cultivar 'Prinqual'[15] but both alleles are different. Results obtained here show that the allele from 'Blanquillo de Cáceres' was associated with low gluten strength, but Nieto-Taladriz et al.[15] showed that the allele of 'Prinqual' had high gluten strength. These results suggest that the D glutenin subunit d_4, present in two different alleles, has a minor effect on quality compared with the effect of the LMW glutenin subunits encoded in the same block.

Acknowledgements

This work was supported by grant AGF97-937 from the Comisión Interministerial de Ciencia y Tecnología (CICYT) of Spain.

References

1. P.I. Payne, *Annu Rev Plant Physiol*, 1987, **38**, 141
2. P.I. Payne, J.A. Seekings, A.J. Worland, M.G. Jarvis and L.M. Holt, *J Cereal Sci*, 1987, **6**, 103.
3. R.B. Gupta and KW Shepherd, in *Proc 7th Wheat Genet Symp,* eds TE Moller and RMD Koebner, Bath Press, Bath UK, 1998, p. 943.
4. R.B. Gupta, J.G. Paul, G.B. Cornish, G.A. Palmer, F. Bekes and A.J. Rathjen, *J Cereal Sci,* 1994, **19**, 9.
5. M.T. Nieto-Taladriz and A. Bouguenec, in *Gluten Proteins 1993*, Association of Cereal Research, Detmold, Germany, 1994, p. 262.
6. M.T. Nieto-Taladriz, M.R. Perretant and M. Rousset, *Theor Appl Genet*, 1994, **88**, 81.
7. M. Rodriguez-Quijano and J.M. Carrillo, *Euphytica,* 1996, **91**:141.
8. N.K. Singh, K.W. Shepherd and G.B. Cornish, *J Cereal Sci*, 1991, **14**, 203.
9. D. Lafiandra and D.D. Kasarda , *Cereal Chem.,* 1985, **62,** 314.
10. P.I. Payne and G.J. Lawrence, *Cereal Res Commun* , 1983, **11**, 29.
11. E.V. Metakovsky, *J Genet & Breed*, 1991, **45**, 325.
12. E.V. Metakovsky, M. Gómez, J.F. Vázquez and J.M. Carrillo, *Plant Breed*, 2000, **119**, 39.
13. L.M. Mansur, C.O. Qualset and D.D. Kasarda, *Crop Sci*, 1990, **30,** 593.
14. F. MacRitchie and D. Lafiandra, in *Food proteins and their applications*, eds S. Damodaran and A. Paraf, Marcel Dekker, Inc., New York,1997, p.293.
15. M.T. Nieto-Taladriz, M. Rodriguez-Quijano and J.M. Carrillo, *Genome*, 1998, **41**, 215.

RELATIONSHIP BETWEEN BREADMAKING QUALITY AND SEED STORAGE PROTEIN COMPOSITION OF JAPANESE COMMERCIAL HEXAPLOID WHEATS (*Triticum aestivum* L.)

H. Nakamura

Tohoku National Agricultural Experiment Station, Morioka, Iwate, 020-0198, Japan

1 INTRODUCTION

The high-molecular-weight (HMW) glutenin subunit designated as the 145kDa subunit was found frequently in Japanese wheat (*Triticum aestivum* L.) varieties.[1] This subunit has identical electrophoretic mobility to the HMW glutenin subunit 2.2 as reported by Payne *et al.*[2] The frequency of varieties with HMW glutenin 145kDa subunit was higher in the southern part of Japan than northern part, and examination of pedigrees shows that the genotypes with and without the 145kDa subunit were preferably selected in each step of the wheat breeding procedures in the southern and northern parts of Japan, respectively.[3] This research reports the relationship between breadmaking quality and seed storage protein composition of Japanese commercial hexaploid wheats.

2 MATERIALS AND METHODS

Endosperm seed storage proteins in 131 Japanese commercial hexaploid wheat (*Triticum aestivum* L.) varieties were fractionated by sodium dodecyl sulphate polyacrylamide gel electrophoresis to identify the alleles for the complex gene loci, *Glu-A1, Glu-B1,* and *Glu-D1* that code for high molecular weight (HMW) subunits of glutenin in Japanese hexaploid wheat varieties. The Norin numbering system has been employed in Japan since 1929 to designate the commercial variety. The varieties studied are the major wheats bred and cultivated in Japan, and are very important for Japanese wheat production. The separation gel contained 1.5M Tris-HCl, pH8.8 and 0.27% SDS. Gels were made with 7.5% (w/v) acrylamide and 0.2% (w/v) bis-acrylamide. The stacking gel contained 0.25M Tris-HCl, pH6.8. Wheat flour (10mg) was suspended in 300 mL 0.25M Tris-HCl buffer (pH6.8) containing 2% (w/v) SDS, 10% (v/v) glycerol, 5% 2-mercaptoethanol and shaken for 2hr at room temperature. The suspension was heated at 95 °C for 3 min. The top portion of the supernatant was collected after centrifugation for 3 min at 12,000 rpm and a portion (30 μL) of the extract was loaded into the gel slot. The electrode buffer was 0.025M Tris-glycine, pH8.3, containing 0.1% (w/v) SDS.

3 RESULTS AND DISCUSSION

A characteristic apparently unique to Japanese commercial wheat varieties is the high frequency of the 145kDa subunit encoded by the *Glu-D1f* allele. The molecular weight of this subunit exceeded that of any other glutenin subunit present in the varieties analyzed. 35.1% (46 of 131 varieties) of the Japanese varieties carried the 145kDa subunit. The high proportion of 145kDa subunit variation contrasts sharply with the situation in 1380 varieties throughout the world.[4] In Japanese commercial varieties, the 145kDa subunit occurs frequently and in some cases occurs in unique combinations. The hardness of wheat flour is correlated with Japanese soft noodle-making quality, with hard wheat varieties having poor quality. Wheat lines ideal for Japanese soft noodle-making quality are of course preferred in Japan and the frequency of the 145kDa subunit in these lines may consequently, be correlated with this character. It is particularly high in southern Japan, but quite low in the northern area.[3] In southern Japan, lines good for Japanese soft noodle-making quality predominate. In the pedigree of the varieties, the many varieties that possess the 145kDa subunit were used by crossing in the Japanese soft noodle wheat breeding program. The breeding areas differ in frequencies of HMW glutenin subunit groups throughout Japan. Subunits 5+10 are seen more frequently in European than Japanese wheat commercial varieties,[5] possibly owing to their correlation with good breadmaking quality, though this is not the case in Japan. Only 1.5% (2 bread wheat varieties, Norin 35 and Haruhikari bred in Hokkaido) of varieties possessed subunits 5+10 encoded by the *Glu-D1d* allele: compared with 41% of 1380 varieties throughout the world.[4] Table 1 gives the *Glu-1* quality scores of the Japanese commercial varieties, which ranged from 5 to 9. The average *Glu-1* quality scores of Japanese wheats have been shown to be lower than those of known quality wheats from Europe, Australia, Canada and the United States.[4] Europe may be considered a bread consumption zone, while Asia is a noodle zone where noodles are made from hexaploid wheats. The pedigrees of the five principal breadmaking varieties of spring or winter wheat bred in the Hokkaido area were investigated. At least half of the crosses made were aimed at improving breadmaking quality in order to increase the proportion of homegrown wheat in the milling grist. The 5 varieties have inherited the good glutenin subunits 1, 17+18, and 5+10 introduced from varieties from other countries. Probably, the most influential factor affecting the composition of *Glu-1* loci is the breeding strategy in relation to breadmaking quality in the Hokkaido district.

It has been revealed that variation in HMW glutenin subunit composition in Japanese hexaploid wheats is very different from that in varieties throughout the world. HMW subunits of glutenin have different properties from other smaller and more abundant subunits[6] and thus allelic variation in HMW glutenin subunits of the Japanese varieties is a matter of considerable importance. In Asian countries, noodles are made from hexaploid wheats rather than tetraploid durum wheats which are preferred for pasta in western countries. Research on the contributions of wheat flour components to noodle quality indicates proteins to be of primary importance in this regard and quantitative and qualitative aspects should be considered in explaining variation in the quality of noodles made from different wheats.[7,8] This matter may be of interest to wheat breeders who consider HMW glutenin subunit alleles when breeding advanced lines of good quality.

Table 1 *Glu-1 quality score of Japanese varieties with respect to high molecular weight glutenin subunit composition*

Subunit composition	*Glu-1* score	Variety
1,7+8,2+12	8	Norin 131
1,7+8,3+12	8	Norin 21, Norin 42
1,7+8,4+12	7	Norin 8, Norin 24, Norin 82,Norin 108
1,7+8,145kDa+12	-	Norin 41, Norin 59, Norin 96
1,7+9,4+12	6	Norin 17, Norin 31, Norin 38, Norin 89
1,6+8,4+12	5	Norin 115
1,17+18,2+12	8	Norin 130
2*,7+8,2+12	8	Norin 87, Norin 91, Norin 107, Norin 119
2*,7+8,145kDa+12	-	Norin 60, Norin 61, Norin 62, Norin 92, Norin 93, Norin 99, Norin 103, Norin 105, Norin 110, Norin 112, Norin 123, Norin 124
2*,7+9,5+10	9	Norin 35
2*,13+19,2+12	-	Norin 111
Null,7+8,2+12	6	Norin 1, Norin 2, Norin 3, Norin 4, Norin 5, Norin 6, Norin 7, Norin 9, Norin 10, Norin 13, Norin 14, Norin 18, Norin 25, Norin 27, Norin 29, Norin 32, Norin 33, Norin 34, Norin 36, Norin 37, Norin 39, Norin 40, Norin 44, Norin 45, Norin 46, Norin 47,Norin 48, Norin 51,Norin 52, Norin 55, Norin 56, Norin 58,Norin 66, Norin 67, Norin 68, Norin 70, Norin 71, Norin 73, Norin 74, Norin 77,Norin 79, Norin 78, Norin 80, Norin 81, Norin 85,Norin 86, Norin 88,Norin 90,Norin 94, Norin 97, Norin 100, Norin 101, Norin 102, Norin 109, Norin 113, Norin 116, Norin 118,Norin 127
Null,7+8,5+10	8	Norin 104
Null,7+8,145kDa+12	-	Norin 15, Norin 19, Norin 20, Norin 22, Norin 23, Norin 26, Norin 28,Norin 30, Norin 43, Norin 49, Norin 50, Norin 53, Norin 54, Norin 57,Norin 63, Norin 64, Norin 65, Norin 72, Norin 95, Norin 106, Norin 117,Norin 120, Norin 122, Norin 129
Null,7+9,2+12	5	Norin 16, Norin 75, Norin 83, Norin 84, Norin 114
Null,7+9,145kDa+12	-	Norin 11, Norin 69, Norin76, Norin 98, Norin 121, Norin 125, Norin 128
Null,20,2+12	-	Norin 12, Norin 126

4 CONCLUSIONS

Twenty-four different major glutenin HMW subunits were identified with each variety containing three to five subunits. Seventeen different glutenin subunit patterns were observed for 14 alleles in Japanese wheat varieties. A catalog of alleles for the complex *Glu-1* loci which code for HMW subunits of glutenin in commercial wheat was compiled. Japanese commercial wheat varieties showed some unique allelic variation in glutenin HMW subunits that was very different from that of foreign wheats. The average *Glu-1* breadmaking quality scores of Japanese commercial hexaploid wheat varieties were lower than those of wheat varieties from Europe, Australia, Canada and the United States.

References

1. H. Nakamura, H. Sasaki, H. Hirano, and Yamashita, A. *Japan. J.Breed.*, 1990, **40**, 485-494.
2. P.I. Payne, L.M. Holt, and G.J. Lawrence, *J. Cereal Sci.*, 1983, **1** , 3-8.
3. H. Nakamura, *Euphytica*, 1999, **136**, 131-138.
4. P.I. Payne, L.M. Holt, E.A. Jackson, and C.N. Law, *Phil. Trans. R. Soc. Lond.*, 1984, **304**, 359-371.
5. A.I. Morgunov, R.J. Pena, J. Crossa, and S.J. Rajarm, *J. Genet. and Breed.*, 1993, **47**, 53-60.
6. P.I. Payne, L.M. Holt, and C.N. Law, *Theor. Appl. Genet.*, 1981, **60**, 229-236.
7. D.M. Miskelly, Proceedings 31st Annual Conference of the Royal Australian Chemical Institute Cereal Chem. Div., Perth., 1981, 61-62.
8. D.M. Miskelly, and H.J. Moss, *J. Cereal Sci.*, 1985, **3**, 379-387.

Acknowledgements

The author expresses his appreciation to Dr. A. Inazu for his advice and comments.

ISOGENIC BREAD WHEAT LINES DIFFERING IN NUMBER AND TYPE OF HIGH M_r GLUTENIN SUBUNITS

B. Margiotta[1], L. Pfluger[2], M.R. Roth[3] , F. MacRitchie[3] and D. Lafiandra[2].

[1]Germplasm Institute, C.N.R., via Amendola 165/a, 70126 Bari, Italy. [2]Dept. of Agrobiology & Agrochemistry, University of Tuscia, 01100 Viterbo, Italy. [3]Dept. of Grain Science, Kansas State University, Manhattan KS, USA.

1 INTRODUCTION

High molecular weight glutenin subunits (HMW-GS) play a key role in affecting gluten viscoelastic properties through their major effect on determining the size distribution of glutenin polymers[1]. These proteins are controlled by genes present at the complex *Glu-1* loci on the long arm of the homoeologous group 1 chromosomes, which have been shown to contain two linked genes encoding a subunit of lower and a subunit of higher M_r termed y- and x-type, respectively, differing in the length and type of repeat motifs of the repetitive domain and number and distribution of cysteine residues[2]. Despite the fact that bread wheats possess six HMW-GS genes, the number of expressed subunits ranges from three to five because of gene silencing processes which have occurred during wheat evolutionary history. In particular, the y-type gene present at the *Glu-A1* locus is always silent in cultivated wheat while the x-type at the same locus and the y-type at the *Glu-B1* locus are expressed only in some cultivars. This results in a variable number of subunits (from three to five) in different bread wheat cultivars. Extensive allelic variation has been detected at each of the encoding loci which has been shown to have different effects on breadmaking performance of different wheat cultivars[3]. Number and allelic type of subunits have been reported to affect bread-making quality through their effect on the amount of large-sized glutenin polymer[4]. Improvement in technological quality of wheat flour can be obtained by increasing the number of genes actively expressing HMW-GS or modifying the allelic composition of HMW-GS[5]. The introduction of HMW-GS genes from wild wheat progenitors in which genotypes expressing both x- and y-type subunits at the *Glu-A1* locus are present, can be one approach to increase the number of glutenin subunits. This has been shown to have incidentally occurred in some Swedish bread wheat lines[6].

Structural characteristics of HMW-GS, such as length of the repetitive domain and number of cysteine residues, have been suggested to be responsible for qualitative differences associated with different allelic subunits. In order to obtain more information on the possible role of these aspects, sets of near isogenic lines have been developed.

2 MATERIALS AND METHODS

2.1 Materials

A set of near-isogenic lines (NIL), using the bread wheat cultivar Pegaso (which possesses glutenin subunit composition: null, 7+9 and 5+10 at the *Glu-A1, Glu-B1* and *Glu-D1* loci) as recipient variety and different bread wheat lines as donors, has been produced. Crossing, backcrossing and electrophoretic selection of lines were carried out as described by Rogers et al.[7].

Similarly, a rare pair of HMW-GS detected at the *Glu-B1* locus in the bread wheat cultivar Cologna[8], designated 26+27, was transferred into Fiorello replacing the pair of subunits 7+8 present at the same locus in this cultivar.

Biotypes of the bread wheat cultivar Halberd differing in HMW-GS at the *Glu-B1* locus were also used.

2.2 Methods

2.2.1 *Electrophoretical analyses.* HMW-GS were extracted from single seeds and analysed by SDS-PAGE as reported by Payne et al.[9].

2.2.2 *Chromatographical analyses (RP-HPLC).* Fractions containing HMW-GS were prepared for RP-HPLC analyses essentially according to the procedure of Marchylo et al.[10] and Margiotta et al.[11]; SE-HPLC, was carried out as reported by Batey et al.[12], but using a Biosep-SEC-S4000 column (Phenomenex).

2.2.3 *Mixographs.* Mixographs were obtained with a computerised 10g mixograph (TMCO) using a Hixsmart software program.

3 RESULTS AND DISCUSSION

3.1 Near Isogenic lines differing in number and type of HMW-GS

Electrophoretic separations of HMW-GS present in the bread wheat cultivar Pegaso together with the nine derived isogenic lines are reported in Figure 1. In particular, lines differ in the number of subunits, ranging from three up to six (lanes 2, 3, 4, 5, 7, 8, 9, 10), and types. In particular, the line with six HMW-GS was obtained using the Swedish bread wheat line W29323 in which both x- and y-type subunits present at the *Glu-A1* locus are expressed[6]. Allelic variants possessing a larger repetitive domain in the Dx or Dy type subunits, such as 2.2*, 2.2 and 12_1 have also been obtained (lanes 2, 3, 4). The material produced is being increased and the qualitative properties will be determined in order to assess the effect of the number of subunits and role of the increased size of the repetitive domain. Preliminary data on the NIL line with six HMW-GS indicate beneficial effects on dough properties resulting from the increased number of HMW-GS.

3.2 Number of cysteine residues and its effect on quality

Chromatographic studies have demonstrated that variation in the number of cysteine residues of HMW-GS is detectable by RP-HPLC. In fact, comparative analyses of reduced and reduced and alkylated subunits, using 4-vinylpyridine as alkylating agent, revealed a

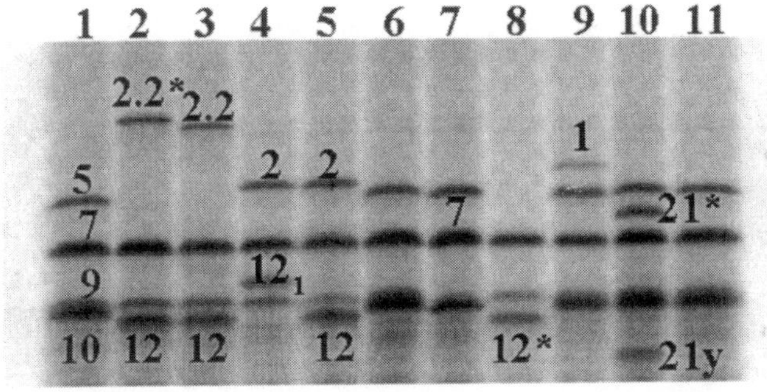

Figure 1. SDS-PAGE of HMW-GS present in Pegaso (1, 6, 11) and derived NILs (2, 3, 4, 5, 7, 8, 9, 10) with different allelic variants introduced.

differential effect of the alkylation on proteins encoded at different loci and on x- and y-type subunits, according to their different number of cysteine residues.

Such analyses carried out on subunits present at the *Glu-B1* locus led to the hypothesis that subunit 20 had a lower number of cysteine residues, on the basis of its chromatographic behaviour compared to subunits 7 and 17 which have been shown to possess four cysteine residues. Subsequently, it was demonstrated that subunit 20 possesses two cysteine residues located one in the N- and the other in the C-terminal region of the molecule and is accompanied by a y-type subunit, termed 20y[11,13]. It has been postulated that differences in the number of cysteine residues, present in subunit 20 compared to 17, are responsible for the larger amount of large-sized polymers associated with the 17+18, when compared to the 20x+20y pair.

Comparative RP-HPLC of HMW-GS present in the bread wheat cultivar Cologna indicates that subunit 26, present together with the y-type subunit 27 at the *Glu-B1* locus in this cultivar, behaves similarly to subunit 20. Subunit 26 was purified, treated with alkylating fluorogenic reagent ABD-F and subsequently digested with trypsin. RP-HPLC separation of the tryptic digest obtained was very similar to that given by subunit 20. Also in this case, only two of the peptides obtained after digestion of the alkylated protein showed fluorescence. The N-terminal amino acid sequences of the fluorescent peptides revealed that they matched the N- and C-terminal regions of HMW-GS and showed the presence of one cysteine residue in each of them, similarly to what was observed for subunit 20 when subjected to the same treatment (data not shown). This confirmed that it is similar to HMW-GS 20.

The allelic pair 26+27 has been transferred into the bread wheat cultivar Fiorello. After repeated steps of backcrossing the NIL produced has been increased and qualitative properties assessed.

Two biotypes, one possessing the subunits 20x+20y and the other the pair 7+9, detected in the Australian bread wheat cultivar Halberd were also used.

SE-HPLC of Fiorello and the NIL carrying the subunit 26+27 indicated no difference in the % of total polymeric proteins (% PI) with similar results being observed for the two Halberd biotypes (Table 1).

Table 1 *SE-HPLC parameters and mixing properties of NIL and biotypes with different alleles at the Glu-B1locus.*

Line	% Peak 1	% Peak 2	%Peak 3	%UPP	PEAK TIME	PEAK HEIGHT	BREAKDOWN SLOPE
Fiorello 1, 7+8, 5* + 12	40.3	48.5	11.3	47.9	2.91	43.28	-2.58
Fiorello 1, 26+27, 5*+12	40.2	50.0	9.8	40.8	1.66	61.20	-4.98
Halberd 1, 7+9, 5+10	36.0	53.6	10.4	51.6	3.62	46.70	-2.99
Halberd 1, 20, 5+10	36.3	53.0	10.7	46.1	2.98	54.80	-4.60

When the percentage of unextractable polymeric proteins (%UPP) was assessed large differences were detected. In fact, Fiorello shows a higher %UPP compared to the derived isogenic line. Similarly, Halberd biotype with subunit 20 shows a lower %UPP compared to the biotype with the pair 7+9.

The rheological properties assessed with the mixograph paralleled the chromatographic data. In fact, the values of peak time were higher in Fiorello compared to the NIL line with the pair of subunits 26+27; similarly the peak time value of the Halberd biotype with 7+9 was higher than that with subunit 20x+20y.

4 CONCLUSIONS

Understanding the genetic/biochemical basis of functionality in wheat requires the development of material with specified protein composition. The production of NIL, differing in the number and type of high-molecular weight glutenin subunits, represents a possible approach to establish structure-function relationships, and clarify the molecular basis of dough rheological properties. Present results confirm the role of the number of cysteine residues in affecting size of glutenin polymers and dough rheological properties. The mechanism by which this happens needs to be clarified.

References

1. F. MacRitchie and D. Lafiandra, "Food Proteins and their Applications", 1996, **10**, p.293.
2. P.R. Shewry and A.S. Tatham, *J. Cereal Sci.,* 1997, **25**, 207.
3. P.I. Payne, *Ann. Rev Plant Physiol.,* 1987, **38**, 141.
4. R.B. Gupta, Y. Popineau, J. Lefebvre, M. Cornec, G.J. Lawrence and F. MacRitchie, *J. Cereal Sci.,* 1995, **21**, 103.
5. D. Lafiandra, S. Masci, B. Margiotta and E. De Ambrogio, Proc 9th Int. Wheat Genet. Symp., A.E. Slinkard Ed., University of Saskatchewan, University Extension Press, 1998, 261
6. B. Margiotta, M. Urbano, G. Colaprico, E. Johansson, F. Buonocore, R. D'Ovidio and D. Lafiandra, *J. Cereal Sci.,* 1996, **23**, 203.
7. W.J. Rogers, P.I. Payne, J.A. Seekings and E.J. Sayers, *J. Cereal Sci.,* 1991, **14**, 209.

8. N.E. Pogna, F. Mellini, A. Beretta and A. Dal Belin Peruffo, *J. Genet. Breed.,* 1989, **43**, 17.
9. P.I. Payne, C.N. Law and E.E. Mudd, *Theor. Appl. Genet.,* 1980, **58**, 113.
10. B.A. Marchylo, J.E. Kruger and D.W. Hatcher, *J. Cereal Sci.,* 1989, **9**, 113.
11. B. Margiotta, G. Colaprico, R. D'Ovidio and D. Lafiandra, *J. Cereal Sci.,* 1993, **17**, 221.
12. I.L. Batey, R.B. Gupta and F. MacRitchie, *Cereal Chem.,* 1991, **68**, 207.
13. F. Buonocore, C. Caporale and D. Lafiandra, *J. Cereal Sci.,* 1996, **23**, 195

Acknowledgements

This research was partly financed by MiPa, Plant Biotechnology project.

QUANTITATIVE ANALYSES OF STORAGE PROTEINS OF AN OLD HUNGARIAN WHEAT POPULATION USING THE SE-HPLC METHOD

A. Juhász[1], F. Békés[2], Gy. Vida[1], L. Láng[1], L. Tamás[3], Z. Bedő[1]

1. Agricultural Research Institute of HAS, Martonvásár, Hungary. 2. CSIRO Plant Industry, Grain Quality Research Laboratory, North Ryde, NSW Australia 3. Department of Plant Physiology, Loránd Eötvös University, Budapest, Hungary

1 INTRODUCTION

Bánkúti 1201, registered in 1931, was perhaps the most famous and most successful variety in the history of Hungarian wheat breeding. It was of outstanding importance among the winter wheat varieties known collectively today as the old Hungarian wheat varieties. Their populations can be classified among the varieties showing good breadmaking quality despite their 2+12[1,2] or 3+12[3] HMW-glutenin subunit composition on chromosome 1D. The technological quality of these old varieties is generally characterised by high gluten content combined with good gluten quality. To discover the reason for this good quality the storage protein composition (HMW-, LMW-GS and gliadin composition) of Bánkúti 1201 was analysed over the last five years and was found to exhibit high genetic variability.

Besides the HMW-GS composition there are other factors, such as the amounts of the individual subunits, the proportions of HMW- to LMW-glutenins and the ratio of insoluble to soluble polymer fractions, which determine breadmaking quality differences among wheat varieties. Size-Exclusion High Performance Liquid Chromatography (SE-HPLC) has been used overall for the quantification of the three main storage protein groups: albumins + globulins, gliadins and polymeric proteins and to determine the molecular size distribution of the polymeric fraction[4,5].

Studies using SE-HPLC indicate that the amount of polymeric proteins and their size distribution correlate positively with technological properties[5,6].

The aim of the present work was to determine the amounts of the individual protein fractions and their ratios for a better understanding of the good breadmaking quality of Bánkúti 1201 variety.

2 MATERIALS AND METHODS

A set of 23 lines selected from a Bánkúti 1201 population based on their HMW-GS type were analysed, glutenin, gliadin, and albumin/globulin contents of the samples were determined in triplicate by SE-HPLC applying the modified method of Batey et al. (1991)[4]. Unextractable polymeric protein (UPP) percentages were determined using the

method of Gupta and MacRitchie (1994)[7]. The relative amounts of polymeric proteins from the two extracts were expressed as UPP%. The RP-HPLC method of Marchylo et al. (1989) was used for the qualitative/quantitative analyses of individual subunits of glutenins, the HMW to LMW GS ratio and the x to y ratios of the individual HMW-glutenin subunits[8].
Statistical analyses were carried out by Analysis of Variance and Analysis of Pearson correlation analysis using the MSUSTAT v 4.1 (Richard E. Lund, Montana State University, Bozeman, MT, USA) and Super Anova v 1.11 (Abacus Concepts Inc., Berkeley, CA, USA).

3 RESULTS AND DISCUSSION

3.1 SE-HPLC

The amounts of the measured protein fractions varied over a wide range in the population. The glutenin values related to the total soluble protein fraction ranged between 34.2 and 41.45 %. Higher relative gliadin amounts were measured, ranging between 47.35 and 55.54 %. The higher gliadin content is characteristic of the old Hungarian wheat varieties and is exhibited in the high gluten spread and extension values.
A high glutenin to gliadin ratio is, according to the literature, one of the features related to good breadmaking quality. In the case of Bánkúti 1201 the value of this ratio was about 0.7. The UPP %, determined as the relative ratio of polymeric protein in the insoluble fraction to the total polymeric protein fraction ranged from 35.8% to 62.1% (mean 48.4%, Std. Dev. 6.62). These values are above or around the average, compared to other wheat varieties possessing HMW-GS 2+12 on chromosome 1D[7,9,10].
Based on the HMW-GS composition 6 different types could be distinguished, confirming the heterogeneous landrace type characteristic of the variety (Table 1). The subunit composition occurring most frequently in the lines examined was 2* 7+9 (2 or 3)+12. Further experiments, for example gene sequencing and 2D-PAGE, will have to be carried out to prove whether HMW glutenin subunit 2 or 3 is the most abundant in Bánkúti 1201. Some lines containing subunits 5+10 could be identified as well. An interesting mutant wheat line was observed, in which the x type HMW-GS on chromosome B was missing.

Table 1. *Allelic variation at the* Glu-1 *loci and protein composition in the six groups of variety Bánkúti 1201*

HMW-GS composition	No.	Protein	Glutenin	Gliadin	Glu/Gli	%UPP
1 7+8 2+12	1	15.20	35.48	54.57	0.65	46.98
2* 7+8 2+12	6	13.75	38.43	50.36	0.76	52.45
1 7+9 2+12	2	14.48	36.95	52.47	0.70	48.97
2* 7+9 2+12	12	14.38	36.29	53.59	0.68	46.37
2* 8 5+10	1	13.70	35.63	52.52	0.68	40.28
2* 7+8 5+10	1	15.50	40.39	48.64	0.83	56.21
Mean	23	14.23	36.93	52.55	0.70	48.37
L.s.d. (p=0,05)		1.18	3.07	3.229	0.08	9.965

Significant differences can be seen for the measured parameters. Glutenin and UPP content were greater in lines carrying subunit pair 5+10 rather than subunits 2+12. These results confirm that subunits 5+10 have stronger effects on the amount and size distribution of the polymeric protein, as suggested earlier[7]. The significantly lower UPP values noticed in group 2* 8 5+10 are probably due to the absence of Bx type HMW-GS. The presence of subunit Bx7 in other wheat lines, having otherwise the same HMW-GS composition, produced significantly higher polymeric protein content.

3.2 RP-HPLC

The relative HMW-glutenin content ranged between 31.45 % and 46.33 % leading to relatively large HMW to LMW ratios (0.46–0.86). The amounts of single subunits were about average, except for subunit Bx7, which was present in higher amounts (max. 52.3 %). Increased Bx type subunit content led to increases in the HMW to LMW and x to y ratios.

Comparing the groups with different HMW-GS compositions, the groups with 2* 7+8 2+12 and 2* 8 5+10 differed significantly in HMW to LMW ratio from the others. The HMW glutenin content observed in lines containing subunits 2* 7+8 2+12 was significantly higher than in other lines. The Bx mutant line showed lower values.

In a 7+8 2+12 background higher amounts of x-type subunits could be observed in contrast to groups containing subunits 7+9 2+12. Although lines possessing 1 7+8 2+12 and 2* 7+8 2+12 expressed subunit Bx7 in larger amounts, the other x- and y-type subunits were expressed in lower amounts, resulting in no change in the proportions of HMW to LMW.

Table 2. *HMW-, LMW-glutenin content and amounts of individual HMW-GSs measured by RP-HPLC*

HMW-GS Composition	No.	HMW%	LMW%	HMW /LMW	Dy	By	Dx	Bx	Ax	x /y
1 7+8 2+12	1	34.44	65.56	0.53	10.51	9.64	16.76	48.50	14.60	3.96
2* 7+8 2+12	6	38.52	61.48	0.64	10.92	9.36	17.44	49.28	13.00	4.02
1 7+9 2+12	2	37.58	62.42	0.60	14.24	12.41	22.60	34.28	16.47	2.76
2* 7+9 2+12	12	37.30	62.70	0.60	14.18	11.97	22.53	34.64	16.68	2.84
2* 8 5+10	1	32.02	67.98	0.47	20.45	18.54	39.99	0.00	21.02	1.56
2* 7+8 5+10	1	38.53	61.47	0.63	12.36	10.73	22.05	44.66	10.20	3.33
Mean	23	37.34	62.66	0.60	13.37	11.46	21.70	37.96	15.52	3.16
L.s.d. (p=0.05)				0.14	0.52	0.68	0.93	3.39	1.77	

4 CONCLUSIONS

High variability in the SE-HPLC data confirm our earlier electrophoretic results indicating high genetic variability in the Bánkúti 1201 population. The moderate content of soluble proteins and high content of insoluble polymeric proteins in the variety could be an explanation for the good breadmaking quality. The high content of polymeric proteins is mainly due to the higher amounts of Bx-type subunit present in the lines examined. The positive effect of the x/y ratio, described in earlier publications was confirmed by the present results, which also provided important information about the quantitative protein composition of the variety Bánkúti 1201. The impact of the amounts of individual fractions on functional properties should be determined. The identification of genotypes showing different amounts of single storage protein fractions and different allele compositions at the *Glu-1* loci and their application as gene sources could be an important resource in breeding programs.

Acknowledgements:

This work was carried out within the framework of cooperation between CSIRO-Plant Industry (Australia) and the Agriculture Research Institute of the HAS and was supported by the Hungarian Scientific Research Fund (OTKA T 32413).

References

1. Z. Bedo, Symp. EUCARPIA Prospectives of Cereal Breeding in Europe, Landquart, Switzerland, 1994, p. 95.
2. Gy. Vida, Z. Bedo, L. Láng, A. Juhász, *Cereal Res. Commun.*, 1998, **26**, 313.
3. M. Kárpáti, Z. Bedo, R. Fata, H. Budai, *Növénynemesítési Tudományos Napok*, 1994, p. 51.
4. I.L. Batey, R.B. Gupta and F. MacRitchie, *Cereal Chem.*, 1991, **68**, 207.
5. R.B. Gupta, K. Khan and F. MacRitchie, *J. Cereal Sci.*, 1993,**18**, 23.
6. T. Dachkevitch and J.C. Autran, *Cereal Chem.*, 1989, **66**, 448.
7. R.B. Gupta and F. MacRitchie, *J. Cereal Sci.*, 1994, **19**, 19.
8. B.A. Marchylo, J.E. Kruger and D.W. Hatcher, *J. Cereal Sci.*, 1989, **9**, 113.
9. M. Ciaffi, D. Lafiandra, L. Dominici, S. Ravaglia, R. Gupta and F. MacRitchie, Gluten 96, Sydney, 1996, p. 35.
10. O. Larroque, M.C. Giannibelli and F. MacRitchie, *J. Cereal Sci.*, 1999, **29**, 27.

IS THE ROLE OF HIGH MOLECULAR WEIGHT GLUTENIN SUBUNITS (HMW-GS) DECISIVE IN DETERMINATION OF BAKING QUALITY OF WHEAT?

R.Lásztity, S.Tömösközi, R.Haraszi, T.Révay and M.Kárpáti

Department of Biochemistry and Food Technology, Budapest University of Technology and Economics, H-15o2 Budapest, Pf.91. Hungary

1 INTRODUCTION

It is generally accepted[1,2] that the properties of the storage proteins of wheat govern its suitability for processing into bread. Among cereals only bread wheats – and to lesser extent triticale- possess storage proteins which interact with water to yield doughs having the necessary cohesiveness and elasticity for making high specific volume leavened breads.

The correlations between the protein content, the protein composition of the gluten complex and the rheological properties of wheat flour doughs has been thoroughly investigated. Some basic conclusions may be summarized as follows :

- Insoluble protein matrix is an essential pre-requisite for the formation of a cohesive dough.
- There must be a sufficient amount of matrix protein to form a continuous protein phase in the presence of starch and water.
- The ratio of high- and low molecular weight gluten proteins has a significant effect on the rheological properties of dough .

Progress in separation techniques and molecular biology has made possible the isolation of individual polypeptides forming the gluten complex and also the identification of genes coding for these proteins. This fact allowed a deeper study of correlations between individual gluten proteins and baking quality. Recently, most attention has been paid to the glutenin components.

According to recent views of specialists, summarized in reviews of Shewry and Miflin[3], Müller *et al*[4], Shewry *et al*[5], and Shewry and Tatham[6], the glutenin is composed of subunits linked by disulphide bonds. The subunits are divided into two groups: (1) high molecular weight subunits (HMW or HMW-GS) and (2) low molecular weight subunits (LMW or LMW-GS). The HMW subunits are numbered according to electrophoretic mobility within the group and according to chromosome coding for individual polypeptide. A catalogue of genes coding for HMW subunits of wheat is given by Payne and Lawrence[7]. The HMW glutenin subunits are classified into two subgroups: x-type and y-type subunits.

Based on worldwide observations relating to correlations between HMW-GS patterns and wheat quality, Payne[8] proposed the use of the so called "*Glu-1* quality score" for prediction of baking quality of wheat varieties. Although this system is used by breeders in many countries, some newer data suggest that its general validity is doubtful and other factors should also be taken in consideration.

2 WHAT MAKES THE CRITICISM JUSTIFIED? WHAT SHOULD BE CLARIFIED?

2.1. The role of absolute quantity of protein

One of the earliest criticisms was expressed by Huifen and Hoseney[9]. "In taking mind all overall view, we must consider that because the total protein is highly correlated with loaf volume potential, then it appears unlikely that quality is controlled by any one or even a small number of proteins or peptides. If that were true, we must assume that the "critical" protein(s) is highly correlated with total protein content of the sample. This appears to us to be an illogical assumption".

In all cases it is certain that any correlation calculated concerning the dependence of wheat quality on the chemical components of grain is influenced by protein quantity. Consequently, any correlation relating to wheat quality is generally valid for a given range of protein content.

2.2. Effect of quantitative ratios of individual HMW-GS

By calculating the "*Glu-1* quality score" only the qualitative distribution of individual HMW-GS is generally considered. It is a logical suggestion that the quantity may also play a role in formation of the gluten complex. It seems that some of first studies on this question (Gupta[10], review papers of Shewry and Tatham[6] and Grosch and Wieser[11]) confirm the importance of quantitative ratios. Hopefully a more detailed study of the amounts of different subunits will help to explain the big differences, in some cases, between the predicted (based on *Glu-1* score) and actual quality of some cultivars e.g. according to the "*Glu-1* score" the combination of subunits 2 and 12 (2+12) or 4+12 is of low quality value. However, newer studies (in Hungary and Croatia) of the old Hungarian variety Bánkuti, known as high quality hard winter wheat, have shown that this wheat contains subunits 2+12. (Other varieties grown in Hungary containing subunits 2+12 have very low scores : 4-6, as expected). It should be also noted that according to Khan *et al*[12], one of the HRS varieties (HY 320) grown in North Dakota with subunits 2+12 had quite a high score of 8.

2.3. Role of LMW-GS and gliadins

It is generally accepted that the LMW-GS play a role in formation of the gluten complex and in determining its rheological properties. Earlier investigations of Gupta[10] and Pogna *et al*[13] showed that LMW-GS quantity and distribution influence the molecular weight distribution and quantity of polymeric proteins and gluten viscoelasticity of durum wheat.

Nevertheless, until now the role of LMW-GS is not yet fully clarified. There are some views and hypotheses concerning the structure of gluten which suggest a secondary role for these subunits. However, data about quantitative ratios of different gluten subunits show that the quantity of LMW-GS and gliadins is about three times higher than that of HMW-GS. The recent work of Grosch and Wieser[11] revealed that the action of low molecular weight thiol compounds, such as reduced glutathione (GSH), is primarily directed at intermolecular - S – S – bonds between LMW-GS and HMW-GS. It is known that GSH may cause drastic changes in consistency of dough.

It is also an important finding that some D-type (coded by chromosome D) LMW-GS contain only one SH- group and can act as terminators in potential polymerization processes (Masci *et al*[14]). The quantity of this component is a factor which should not be ignored.

The role of gliadin polypeptides also needs further study. It seems that the hypothesis that gliadins act as fillers, plasticizers in the gluten complex, should be modified. It was shown e.g. by Keck *et al*[15] that some peptides obtained by enzymic hydrolysis of purified glutenins contain gamma-gliadin components.

2.4. The effect of dough formation and mixing

It is important to appreciate that under conditions of bread making technology, the disulphide bond system of proteins may be viewed as a dynamic system. Their number and distribution may change depending on mixing conditions, the presence of low molecular weight thiol (disulphide) compounds, enzymes etc. (Grosch and Wieser[11]). For example, the breakage of disulphide bonds during mixing of dough and their reformation during the resting period has been shown experimentally by several researchers. Low molecular weight thiol compounds e.g. GSH may cause disulphide-thiol interchange by formation of protein-S-S-G molecules. This disruption of interprotein bonds may cause weakening of the dough structure.

2.5. The mechanism of disulphide bond formation and polymerization of polypeptides

One of the open questions concerning the disulphide bond system of the gluten complex is the mechanisn of disulphide bond formation *in vivo*. Earlier experiments have shown[1] that using the same mixture of polypeptides (produced by reducing the disulphide bonds of gluten by 2-mercaptoethanol) fractions with quite different rheological properties may be produced by reoxidation under different conditions (concentration, pH, presence of urea etc). These results suggest that the process of formation of disulphide bonds and probably the polymerization process of glutenin subunits may be regulated. At present we know that all wheat gluten proteins are synthetized on the rough endoplasmic reticulum (RER), with a signal peptide that is cleaved as it directs the nascent polypeptide into the RER lumen. Protein folding and disulphide bond formation then occur within the RER. Among enzymes playing a role in post-translational modification of proteins, protein disulphide isomerase (PDI) may be of interest in elucidation of the mechanism of disulphide bond formation in gluten-forming polypeptides. This enzyme is located within the lumen of ER and has been demonstrated to catalyze disulphide bond formation in secretory proteins in a range of biological systems[16]. The presence of PDI in wheat endosperm, aleurone layer and embryo was confirmed by several researchers[17-19]. Bulleid and Friedman[20,21] and Bulleid *et al*[22] showed that PDI is able to catalyse the formation of intrachain disulphide bonds in gamma-gliadin synthesised *in vitro* and also in some HMW-GS and LMW-GS. However, none of the proteins were observed to form stable disulphide-linked oligomers. Experiments made with procollagen[23] also showed that the formation of disulphide-stabilised trimer is much faster in the presence of PDI. Consequently, the role of PDI in the formation of interchain disulphide bonds in gluten cannot be excluded. Bearing in mind that wheat grain contains a lot of other redoxy-enzymes and the possible role of molecular chaperones, it is clear that a lot of further investigations are needed to give a final answer to the questions raised in this paper.

3 CONCLUSIONS

The prediction of wheat quality based on calculation of the "*Glu-1* score" has been widely applied in recent breeding strategies in several countries. Nevertheless, the level of significance between predicted and actual quality is, in many cases, low and some contradictory data were recently obtained. These facts suggest that further investigations of additional factors such as protein quantity, the role of LMW-GS and gliadins, the involvement of enzyme(s) and molecular chaperons, are needed.

References

1. R.Lásztity, 'The Chemistry of Cereal Proteins', CRC Press Inc., Boca Raton, 1996.
2. R.Lásztity, 'Cereal Chemistry', Akademiai Kiadó, Budapest, 1999,p.69.
3. P.R.Shewry and B.J.Miflin, Seed storage proteins of economically important cereals', Advances in Cereal Science and Technology Vol.7., AACC, St.Paul, 1985, p.1.
4. S.Müller, W.H.Vensel, D.D.Kasarda, E.Köhler and H.Wieser, *J.Cereal Sci.*,1998, **27**, 109
5. P.R.Shewry, N.G.Halford and A.S.Tatham, *J.Cereal Sci*, 1992 ,**15**, 105
6. P.R.Shewry and A.S.Tatham, *J.Cereal Sci.*,1997, **25**,207
7. P.I. Payne and G.J.Lawrence, *Cereal Res.Comm.* 1983, **11**,29
8. P.I.Payne, 1987, *Aspects Appl.Biol.*, **15**, 79
9. Huifen He and R.C.Hoseney, 'Gluten, a theory of how it controls bread-making quality', Gluten Proteins 1990 AACC, St Paul, 1991,p.1.
10. R.B.Gupta, Qualitative and quantitative variation in LMW and HMW glutenin subunits: Their effect on molecular size of proteins and dough properties' Gluten Proteins 1993', Detmold, 1993, p.151
11. W.Grosch and H.Wieser, *J.Cereal Sci.*,1999, **29**,1
12. K.Khan, J.Figueroa and K.Chakraborty, Relationship of gluten protein composition to breadmaking quality of HRS wheat grown in North Dakota, Gluten Proteins 1990, AACC, St Paul, 1991,p.81
13. N.Pogna, D.Lafiandra, P.Feillet and J.C.Autran, *J.Cereal Sci.*,1988, **7**,211
14. S.Masci, T.A.Egorov, C.Ronchi, D.D.Kuzmicky and D.D.Kasarda, *J.Cereal Sci.*, 1999, **29**, 17
15. B.Keck, P.Köhler and H.Wieser, *Z.Lebensm.Untersuch. Forsch.*, 1995, **200**, 432
16. N.J.Bulleid, 'Protein disulphide isomerase: Role in biosynthesis of secretory proteins', Advances in Protein Chemistry Vol.44.,1992, p.125
17. L.T.Roden, B.J.Miflin and B.B.Freedman, *FEBS Lett*, 1982, **138**, 121

18. Y.Shimoni, X-Z.Zhu, H.Levonony, G.Segal and G.Galili, *Plant Physiol.*,1995, **108**, 327
19. M.A.Livesley, N.J.Bulleid and C.M.Bray, *Seed Sci. Res.,* 1992, **2**, 97
20. N.J.Bulleid and R.B.Friedman, *Nature*, 1988, **335**, 649
21. N.J.Bulleid and R.B.Friedman, *Biochem. J.*, 1988, **254**, 805
22. N.J.Bulleid, P.R.Shewry and R.B.Friedman, 'Plant Protein Engineering', Cambridge University Press, Cambridge, 1992, p. 201
23. J.Koivu and R.Myllyla, *J.Biol.Chem.*1987, **262**, 6159

Acknowledgements

Funding for the work by OTKA Res. Project T 029712, is gratefully acknowledged.

LOW MOLECULAR WEIGHT GLUTENIN SUBUNIT COMPOSITION AND GENETIC DISTANCES OF SOUTH AFRICAN WHEAT CULTIVARS

H. Maartens and M.T. Labuschagne

Department of Plant Breeding, University of the Orange Free State, Bloemfontein 9300, South Africa.

1 INTRODUCTION

It is becoming increasingly difficult to distinguish between newly released varieties. The accurate identification of plant breeding material is very important for the protection of plant breeders' rights and for effectiveness in breeding programmes. In South Africa, the HMW glutenins are mainly used to distinguish between wheat cultivars[2]. This study has proved, however, that the HMW glutenins are not reliable for cultivar identification since many cultivars in South Africa have the same HMW banding combinations. The LMW glutenins have proved to be very effective for cultivar identification, since all the cultivars had different banding patterns. Research on the LMW glutenin subunits was limited in the past, since they fractionated inadequately on SDS-PAGE. The introduction of a simplified one-step, one-dimensional SDS-PAGE procedure[1] provided a rapid method for analysing a large number of samples on a single gel in a gliadin-free background. The aim of this study was to determine the LMW glutenin subunit composition of South African wheat cultivars and to use this data to determine the genetic distances between cultivars.

2 MATERIALS AND METHODS

Forty-five South African cultivars were screened for their HMW and LMW glutenin subunit composition. Glutenins were extracted[1]. The HMW and LMW glutenins were evaluated on the same gel. Due to their overlap with the LMW glutenins, the gliadins were extracted and discarded. Forty µl of the samples were loaded on a 10% SDS-PAGE gel. It was run at 66 mA for three hours at 15^{0}C. The gel was fixed overnight in acetic acid and stained with trichloroacetic acid and Coomassie Blue. After it was destained with distilled water, it was analysed with the "Molecular Analyst Fingerprinting" software of Biorad.

Table 1 *LMW glutenin bands of some of the near isogenic lines which were tested*

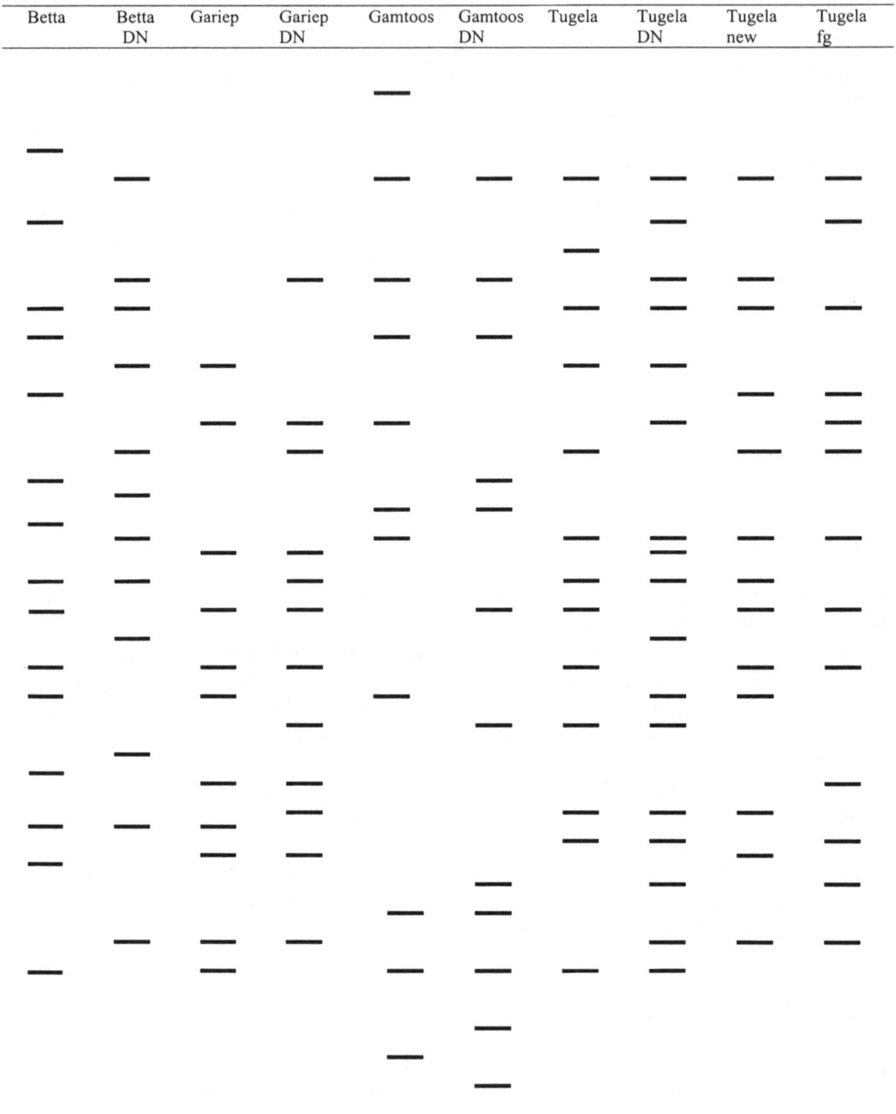

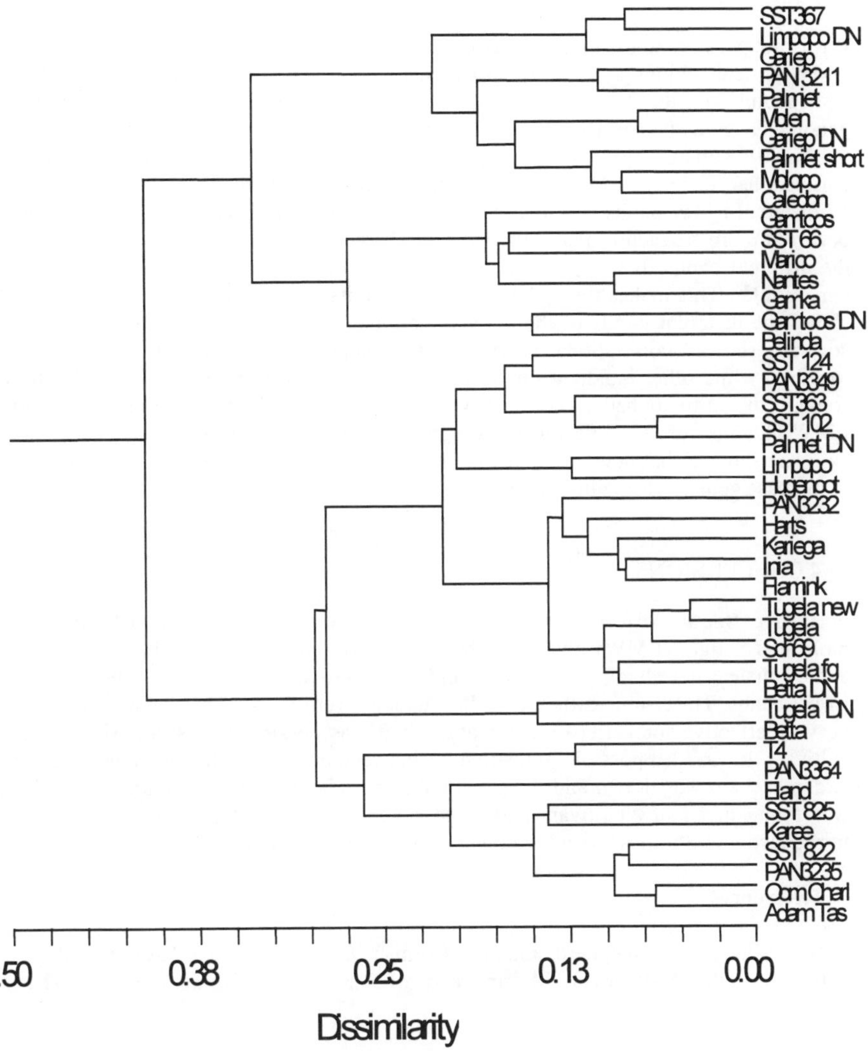

Figure 1 *A dendrogram of the South African wheat cultivars tested.*

3 RESULTS AND DISCUSSION

It was not possible to distinguish between cultivars with a close genetic relationship using the HMW glutenin subunits. Betta, Betta DN, Gariep, Gariep DN, Gamtoos and Gamtoos DN all had subunit 1 on the A genome, subunits 7+9 on the B genome and subunits 5+10 on the D genome. Tugela, Tugela DN, Tugela new and Tugela fast-growing all had subunit 2* on the A genome, subunits 7+8 on the B genome and subunits 5+10 on the D genome. All the cultivars had different LMW glutenin banding patterns. The cultivars had between nine and 17 different LMW bands and there were 39 different LMW bands. Some of the LMW bands occurred more frequently than other bands. There were also differences in the intensities of the different bands. It was possible to distinguish between cultivars which were genetically very similar. This makes the LMW glutenin subunits ideal for cultivar identification. Table 1 shows the different bands of the above-mentioned cultivars. Analysis with the NCSS 2000 software showed correlations between certain bands. Some combinations always occurred together, while other bands never occurred together. A dendrogram (Figure 1) showed that some of the cultivars had a very close genetic relationship. It can therefore be concluded that LMW glutenins are suitable for cultivar identification, but their expression in South African cultivars showed that new germplasm must be introduced into existing breeding programmes to increase genetic variability in wheat cultivars.

4 CONCLUSIONS

Forty-five South African bread wheat cultivars were screened to determine their low molecular weight (LMW) glutenin subunit composition. The LMW subunits were extracted and evaluated in a gliadin-free background[1]. The cultivars had between nine and 17 different LMW bands. Thirty-nine different LMW subunits were found. The LMW subunits proved to be very effective for variety identification as it was possible to distinguish between all the cultivars. It was also possible to distinguish between cultivars which were genetically similar. A dendrogram was drawn and the genetic distances between the cultivars were calculated. It was found that many cultivars had a close genetic relationship. New germplasm must be introduced into South African bread wheat breeding programmes.

References

1. N.K. Singh, K.W. Shepherd, and G.B. Cornish, *J. Cereal Science,* 1991, **14,** 203.
2. P.G. Randall, M Manley, L. Meiring, and A.E.J. McGill, *J. Cereal Science,* 1992,**16,** 211.

A NEW LMW-GS NOMENCLATURE FOR SOUTH AFRICAN WHEAT CULTIVARS

H. Maartens and M.T. Labuschagne

Department of Plant Breeding, UOFS, P.O.Box 339, Bloemfontein 9300, South Africa.

1 INTRODUCTION

Wheat is one of the world's most important crops and the need for variety identification is probably far greater than for any other cereal grain. The endosperm is the largest tissue in the grain and it contains the majority of proteins. The gliadins and glutenins are the most important proteins. Research on low molecular weight glutenin subunits (LMW-GS) has been limited, because they do not fractionate adequately by SDS-PAGE. The introduction of a one-step, one-dimensional SDS-PAGE procedure[1] provided a rapid method for analysing a large number of samples in a single gel in a gliadin free background. A nomenclature system was developed for LMW-GS in Australia[2] but this system was not suitable for South African wheat cultivars, as many did not fit the system. The aim of this study was therefore to use the LMW-GS composition of South African bread wheat cultivars and correlations between the bands, and the analysis of monosomics to develop a new nomenclature system that will include all the cultivars.

2 MATERIALS AND METHODS

The materials consisted of a total of 101 South African bread wheat cultivars (all the old and new commercial wheat cultivars were included) and 46 near isogenic lines (NIL's) as well as monosomic lines of Chinese Spring, Flamink, Inia and SST3. Protein extraction was done[1], and samples were then run with SDS-PAGE[3] followed by a staining procedure[4]. Six replications of each entry were analysed. The "Molecular Analyst Fingerprinting" software was used for gel analysis. A low range SDS-PAGE marker of BIORAD was used to establish molecular weights of subunits. Band intensities were also calculated by dividing bands into classes from very light (1) to very dark (5). Correlations between bands were determined using the Spearman-rank correlation matrices of the NCSS 2000 programme. A combination of the correlations and the monosomics were used to determine the chromosomes which code for specific LMW-GS bands.

3 RESULTS AND DISCUSSION

Thirty-nine different LMW-GS were identified. The different bands were divided into class intervals. All the entries had specific banding patterns. A detailed description of banding patterns is given in another paper in these proceedings. The 39 bands were used to test for correlations between them. Significant negative correlations between bands indicated that the presence of one band was often associated with the absence of another, while significant positive correlations indicated that some band combinations occurred together. Table 1 gives a summary of all the LMW glutenin bands present in Chinese Spring, Flamink, Inia and SST3 and their monosomic lines. A combination of the data from the monosomic analysis and the correlations was used to determine chromosomes responsible for LMW-GS. Band 1 was coded by 1B and 1D, bands 2, 3 and 4 were coded by 1A, 1B and 1D. Bands 5 and 6 were coded by 1A and 1D, band 7 is coded by 1A, and band 8 by 1A, 1B and 1D. Band 9 was coded by 1A and 1B, and band 10 by 1A, 1B and 1D. Bands 11 and 12 were coded by 1B and 1D. Band 13 was coded by 1A, 1B and 1D, and band 14 by 1B and 1D. Band 15 was coded by 1A and 1B, and band 16 by 1A, 1B and 1D. Band 17 was coded by 1A, 1B and 1D, and band 18 by 1A and 1D. Band 19 was coded by 1B and 1D, and band 20 by 1B. Band 21 was coded by 1A and bands 22 and 23 by 1B and 1D. Band 24 was coded by 1A, 1B and 1D, and band 25 by 1B and 1D. Band 26 was coded by 1A and 1D, and bands 27, 28 and 29 by 1A, 1B and 1D. Band 30 was coded by 1B, and band 31 by 1A and 1B. Band 32 was coded by 1B and 1D, and band 33 by 1A and 1B. Band 34 was coded by 1B and 1D and band 35 by 1B. Band 37 was coded by 1B and 1D and band 38 by 1A and 1B. Band 39 was coded by 1B.

Definite banding combinations were found (102 pairs), and it can be assumed that these combinations will always occur together. The least number of bands was controlled by chromosome 1A, which confirmed results reported by other authors[5]. A minimum of 3 (in SST367) and maximum of 13 bands (Tugela and Tugela DN) were controlled by 1B. A minimum of 2 bands (in SST367) and maximum of 13 bands (in SST66, SST101, SST822 and T4) were controlled by chromosome 1D.

The monosomic lines proved to be effective to determine which chromosomes are responsible for the expression of specific bands or subunits. With the help of the calculated correlations between the bands, band combinations were determined, which were expressed by specific chromosomes. With the new nomenclature system, all the cultivars could be distinguished, which was not the case when the Australian system[2] was used. With their system Belinda, Betta, Limpopo, Molopo and Nantes had no matching combinations in group 2, and Molen, Molopo and Nantes had no matching combinations in group 3. In four of 30 cultivars tested previously[6] banding combinations c and f (group 1) appeared together in the same South African cultivars, and these combinations were therefore not alternatives to each as other as reported by other authors[2]. Another problem was that many bands, which were reported by them as being faint, were dark bands in this study (for example Tugela with combination, group 2 was supposed to have faint bands, but in our study the bands were dark). This raises the question whether there are differences in this regard between South African and Australian wheats, or whether different genes code for these bands.

Tugela and Karee NIL's, which are very closely related, could be distinguished by this nomenclature system. In some cases, bands in the same group were expressed by all

Table 1 *The different LMW glutenin bands in the cultivars and their monosomic lines.*

	CS	1A	1B	1D	Fla	1A	1B	1D	In	1A	1B	1D	Sst	1A	1B	1D
1																
2	—	—	—	—	—	—	—	—		—	—	—		—	—	—
3									—				—			—
4		—		—	—			—	—	—	—	—	—		—	
5	—		—											—		
6		—		—	—	—	—			—			—		—	—
7					—				—		—	—				
8	—					—	—			—				—		—
9		—		—				—			—	—			—	
10	—		—	—	—	—			—			—			—	—
11											—		—	—		
12		—							—	—						
13			—		—	—	—					—	—			—
14	—	—	—	—		—		—			—		—	—	—	
15																
16	—							—		—	—	—	—	—	—	
17		—	—	—	—	—			—							
18							—	—		—				—		
19	—	—	—	—	—	—				—	—	—			—	—
20									—				—	—		—
21								—	—		—	—			—	—
22	—	—	—		—	—	—	—		—			—	—		
23				—	—	—			—	—		—		—		—
24	—	—					—		—		—	—	—		—	—
25			—			—		—		—						—
26					—		—	—				—	—	—	—	
27		—	—	—					—							—
28							—	—		—					—	
29	—	—	—	—	—	—					—	—	—	—		
30													—	—		—
31	—			—		—	—	—	—			—				
32					—					—	—					
33		—	—	—				—	—			—		—	—	—
34				—	—	—	—	—			—		—	—		
35										—		—				
36	—		—	—					—		—		—	—	—	—
37	—	—		—	—	—	—		—	—	—	—				
38	—	—	—	—	—	—	—	—		—	—	—	—	—	—	—
39													—			

CS = Chinese Spring, Fla = Flamink, In = Inia, Sst = SST3

three chromosomes. These bands could have the same molecular weight but a different amino acid composition or quality effects. We are planning to shortly look at the relationship between gliadins and LMW-GS, and how these fractions can be used in combination for cultivar identification and quality assessment.

Table 2 *LMW-GS composition of Tugela NIL's, indicating different subunit composition*

	Chromosome 1A	Chromosome 1B	Chromosome 1D
Tugela	10,15,18,21,28,29	10,12,15,20,25,28,29	10,12,18,25,28,29
Tugela DN	7,9,10,18,24,28,29,33	9,10,12,14,19,20,22,26, 25,28,29,33,34	10,12,14,18,19,22,24, 25,28,29,34
Tugela fast growing	5,13,15,18,27,29	13,14,15,23,27,29	5,13,14,18,23,27,29
Tugela selection	9,10,15,21,24,28	9,10,15,20,23,24,28	10,23,24,28

References

1. N.K.Singh, K.W. Shepherd and G.B. Cornish,. *J. Cereal Science*, 1991, **14**, 203.
2. R.B.Gupta and K.W. Shepherd,. TAG, 1990, **80,** 65.
3. H. Maartens,. Ph.D. thesis, 1999, UOFS, Bloemfontein, South Africa.
4. C.W. Wrigley, 1992. In: *Advances in Cereal Science and Technology*, vol.5, ed. Y. Pomeranz. American Association of Cereal Chemists, St. Paul, MN.
5. F.MacRitchie,. *Advances in Food and Nutritional Research*, 1992, **36**, 1.
6. H. Maartens,. MSc Thesis,1997, UOFS, Bloemfontein, South Africa.

INTRODUCTION OF THE D-GENOME RELATED HIGH- AND LOW-M_r GLUTENIN SUBUNITS INTO DURUM WHEAT AND THEIR EFFECT ON TECHNOLOGICAL PROPERTIES

Lafiandra D.[1], Margiotta B.[2], Colaprico G.[2], Masci S.1, Roth, M.R.[3], MacRitchie F.[3]

1. Dept. of Agrobiology & Agrochemistry, University of Tuscia, Viterbo (Italy). 2. Germplasm Institute, C.N.R., Bari (Italy). 3. Dept. of Grain Science, Kansas State University, Manhattan (KS), USA.

1 INTRODUCTION

High- and low-M_r glutenin subunits are of prime importance in determining technological properties, both in bread and durum wheat, through their effect in modulating glutenin polymer size[1,2]. Although durum wheat is mostly used for pasta production, its use for the preparation of bread is also widespread, expecially in many Mediterranean countries, in spite of the fact that durum wheat breadmaking quality is inferior to that of bread wheat. The poor performance of durum wheat in breadmaking has been attributed to the absence of the D-genome related proteins. In fact, analyses of D-genome disomic substitution lines of durum wheat cultivar Langdon, carried out by Liu *et al.*[3], have demonstrated that the chromosome 1D substitutions have large effects on the amount of glutenin, SDS sedimentation value, mixing time and peak resistance value.

More recently chromosomal segments containing genes encoding D-genome related gliadins and glutenins have been transferred into durum wheat[4-7]. In particular, Lukaszewsky and Curtis[6,7], through chromosome engineering, have been able to transfer chromosome segments carrying the pairs of high-M_r glutenin subunits 5+10 or 2+12, encoded by genes present at the *Glu-D1* locus of bread wheat, onto chromosomes 1R and 1A of triticale and subsequently to the durum wheat cultivars Monroe, Turbo and WPB 881, replacing the null allele present at the *Glu-A1* locus[8]. Pogna *et al.*[9] have used the bread wheat cultivar Perzivan, which contains a translocated segment on the short arm of chromosome 1A containing the genes encoding gliadins and low molecular weight glutenin subunits, to introduce the proteins normally present at *Gli-D1/Glu-D3* on the short arm of chromosome 1D into durum wheat.

Several isogenic lines carrying the pairs of subunits 5+10 or 2+12 have been obtained by Lafiandra *et al.*[1], using Italian durum wheat cultivars as recurrent parents. Preliminary quality data on one of such lines are reported together with the production of a durum wheat line carrying the entire set of gliadin and glutenin components normally associated with chromosome 1D of bread wheat.

2 MATERIALS AND METHODS

Durum wheat lines carrying the 1D/1A translocation containing genes encoding high-M_r glutenin subunits 5+10 were supplied by A. Lukaszewsky. These materials have been crossed with the Italian durum wheat cultivar Svevo. Similarly, the bread wheat cultivar Perzivan carrying a translocated segment of the short arm of chromosome 1D containing the *Gli-D1/Glu-D3* loci on the short arm of chromosome 1A, was crossed with the same durum wheat cultivar. After repeated backcrosses using Svevo as recurrent parent, two lines were obtained carrying the genes encoding high-M_r subunits 5+10 on the long arm of chromosome 1A and the genes encoding *Gli-D1/Glu-D3* proteins on the short arm of the same chromosome. These were crossed and a line expressing both subunits 5+10 and the *Gli-D1/Glu-D3* proteins was isolated.

2.2.1 *Electrophoretic analyses.* Two dimensional electrophoretic analyses (A-PAGE SDS-PAGE) were carried out as reported by Redaelli *et al.*[10].

2.2.2 *Chromatographic analyses (SE-HPLC).* SE-HPLC was carried out as reported by Batey *et al.*[11], but using a Biosep-SEC-S4000 column (Phenomenex).

2.2.3 *Mixographs.* Mixographs were obtained with a computerised 10g Mixograph (TMCO) using a Mixsmart software program.

3 RESULTS

The qualitative properties of a near isogenic line (nil) possessing high-M_r glutenin subunits 5+10 together with subunits 7+8, derived from the durum wheat cultivar Svevo, which represents one of the most currently widely grown cultivars in Italy, have been evaluated and are reported in Table 1.

Table 1 *SE-HPLC parameters and mixing properties of the durum wheat cultivar Svevo and a derived line with the subunit-pair 5+10.*

High M_r Composition	%Peak 1	%Peak 2	%Peak 3	%UPP	Mixographic peak dev. time (min)
7+8	47.2	41.9	10.9	54.1	5.1
7+8/5+10	49.8	40.0	10.2	62.1	15.0

The chromatographic parameters (SE-HPLC) of the durum wheat cultivar Svevo do not show great difference in the % of total polymeric proteins (% Peak 1) from the derived nil which also has subunits 5+10. In contrast, when the percentage of unextractable polymeric proteins (%UPP) was determined, a large difference was found. In fact, a higher amount of unextractable polymeric proteins was present in the durum wheat line possessing subunits 5+10 compared to Svevo, indicating that the molecular weight distribution was shifted upwards. This was consistent with the mixing properties which showed an appreciable increase in dough strength in the subunit 5+10 line as reflected in the much longer dough development time.

Crossing the durum wheat cultivar Svevo with the bread wheat cultivar Perzivan allowed the replacement of the low-M_r glutenin subunits encoded by *Glu-A3* with homoeoallelic subunits normally present at the *Glu-D3* locus in bread wheat.

Subsequently, crossing of this with the line carrying subunits 5+10 resulted in the production of a durum wheat genotype having both high- and low-M_r glutenin subunits associated with the D-genome (Fig. 1). This material is currently being increased for quality evaluation.

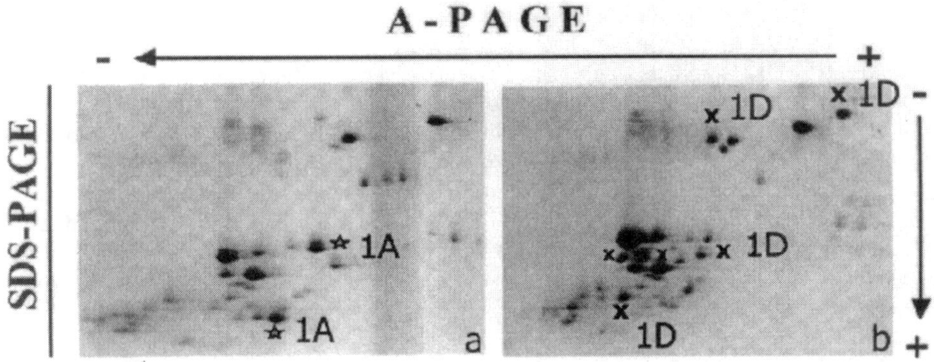

Figure 1. *Two-dimensional electrophoretic separation of glutenin subunits present in Svevo (a) and a derived nil possessing D-genome associated high- and low-*M_r *glutenin subunits (x). Stars indicate 1A encoded proteins of durum wheat Svevo replaced by 1D encoded low-*M_r *glutenin subunits.*

4 CONCLUSIONS

In bread wheat it has been suggested that an increase in the number of genes actively expressing high-M_r glutenin subunits can lead to improvement of flour breadmaking properties as a result of increases in the amount of the large polymeric glutenins.

The introduction of specific subunits that have been associated with dough strength (e.g., 5+10) may prove particularly effective. This can be simply done by replacement of the null allele present at the *Glu-A1* locus with alleles expressing only the x- or both x- and y-type subunits. This latter combination is quite widespread in diploid and tetraploid wild wheats[13]. A similar approach can be taken for durum wheat; indeed, durum wheat lines with an increased number of high-M_r glutenin subunits have already been produced. In particular, Ciaffi *et al.*[12] demonstrated that durum wheat lines possessing both x- and y-type subunits at the *Glu-A1* locus had high dough strength and baking performance, as good as those of the bread wheat cultivars used as controls.

Wheat chromosome engineering, which consists of the transfer of chromosomal segments between wheat and related *Triticeae* species through manipulation of the homoeologous pairing process[4], has been successfully employed in manipulating the protein composition of triticale and wheat; subunits 5+10 or 2+12, and also gliadins and low-M_r glutenin subunits normally associated with the chromosome 1D of bread wheat, have been transferred to chromosome 1A of durum wheat using this approach and quality studies have been already carried out.

For example, Ammar *et al.*[8] have produced translocated durum wheat lines carrying subunits 5+10 encoded by genes present at the *Glu-D1* locus in the bread wheat cultivar Wheaton. A large increase in SDS-unextractable polymeric proteins and a relative increase in the SDS sedimentation volume were observed although the increase was dependent on the different durum cultivars used.

Similarly, Pogna *et al.*[9] have demonstrated the positive influence of the *Gli-D1/Glu-D3* proteins in influencing the breadmaking properties of durum wheat semolina. In particular, they found that these proteins reduced dough tenacity and increased extensibility improving the breadmaking properties of durum wheat.

Durum wheat transformation has recently been used to introduce a gene corresponding to subunit 1Dx5[14], but this approach still has severe limitations before practical use in breeding is feasible. The material developed by using chromosome engineering does not suffer from these problems and permits manipulation of the protein composition of durum wheat by introducing useful proteins from genomes different to A and B.

The introduction of D-genome related high-M_r glutenin subunits further increases the limited genetic variation existing for these components in durum wheat and allows exploration of their effects on technological properties with a view to diversify semolina end uses.

References

1. D. Lafiandra, S. Masci, C. Blumenthal, C.W. Wrigley, *Cereal Foods World*, 1999, **44**, 572.
2. M.C. Gianibelli, M. Ruiz, J.M. Carrillo, F. MacRitchie, 'Wheat Structure Biochemistry and Functionality', I.P. Schofield Ed., Royal Society of Chemistry, 1995, p. 146.
3. C.Y. Liu, A.J. Rathijen, K.W. Shepherd, P.W. Gras, L.C. Giles, *Plant Breeding*, 1995, **114**, 34.
4. C. Ceoloni, M. Ciaffi, D. Lafiandra, B. Giorgi, In: Proc. 8th Int. Wheat Genet. Symp. Z.S. Li and Z.Y Xin Eds. 1993, p. 159.
5. F. Vitellozzi, M. Ciaffi, L. Dominici, C. Ceoloni, *Agronomie*, 1997, **17**, 413.
6. A.J. Lukaszewski, C.A. Curtis, *Plant Breeding*, 1992, **109**, 203.
7. A.J. Lukaszewski, C.A. Curtis, *Plant Breeding*, 1994, **112**, 177.
8. K. Ammar, A.J. Lukaszewski, G.M. Banowetz, *Cereal Foods World*, 1997, **42**, 610.
9. N.E. Pogna, M. Mazza, R. Redaelli, P.K.W. Ng, 'Proc. 6th Int. Gluten Workshop', C.W. Wrigley Ed. 1996, p. 18.
10. R. Redaelli, M.H. Morel, J.C. Autran, N.E. Pogna, *J. Cereal Sci.*,1995, **21**, 5.
11. I.L. Batey, R.B. Gupta, F. MacRitchie, *Cereal Chem.*, 1991, **68**, 207.
12. D. Lafiandra, S. Masci, B. Margiotta, E. De Ambrogio, 'Proc 9th Int. Wheat Genet. Symp.', A.E. Slinkard Ed., University of Saskatchewan, University Extension Press, 1998, p. 261
13. M. Ciaffi, D. Lafiandra, T. Turchetta, S. Ravaglia, H. Bariana, R. Gupta, F. MacRitchie, *Cereal Chem.*, 1995, **72**, 465.
14. G.Y. He, L. Rooke, S. Steele, F. Békés, P. Gras, A.S. Tatham, R. Fido, P. Barcelo, P.R. Shewry, P.A. Lazzeri, *Molecular Breeding*, 1999, **5**, 377.

Acknowledgements

We wish to thank A. Lukaszewsky for supplying durum wheat lines carrying the 1D-1A translocation. The research was supported by the Italian "Ministero dell'Università e della Ricerca Scientifica e Tecnologica (ex 40%).

EFFECTS OF HMW GLUTENIN SUBUNITS ON SOME QUALITY PARAMETERS OF PORTUGUESE LANDRACES OF *TRITICUM AESTIVUM* SSP. *VULGARE*

C. Brites[1], A. S. Bagulho[1], M. Rodriguez-Quijano[2], J.M. Carrillo[2]

1.Estação Nacional de Melhoramento de Plantas, Apartado 6 7351 Elvas Codex, Portugal.
2.Departamento de Genética, ETSIA-Agrónomos, Universidad Politécnica, E-28040 Madrid, Spain

1 INTRODUCTION

Old cultivars can be useful for breeding purposes as a source of protein variation[1]. For this, it is necessary to know the variability in protein composition and its relationship with technological quality of the germplasm.

Wheat storage proteins are known to play a major role in determining dough technological quality and many studies[2,3] have demonstrated the influence of allelic variation in HMW-glutenin subunits controlled by the *Glu-1 loci* on quality differences in bread wheat varieties. The HMW glutenin subunit composition of Portuguese collections of hexaploid wheats has been characterized[4,5]. However, no information was available about the technological quality of these lines and their relationship with the protein composition. The objective of present study was to investigate the relationship between HMW glutenin composition and quality parameters in a collection of bread wheat landraces.

2 MATERIALS AND METHODS

2.1 Plant material

The 39 Portuguese landraces of *Triticum aestivum* ssp. *vulgare* used in this study were taken from the collection[6] kept at the National Plant Breeding Station, Elvas, Portugal. The material was grown in a randomised complete block design with three replicates in 1999.

2.2 Methods

2.2.1 Glutenin composition. The HMW glutenin subunits have been partially characterised in a previous work[4] and reanalysed in milled flour by the same methods. The same nomenclature for HMW glutenin subunits was used.

2.2.2 Quality evaluation. Samples of grains from each replicate were tempered to 14% moisture during 24 hours and milled using a Cyclotec mill (Tecator, Sweden) equipped with a 0.5 mm sieve. The protein content and grain hardness were determined by NIR[7,8] and the SDS-sedimentation test[9] was performed. To assess the mixing properties, wholemeal was fractionated to obtain a particle size below 250 μm and then used for the 10 g Mixograph[8] with modification of absorption water according to protein content and grain hardness[10]. The following parameters were measured: maximum peak height (MPH), time to MPH (mixing development time-MDT) and the difference in percentage between MPH and height at 3 min after the peak of the curve (resistance breakdown - BDR).

In 25 varieties with sufficient grain, the three replicates were bulked and the micro alveograph test was performed on 50 g flour produced on a Chopin CD1 mill, according the manufacture's instructions and the gluten strength-W, tenacity–P and extensibility–L were determined.

2.2.3 Statistical analysis. Analysis of variance (general linear model procedure[11]) was used to study the effects of different HMW glutenin alleles on mean technological values. The allelic variation at *Glu-A1*, *Glu-B1* and *Glu-D1 loci* were considered as effects. The F-test for variance significance was derived from the type III sum of squares[11] according to fixed effect models. Duncan test was used for allele means comparisons. The relationships among the quality tests were examined by Pearson correlation coefficients.

3 RESULTS AND DISCUSSION

3.1 Variation in HMW–Glutenin Subunits

The SDS-PAGE patterns showed (Figure 1, Table 1) that some genotypes were not identical to those previously characterised[4]. At *Glu-A1*, some variation was detected: the HMW 2* (allele *b*) was identified in four genotypes with the assigned null allele; in seven genotypes a HMW subunit was detected with mobility between subunits 1 and 2* which may be identical to the 2¨ subunit (*Glu-A1 f* allele) previously detected[12] in populations of 'Barbela' wheat. At the *Glu-D1* locus, HMW glutenin subunits 2+12 (allele *a*) were largely predominant being expressed in 36 lines out of 39, only one genotype was found to have HMW glutenin subunits 5+10.

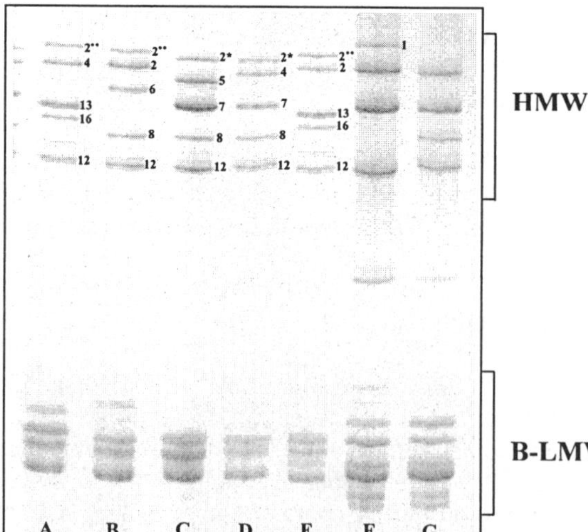

Figure 1: SDS-PAGE patterns of HMW glutenin subunits in some Portuguese wheat varieties and varietal standards:
A – 'Barbela'
B – 'Ribeiro'
C – 'Ablaca' (standard)
D – 'Tremês Arroxeado'
E – 'Viloso Mole'
F – 'Flinor' (standard)
G – 'Champlein' (standard)

3.2 Variation in Quality Parameters

There was a high concentration of protein among the landraces (Table 1, higher 13%) in agreement with other characterised collections[1]. Large variation was observed for hardness and SDS sedimentation, but in general the dough was weak with low P/L values corresponding to extensible gluten type.

Table 1 *Varieties analysed,* Glu-1 *alleles and mean values for grain hardness, protein content, SDS sedimentation test (SDS), mixograph (MDT, MPH and BDR) and alveograph (W and P/L) parameters*

Varieties	*Glu-1* alleles†			Hardness	Protein	SDS	MDT	MPH	BDR	W	P/L
	Glu-A1	*Glu-B1*	*Glu-D1*		(%)	(mm)	(min)	(mm)	(%)	(10^{-4}J)	
'Alentejano'	*f*	*f*	*a*	46	14,7	105	1,6	95	20,1	151	0,60
'Almadense'	*b*	*f*	*a*	46	15,5	36	0,5	81	42,9	32	0,44
'Barbela'	*f*	*f*	*c*	47	14,3	67	1,1	81	25,5	118	0,43
'Bejense'	*b*	*e*	*a*	97	16,9	53	0,8	91	48,4	-	-
'Belém'	*b*	*a*	*a*	37	15,7	57	0,7	92	39,9	86	0,35
'Eborense'	*b*	*f*	*a*	43	16,3	97	1,5	97	30,0	-	-
'Egípcio'	*f*	*d*	*a*	81	16,9	55	1,1	99	42,4	-	-
'Fronteiriço'	*c*	*e*	*a*	43	14,0	28	0,6	82	40,5	51	0,62
'Funchal'	*b*	*e*	*a*	38	18,0	74	0,8	99	33,3	-	-
'Galego barbado'	*b*	*e*	*a*	47	15,7	38	0,6	88	45,7	54	0,47
'Galego rapado'	*a*	*e*	*a*	39	13,3	89	1,6	87	23,0	-	-
'Grécia'	*b*	*e*	*a*	43	15,0	58	0,8	90	40,1	54	0,37
'Grécia ruivo'	*b*	*e*	*a*	31	15,5	48	0,6	90	46,6	45	0,57
'Ideal'	*b*	*a*	*a*	40	17,2	43	0,5	95	43,2	-	-
'Liz'	*b*	*e*	*a*	78	19,0	61	0,8	89	38,2	-	-
'Magueija'	*b*	*e*	*a*	95	16,7	47	1,1	100	42,7	94	0,82
'Mestiço'	*b*	*e*	*a*	91	18,4	65	0,9	99	39,4	-	-
'Mocho cabeçudo'	*b*	*f*	*a*	43	15,9	91	1,4	92	26,1	174	0,35
'Mocho espiga branca'	*b*	*f*	*a*	40	13,0	72	1,4	82	26,8	148	0,41
'Mocho espiga ruiva'	*b*	*f*	*a*	39	13,9	74	1,5	83	21,3	129	0,47
'Mocho rapado'	*b*	*f*	*d*	34	15,8	84	1,3	94	28,2	163	0,32
'Mole Algarvio'	*c*	*e*	*a*	30	14,5	28	0,6	89	42,3	43	0,74
'Português'	*b*	*e*	*a*	34	15,1	44	0,7	95	38,2	71	0,53
'Poveiro'	*b*	*e*	*a*	22	14,9	69	0,8	93	32,2	108	0,90
'Precoce'	*b*	*e*	*a*	27	15,1	50	0,8	94	41,0	95	0,48
'Ribatejano'	*f*	*d*	*a*	39	16,5	98	1,2	80	23,8	-	-
'Ribeiro'	*f*	*d*	*a*	23	13,4	88	1,2	85	25,8	171	0,57
'Ruivo'	*b*	*e*	*a*	46	14,9	61	0,9	97	33,8	73	0,93
'Sacho'	*f*	*d*	*a*	46	15,0	71	1,0	94	36,8	118	0,50
'Sado'	*b*	*e*	*a*	35	15,6	65	0,4	74	33,8	-	-
'Saloio'	*b*	*e*	*a*	96	15,5	53	0,9	94	45,5	78	0,89
'Temporão de Coruche'	*b*	*e*	*a*	45	16,9	60	0,5	95	44,4	-	-
'Transmontano'	*b*	*e*	*a*	82	15,6	64	0,8	77	36,4	-	-
'Transtagano'	*b*	*b*	*a*	34	18,1	99	0,8	88	29,5	-	-
'Tremês arroxeado'	*b*	*b*	*c*	39	16,5	98	1,3	100	26,0	-	-
'Tremês branco'	*b*	*d*	*a*	45	15,1	76	0,9	93	32,8	148	0,56
'Tremês de Tavira'	*b*	*e*	*a*	40	14,8	54	0,6	83	39,2	57	0,47
'Tremês ruivo'	*b*	*e*	*a*	35	14,4	75	0,8	97	34,0	59	2,17
'Viloso mole'	*f*	*f*	*a*	42	13,3	99	1,3	85	22,7	139	0,38

†-For corresponding HMW glutenin subunits, see table 3

These results show that some varieties ('Barbela','Belém', 'Mocho espiga branca', 'Precoce', 'Ribeiro') are suitable for biscuit production or for improving the extensibility of commercial varieties with tenacious gluten type.

3.3 Relation between Quality Tests

No significant relation between protein content and gluten strength-related parameters (SDS, MDT, W) was observed (data not presented) in agreement with other studies[13]. SDS sedimentation volume, mixing time (MDT), dough strength (W), tenacity (P), extensibility (L) are all interrelated, which was previously shown[10]. Flour protein was correlated with mixograph peak height (MPH) and breakdown (BDR).

3.4 Relationship between HMW glutenin subunits and quality characteristics

The allelic variation at the *Glu-1 loci* had no influence on flour protein content, hardness, mixograph peak height and dough tenacity. HMW glutenins controlled by *Glu-A1* and *Glu-B1 loci* had significant effects on the SDS sedimentation value, mixograph development time and resistance breakdown (Table 2).

Table 2 *Analysis of variance for SDS sedimentation, mixograph (MDT, BDR) and alveograph (W, L) values*

Source of Variation	df	Mean Squares			df	Mean Squares	
		SDS	MDT	BDR		W	L
Glu-A1	3	1108**	0,25**	110*	2	739	130
Glu-B1	4	1113**	0,32**	168**	3	3498*	1987*
Glu-D1	2	212	0,00	0,13	2	812	418
residual	29	200	0,05	34	17	1029	544
r square		0,65	0,65	0,60		0,64	0,56

df=degrees of freedom; *, ** significant at the 5% and 1% levels, respectively

The following remarks can be made about the mean values for quality characteristics of varieties grouped according to the *Glu-1* HMW glutenins (Table 3): i) the varieties with the null allele at *Glu-A1* have quality mean values significantly lower than those with subunits 1, 2* or 2¨, which is consistent with previous work[14]; ii) there were no significant differences between mean values of varieties having subunits 7+8, 6+8 or 13+16 at the *Glu-B1 locus*; iii) the quality mean values (excluding the extensibility) of varieties with subunits 20+20y or 7 at the *Glu-B1 locus* were lower than those with subunits 7+8, 6+8 or 13+16, confirming previous results[15]. The extensibility of dough was significantly different in varieties with the HMW subunit 7 to those with HMW subunits 20+20y; iv) in contrast with the expected[3,13], no significant differences at *Glu-D1* locus were observed due to the low allelic variation present.

4 CONCLUSIONS

In the modern commercial cultivars the presence of HMW glutenin subunit alleles giving high gluten strength is a main objective. Consequently, the Portuguese bread wheat collection is not of value to improve the variation for this character.

However, LMW-glutenins and other proteins remain to be explored. Those proteins could influence the grain hardness and extensibility, important parameters which define different classes of wheats, with great variation present in the collection.

Table 3 *Mean values for quality characteristics (SDS, MDT, BDR, W and L) of 39 varieties grouped according to* Glu-1 *HMW glutenin subunit composition and classification of means using Duncan test*

Locus	***allele***	**N° varieties**	**Subunit**	**Means**				**Means**	
				SDS	**MDT**	**BDR**	**N°**	**W**	**L**
Glu-A1	*a*	1	1	89^{a}	$1,56^{a}$	$23,0^{c}$	-	-	-
	b	29	2*	64^{a}	$0,87^{bc}$	$36,5^{ab}$	18	93^{ab}	88^{ab}
	c	2	null	28^{b}	$0,58^{c}$	$41,4^{a}$	2	47^{b}	60^{b}
	f	7	2··	83^{a}	$1,23^{ab}$	$28,2^{bc}$	5	139^{a}	106^{a}
Glu-B1	*a*	2	7	50^{c}	$0,57^{c}$	$41,6^{a}$	1	86^{ab}	132^{a}
	b	2	7+8	99^{a}	$1,06^{ab}$	$27,8^{b}$	-	-	-
	d	5	6+8	78^{ab}	$1,09^{ab}$	$32,3^{ab}$	3	146^{a}	104^{ab}
	f	9	13+16	81^{a}	$1,29^{a}$	$27,1^{b}$	8	132^{a}	110^{ab}
	e	21	20+20y	56^{bc}	$0,78^{bc}$	$39,0^{a}$	13	68^{b}	71^{b}
Glu-D1	*a*	36	2+12	65^{a}	$0,92^{a}$	$35,6^{a}$	23	95^{a}	88^{a}
	c	2	4+12	83^{a}	$1,17^{a}$	$25,8^{a}$	1	118^{a}	111^{a}
	d	1	5+10	84^{a}	$1,28^{a}$	$28,2^{a}$	1	163^{a}	137^{a}

References

1. A. Blum, B. Sinmena, G. Golan, and J. Mayer, *Plant Breed,* 1987, **99**, 226

2. P. I. Payne, K. G. Corfield and J. A. Blackman, *Theor. Appl. Genet.*, 1979, **55,** 153

3. P. I. Payne, M. A. Nightingale, A. F. Krattiger and L. M. Holt,. *J. Sci. Food Agric.*, 1987, **40,** 51

4. M. Rodriguez-Quijano, J.F. Vazquez, C. M. Brites and J.M. Carrillo, *J. Genet. & Breed.* 1998, **52,** 95

5. G. Igrejas, H. Guedes-Pinto, V. Carnide and G. Branlard, *Plant Breed.,* 1999, **118,** 297

6. J. C. Vasconcelos, *Trigos portugueses ou de há muito cultivados no país*, I série ,Sep. Bol. Agric. I, Lisboa, 1933, N° 1 e 2, p.134

7. ICC Standard N°. 159.1995, "Determination of protein by Near Infrared Reflectance (NIR) spectroscopy". ICC 4 p.

8. American Association of Cereal Chemist. Approved method of the AACC. Methods 54-40A approved in 1995 and 39-70A approved in 1997. AACC, St. Paul, Minnesota.

9. J. W. Dick and J. S Quick, *Cereal Chem.* 1983, **60,** 315

10. J. P. Martinant, Y. Nicolas, A. Bouguennec, Y. Popineau, L. Saulnier and G. Branlard,. *J. Cereal Sci.*, 1998, **27:**179

11. SAS Institute Inc. SAS/STAT® User`s guide-release 6.03 edition. U.S.A.: SAS Institute Inc. Cary, NC., 1988, 1028p.

12. G. Igrejas, G. Branlard, V. Carnide, I. Gateau and H. Guedes-Pinto, *J. Genet. & Breed.*, 1997, **51**, 167
13. H. Dong, R. G. Sears, T. S. Cox, R. C. Hoseney, G. L. Lookhart and M. D. Shogren, *Cereal Chem.*, 1992, **69**, 132
14. N. G. Halford, J. M. Field, H. Blair, P. Urwin, K. Moore, L. Robert, R. Thompson, R. B. Flavell, A. S. Tatham and P. R. Shewry, *Theor Appl. Genet.*, 1992, **83**, 373
15. W. J. Rogers, P. I. Payne, J. A. Seekings and E. J. Sayers, *J. Cereal Sci.*, 1991, **14**, 209

Acknowledgements

Financial support for this study was provided by projects PCNA/BIO/0703/96 from PRAXIS XXI programme and PIDDAC 409/99 from INIA, Portugal.

GENETIC ANALYSIS OF DOUGH STRENGTH USING DOUBLED HAPLOID LINES

O. M. Lukow

Cereal Research Centre, Agriculture and Agri-Food Canada, 195 Dafoe Road, Winnipeg, Manitoba, Canada R3T 2M9.

1 INTRODUCTION

Double haploidy is an effective means of generating homozygous populations for the genetic analysis of wheat quality. This technique has been used extensively at the Cereal Research Centre to produce a number of doubled haploid populations, and each designed to address specific questions about quality. The most extensive evaluation to date has been on the lines derived from crossing Glenlea (semi-dwarf type) to AC Domain. The Canadian cultivar, Glenlea, is recognized for its extra-strong dough mixing properties, over-production of HMW-glutenin subunit 7 and, more recently, a rare B-zone LMW-glutenin subunit pattern. AC Domain is a Canadian bread wheat with moderately strong dough mixing properties. The HMW- and LMW-glutenin subunit composition of AC Domain is typical of other bread wheats. Doubled haploid lines from the cross were used to determine the key HMW- and LMW-glutenin subunits responsible for dough strength.

2 MATERIALS AND METHODS

Doubled haploid lines were generated by the maize pollen technique developed and implemented at the Cereal Research Centre. One hundred and eight-two lines, and the parent cultivars, Glenlea and AC Domain, were grown in three locations in Manitoba in 1998 and bulked. Composite seed samples were milled on a Buhler pneumatic laboratory mill into straight-grade flour.

The HMW- and LMW-glutenin compositions of parents and progeny were evaluated electrophoretically by SDS-PAGE[1,2]. The progeny had HMW-subunits 7+8 or 7+9 and one of 8 (ie. 2^3) possible LMW-glutenin banding patterns. Over-production of subunit 7 was associated with subunit 8. The genetic variation between the parents was as follows:

	HMW-glutenin subunits			LMW-glutenin subunits		
Chromosome	1A	1B	1C	1A	1B	1C
Glenlea	2*	7+8	5+10	50	8	10
AC Domain	2*	7+9	5+10	Null	51,53	13,32,38

Gluten strength and dough mixing properties were determined by the SDS-sedimentation test (AACC method 56-70) and the 10g computerized mixograph using constant flour weight and absorption (62%)[3]. Protein content was analyzed by NIR.

Data were statistically evaluated by ANOVA.

3 RESULTS AND DISCUSSION

Earlier we reported on the effects of the HMW- and LMW-glutenin subunits on protein content, dough strength and baking properties based on smaller doubled haploid populations (ie. 35 and 77) from the same cross[4,5,6,7]. The doubled haploid population was increased to182 lines in order to facilitate an extensive statistical analysis of all possible HMW/LMW combinations. A typical SDS-PAGE separation of the LMW-glutenin subunits of the doubled haploid lines is shown in Fig.1.

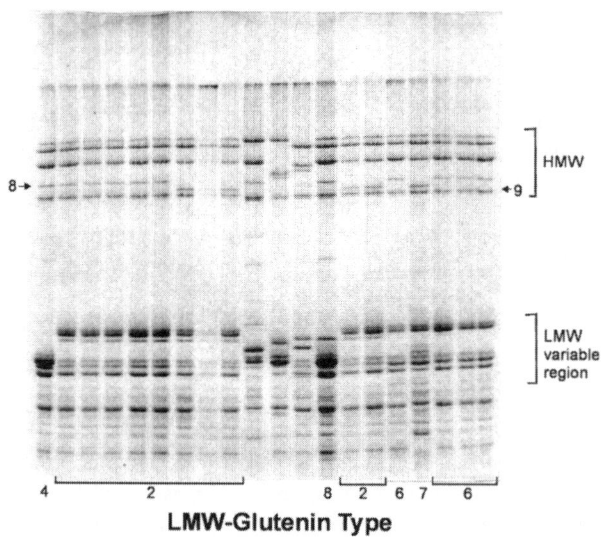

Figure 1 *SDS-PAGE analysis of HMW- and LMW-glutenins.*

Eight LMW-glutenin types were identified based on the mobility of 10 different bands; two types were identical to the parents, Glenlea and AC Domain[8,9] (Fig. 2). LMW-glutenin subunits were assigned to group 1 chromosomes by their linkage with D-zone omega gliadin. All LMW-glutenin subunits were assigned with the exception of bands 20 and 30, which were common to both parents. LMW-glutenin bands coded by one locus were determined by co-segregation in the doubled haploid lines.

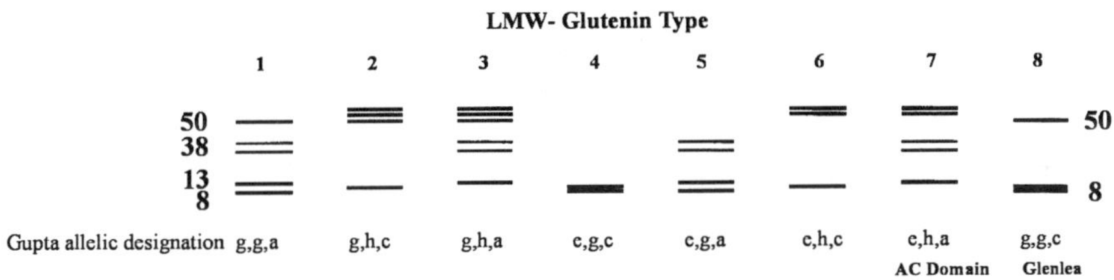

PATTERN	BANDS
1	8, 13, 32, 38, 50
2	10, 50, 51, 53
3	13, 32, 38, 50, 51, 53
4	8, 10
5	8, 13, 32, 38
6	10, 51, 53

Figure 2 *LMW-glutenin subunits that differ in the Glenlea X AC Domain doubled haploids.*

The effect of HMW-glutenin subunit composition on dough strength was determined (Table1). SDS-sedimentation value and mixograph development time, energy to peak, peak band width, total energy and band width energy were significantly greater in doubled haploid lines containing 7+8 than 7+9. This result confirmed our earlier findings on this population and is in agreement with the generally accepted view on the greater contribution to strength of 7+8 than 7+9. Unlike our earlier studies, there was a slight increase in flour protein content in the 7+9 lines.

The effect of the LMW-glutenin subunits on protein content and strength was determined by comparing the LMW alleles from each chromosome in either the 7+9 or 7+8 backgrounds (Table 2). The LMW-glutenin subunits derived from Glenlea significantly increased some dough strength parameters compared to the AC Domain allele. The strengthening effect of the Glenlea LMW-glutenin subunits was most pronounced in the 7+8 lines. LMW-glutenin subunit 8, coded by chromosome 1B, had the greatest effect on dough strength.

4 CONCLUSIONS

This study confirms our earlier finding that the extra-strong mixing characteristics of Glenlea wheat are a result of the additive effects of the HMW- and LMW-glutenins. Maximum dough strength properties are evident when Glenlea-type HMW- and LMW-glutenin protein subunits are present. These protein subunits are used currently as effective makers in early generation screening in the extra-strong wheat breeding program at the Cereal Research Centre.

Table 1 *Effect of HMW-glutenin subunits on quality trait means*

HMW-Glutenin Subunits	Flour Protein Content (%)	Sedimentation Value (ml)	Mixograph Parameters[a]					
			PHG	MDT	ETP	PBW	TEG	BWE
7+8[b]	14.1	83	0.153	4.0	29.5	0.108	91.9	61.7
7+9	14.3*[c]	77***	0.156ns	3.0**	22.0**	0.102*	99.5*	54.2***

[a] mixograph: PHG = peak height, MDT = development time, ETP = energy to peak, PBW = peak band width, TEG = total energy, BWE = band width energy.

[b] 92 lines with 7+8, 90 lines with 7+9.

[c] ns = non-significant, * = P < 0.05, ** = P < 0.01, *** = P < 0.001.

Table 2 *Effect of LMW-glutenin subunits on quality trait means*

LMW-Glutenin Subunits	HMW-Glutenin Subunits 7+8					HMW-Glutenin Subunits 7+9
	Flour Protein Content (%)	Sedimentation Value (ml)	Mixograph Parameters[a]			Sedimentation Value (ml)
			MDT	ETP	BWE	
Null[b]	14.0	84	3.9	29.6	61.1	76
50	14.1ns[c]	82*	4.0ns	29.4ns	62.2ns	78ns
51, 53	14.1	82	3.7	28.1	59.9	76
8	14.0ns	85***	4.2*	30.9*	63.4*	78*
13, 32, 38	14.3	84	3.7	28.0	61.0	76
10	13.8*	83ns	4.3*	31.0	62.3ns	77ns

[a] mixograph: MDT = development time, ETP = energy to peak, BWE = band width energy.

[b] 91 lines with Null, 91 lines with 50; 86 lines with 51, 53, 96 lines with 8; 100 lines with 13, 32, 38, 82 lines with 10.

[c] ns = non-significant, * = P < 0.05, ** = P < 0.01, *** = P < 0.001.

References

1. O.M. Lukow, A. Hussain and G. Branlard, *Cereal Foods World*, 1994, **39**, 617.
2. O.M. Lukow and K. Kidd, in *Gluten Proteins 1990*, ed. W. Bushuk and R. Tkachuk, AACC, St. Paul, MN, 1991, p.491.
3. O.M. Lukow, *Cereal Foods World*, 1991, **36**, 487.
4. O.M. Lukow and F. Townley-Smith, *Cereal Foods World*, 1995, **40**, 668.
5. O.M. Lukow and F. Townley-Smith, *Cereal Foods World*, 1996, **41**,567.
6. C. Perron, O.M. Lukow and F. Townley-Smith, *Cereal Foods World*, 1997, **42**, 618.
7. C. Perron, O.M. Lukow and F. Townley-Smith, in *Proceedings of the 9th International Wheat Genetics Symposium*, Vol. 4, ed. A. Slinkard, University Extension Press, Saskatoon, SK, 1998, p.276.
8. O.M. Lukow, A. Hussain and G. Branlard, *Cereal Foods World*, 1994, **39**, 617.
9. R.B. Gupta and K.W. Shepherd, *Theor. Appl. Genet.*, 1990, **80**, 65.

Acknowledgments

The technical assistance of the Wheat Quality Research Team is gratefully acknowledged.

RELATIONSHIP BETWEEN ALLELIC VARIATION OF *GLU-1*, *GLU-3* AND *GLI-1* PROLAMIN LOCI AND BAKING QUALITY IN DOUBLED HAPLOID WHEAT POPULATIONS

B. Killermann and G. Zimmermann

Bayerische Landesanstalt für Bodenkultur und Pflanzenbau, Vöttingerstraße 38,
D-85354 Freising, Germany

1 INTRODUCTION

The storage proteins in wheat kernels (*T. aestivum*, L.) are composed mainly of high and low molecular weight glutenin subunits (HMW-GS and LMW-GS, respectively) and gliadins. Their uniqueness and importance in dough mixing and bread making due to the formation of the three dimensional protein network called gluten is well known. Genetic variation of gliadins and glutenin subunits enables plant breeders to rank particular proteins, or alleles encoding these proteins, in terms of their contribution to the complex baking quality[1,2]. These proteins are encoded by gene clusters located on chromosomes of groups 1 and 6, with *Gli-1* and *Gli-2* encoding for gliadins, and *Glu-1* and *Glu-3* encoding for glutenins. In numerous studies the impact of the different alleles at these gene loci on the components of baking quality has been investigated. In most cases only a selection from the great number of quality parameters has been included in the investigations e.g. SDS and Zeleny sedimentation values, wet gluten content, dough properties elaborated by Extensograph, Mixograph or Alveograph depending on the country where the study was carried out. Similar multiplicity can be found with regard to the plant material used in the experiments: registered cultivars from various countries, recombinant inbred lines, near isogenic lines, substitution and translocation lines. Very often only a limited number of test lines could be analysed with a limited number of quality test methods. In Germany the main characteristic for quality classification of wheat cultivars is the bread volume determined by the Rapid Mix Test (RMT). For this feature varying correlations are found with other attributes used for quality characterization and selection depending on the genetic background of the investigated material.

The aim of this work was to determine the allelic variation of glutenins and gliadins in four doubled haploid wheat populations homozygous for these loci and to investigate the individual and combined effects of alleles in a common background. Doubled haploid populations with a sufficient number of lines per population seemed to be the material best suited for this purpose. The whole spectrum of quality features used in Germany for the determination of dough properties and baking quality was analysed. In this presentation the results of baking volume (RMT) are shown in detail. The four doubled haploid wheat populations represent a wide range of genetic background of the central and west European gene pool. The parents used in the crosses were distinct with respect to their

Glu-1 alleles and to some extent to their *Glu-3* and *Gli-1* loci and they were selected to cover a wide range with respect to the different quality features.

2 MATERIALS AND METHODS

The plant material studied comprised 495 doubled haploid lines derived from anther culture from four different F1 hybrids of winter wheat. The lines together with the parental cultivars were grown in field trials in 1998 (Saatzucht Schweiger, Feldkirchen Bavaria). The parental cultivars and breeding lines (CWW 92-6 and W 84332) as well as their quality characteristics and allelic combinations are listed in Table 1.

Table 1 *Allelic combinations of parental cultivars and their quality characteristics together with mean values of the four doubled haploid wheat populations*

DH-Pop (DH-lines)	Parental cultivars	*Glu-1* (HMW-GS) *A1*	*B1*	*D1*	*Glu-3* (LMW-GS) *A3*	*B3*	*D3*	*Gli-1* (gliadins) *A1*	*B1*	*D1*	Baking volume (ml/ 100g fl.)	Grain protein content %	Zeleny sedim. (units)	Kernel hardness (index)
1	Atlantis	*a*	*d*	*a*	*a*	*j*[1]	*c*	*b*	*l*[1]	*b*	584	13.1	21	35
(N=142)	Bovictus	*c*	*c*	*d*	*e*	*j*[1]	*c*	*b*	*l*[1]	*b*	571	12.6	25	34
	Mean of pop 1										612	13.4	25	35
2	Flair	*c*	*d*	*d*	*e*	*g*	*c*	*m*	*f*	*b*	534	12.6	29	47
(N=103)	CWW 92-6	*a*	*i*	*a*	*e*	*g*	*c*	*b*	*f*	*b*	650	12.9	38	45
	Mean of pop 2										577	13.5	34	51
3	Atlantis	*a*	*d*	*a*	*a*	*j*[1]	*c*	*b*	*l*[1]	*b*	584	13.1	21	35
(N=150)	Lindos	*c*	*c*	*d*	*d*	*g*	*c*	*o*	*f*	*b*	575	12.6	36	42
	Mean of pop 3										618	13.3	28	42
4	W 84332	*c*	*d*	*a*	*e*	*g*	*c*	*f*	*c*	*b*	511	13.1	27	51
(N=100)	Bussard	*a*	*c*	*d*	*a*	*g*	*c*	*f*	*c*	*b*	723	14.0	59	52
	Mean of pop 4										620	14.0	43	51

[1] *Glu-B3j, Gli-B1l* = rye alleles (cultivars containing the 1BL/1RS wheat-rye-translocation)

The samples were prepared for electrophoresis as crushed half single kernels. Gliadins were extracted with 70% (w/v) ethanol and fractionated by electrophoresis at acid pH as described by Jackson et al.[3]. Protein patterns obtained were classified according to the nomenclature proposed by Jackson et al.[3]. Glutenins were extracted using the procedure proposed by Singh et al.[4]. After reduction and alkylation they were separated in 12% polyacrylamide gels containing SDS. The HMW-GS were classified according to the numbering system of Payne and Lawrence[5]. The LMW-GS are designated according to the numbering system proposed by Jackson et al.[3]. For quality evaluation, kernel hardness, protein content, Zeleny sedimentation value, rheological characteristics from Brabender Extensograph and Farinograph as well as the Rapid Mix Test (RMT) baking volume[6] were determined. The effects of allelic variation at gliadin and glutenin subunit loci on baking volume of this unbalanced data set were studied separately for each of the populations using the GLM procedure of the SAS statistical package[7]. All main and interaction effects were entered to the models and least squares means of the effects were estimated by the GLM procedure.

3 RESULTS AND DISCUSSION

The parents show allelic variation at all three *Glu-1* loci and only at some of the *Glu-3* and *Gli-1* loci. Electrophoretic profiles of HMW-GS and LMW-GS for each parental cultivar are shown in Figure 1. Table 2 lists the results of ANOVA with the variable RMT volume.

Table 2 *GLM-ANOVA with dependent variable RMT volume. F-Values of all main effects and of interloci interactions significant at the 0.1 level*

Effects	F-Values			
	pop 1	pop 2	pop 3	pop 4
Glu-A1	17.45	0.25	0.01	0.45
Glu-B1	10.83	20.17	5.43	1.19
Glu-D1	59.82	39.07	3.39	18.22
*Glu-B1*Glu-D1*	3.49	n.s.	12.40	7.93
*Glu-A1*Glu-D1*	n.s.*	4.61	3.43	n.s.
Glu-A3	4.87	n.s.	0.08	1.43
Glu-B3	n.s.	n.s.	6.55	n.s.
Gli-A1	n.s.	1.98	n.s.	n.s.
*Glu-A1*Glu-A3*	5.60	n.s.	n.s.	n.s.
*Glu-D1*Gli-A1*	n.s.	2.82	n.s.	n.s.
*Glu-B1*Glu-A3*	n.s.	n.s.	3.46	n.s.
*Glu-D1*Glu-A3*	n.s.	n.s.	7.34	n.s.
*Glu-A1*Glu-B3*	n.s.	n.s.	2.80	n.s.
*Glu-A1*Glu-B1*Glu-D1*	n.s.	6.88	n.s.	n.s.
	R^2 of model (*Glu-1, Glu-3, Gli-1*)			
	0.57	0.55	0.32	0.39
	R^2 of model (only *Glu-1*)			
	0.51	0.49	0.21	0.35

* n.s. = not significant

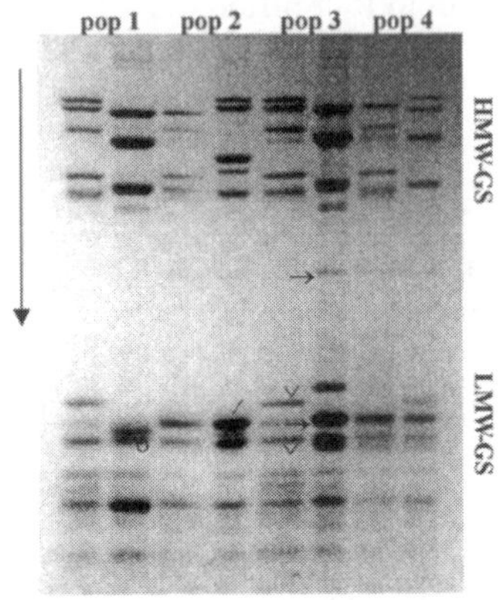

Figure 1: SDS-PAGE patterns of HMW-GS and LMW-GS of the parental cultivars and breeding lines Atlantis, Bovictus, Flair, CWW 92-6, Atlantis, Lindos, W 84332 and Bussard. The alleles *Glu-A3a*, *Glu-A3d*, *Glu-B3g* and *Glu-D3c* controlled by genes on chromosome 1AS, 1BS and 1DS are marked by /, v, →, and υ, respectively.

The R-squares of the full model (all main and interaction effects) are highest in pop 1 and pop 2 where 57% and 55% of the variation in RMT volume can be accounted for by the prolamin loci. Much lower values were found in pop 3 and pop 4 with 32% and 39%, respectively. The predominating effect of the *Glu-1* alleles becomes obvious when calculating R-squares separately for these alleles (see Table 2). The small effects of *Glu-A3* and *Gli-1* alleles must be ascribed to the low heterogeneity of the parent cultivars at these loci. In Figure 2a – 2f the impact of the individual alleles on the RMT volume and the most important interaction effects are shown. The *Gli-1* alleles of pop 3 have not yet been electrophoretically determined and have therefore not been taken into account in the statistical analysis. As repotrd in many studies the most favourable effect on baking quality was found for the allele *Glu-D1d*, but with clear differences between the populations (Figure 2c). It was lowest in pop 3 where two parental cultivars were combined, which are similar in baking volume but differ widely in sedimentation value, kernel hardness and dough properties, moreover one parent (Atlantis) carries the 1BL/1RS translocation. It is noteworthy that the *Glu-B3j* allele had a favourable effect in this population. The lines with and without this allele showed mean RMT volumes of 622 and 577 ml respectively (not shown in Figure 2). Negative effects were found for allele *Glu-B1d* without exception (Figure 2b). At this locus *Glu-B1i* in pop 2, which has been found very seldom in German cultivars, was most advantageous. At the *Glu-A1* locus the allele *a* showed a significant positive effect only in pop 1. The small main effects of the *Glu-3* and *Gli-1* loci are shown in Figure 2d. The most distinct interaction between *Glu-1* loci was found for *Glu-B1*Glu-D1* as shown in Figure 2e. In pop 3 the *Glu-D1d* allele, normally found as favourable even had a negative effect on RMT volume in combination with the favourable *Glu-B1c* allele. Pop 2 did not show the *Glu-B1*Glu-D1* interaction like pop 1, pop 3 and pop 4. Instead the *Glu-D1* locus reacted in a similar way with the *Glu-A1* locus in this population. Only a few interactions were found between *Glu-1, Glu-3* and *Gli-1* loci and they were not consistent over populations.

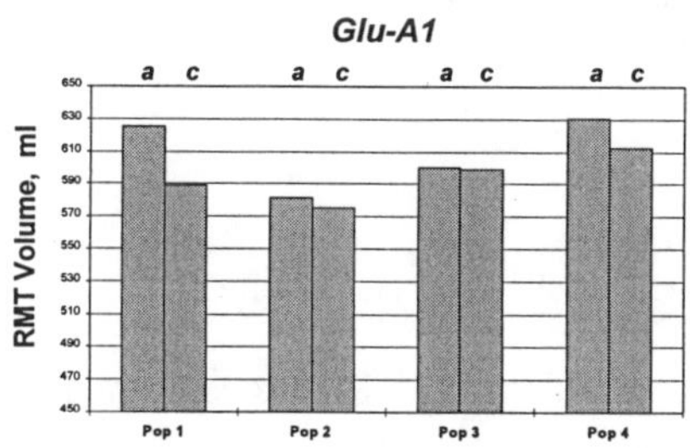

Figure 2a

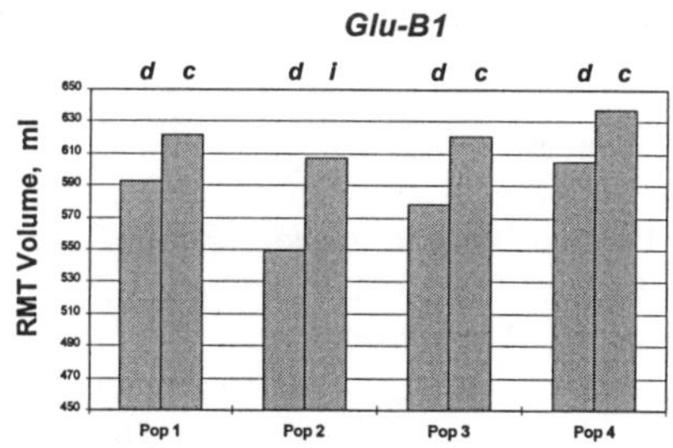

Figure 2b

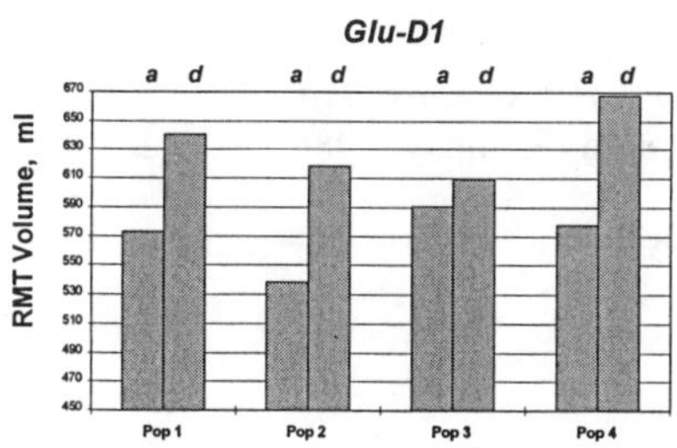

Figure 2c

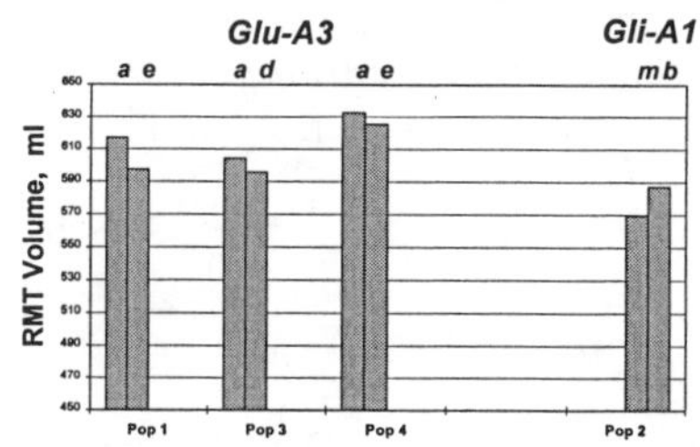

Figure 2d

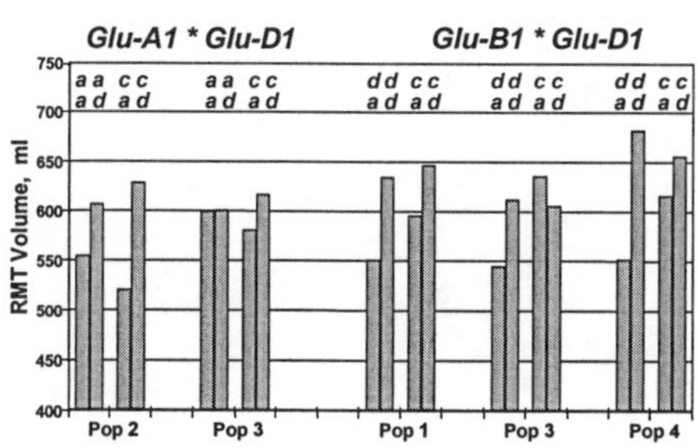

Figure 2e

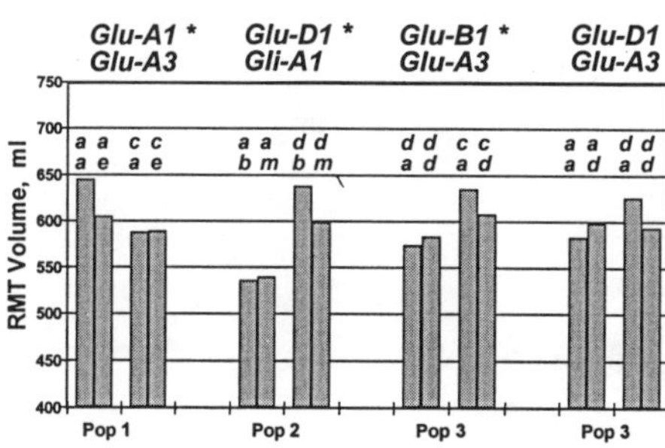

Figure 2f

Figures 2a – 2f *Average RMT volume of lines carrying the individual alleles on gene loci Glu-A1, Glu-B1, Glu-D1, Glu-A3 and Gli-A1 (main effects) and of allele combinations Glu-B1*Glu-D1, Glu-A1*Glu-D1, Glu-A1*Glu-A3, Glu-D1*Gli-A1, Glu-B1*Glu-A3 and Glu-D1*Glu-A3 (interaction effects).*

4 CONCLUSIONS

From these results it can be concluded that the impact of the prolamin loci on the complex feature RMT volume is strongly dependent on the genetic background of the investigated material. This must be kept in mind by breeders when combination and selection strategies are considered. Additive effects and interactions of supposedly favourable alleles can vary considerably in progeny of different crosses. In particular, diverse dough properties of the parents make it necessary to modify the rating of favourable and unfavourable alleles. Positive effects on one or the other component feature are not necessarily confirmed with regard to the influence on the complex baking qualitiy.

References

1. P.I. Payne, *Annu. Rev. Plant Physiol.*, 1987, **38**, 141.
2. R.B.Gupta , J.G. Paul, G.B.Cornish, G.A. Palmer, F. Bekes and A.J. Rathjen, *J. Cereal Sci.*, 1994, **19**, 9.
3. E.A. Jackson, M.-H. Morel, T. Sontag-Strohm, G. Branlard, E.V. Metakovsky and R. Redaelli, *J. Genet. & Breed.*, 1996, **50,** 321.
4. N.K. Singh, K.W. Shepherd and G.B. Cornish, *J. Cereal Sci.*, 1991, **14,** 203.
5. P.I. Payne, G.J. Lawrence, *Cereal Res. Commun.*, 1983, **11,** 29.
6. ICC-Standards, *Standard-Methoden der Internationalen Gesellschaft für Getreidechemie (ICC)*, Verlag Moritz Schäfer, Detmold Germany, 1986, ISBN 3-87696-010-X
7. *SAS Procedure Guide*, 3rd Edn., SAS Institute Inc, 1990a, Vers 6, Cary, NC

Acknowledgements

We are grateful to Dr. G. Daniel (Bayerische Landesanstalt für Bodenkultur und Pflanzenbau) for producing the doubled haploid populations. We also thank the GFP (Gemeinschaft zur Förderung der privaten deutschen Pflanzenzüchtung e.V.) for financial support (project number G 76/97 HS).

Biotechnology

IMPROVEMENT OF WHEAT PROCESSING QUALITY BY GENETIC ENGINEERING

P.R. Shewry[1], H. Jones[2], G. Pastori[2], L. Rooke[2], S. Steele[2], G. He[2], P. Tosi[1,2,3], R. D'Ovidio[3], F. Békés[4], H. Darlington[1], J. Napier[1], R. Fido[1], A.S. Tatham[1], P. Barcelo[5] and P. Lazzeri[5]

1. IACR-Long Ashton Research Station, Department of Agricultural Sciences, University of Bristol, Bristol BS41 9AF, UK. 2. IACR-Rothamsted, Harpenden, Herts, AL5 2JQ, UK. 3. Dipartimento Di Agrobiologia de Agrochimica, Via San Camillo de Lellis, Viterbo 01100, Lazio, Italy. 4. Plant Science CRC, CSIRO, North Ryde, NSW 2113, Australia. 5. DuPONT CIC, IACR-Rothamsted, Harpenden, Herts AL5 2JQ, UK.

1 INTRODUCTION

It is little more than 10 years since the first transgenic plants of cereals (rice and maize) were reported based on transformation and regeneration of protoplasts[1,2] with the first confirmed transgenic plants of wheat being produced as late as 1992.[3] The latter work used micorprojectile bombardment to deliver DNA into regenerable tissue (immature embryos) and this approach has since been adopted in a number of laboratories worldwide.[4,5] The first target to be identified for improvement by genetic engineering of wheat has been gluten quality, and in particular the amount and composition of the high molecular weight (HMW) subunits of wheat glutenin.[6-9] In the present paper we briefly review work within IACR (Rothamsted and Long Ashton) on the transformation of commercial cultivars of wheat to improve grain quality.

2 RESULTS AND DISCUSSION

2.1 Transformation of model wheat lines with subunits 1Ax and 1Dx5

Two near-isogenic lines of spring wheat were initially selected for transformation with genes for subunits 1Ax1 and 1Dx5 : L88-31 which expresses only two subunits (1Bx17, 1By18) and L88-6 which expresses five HMW subunits (1Ax1, 1Dx5, 1Dy10, 1Bx17, 1By18).[10] Several lines were isolated which expressed subunit 1Dx5 in the L88-6 background and 1Ax1/1Dx5 in L88-31, as described by Barro *et. al.* (1997).[8] These included B73-6-1 which exhibits a massive over-expression of 1Dx5 in L88-6, resulting in increases in subunit 1Dx5 from 2.7 to 10.7% of the total gluten proteins and in total HMW subunits from 12.7 to 20.5%.[11] Flour from this line failed to form a normal dough when mixed but blending with flour from a normal cultivar resulted in increased dough strength as measured by Mixograph mixing time.[11]

Several of the lines have been grown in replicate field trials over two years (1998, 1999) at two sites in the UK (Rothamsted and Long Ashton), providing material for detailed analyses of gluten composition and mixing properties. Preliminary results from this are reported elsewhere in this volume.

Detailed studies have also been made of transgene inheritance and stability in six selected lines.[12] Transgene insertion number was shown to range from one up to 20, with some copies being rearranged, truncated or arranged in concatamers. Transgenes were often located at disperse unlinked sites leading in two cases to segregation between the HMW subunit transgenes and the *bar* (Basta resistance) selectable marker genes.

2.2 Transformation of commercial genotypes and breeding lines

Work at present is focusing on the transformation of lines for plant breeding, using whole plasmids containing 1Ax1, 1Dx5 and 1Dy10 HMW subunit genes or "clean" transgene fragments (i.e. lacking the plasmid sequences which include the Amp^R gene). A number of lines have been produced by transforming the cultivars Cadenza and Canon with the whole HMW subunit 1Ax1 plasmid and expression of the transgenes confirmed by SDS-PAGE of seed proteins (unpublished results of G. Pastori, S. Steele and P.R. Shewry). Transformation with the other genes and clean fragments is in progress.

2.3 Transformation of model wheats with mutant HMW subunit genes

Mixograph studies carried out using *in vitro* incorporation of heterologously expressed HMW subunits[13] have indicated that longer HMW subunits have a greater impact on gluten elasticity than shorter subunits. To test this hypothesis we have transformed the model line L88-31 (containing only HMW subunits 1Bx17 and 1By18) with the 1Dx5 gene and mutant forms in which the subunit repetitive domain has been changed in length from 696 to 853, 576 and 441 residues .[14]

Lines expressing all four genes have been produced and are currently being used to isolate homozygous lines for functional analysis and for incorporation into near isogenic series (unpublished results of G. He, R. D'Ovidio, P. Lazzeri, R. Fido and P.R. Shewry in collaboration with O.D. Anderson, USDA, Albany).

2.4 Transformation of durum wheat

Dough strength is an important quality parameter for durum wheat used to make noodles or bread.[15] We have shown that transformation of durum wheat lines with genes for subunits 1Ax1 or 1Dx5 from breadwheat results in increased dough strength, provided that the transgenes are expressed at similar levels to the endogenous HMW subunit genes.[16] However, over-expression of the subunit 1Dx5 transgene resulted in unusual mixing characteristics as determined for the bread wheat line B73-6-1 (see above).

Processing quality in pasta wheat is also associated with the expression of LMW subunits.[17] We are, therefore, also transforming pasta wheat with PCR-amplified genomic sequences encoding two LMW subunits,[18] under the control of the HMW subunit 1Dx5 gene promoter. The proteins have also been expressed in native and "epitope tagged" forms, the latter containing a short sequence at their *C*-termini to allow detection using a commercially available monoclonal antibody (unpublished results of P. Tosi, J.A. Napier, R. D'Ovidio and P.R. Shewry).

2.5 Other quality targets

The strong endosperm-specific expression of the HMW subunit gene promoters[19] means that they are an important tool to drive other transgenes in wheat endosperms. We are interested in the control of grain texture, including both hardness and vitreousness. The γ-zeins of maize are of interest in this respect as their expression in maize endosperms is thought to determine whether the endosperm is floury or vitreous.[20] We have, therefore, transformed breadwheat with genes for γ-zein, either under control of the HMW subunit 1Dx5 gene promoter or the endogenous γ-zein gene promoter. Preliminary results have shown high endosperm-specific expression when the 1Dx5 gene promoter was used but little or no expression with the γ-zein gene promoter (unpublished results of G. He and P.R. Shewry).

3 CONCLUSIONS

Wheat transformation is an attractive system to explore and manipulate aspects of wheat grain structure, composition and functionality, allowing the insertion of wild type and mutant forms of endogenous wheat genes and genes from other sources. The 1Dx5 gene promoter used here and described in detail elsewhere in this volume provides an excellent tool for this, conferring high levels of starchy endosperm-specific expression to endogenous and exogenous transgenes.

References

1. C. A. Rhodes, D. A. Pierce, I. J. Mettler, D. Mascarenhas and J. J. Detmer, *Science*, 1988, **240**, 204.
2. K. Shimamoto, R. Terada, T. Izawa and H. Fujimoto, *Nature*, 1989, **338**, 274.
3. V. Vasil, A. M. Castillo, M. E. Fromm and I. K Vasil, *Bio/Technol.*, 1992, **10**, 667.
4. J. T. Weeks, O. D. Anderson and A. E. Blechl, *Plant Physiol.*, 1993, **102**, 1077.
5. P. Barcelo and P. A. Lazzeri, *Methods in Molecular Biology - Plant Gene Transfer and Expression protocols*, ed. H. Jones, Humana Press Inc., Totowa, NJ, 1995, **49**, p. 113.
6. A. E. Blechl and O. D. Anderson, *Nature Biotech.*, 1996, **14**, 875.
7. F. Altpeter, V. Vasil, V. Srivastava and I. K. Vasil, *Nature Biotech.*, 1996, **14**, 1155.
8. F. Barro, L. Rooke, F. Békés, P. Gras, A. S. Tatham, R. Fido, P. A. Lazzeri, P. R. Shewry and P. Barcelo, *Nature Biotech.*, 1997, **15**, 1295.
9. M. L. Alvarez, S. Guelman, N. G. Halford, S. Lustig, M. I. Reggiardo, N. Ryabushkina, P. R. Shewry, J. Stein and R. H. Vallejos, *Theor. Appl. Genet.*, 2000, **100**, 319.
10. G. J. Lawrence, F. Mac Ritchie and C. W. Wrigley, *J. Cereal Sci.*,1988, **7**, 109.
11. L. Rooke, F. Békés, R. Fido, F. Barro, P. Gras, A. S. Tatham, P. Barcelo, P. Lazzeri and P. R. Shewry, *J. Cereal Sci.*, 1999, **30**, 115.
12. L. Rooke, S. H. Steele, P. Barcelo, P. R. Shewry and P. A. Lazzeri, 2000, Submitted.
13. F. Békés, P. W. Gras and R. B. Gupta, in *Cereals '95*, eds Y. A. Williams and C. W. Wrigley, Royal Australian Chemical Institute, North Melbourne, Australia, 1995, p. 92.
14. R. D'Ovidio, O. D. Anderson, S. Masci, J. Skerritt and E. Prceddu, *J. Cereal Sci.*, 1997, **25**, 1.
15. C.-Y. Liu, K. W. Shepherd and A. J. Rathjen, *Cereal Chem.*, 1996, **73**, 155-166.
16. G. Y. He, L. Rooke, S. H. Steele, F. Békés, P. Gras, A. S. Tatham, R. Fido, P. Barcelo, P. R. Shewry and P. A. Lazzeri, *Molec. Breeding*, 1999, **5**, 377.
17. N. E. Pogna, J.C. Autran, F. Mellini, D. Lafiandra and P. Feillet, *J. Cereal Sci.*, 1990, **11**, 15.

18. R. D'Ovidio, M. Simeone, C. Marchitelli, S. Masci and E. Porceddu, in *Proceedings of the 6th International Gluten Workshop*, ed. C. W. Wrigley, 1996, p. 81.
19. C. Lamacchia, P. R. Shewry, N. Di Fonzo, N. Harris, P. A. Lazzeri, J. A. Napier, N. G. Halford and P. Barcelo, 2000, submitted.
20. M. A. Lopes and B. A. Larkins, *Crop Sci.*, 1991, **31**, 1655.

Acknowledgement

IACR receives grant-aided support from the Biotechnology and Biological Sciences Research Council of the United Kingdom. We wish to thank all of our colleagues who have collaborated in this work, including Dr. O.D. Anderson (USDA, Albany, Ca) and Dr. D. Ludevid (Barcelona) who provided the γ-zein cDNA and gene.

EXPRESSION OF HMW GLUTENIN SUBUNITS IN FIELD GROWN TRANSGENIC WHEAT.

R.J. Fido[1], H.F. Darlington[1], M.E. Cannell[2], H. Jones [2], A.S. Tatham[1], F. Békés [3] and P.R. Shewry[1].

1. IACR-Long Ashton Research Station, University of Bristol, Long Ashton, Bristol BS41 9AF, 2. IACR-Rothamsted, Biochemistry and Physiology Department, Herts, AL5 2JQ, 3. Plant Science CRC, CSIRO , North Ryde, NSW 2113, Australia.

1 INTRODCTION

The high molecular weight (HMW) subunits of glutenin together account for up to about 10-12% of the total grain proteins of wheat (with each accounting for some 2%) [1,2] and are the major determinants of the visco-elastic properties of gluten. These properties enable wheat flour to be used to make bread, pasta as well as a range of other food products. The HMW subunits show both quantitative and qualitative effects on the quality of the grain, with the former related to differences in the number of expressed HMW sub-unit genes.

Because of the functional and economic importance of the HMW subunits and the identification of quantitative effects on the quality it is not surprising that the HMW subunit genes have been identified as a target for expression in transgenic wheat.

We have therefore transformed Australian spring bread wheat lines with genes for HMW subunits 1Ax1 and 1Dx5 by particle bombardment in order to increase the amount of HMW subunits and to determine the effects of this on the functional properties of the flour.

2 MATERIALS AND METHODS

The Australian spring bread wheat lines L88-6 and L88-31 form part of a series of near-isogenic lines derived from crossing mutants of the cultivars Olympic and Gabo with null (silent or absent) alleles at the *Glu-1* (HMW subunit) loci[3]. L88-6 expresses genes encoding five HMW subunits (1Ax1, 1Bx17, 1By18, 1Dx5 and 1Dx10) while L88-31 only expresses the chromosome 1B encoded subunits 1Bx17 and 1By18. These two lines were transformed with genes for HMW subunits 1Ax1 and 1Dx5 using gold particle bombardment in order to increase the proportions of HMW subunits present[4].

To fully evaluate the transformed lines, replicate field trials, with 84 one meter square plots containing five transgenic wheat lines (plus two controls) were grown at two UK sites (Rothamsted and Long Ashton) over 1998, 1999 and will be grown again during 2000.

The plants and mature grain were used to assess the effects of the transformation on the agronomic performance and grain end use quality.

3 RESULTS AND DISCUSSION

3.1 Expression of Additional HMW Subunits

Total proteins in the transgenic lines were separated by SDS-PAGE using a Tris-borate gel system with 10% w/v acrylamide gels[5], and HMW subunit expression levels estimated from the protein bands corresponding to subunits 1Ax1 and 1Dx5 (Table 1). The lines selected for field trials were homozygous and showed medium to high expression of the HMW subunit genes[4].

Table 1. *Transgenic Wheat Lines and Expression levels*

Genotype	Line	Subunit	1Ax1	1Dx5
L88-31	B72-8-11a	null	null	null
	B72-8-11b	1Dx5	n/a	medium
	B102-1-1	1Ax1	medium	n/a
	B102-1-2	1Ax1	medium	n/a
L88-6	B73-6-1	1Dx5	n/a	high

3.2 Effects of Transformation on Agronomic Performance.

The field grown transgenic lines showed some reduction in performance when compared with the control plants, with lower establishment and yields for the 1998 season (Table 2). The nitrogen content and thousand grain weight values of the transgenic lines in L88-31 did not vary markedly from the control line, but B73-6-1 had a higher nitrogen content and lower grain weight than the L88-6 control (Table 2).

Table 2. *Transgenic Field Grown Wheat Samples 1998 Season (IACR-LARS)*

Line	subunit	1000 grain weight	Establishment %	Yield wt/plot	Nitrogen % Fr. Wt.
B72-8-11a	null	31	60	552	2.26
B72-8-11b	1Dx5	29	27	366	2.33
B102-1-1	1Ax1	32	40	269	2.47
B102-1-2	1Ax1	29	29	304	2.38
B73-6-1	1Dx5	33	75	566	2.59
L88-6	control	40	71	704	2.48
L88-31	control	33	81	728	2.33

3.3 Effects of Additional HMW Subunits on Grain End Use Quality.

Detailed studies of the rheological and functional properties of doughs and gluten fractions from the lines are being carried out in collaboration with Mr Y. Popineau (see

Popineau, Y. and Deshayes, G. this volume) (INRA, Nantes) and Dr F. Békés (CSIRO, North Ryde).

4 CONCLUSIONS

Wheat lines transformed to increase the proportion of HMW subunit proteins were successfully grown in field trial experiments over two successive growing seasons (1998 and 1999). Results from the first growing season (1998) showed that transformation with additional HMW subunit genes had no dramatic effects on agronomic performance. Small scale testing using a 2g Mixograph[6] showed impacts of the transgenic subunits on dough mixing properties, with either improved quality or "overstrong" characteristics being observed[7] The field trial experiments are ongoing and the results from these studies are providing new information on the roles of the individual subunits in glutenin polymer structure and functionality.

References

1. W. Seilmeyer, H.-D. Belitz, and H. Wieser, *Lebensm. Unters. Forsch.* 1991, **192**, 124.
2. N.G. Halford, J.M. Field, H. Blair, P. Unwin, K. Moore, L. Robert, *Theor. Appl. Genet.*, 1992, **83**, 373.
3. G.J. Lawrence, F. MacRitchie and C.W. Wrigley, *J. Cereal Sci.*, 1988, **7**, 109.
4. F. Barro, L. Rooke, F. Békés, P. Gras, A.S. Tatham, R.J. Fido, P. Lazzeri, P.R. Shewry, and P. Barcelo, *Nature Biotechnology.,* 1997 **15**, 1295.
5. R.J. Fido, A.S. Tatham, P.R Shewry, in *Methods in molecular biology – plant gene transfer and expression protocols*, vol 49, ed H. Jones, Humana Press Inc, Totowa, NJ 1995, p423
6. F. Békés and P.W. Gras, *Cereal Chemistry*, 1992, **69**, 229.
7. L. Rooke, F. Békés, R. J. Fido, F. Barro, P. Gras, A.S. Tatham, P. Barcelo, P.Lazzeri, and P.R. Shewry, *J. Cereal Sci*, 1999. **30**, 115-120.

Acknowledgments

IACR receives grant-aided support from the Biotechnology and Biological Sciences Research Council of the United Kingdom. We are grateful to Dr Rudi Appels (CSIRO Canberrra) for providing the L88-6 and L88-31 lines. Francisco Barro, Lee Rooke (IACR Rothamsted), Pilar Barcelo and Paul Lazzeri (Du Pont CIC, Rothamsted) for providing the transgenic lines. Experimental husbandry staff at Long Ashton and Rothamsted for carrying out the field experiments.

PROLAMIN AGGREGATION AND MIXING PROPERTIES OF TRANSGENIC WHEAT LINES EXPRESSING 1Ax AND 1Dx HMW GLUTENIN SUBUNITS TRANSGENES

Y. Popineau[1], G. Deshayes[1], R. Fido[2], P.R. Shewry[2] and A.S. Tatham[2].

1. INRA, Unité de Biochimie et de Technologie des Proteines, B.P. 71627, Rue de la Géraudiere, 44316 Nantes Cedex 03, France. 2. IACR-Long Ashton Research Station, Department of Agricultural Sciences, University of Bristol, Long Ashton, Bristol BS41 9AF, United-Kingdom.

1 INTRODUCTION

The role of HMW glutenin subunits in determining glutenin aggregation and gluten rheological behaviour has been assessed by comparing the properties of wheat varieties[1] or those of near-isogenic lines,[2] and by addition of purified subunit to flours.[3] Creation of transgenic lines differing in their HMW subunit composition allows us to study further the effect of subunit structure on glutenin association and functionality.[4] In this work, we compared the effect of expression of trangenes coding for subunits 1Ax1 (2 cysteines available for intermolecular bonds) and 1Dx5 (3 cysteines available for intermolecular bonds) on glutenin aggregation and the mixing properties of doughs.

2 MATERIAL AND METHODS

Flours were milled from control and transformed wheat lines grown at IACR-Long Ashton Research Station in 1998 (Table 1). Protein contents of flours were determined by Kjeldhal digestion and colorimetric determination of N content. Glutenin subunit composition was determined by SDS-PAGE in the presence of reducing agent and by capillary electrophoresis (analyses by ULICE, Riom, France).

The extractability, size distribution and aggregation of prolamins in flours were analysed by SE-HPLC on a Superose 6 column (1 x 30 cm; flow rate 0.3 ml / min;

Table 1. *Characteristics of wheat lines grown at IACR-Long Ashton Research Station*

Genotype	Line	HMW SU composition	Expressed subunit	Number of plots
L 88-31	control	17+18		4
	B72-8-11b	17+ 8	1Dx5	3
	B102-1-1	17+18	1Ax1	2
	B102-1-2	17+18	1Ax1	4
L 88-6	control	1, 17+18, 5+10		4
	B73-6-1	1, 17+18, 5+10	1Dx 5	4

detection at 220 nm) equilibrated in 0.0125 M Na borate buffer, 0.1% SDS. Samples were prepared by a three-step extraction. The first step comprised stirring in 0.0125 M Na borate buffer, pH 8.5, 0.5 % SDS. Residual proteins were sonicated under controlled conditions (6W, 30 s) in 0.0125 M Na borate buffer with 2% SDS. In addition, a third extraction step was made on the last pellet with 0.0125 M Na borate buffer, 2% SDS containing 1% DTT (dithiothreitol). All three protein extracts were analysed by SE-HPLC. Chromatographic patterns were divided into three peaks (P1, P2, P3) corresponding, respectively, to large-size glutenin polymers (MW>500k), medium-size glutenin polymers (500k<MW<70k) and gliadin monomers (or glutenin subunits in the case of the third reduced extract). Peak were quantified by measuring surface areas.

Mixing tests were performed with a 2 g Micromixograph (National Manufacturing Division) : 2g of flour were hydrated with 1.2 ml of distilled water and mixed for 10 minutes at 88 rpm and 20°C.

3 RESULTS AND DISCUSSION

3.1 Glutenin Subunit Composition

SDS-PAGE confirmed that proteins encoded by transgenes were expressed and that the apparent molecular weights of the expressed subunits were identical to those of wild type proteins. The total glutenin content depended on the number of HMW subunits expressed in the lines (Table 2). The 1A, 1D null control line (L88-31) contained less glutenin than the L88-6 control line, which contained 1A, 1B and 1D subunits. As expected, insertion of trangenes increased the proportions of HMW subunits. However, the total protein contents were not modified. The 1Ax1 and 1Dx5 subunits encoded by the transgenes accounted for about 50% and over 70% of the total HMW subunits, respectively, in the transformed lines.

Table 2. *Glutenin subunit composition of transgenic lines of wheat determined by capillary electrophoresis.*

Line	total glutenin % total protein	HMW GS % total glutenin subunits	Glu 1A % HMW GS	Glu 1B % HMW GS		Glu 1D % HMW GS	
			x	x	y	x	y
L88-6 Control	44	36	16	33	16	26	10
L88-6 + 1Dx5	53	44	4	8	4	73	4
L88-31 Control	33	18	0	75	25	0	0
L88-31 + 1Ax1	37	31	50	38	12	0	0
L88-31 + 1Dx5	44	28	0	21	8	71	0

3.2 Extractability and Aggregation of Prolamins

Little difference was observed between plots of the same line and only mean values of plots will be considered below. The effects of over-expression of subunits 1Ax1 or 1Dx5 in two different genotypes are compared in Table 3. When only subunits 17+18 were present (L88-31 control) glutenin polymers were less aggregated than when five subunits were expressed (L88-6 control). Over-expression of subunit 1Ax1 increased the proportion of glutenin extractable with 2% SDS + U.S. However, the amount extractable with DTT did not change. This indicated the presence of more aggregated glutenin polymers, but that these had properties similar to those of the wild type line.

Over-expression of subunit 1Dx5 modified glutenin aggregation more extensively. The proportion of proteins extractable with 0.5 % and 2% SDS decreased, but the amount of DTT-extracted proteins increased considerably. Subunit 5 was concentratred in the latter fraction. This change was greater when subunit 5 was expressed in the L88-31 background which was probably due to the different amounts of the subunit 5 in the glutenins from the two genotypes. This indicates the ability of subunit 5, which contains an additional cysteine residue, to promote the formation of highly crosslinked (by SS bonds) and aggregative glutenin polymers.

3.3 Mixing Properties

For a given line the mixing behaviour was almost independant of the plot considered. The control line L88-31 exhibited weaker mixing properties than the control line L88-6 : lower peak time, peak value and peak width, lower width at 10 min (Table 4). This is related to the difference in HMW subunit composition and glutenin aggregation. Expression of subunit 1Ax1 enhanced the mixing properties of L88-31. The effect of the insertion of 1Dx5 subunit was unexpected. In line L88-31, this increased the peak time but decreased all other mixogram characteristics. Insertion in line L88-6, originally much stronger than L88-31, decreased all mixogram characteristics. In fact, the transgenic 1Dx5 lines failed to form a cohesive dough under hydration and mixing.

4 DISCUSSION

Glutenin aggregation depended on the number of HMW subunits expressed by the lines.

Table 3. *SE-HPLC analysis of proteins extracted from control and transgenic lines.*

Genotype	Line	Inserted subunit	0.5 % SDS			2% SDS + US	2% SDS + DTT
			P1 % tot. prot.	P2 % tot. prot.	P3 % tot.prot.	% tot. prot.	% tot.prot.
L88-31	Control		7.5	13.6	66.3	10.6	2.0
	B72-8-11b	1Dx5	4.8	11.1	57.5	8.3	18.4
	B102-1-1	1Ax1	4.7	13.0	65.1	15.9	2.2
	B102-1-2	1Ax1	8.0	15.0	63.3	11.8	2.0
L88-6	Control		3.2	9.7	53.4	29.9	3.1
	B73-6-1	1Dx5	2.8	8.7	47.0	12.1	29.4

Table 4. *Mixograph analysis of flours from control and transgenic lines.*

Genotype	Line	Inserted subunit	Peak time (min)	Peak value torque %	Peak width torque %	Midline at 10min torque %	Width at 10 min torque %
L88-6	Control		4.9	43.7	20.6	34.3	8.6
	B73-6-1	1Dx5	1.3	16.8	21.0	13.4	7.9
L88-31	Control		2.6	27.7	11.2	20.0	5.3
	B72-8-11b	1Dx5	4.5	16.0	7.8	8.7	4.9
	B102-1-1	1Ax1	2.6	39.3	18.8	27.9	6.6
	B102-1-2	1Ax1	3.1	36.3	16.5	26.5	6.5

Insertion of subunit 1Ax1 increased glutenin aggregation, but may not have given any increase in crosslinking by SS bonds. Thus, only the average size of the glutenin polymers may have been increased. This increased peak mixing time and dough consistency. Insertion of subunit 1Dx5 increased glutenin aggregation considerably, possibly through covalent crosslinking between aggregates, generating very large and insoluble aggregates.This difference was attributed to the presence of an additional cysteine residue available for intermolecular crosslinking in subunit 1Dx5. This resulted in abnormal mixing behaviour, such as absence of real peak of torque and very low torque. It can be postulated that an excess of subunit 1Dx5 modified the glutenin (gluten) structure and hindered the formation of an homogeneous protein network. Subunit 1Dx5 is always expressed as a pair with 1Dy10 and there is evidence that dimers between these two subunits, and between other x-type and y-type subunits, are present as "building blocks" in the glutenin polymers. Over-expression of subunit 1Dx5 in the absence of additional subunit 1Dy10 could therefore result in extensive restructuring of the glutenin polymers with important consequences for the mixing and baking properties. Thus, the results show that transformation can be used to modify the technological properties of gluten proteins. Furthermore, the drastic effects obtained by expression of subunit 1Dx5 may facilitate the development of new uses of wheat, either in the food industry or non-food applications.

References

1. P.I. Payne, K.G. Corfield and J.A. Blackman, *Theoret. Appl. Genet.*, 1979, **55,** 153.
2. Y. Popineau, M. Cornec, J. Lefebvre and B. Marchylo, *J. Cereal Sci.*, 1994, **19,** 231.
3. F. Békés, P.W. Gras, R.B. Gupta, D.R. Hickman and A.S. Tatham, *J. Cereal Sci.*, 1994, **19,** 3.
4. I. Rooke, F. Békés, R. Fido, F. Barro, P. Gras, A.S. Tatham, P. Barcelo, P. Lazzeri, P.R. Shewry, *J. Cereal Sci.*, 1999, **30,** 115.

Acknowledgements

Part of this research was supported by the European Community. FAIR CT96-1170 Eurowheat. IACR receives grant-aided support from the Biotechnology and Biological Sciences Research Council of the United Kingdom.

MODIFICATION OF STORAGE PROTEIN COMPOSITION IN TRANSGENIC BREAD WHEAT

G.Y. He[1], R. D'Ovidio[2], O.D. Anderson[3], R. Fido[4], A.S. Tatham[4], H.D. Jones[1] P.A. Lazzeri[5], and P.R. Shewry[4]

[1]Biochemistry and Physiology Department, IACR-Rothamsted, Harpenden, Herts, AL5 2JQ, UK; [2]Dipartimento di Agrobiologia e Agrochimica, Università della Tuscia, Via S. Camillo de Lellis, 01100 Viterbo, Italy; [3]U.S. Department of Agriculture-Agricultural Research Service, Western Regional Research Centre, 800 Buchanan Street, Albany, CA 94710, U.S.A.; [4]IACR-Long Ashton Research Station, Department of Agricultural Sciences, University of Bristol, Long Ashton, Bristol, BS41 9AF, UK; [5]DuPont Wheat Transformation Laboratory, C/O IACR-Rothamsted, Harpenden, Herts, AL5 2JQ, UK.

1 INTRODUCTION

Although the functional properties of bread wheat dough are determined by many seed components (proteins, carbohydrates and lipids) and their interactions, gluten storage proteins are of greatest importance. Of these, the HMW subunits of glutenin are major determinants of dough elasticity, with both quantitative effects related to subunit number and qualitative effects related to allelic variation in subunit properties.[1-3] We have therefore produced transgenic wheat plants with specific differences in their gluten structure by the addition of genes for wild type or mutant proteins in order to explore the relationships between HMW subunit protein structure and functional properties.

2 MATERIALS AND METHODS

2. 1 Materials

The spring wheat line L88-31 is null for the 1A and 1D subunit genes and expresses only subunits 1Bx17 + 1By18 while the near isogenic sister line L88-6 contains HMW subunits 1Ax1, 1Bx17, 1By18, 1Dx5 and 1Dy10 encoded by the *Glu-A1*, *Glu-B1* and *Glu-D1* loci, respectively. Donor plants were grown in controlled environment chambers under 16/8 h light/dark photoperiod with artificial light of 350 mol $m^{-2}.s.^{-1}$ at 18°C/16°C.

2.2 Methods

2.2.1 Plasmid. Plasmid pHMW1Dx5 contains an 8.7kb *Eco*RI fragment containing the complete coding region of the 1Dx5 gene flanked by about 3.8kb and 2.2kb of 5' and 3' flanking sequence, respectively.[4] This was used to construct an expression cassette (pLRPT) containing about 1.3kb of the 5' flanking region of the 1Dx5 gene with a 650bp fragment of the 3' flanking region. This cassette was used for the expression of three mutant forms of subunit 1Dx5 (pLRPTDx5-R441, pLRPTDx5-R576 and pLRPTDx5-R853).

The plasmid pCaI-*neo* contains a *neo* gene driven by CaMV 35S promoter plus *adh1* intron, and geneticin sulphate (G418) or paromomycin were applied for selection of transformed cultures. Plasmid pAHC25 contains *gus* and *bar* genes each driven by the ubiquitin promoter (plus intron) and Bialaphos or PPT were applied for selection.

2.2.2 Transformation. The genes of interest were delivered into target tissue in combination with the selectable marker gene. Scutella of L88-31 and L88-6 were isolated one-day before bombardment and cultured on MS-based induction medium with either 1mg/l 2,4-D or 2mg/l picloram. Sub-micron gold particles were used for DNA delivery under high pressure, using the Helium-Driven PDS-1000/He Biolistic® Particle Delivery System. Cultures were maintained in darkness for three weeks for embryogenic callus induction.

2.2.3 Analysis of transgene integration and expression. Genomic DNA was isolated from leaves using a CTAB extraction method.[5] PCR analysis was used to determine transgene integration while SDS-PAGE, dot blotting and western blotting analyses of seed protein were carried out to detect the presence of HMW subunit proteins.[6-7]

2.2.4 Progeny analysis of transgenic plants. The T_0, T_1 and T_2 transgenic wheat lines containing the wild type and mutant HMW subunit genes were grown under glasshouse conditions to provide bulk seed samples, and to identify homozygous lines of transgenic plants.

3 RESULTS AND DISCUSSION

3.1 Plasmid construction

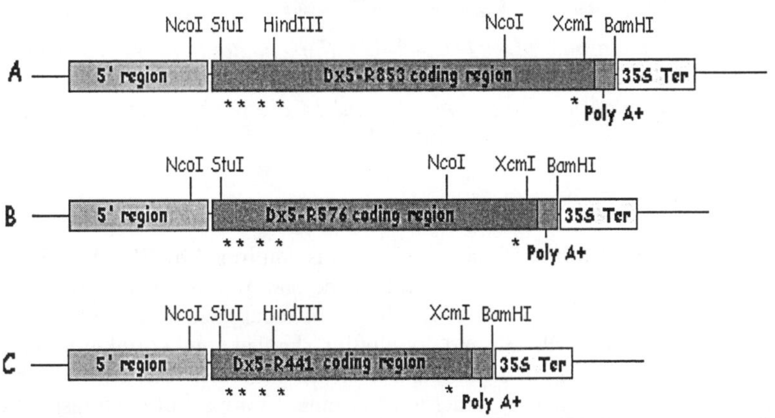

Figure 1. *Diagrams of wild type and mutant subunit 1Dx5 constructs with the positions of restriction sites used for preparing the constructs indicated. The 5' region and Poly A+ region are derived from the 1Dx5 gene. The * indicates the positions of cysteine resides.* ***A****. Plasmid pLRPTDx5-R853;* ***B****. Plasmid pLRPTDx5-R576;* ***C****. Plasmid pLRPTDx5-R441.*

The pLRPTDx5-R441, pLRPTDx5-R576 and pLRPTDx5-R853 constructs are under control of the 1Dx5 promoter region and encode modified 1Dx5 subunits with repetitive domains about 34% and 17.2% shorter, and 22.5% larger than the native 1Dx5 protein, respectively (Figure 1). Comparison of transgenic plants expressing them can therefore contribute to defining the importance of the repetitive domain in the viscoelastic properties of gluten.

3.2 Confirmation of transgenic plants

A number of healthy putative transgenic plants were recovered from bombarded scutella transformed with all of the subunit 1Dx5 gene constructs. PCR analysis for construct pLRPT-Dx5-R576 showed 14 positive T_0 plants in the wheat line L88-6 that contains the endogenous 1Dx5 subunit gene. Consequently, PCR products of the short mutant (Dx5-R576) gene could not be distinguished from the native 1Dx5 product and the presence of the 35S terminator of the construct was therefore used to identify trangenic plants. Similarly, PCR analysis was carried out to confirm transformation events with the other subunit 1Dx5 constructs (pLRPT-Dx5-R853, pLRPT-Dx-441, pHMW1Dx5) with a total of 31 transgenic plants were recovered.

Further studies demonstrated that the Dx5-R853 gene was stably inherited in all 27 T_1 plants derived from 5 independent lines (Figure 2).

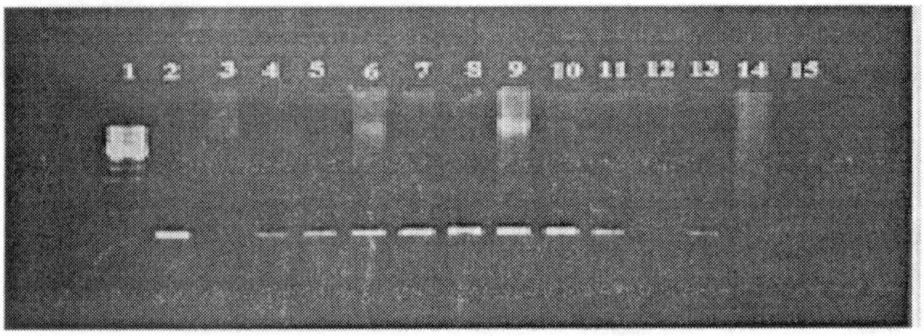

Figure 2. *PCR analysis of plasmid pLRPTDx5-R-853 (long mutant) in T_1 transgenic plants. Lane 1 is a 1kb marker; Lane 2 is plasmid control; Lanes 3-11 inclusive are T_1 plants; Lane 12 is negative plant control; Lanes 13 and 14 are positive plant controls; Lane 15 is a water control.*

3.3 Transgene Expression

The expression of the mutant forms of subunit 1Dx5 was confirmed by SDS-PAGE of seed proteins over three generations for the Dx5-R853 gene and over two generations for 1Dx5-R576 (Figure 3). Transgene expression levels in different generations of wheat transgenic lines determined by SDS-PAGE and dot blotting showed good correlation with the results of PCR analysis. Small-scale Mixograph analyses and other tests will now be applied to flour from the transgenic lines in order to determine the impact of the transgenes on functional properties.

4 CONCLUSIONS

The structures and interactions of the HMW glutenin proteins are the major determinants of dough strength (elasticity) which is an important determinant of the end-use properties of wheat. The application of reliable and robust transformation methods allows the production of wheat plants with specific differences in their gluten structure, by the introduction of additional genes for wild type or mutant proteins. It therefore allows the relationships between protein composition and grain functionality to be established.

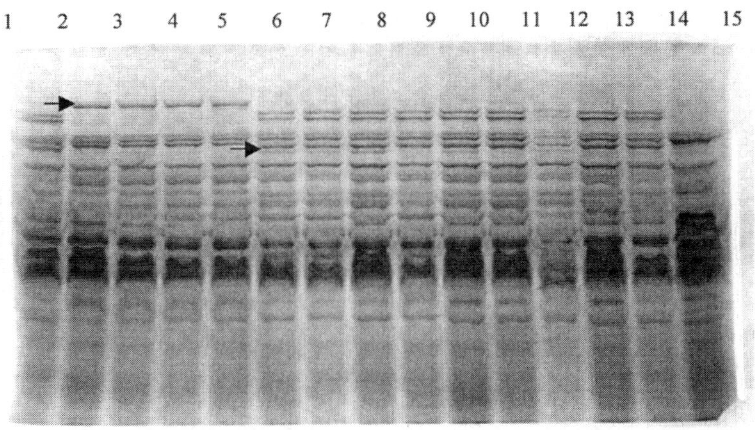

Figure 3. *Expression of plasmid pLRPT-Dx5-R853 in T_3 plants of wheat line L88-31 (lanes 2-5) and pLRPT-Dx5-R576 in T_2 plants of line L88-6 (lanes 6-13). Proteins from non-transformed control seeds were also included (line L88-6 in lanes 1 and 14 and L88-31 in lane 15). Transgene products are indicated by arrows in lanes 2 and 6.*

References

1. P.R. Shewry, P.R. Tatham, F. Barro, P. Barcelo, P.A. Lazzeri, *Bio/Technology*, 1995a, **13**:1185-1190

2. P.R. Shewry, P.R. Tatham, P.A. Lazzeri, Shewry, *J. Sci. Food Agric.*, 1997, **73**, 397.

3. N.G. Halford, J.M. Field, H. Blair, P. Urwin, K. Moore, L. Robert, R. Thompson, R.B. Flavell, A.S. Tatham, and P.R. Shewry, *Theor. Appl. Genet.* 1992, **83**, 373.

4. N. G. Halford, J. Forde, P.R. Shewry, and M. Kreis, 1989, *Plant Sci.* **62**, 207.

5. J. Stacey, and P.G. Isaac, in *Methods in Molecular Biology-Protocols for Nucleic Acid Analysis by Nonradioactive Probes,* ed. P. G. Isaac, Humana Press Inc., Totowa, NJ, 1994, **28** p. 9.

6. P.R. Shewry, A.S. Tatham, and R.J. Fido, in *Methods in Molecular Biology-Plant Gene Transfer and Expression Protocols*, ed. H. Jones, Humana Press Inc., Totowa, NJ, 1995, **49**, p. 399.

7. R.J. Fido, A.S. Tatham, and P.R. Shewry, in *Methods in Molecular Biology-Plant Gene Transfer and Expression Protocols*, ed. H. Jones, Humana Press Inc., Totowa, NJ, 1995, **49**, p. 423.

Acknowledgement

The Institute of Arable Crops Research (IACR) receives grant-aided support from the Biotechnology and Biological Sciences Research Council of the United Kingdom. Part of the work is funded by European Commission in the project "Improving the Quality of EU wheats for use in the Food Industry 'EUROWHEAT'" (Fair CT96-1170).

TRANSFORMATION OF COMMERCIAL WHEAT VARIETIES WITH HIGH MOLECULAR WEIGHT GLUTENIN SUBUNIT GENES

G.M. Pastori[1], S.H. Steele[1], H.D. Jones[1] and P.R. Shewry[2]

1. Biochemistry and Physiology Department, IACR-Rothamsted Experimental Station, Harpenden, Herts AL5 2JQ, UK. 2. IACR-Long Ashton Research Station, Long Ashton, Bristol BS41 9AF, UK.

1 INTRODUCTION

The high molecular weight (HMW) subunits of wheat glutenin are major determinants of the elastic properties of gluten that allow the use of wheat doughs to make bread, pasta and a range of other foods[1]. There are both quantitative and qualitative effects of HMW subunits on the quality of the grain, the former being related to differences in the number of expressed HMW subunit genes[1]. Mixograph analysis of wheat isogenic lines L88-6 and L88-31 transformed with HMW subunits 1Ax1 and 1Dx5 revealed a significant increase in dough strength when compared with non-transformed lines, being highest in transgenic lines expressing 1Ax1+1Dx5 subunits[2]. Similarly, expression of additional HMW subunits 1Ax1 and 1Dx5 in pasta wheat (*Triticum turgidum* var. Durum) resulted in improved dough strength and stability[3]. However, dough from the transgenic lines expressing 1Dx5 was too strong for conventional mixograph analysis and blending was required[3], suggesting that high levels of expression of the allelic subunit 1Dx5 without the equivalent expression of 1Dy10 may result in modification of the gluten network and altered physical properties.

The aim of this work is to transfer additional HMW subunit genes to commercially grown wheat varieties differing in their quality for breadmaking and to determine the effects of the resulting changes in the number of actively expressed HMW subunit genes on the amount, composition and biophysical properties of gluten and doughs.

2 MATERIALS AND METHODS

2.1 Materials

Wheat (*Triticum aestivum* L.) plants were grown in controlled environment chambers at 18°C/15°C (day/night), with a 16h/8h-photoperiod and a photosynthetic photon flux density of 700 μmol m^{-2} s^{-1} at a relative humidity of 80%. Wheat plants from varieties Cadenza, Canon and Imp were grown for 10-13 weeks until explants were collected for bombardment.

2.2 Methods

2.2.1 Transformation procedure. Scutella from immature embryos were transformed by particle bombardment[4,5] with plasmids containing the HMW glutenin subunit genes 1Ax1, 1Dx5 and/or 1Dy10 (Figure 1) and the selectable/scorable marker plasmid pAHC25 containing the *uidA* and *bar* genes under the control of maize ubiquitin 1 promoter. Embryogenic calli were regenerated and plants selected with the herbicide phosphinothricin.

2.2.2 DNA analysis. Leaf genomic DNA from control and putative transgenic plants was extracted[5] and the presence of the transgenes was determined by PCR.

2.2.3 Protein analysis. HMW glutenin subunits were analysed by SDS-PAGE using the Tris-borate buffer system[2].

Figure 1 *HMW glutenin subunits gene constructs*

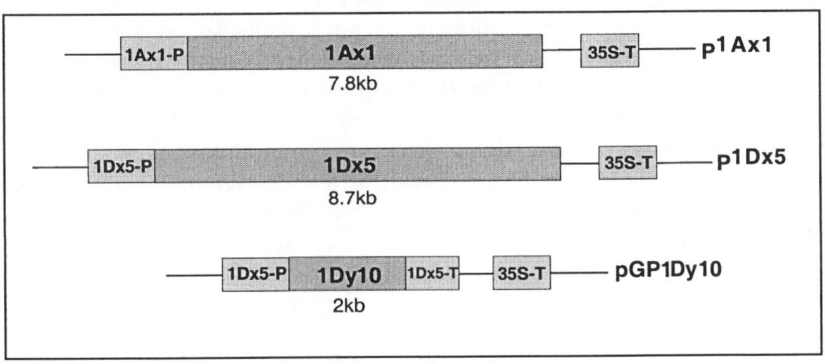

3 RESULTS AND DISCUSSION

3.1 Production of transgenic lines

A total of 1749 immature embryos isolated from varieties Cadenza and Canon were bombarded with plasmids containing the HMW subunit 1Ax1 (p1Ax1) and the selectable/scorable marker genes (pAHC25). The embryogenic capacity of non-bombarded calli of both varieties was 70%, which was maintained in bombarded tissues of Cadenza and Canon, both in controls bombarded with gold only and in treatments bombarded with gold plus DNA. There was no correlation between transformation frequency (number of transgenic lines / number of bombarded embryos x 100) and the embryogenic capacity in either variety (r^2= 0.0106 for Cadenza; r^2= 0.0179 for Canon).

The efficiency of the selection procedure was demonstrated by a low percentage of escapes, i.e. number of plants surviving herbicide-selection but lacking the transgene. In Cadenza, 68.5% of the bombardments resulted in less than 30% escapes, and 43% of these bombardments gave no escapes. Similar values were achieved in the variety Canon in which 41% of bombardments had no escapes. There were no significant differences between the two varieties in terms of their ability to survive/escape the herbicide. The frequencies of transformation were high in both varieties. Mean frequencies of 4.9% and 3.9% were achieved for Cadenza and Canon, respectively, with the best bombardments giving up to 7% in both varieties. A high correlation between transformation frequency and the age of wheat donor plants was observed in varieties Cadenza and Canon (r^2= 0.946)[5]. Similar results were observed in the variety Imp, although the highest frequency obtained was 2.5% from 300 bombarded embryos, probably due to the lower regeneration capacity of the explants when compared with Cadenza and Canon. A total of 79 transgenic lines containing the p1Ax1 construct were obtained from the three varieties and analysed for protein expression.

3.2. Protein analysis

The pattern of HMW glutenin subunits was analysed in control and transgenic lines of the three varieties by separating total protein fractions from 5-10 individual T_1 half seeds by SDS-PAGE. Cadenza contains the endogenous subunits 1Dx5+1Dy10 and 1Bx14+1By15 while the HMW subunit pattern of Canon consists of 1Ax2*, 1Dx2+1Dy12 and 1Bx7+1By9 (Figure 2). The presence of the additional subunit 1Ax1 was clearly detected in the HMW subunit patterns of transgenic lines of both varieties. The transgene segregated in a Mendelian fashion (3:1) in almost all regenerated plants. Transgenic lines (selected T_1 half seeds) with different levels of expression are being grown and analysed for protein expression. Selected homozygous lines will be multiplied and tested for gluten composition and functional properties.

More transgenic plants are being generated expressing the HMW subunits 1Ax1, 1Dy10 and 1Dx5+1Dy10 in the varieties Cadenza, Imp and Buster. Bombardments are carried out using either whole plasmids or purified DNA fragments encoding only for the gene of interest and, therefore, free of the ampicillin-resistance gene. From the 32 transgenic lines produced until now containing DNA fragments encoding subunits 1Ax1 or 1Dx5, only one contains the ampicillin-resistance gene, indicating the relatively high efficiency of purification of the DNA fragments (data not shown). Transgenic plants overexpressing 1Ax1, 1Dx5 and 1Dx5+1Dy10 genes are being characterized by Southerns to determine copy number, insertion pattern and inheritance of the transgenes, and analysed for protein expression by SDS-PAGE. Selected homozygous lines will be analysed for effects on gluten composition and functional properties.

Figure 2 *HMW glutenin subunit pattern in varieties Cadenza and Canon transformed with the HMW subunit 1Ax1. Protein fractions from control (C) and transgenic plants were separated by SDS-PAGE. Arrows indicate the presence of the additional subunit 1Ax1.*

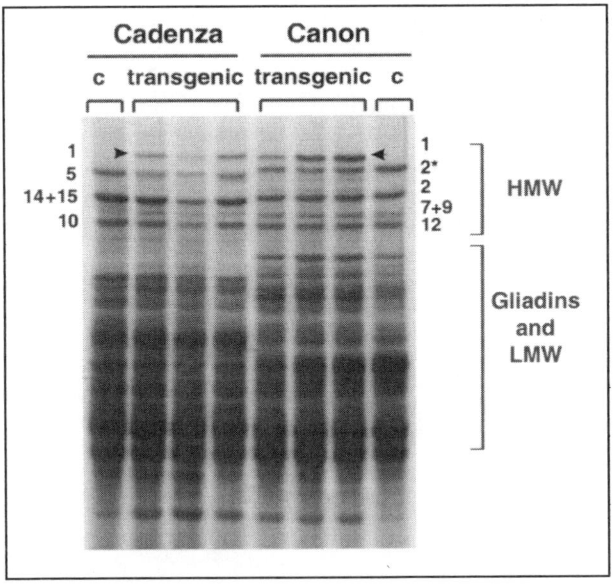

4 CONCLUSIONS

A significant number of transgenic lines of elite wheat varieties containing the HMW subunit gene 1Ax1 have been generated at relatively high frequencies. Other transgenic lines are being produced expressing subunits 1Dx5, 1Dx5+1Dy10 and 1Dy10. The availability of a range of transgenic lines expressing additional HMW subunits, either alone or in allelic pairs, using whole plasmids or purified DNA fragments offers a unique opportunity to determine the effects of HMW subunits on gluten structure and functionality, and to dissect the pattern of insertion and inheritance of HMW transgenes in successive generations.

References

1. P.R. Shewry, A.S. Tatham, F. Barro, P. Barcelo and P. Lazzeri, *Bio/Technology*, 1995, **13**, 1185

2. F. Barro, L. Rooke, F. Bekes, P. Gras, A. Tatham, R. Fido, P. Lazzeri, P.S. Shewry and P. Barcelo, *Nature Biotech.*, 1997, **15**, 1295

3. G.Y. He, L. Rooke, S.H. Steele, F. Bekes, P. Gras, A.S. Tatham, R. Fido, P. Barcelo, P.R. Shewry and P. Lazzeri, *Mol. Breed.*, 1999, **5**, 377

4. P. Barcelo and P. Lazzeri, 'Methods in Molecular Biology-Plant Gene Transfer and Expression Protocols', Humana Press Inc. Totowa, NJ, 1995, **49**, p.113

5. G.M. Pastori, M.D. Wilkinson, S.H. Steele, C.A. Sparks, H.D. Jones and M.A.J. Parry, submitted to *Mol. Breed.*, 2000

Acknowledgements

IACR receives grant-aided support from the Biotechnology and Biological Sciences Research Council (BBSRC) of the United Kingdom. G.M. Pastori acknowledges financial support from BBSRC. This work forms part of a LINK project supported by BBSRC and Biogemma.

MODIFICATION OF THE LMW GLUTENIN SUBUNIT COMPOSITION OF DURUM WHEAT BY MICROPROJECTILE-MEDIATED TRANSFORMATION

P. Tosi[1,2], J.A. Napier[1], R. D'Ovidio[3], H.D. Jones[2], P.R. Shewry[1].

1.IACR-Long Ashton, Long Ashton, Bristol BS41 9AF, UK. 2.IACR-Rothamsted, Harpenden, Herts, AL5 2JQ, UK. 3.Universita' degli Studi della Tuscia, Facolta' di Agraria, 01100 Viterbo, Italy.

1 INTRODUCTION

Durum wheat is the hardest of all wheats, making the semolina obtained by grinding of durum wheat seeds the most suitable material for manufacturing food products such as pasta and couscous. The viscoelastic properties of wheat dough are primarily due to the gluten proteins present in the starchy endosperm of wheat seeds, the so-called prolamins[1]. For semolina dough, in particular, performance has been correlated with the specific composition of the low molecular weight glutenin subunits (LMW-GS)[2], a class of highly polymorphic proteins participating in the formation of the extensive disulphide-linked polymers of the endosperm. At the molecular level, LMW-GSs consist mainly of two non-repetitive domains flanking a central domain composed of short amino-acid motifs repeated in tandem series.

In this work, transgenic wheat technology has been used to modify the LMW glutenin composition of the commercial durum wheat cultivars Svevo and Ofanto by over-expression of three particular LMW-GS genes. The genes used were *lmw*1A3[3,4], *lmw*1B[3,4] and its mutant form *lmw*1B- and they differed in the length and regularity of the internal repetitive domain and/or in the number and position of cysteine residues. Since these differences are considered to be important in determining dough quality, these transgenic plants represent a valuable means to study the role of such structural features of LMW subunits in the complex dough system. The addition of an epitope tag at the 3' end of the proteins encoded by the introduced genes facilitated their identification and opens new opportunities to study their trafficking and deposition.

2 MATERIAL AND METHODS

2.1 Materials

Donor plants of durum wheat (*Triticum turgidum* L. ssp durum) cultivar Svevo and Ofanto were grown in glasshouse at 18-20°C day and 10-12°C night temperatures, with a 16h photoperiod provided by banks of fluorescent tubes and incandescent bulbs.

2.2 Methods

2.2.1 *Plasmid DNAs.* Plasmids pHMW1Dx5[5]and pLRPT were used to construct the expression cassette $pRDPT_5$. Plasmid RMBS was used to add a c-myc tag, recognised by the monoclonal antibody 9.E10, to the 3'end of the LMW-GS genes.
Plasmid pLDNLMW1B- was the source of the mutant gene lmw1B-, the encoded protein lacking the first cysteine residue present in related LMW subunit proteins.
Plasmid pAHC25, containing the *uid* and *bar* genes under the control of the maize ubiquitin *Ubil* promoter, was used as selectable marker.
2.2.2 *Transformation procedure.* Explants prepared from immature inflorescences[6] were co-transformed by particle bombardment with a plasmid containing the gene of interest and plasmid pAHC25. Plants were regenerated and selected under the herbicide phosphinothricin[7].
2.2.3 *DNA analysis.* Plants surviving selection were grown in greenhouses and DNA extracted from their leaves analysed by PCR to assay transgene integration.
2.2.4 *Protein analysis of the transgenic plants.* LMW glutenin subunit composition was characterised by SDS-PAGE[8] and by western and dot blotting[9] using the monoclonal antibody 9.E10 (AgrogenBioclear). The transgenic lines were multiplied and their T_2 seeds analysed in order to identify homozygous plants.

3 RESULTS AND DISCUSSION

3.1 Preparation of constructs for bombardment

A total of six constructs containing LMW glutenin subunit genes with or without a 3' epitope-tag were generated as shown in Figure 1. The epitope was added in order to facilitate the analysis of the transgenic plants; however, since it may influence the expression or the functionality of the proteins, constructs lacking the epitope-tag were also prepared.

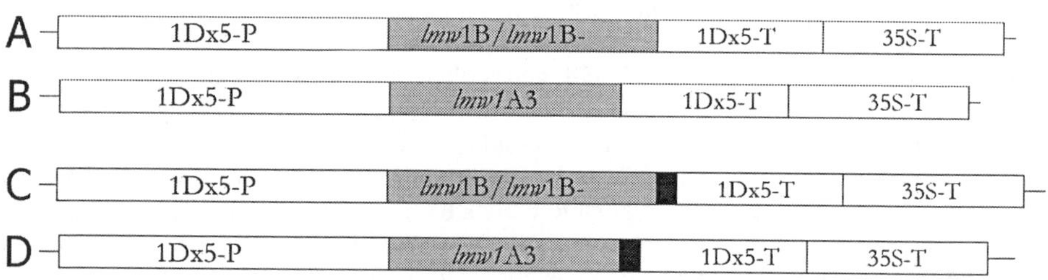

Figure 1. *Constructs used for transformation.*
In order to generate constructs with regulatory sequences resembling as much as possible those of glutenin genes, 650 bp of the 3' untranslated region of the HMW subunit 1D X5 (1D X5-T) were amplified from plasmid pHMW1DX5 and cloned into the vector pLRPT already containing a 1.3kb promoter fragment of the same gene (1DX5-P) and a CaMV 35S terminator sequence (35S-T); the resulting transformation vector was named $pRDPT_5$. The coding regions of the lmw1A3, lmw1B and lmw1B- genes were cloned in the RMBS cassette and recovered together with the epitope-tag (▮) to be inserted into $pRDPT_5$.
A *represents the constructs $pRDPT_51B$ and $pRDPT_51B$- containing lmw1B or lmw1B- inserted into $pRDPT_5$.*

B *represents the construct pRDPT$_5$1A3 containing lmw1A3 inserted into pRDPT$_5$.*
C *and* **D** *represent the constructs pRDPT$_5$1B*, pRDPT$_5$1B*- and pRDPT$_5$1A3*, containing the same three lmw-gs genes with the c-myc tag added at the 3' end.*

3.2 Transgene expression analysis

PCR positive lines were obtained for each of the constructs containing the epitope and for the constructs without the epitope carrying *lmw*1A3 or *lmw*1B. The PCR results were confirmed by Southern blot analysis. Transformants were grown to maturity and their seeds analysed to detect transgene expression.
Detection of transgene expression in lines transformed with constructs containing the epitope was simplified by using dot immunoblotting to screen up to 30 seeds of each T_0 transgenic line. All lines examined so far in this way expressed the epitope-tagged transgenes. In order to localise the transgenic protein among the endogenous proteins, protein extracts of seeds expressing the transgenes and of seeds of control untransformed plants were separated by Tris-borate SDS-PAGE. In the lines transgenic for *lmw*1B* and *lmw*1B*-, the novel LMW subunit proteins were easily detected among the native proteins (Figure 2, arrow). However, attempts to identify the transgenic protein encoded by *lmw*1A3 have so far been unsuccessful. The size of the bands corresponding to the transgenic proteins, detected either by SDS-PAGE or western blot analysis, are consistent with the inserted genes not having undergone rearrangements. However, SDS-PAGE of the only line so far analysed which expresses pRDPT$_5$1B*- indicated that the transgenic protein migrated slightly faster than the corresponding transgenic protein from the pRDPT$_5$1B* lines. The reason for this is not known but it should be resolved by analysis of other independent lines transgenic for the construct.

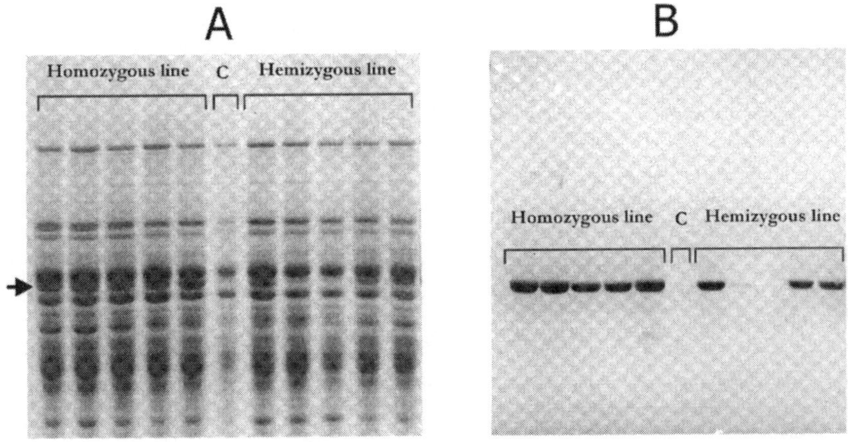

Figure 2. *Tris-borate SDS-PAGE (A) and western blot (B) analysis of T_2 seeds from two transgenic lines of cv. Ofanto expressing pRDPT$_5$1B*. The position of the transgenic protein on the gel is indicated with an arrow.*

Of the lines transformed with constructs without the epitope tag, protein analysis has only been conducted so far on lines transgenic for *lmw*1A3. SDS-PAGE failed to identify the transgenic protein and western blot analysis was not an option due to the unavailability of an antibody specific for the transgenic subunit. For this reason it has not been possible to determine if the transgene actually expressed the protein, but at levels not detectable on gel, or if it had undergone rearrangements that resulted in lack of expression. It is also possible, in view of the results obtained with the lines transgenic for pRDPT$_5$1A3*, that the level of expression was sufficiently high but the transgenic protein could not be resolved because it co-migrated with the native proteins. Alternative protein separation methods will be therefore used to determine if this is the case.

4 CONCLUSIONS

In this work, the generation of lines transgenic for a series of LMW subunit genes is reported. The resulting modifications in the gluten protein composition will be investigated for their impact on dough quality. The c-myc epitope-tag has been used to unambiguously identify and follow the expression of the transgenic LMW subunits. This is the first time an endosperm storage protein has been specifically tagged and the lines therefore represent a valuable resource to investigate aspects of gluten protein deposition and trafficking.

5 REFERENCES

1 P.R. Shewry and B.J. Miflin, 1985, *Advances in Cereal Science and Technology*, **7**, 1.

2 N.E. Pogna, D. Lafiandra, P. Feillet, J.C. Autran, 1988, *J. Cereal Science*. **7**, 211.

3 R. D'Ovidio, M. Simeone, C. Marchitelli, S. Masci, E. Porceddu, 1996, *Proc 6th Int Gluten Workshop* , 81.

4 R. D'Ovidio, S. Masci, C. Mattei, P. Tosi, D. Lafiandra, and E. Porceddu, (These Proceedings).

5 N.G. Halford, J. Forde, P.R. Shewry, and M. Kreis, 1989, *Plant Sci.* **62**, 207.

6 P. Barcelo and P. Lazzeri, 1995, *Methods in Molecular Biology-Plant Gene Gransfer and Expression Protocols* (Vol.**49**) Jones, H. (ed). Humana Press Inc., Totowa, NJ.

7 G.M. Pastori, M.D. Wilkinson, S.H. Steele, C.A. Sparks, H.D. Jones and M.A.J. Parry, submitted to *Mol. Breed*, 2000.

8 P.R. Shewry, A.S. Tatham, and R.J. Fido, 1995b, *Methods in Molecular Biology-Plant Gene Transfer and Expression Protocols* (Vol.**49**), 399, Jones, H. (ed). Humana Press Inc., Totowa, NJ.

9 R.J. Fido, A.S. Tatham, and P.R. Shewry, 1995b, 433 *Methods in Molecular Biology-Plant Gene Transfer and Expression Protocols* (Vol.**49**), 423, Jones, H. (ed). Humana Press Inc., Totowa, NJ.

GENETIC MODIFICATION OF THE TRAFFICKING AND DEPOSITION OF SEED STORAGE PROTEINS TO ALTER DOUGH FUNCTIONAL PROPERTIES

C. Lamacchia[1,2,3], N. Di Fonzo[1], N. Harris[4], A.C. Richardson[5], J.A. Napier[3], P.A. Lazzeri[6], P.R. Shewry[3] and P. Barcelo[6],

1. Istituto Sperimentale per la Cerealicoltura, Foggia, SS 16 Km 675, Italy. 2. IACR-Rothamsted, Harpenden, Herts, AL5 2JQ, UK. 3. IACR-Long Ashton, Long Ashton, Bristol, BS18 9AF, UK. 4. Quality Assurance for Higher Education, Southgate House, Glocester GL1 1UB, U.K.. 5. Department of Biological Sciences. University of Durham, Durham, DH1 3LE, UK. 6. DuPont CIC-Wheat Transformation Laboratory, c/o IACR-Rothamsted, UK.

1 INTRODUCTION

The gluten proteins are major determinants of the functional properties of the wheat grain, conferring visco-elastic properties which allow dough to be processed into a range of food products including bread (unleavened and leavened), pasta and noodles. In particular, the glutenin polymers appear to determine dough strength, forming an elastic network which interacts with the gliadins by non-covalent forces, principally hydrogen bonds[1]. Protein-protein interactions occur in the secretory pathway during the trafficking and deposition of gluten proteins in protein bodies and this process remains incompletely understood. Currently available evidence indicates that two routes may exist.[2] All gluten proteins are initially synthesised on ribosomes bound to the rough endoplasmic reticulum (ER) and translocated into the lumen with the cleavage of a short (≅ 20 amino acid) signal peptide. Within the lumen the proteins fold and disulphide bonds are formed, possibly assisted by lumenal proteins: molecular chaperones and protein disulphide isomerase, respectively. Some gluten proteins are then transported via the Golgi apparatus to the vacuole where they form protein bodies. In contrast, other gluten proteins appear to remain within the ER lumen where they accumulate to form a second population of protein bodies of ER origin. These two populations of protein bodies may subsequently fuse to give rise to a continuous protein matrix in the mature endosperm cells.

The factors which determine whether gluten proteins are transported to the vacuole or remain in the ER are not known and no specific targeting or retention sequences have been detected. However, the suggestion that glutenins tend to be retained within the ER and gliadins transported to the vacuole indicates that the solubility and polymerisation of the proteins may be important. The trafficking of the gluten proteins may also be important in establishing protein:protein interactions which affect the structure and functional properties of the gluten in the mature grain and in derived flours.

The transport of vesicles from the ER to the Golgi apparatus is mediated in yeast and mammalian systems by the Rab1 class of small GTP-binding proteins.[3,4] We have, therefore, attempted to up-and down-regulate this step to determine the impact on the trafficking, interactions and properties of the gluten proteins. Wheat cultivars Ofanto and Svevo have therefore been transformed with a tobacco *rab1* cDNA and with a transdominant mutant form[5] of the same cDNA, both under the control of the endosperm-specific HMW glutenin subunit promoter.[6]

2 MATERIALS AND METHODS

2.1 Wheat Transformation

Transformation procedures were as published by Barcelo and Lazzeri (1995). Two durum wheat varieties were used: Ofanto and Svevo.

2.2 Plasmid DNA

The plasmid pAHC25[8] containing marker genes was delivered in co-transformation at equimolar ratio with two different plasmid: pCWL24 containing the wild type tobacco *rab*1 gene[5] or pCML24 containing the mutant tobacco *rab* 1 gene,[5] both driven by the endosperm-specific HMW 1Dx5 glutenin subunit promoter.[6]

2.3 Analysis of transgenic plants

PCR and Southern analyses were performed as described by Barro *et al.* (1997). Endosperms for EM analysis were prepared by standard techniques and embedded in Spurr Hard resin for conventional EM and LR White resin for immunogold labelling. Wheat grain samples for immunolocalisation of seed proteins were prepared according to Leitch *et al.* (1994).

Total proteins were extracted from single half grains and separated by SDS-PAGE using Tris-borate buffer system with 10% (w/v) acrylamide gels.[1]

3 RESULTS AND DISCUSSION

3.1 Transformation Experiments

The pCWL24 and pCML24 constructs were each used to bombard explants of immature inflorescences and embryos of durum wheat cultivars Ofanto and Svevo, in co-transformation with the pAHC25 plasmid.[8] Six bombardments of one humdred and twenty explants each, followed by selection on medium containing 3 mg l^{-1} bialaphos, gave rise to thirteen herbicide resistant plants. PCR analysis confirmed that nine plants contained the genes of interest, either wild type *rab*1 (five plants) or mutant *rab*1 (four plants).

3.2 Molecular Analysis

Two T_1 transgenic lines (one for each construct) were selected for further analysis, after confirming the presence of the tobacco wild type and mutant *rab*1 transcripts by RT-PCR. This analysis was performed on samples extracted from developing seeds (from 12 to 18 days post anthesis) and from full expanded leaves. In all four lines cDNA from endosperm tissues showed RT-PCR products for the genes of interest (≅ 600 bp band for both wild type and mutant *rab*1 genes) while no RT-PCR products were detected in leaf samples.

Southern blot analysis was performed on the two RT-PCR-positive plants, and on two negative controls. The detection of the 1.9 kb *rab*1 expression cassette (containing the gene coding region and HMW 1Dx5 glutenin subunit promoter) showed the presence of at least one intact expression cassette in both transgenic lines.

3.3 EM Analysis

3.3.1. Conventional EM examination. Caryopses from transgenic homozygous lines (expressing the wt or mutant *rab*1 genes) were harvested between 15 and 18 d.p.a., then fixed and embedded for conventional EM examination. The expression of the wt and mutant *rab*1 genes did not appear to have any obvious effects on the structures of the starchy endosperm cells (ER or Golgi apparatus) or in the amount of protein bodies in the same tissue.

3.3.2 Immunogold labelling experiments. Sections of the starchy endosperm were cut and labelled using either the polyclonal antibody anti-R2-HMW[11] for HMW glutenin subunits or the wide specificity monoclonal antibody IFRN 0610[12] for total prolamins. Expression of the wild type *rab*1 gene did not have any obvious effects on seed storage protein deposition in protein bodies compared to the control plants. However differences could be detected between the line expressing the mutant *rab*1 gene and the control line. Most of the prolamins in the starchy endosperm sections of the transformed lines appeared to be retained in the ER.

3.4 Immunological Localisation of Seed Proteins

One hundred and twenty developing grains (forty for each antibody) from transgenic homozygous lines expressing the wt or mutant *rab*1 genes were examined for the distribution of glutenin and gliadins by immunological localisation with the following antibodies all provided by Dr. Denery-Papini INRA Nantes; AB-HMG-2 for the HMW glutenins, AB-ABG-1 for the α-, and β-gliadins and AB-OMG-5 for ω-gliadins. Expression of the *rab*1 wild type gene did not appear to have any effects on the amount and distribution of HMW glutenin subunits or α-, β-, and ω-gliadins when compared with the control plants. However, expression of the *rab*1 mutant gene appeared to result in small differences in the amount of HMW glutenins when compared with the control plants, the amount being lower in the ventral part of the starchy endosperm. Expression of the *rab*1 mutant gene also appeared to result in differences in the amount and distribution of α- and β-gliadins in comparison with the control; these proteins being more abundant across the starchy endosperm of the transgenic seeds.

3.5 SDS-PAGE of Proteins

Total seed proteins from the two transgenic homozygous lines were extracted and separated by SDS-PAGE. A 1% (w/v) solution of each sample was loaded three times with different volumes to determine any differences between the transgenic and the control lines. The gel was scanned and the intensities of the bands (OD x mm^2) corresponding to HMW glutenin subunits, S-poor prolamins and S-rich prolamins calculated as a proportion of total protein content (100%). This was carried out for each of the threee volumes and the mean for each proportion calculated. No clear differences could be detected between the transgenic line containing the wild type *rab*1 gene and the control line. However, decreases in HMW glutenin subunits (from 7.5% to 5.1%) and S-poor prolamins (from 10.1% to 8.8%) together with an increase in S-rich prolamins (from 82.44% to 86.27%) occurred in the transgenic line containing the mutant *rab*1 gene in comparison with the control line.

4 CONCLUSION

The disruption of the *Rab*1-dependent transport of vesicles from the ER to the Golgi appparatus by the mutant *rab*1 gene appeared to have effects on the amount and distribution of the total prolamins. Decreases in the proportion of HMW glutenin subunits and S-poor prolamins (ω-gliadins) occurred together with an increase in the proportion of S-rich prolamins (α-, β-, γ-gliadins and LMW glutenins). These could be explained by the action of a feed-back mechanism of the prolamins that are normally deposited in the vacuole-derived protein bodies. The failure to deposit these proteins (some or all of which may be S-rich prolamins) in the vacuole-derived protein bodies may induce an increase in their synthesis at the expense of other prolamins which do not follow this targeting pathway (HMW glutenin subunits and possibly also S-poor prolamins).

Small scale rheological testing of doughs from the transformed plants will be carried out and the effects of changes in total amount, composition and assembly of prolamins on the visco-elastic properties of the dough will be determined by comparison with the control plants. Since the rheological properties of durum wheat dough are affected by changes in the amount and composition of the prolamins,[13] differences in dough quality between the mutant *rab*1 transformed line and the control would be expected.

References

1. P.R. Shewry, A.S. Tatham, F. Barro, P. Barcelo and P.A. Lazzei, *Bio/Technol.*, 1995, **13**, 1185.
2. R. Rubin, H. Levanony, and G. Galili, *Plant Physiol.,* 1992 **99**, 718.
3. J. Downward, *Trends Biochem. Sci.,* 1990, **15**, 469.
4. J.E. Rothman, and L. Orci, *Nature*, 1992, **355**, 409.
5. V.A. Andreeva, D.E. Evans, C.R. Hawes, and J.A. Napier, *Russian J. Bioorganic Chem.,* 1997, **23**, 164.
6. N.G. Halford, J. Forde, P.R. Shewry and M. Kreis, *Plants Sci.,* 1989, **62**, 207.
7. P. Barcelo and P.A. Lazzeri, in *Methods in Molecular Biology: Plant Gene Transfer and Expression Protocols.,* ed. H. Jones, Umana Press Inc, Totowa NJ, 1995, p. 113.
8. A.H. Christensen and P.H. Quail, *Transgenic Research*, 1996, **5**, 231.
9. F. Barro, L. Rooke, F. Békés, P. Gras, A.S. Tatham, R. Fido, P.A. Lazzeri, P.R. Shewry and P. Barcelo, *Nature Biotechnol.*, 1997 **15**, 1295.
10. A.R. Leitch, T. Schwarzacher, D. Jackson, I.R. Leitch, In situ hybridization: practical guide. *Bios. Scientific Publisher*, Oxford, 1994.
11. S. Denery-Papini, Y. Popineau, L. Quillien and M.H.V. Van Regenmortel, *J Cereal Sci.,* 1996, **23**, 133.
12. G.M. Brett, E.N.C. Mills, B.J. Goodfellow, R.J. Fido, A.S. Tatham, P.R. Shewry and M.R.A. Morgan, *J. Cereal Sci.,* 1999, **29**, 117.
13. G. Galterio, E. Biancolatte and J.-C. Autran, *Genet. Agr.,* 1987 **41**, 461.

PRODUCTION OF TRANSGENIC BREAD WHEAT LINES OVER-EXPRESSING A LMW GLUTENIN SUBUNIT

R. D'Ovidio[1], R. Fabbri[1], C. Patacchini[1], S. Masci[1], D. Lafiandra[1], E. Porceddu[1], A.E. Blechl[2] and O.D. Anderson[2]

1. Dipartimento di Agrobiologia e Agrochimica, Università della Tuscia, Via San Camillo De Lellis, 01100 Viterbo, Italy. 2. Agricultural Research Service- USDA, 800 Buchanan Street, Albany, CA 94710, U.S.A.

1 INTRODUCTION

Low molecular weight glutenin subunits (LMW-GS) are important components of the glutenin polymer structure. Their relative amount and/or the presence of specific components can influence the visco-elasticity of gluten dough which is correlated with the end-use properties of wheat flour[1,2].

On this basis, we are investigating the possibility to manipulate gluten dough strength and elasticity by altering the relative ratio between glutenin subunits and gliadins by increasing the LMW-GS. We are pursuing this goal by transforming bread wheat with a LMW-GS gene driven by its own promoter or by the promoter of the high-molecular weight glutenin subunit Dy10.

2 MATERIALS AND METHODS

2.1 Materials

2.1.1 Plant material. Immature embryos of the bread wheat cultivar Bobwhite were used for transformation experiments.

2.1.2 DNA constructs. UBI:BAR[3], pLMWF23A[4] and pDy10BK/F23 plasmid DNA clones were used for wheat transformation. The pDy10BK/F23 construct is a derivative of the pDy10BK1 and the pLMWF23A clones.

2.2 Methods

2.2.1 Genetic transformation. 12.5 μg/μl of each UBI:BAR and pLMWF23A or pDy10BK/F23 plasmid DNA clones were cotransformed into wheat using microprojectile bombardment as described by Blechl and Anderson[5]. Two different transformation experiments were performed and 1200 immature embryos were bombarded for each experiment.

2.2.2 SDS-PAGE analysis. Half seeds without embryos, derived from putative transformed plants, were crushed and extracted with SDS sample buffer. Aliquots of each were loaded on an SDS-PAGE (T=12, C=1.28) gel. For each gel the endosperm protein extracted from cultivars Bobwhite and Cheyenne were added as reference genotypes.

In order to determine if the transgenic product was incorporated into the polymeric fraction, seeds from two samples belonging to the T_2 were halved, crushed, and the monomeric and oligomeric fraction was extracted by using 50% propan-1-ol. Aliquots corresponding to the soluble fraction were analysed by SDS-PAGE as described above, both in reducing and non-reducing conditions. The residue remaining after removal of the soluble fraction (insoluble fraction) was extracted with the SDS sample buffer and aliquots loaded on the same SDS-PAGE gel used for the soluble samples.

2.2.3 Size Exclusion-High Performance Liquid Chromatography (SE-HPLC). The propan-1-ol soluble fraction from half transgenic seeds, and from seeds of cultivar Bobwhite, were dried down and resuspended in elution buffer (50 mM Tris-HCl, pH 6.8, 2% SDS, 4M Urea). The insoluble fraction was extracted with the elution buffer and aliquots of both soluble and insoluble fractions loaded onto a TSK4000 column. Separation was performed in 30 min at 0.7 ml/min (280 nm wavelength).

2.2.4 Reversed Phase-High Performance Liquid Chromatography (RP-HPLC). Sample extraction of the insoluble fraction and running conditions were as described in Masci et al[6]. Peaks were collected and analysed by SDS-PAGE as above.

3 RESULTS AND DISCUSSION

3.1 Genetic transformation

In order to obtain a strong expression of a LMW-GS we have followed two different strategies. The first included the use of a LMW-GS gene under control of its own 5' and 3' regulatory regions, whereas the second consisted in the use of a LMW-GS gene under control of the 5' and 3' regulatory regions of the Dy10 HMW-GS gene. For the first strategy a genomic clone (pLMWF23A) isolated from the bread wheat cv. Cheyenne was used, whereas for the second one, the construct pDy10BK/F23 was prepared from the pDy10BK1 and pLMWF23A clones. The pDy10BK1 clone contains the coding region of the Dy10 high-molecular-weight glutenin subunit (HMW-GS) flanked by 2937 bp and 1585 bp of the 5' and 3' regions, respectively. The pLMWF23A clone[7] contains the coding region of a LMW-GS gene flanked by about 1200 bp and 1600 bp of the 5' and 3' regions, respectively. The pDy10BK/F23 was prepared by replacing the coding region of the Dy10 with the coding region of the pLMWF23A in pDy10BK1.

The pLMWF23A plasmid was cotransformed with UBI:BAR into immature embryos of cultivar Bobwhite by microprojectile bombardment. Eleven independent lines were selected based on their ability to grow on culture media containing 3mg/L of bialaphos. All the selected lines were checked for the presence of the transgenic LMW-GS by SDS-PAGE analysis of the proteins extracted from endosperm tissue of T1 seeds. Of the 11 selected plants only one showed expression of the LMW-GS encoded by the pLMWF23A transgene (Fig. 1).

Similar genetic transformation experiments were also performed by cotransforming the pDy10BK/F23 and UBI:BAR into immature embryos of cv. Bobwhite. Forty five independent lines were recovered after selection on 3mg/L of bialaphos. Endosperm proteins of T1 seeds from twenty selected lines were checked by SDS-PAGE for the presence of the transgenic LMW-GS, but none of them contained detectable levels of LMW-GS encoded by the pDy10BK/F23 transgene.

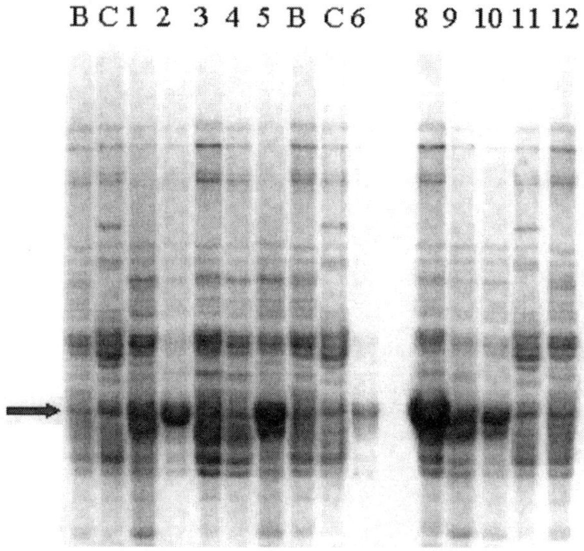

Fig. 1. *SDS-PAGE analysis of T1 seed endosperm protein. The arrow indicates the over-expressed LMW-GS encoded by the pLMWF23A transgene. B, cv. Bobwhite; C, cv. Cheyenne.*

3.2 Protein analysis

SDS-PAGE analysis showed a marked accumulation of the LMW-GS encoded by the pLMWF23A transgene (Fig. 1). Densitometric analysis showed a twelve fold increase in the transgenic product compared to the corresponding native one. The higher amount of LMW-GS present in the transformed genotypes was confirmed by RP-HPLC, which showed the presence of a major peak in the LMW-GS region, corresponding to the transgenic LMW-GS.

The transgenic LMW-GS was also analysed to determine crosslinking and participation in the glutenin polymer. SDS-PAGE analysis of soluble and insoluble fractions showed that the encoded transgenic LMW-GS is incorporated mainly in the glutenin polymers having the highest molecular weight. In fact, the transgenic LMW-GS is not present as a monomer in the unreduced soluble fraction, but is observed in low amount in the reduced soluble sample. In contrast, the transgenic LMW-GS is abundant in the insoluble fraction. To test whether the large production of the LMW-GS changed the molecular size distribution of the glutenin polymer, SE-HPLC was performed both on the soluble and insoluble fractions, and compared to cultivar Bobwhite. These analyses showed that the transformed genotypes contained a higher amount of glutenin polymers, and apparently a lower amount of gliadins, compared to the cultivar Bobwhite. Moreover, a slight shift in retention time of peak corresponding to the excluded volume (polymeric fraction) might be an indication that the glutenin polymers present in the transformed genotypes are of a slightly higher molecular size.

4 CONCLUSIONS

In planning these experiments, we anticipated that we might encounter difficulties in identifying the expression of the LMW-GS transgene, because of the high number and similarity of LMW-GS already present. Although a molecular tag could facilitate the identification of transgenic LMW-GS, we decided to attempt to isolate transgenic lines expressing a completely naturally-occurring LMW-GS sequence because i) our goal was to produce major changes in the glutenin/gliadin ratio to increase the probability of detecting effects, and ii) of possible influences of the tag itself on gluten properties.

Due to the difficult in identifying the transgenic protein, some of the BAR-resistant lines selected do not show over-expression, but might still express the transgene at a low level. We were able to obtain one transgenic line over-expressing the pLMWF23A transgene. It is noteworthy that the transgenic LMW-GS is totally incorporated into the glutenin polymer, and that its molecular size distribution may exert an influence on dough visco-elastic properties.

Finally, increasing LMW-GS in the transgenic genotypes seems to be associated with a decrease in gliadin synthesis. This effect might indicate either a regulation on the level of protein synthesis or a limitation on cellular resources for protein synthesis.

References

1. J.C. Autran, B. Laignelet, M.H. Morel, 1987, *Biochimie*, **69**, 699
2. R.B. Gupta, J. G. Paul, G.B.Cornish, G.A. Palmer, F. Bekes, A.J. Rathjen, *J. Cereal Sci.*, 1994, **19**, 9
3. M.-J. Cornejo, D. Luth, K.M. Blankenship, O.D. Anderson, A.E. Blechl,. *Plant Mol. Biol.*, 1993, **23**, 567
4. B.G. Cassidy, J. Dvorak and O.D. Anderson, *Theor. Appl. Genet.*,1998, **96**, 743
5. A.E. Blechl and O.D. Anderson, *Nat. Biotechnol.,*1996, **14**, 875
6. Masci S., E.J.L. Lew, D. Lafiandra, E. Porceddu and D.D. Kasarda, *Cereal Chem.*,1995, **72**, 100
7. R. D'Ovidio, O.A. Tanzarella and E. Porceddu, *J. Genet. & Breed.*, 1992, **46**, 41S.

Acknowledgements

Research supported by the Italian National Research Council, Grant n. 97.04356.CT14, to RD.

PCR AMPLIFICATION AND DNA SEQUENCING OF HIGH MOLECULAR WEIGHT GLUTENIN SUBUNITS 43 AND 44 FROM *TRITICUM TAUSCHII* ACCESSION TA2450

M. Tilley[1], S.R. Bean[2], P.A. Seib[2], R.G. Sears[3*] and G.L. Lookhart[1].

1.USDA-ARS GMPRC, Manhattan, KS 66502. 2. Kansas State Univ., Dept. of Grain Science Industry, Manhattan, KS 66502. 3. Kansas State Univ., Dept. of Agronomy, Manhattan, KS 66502.*Current address: AgriPro Wheat, Junction City, KS.

1 INTRODUCTION

Triticum tauschii, the diploid origin of the D genome of hexaploid wheat, is represented in modern bread wheat by a small pool of genetic diversity, thus serving as a valuable resource for improvement of pest resistance[1]. Additionally, *T. tauschii* germplasm may serve as a source of novel high molecular weight glutenin subunit (HMW-GS) alleles[2]. As the HMW-GS encoded at the *Glu-D1* loci appear to be significant in determination of bread wheat quality, the discovery of novel HMW-GS and introgression into current cultivars is an attractive means for enhancing quality. Preliminary breeding work led to the production of new bread wheat lines that have been used in crosses with established cultivars to develop plants bearing disease and insect resistance, and new combinations of HMW-GS[3]. Some of the resulting lines exhibited shorter mixing times and improved milling and baking characteristics when compared to parental hexaploid lines[4]. *T. tauschii* lines that contain the novel HMW-GS 43 and 44 were studied to compare the properties of the gluten proteins of the *T. tauschii* lines to those found in typical U.S. hard red winter wheat cultivars.

2 MATERIALS AND METHODS

2.1 Materials

Triticum tauschii accession TA2450 was obtained from the Wheat Genetics Research Center, KSU, Manhattan KS.

2.2 Methods

2.2.1 SDS-PAGE. SDS-PAGE analysis was performed[5] using the Novex system.

2.2.2 Polymerase chain reaction and cloning. Genomic DNA was extracted from ground kernels[6] and coding regions of the Dx and Dy HMW-GS were amplified using the polymerase chain reaction[7,8,9]. Reactions were amplified using a GeneAmp 2400 thermal cycler (Perkin Elmer Foster City CA) using Platinum *Taq* DNA Polymerase High Fidelity (Life Technologies Rockville, MD). Bands were excised from low melt agarose gels prepared and run with 1X TAE, gel purified, and cloned in the pST Blue-1 vector (Novagen). Following blue/white selection, clones of interest were isolated, analyzed by restriction analysis and verified by Southern blot analysis using digoxygenin (DIG) labeled pHMWDx4 clone.

2.2.3 DNA Sequencing. Due to the repetitive nature of HMW-GS, primer walking was not a possible sequencing strategy. Subclones were prepared from restriction fragments yielding overlapping products between 500 and 1,200 base pairs. Dideoxy sequencing was performed using an ABI 3700 automated DNA sequencer (Kansas State University). A minimum of 4 clones for each subclone were fully sequenced to compensate for possible misincorporations due to DNA polymerase infidelity. Both strands were fully sequenced and data was analyzed using Lasergene software (DNAstar).

3 RESULTS AND DISCUSSION

3.1 SDS-PAGE

Triticum tauschii accession TA2450 contained two HMW-GS designated 43 and 44[4] (Figure 1). In comparison to standard cultivars TAM-105 and Karl 92, subunit 43 had a mobility similar to that of HMW-GS Dx2 and subunit 44 had a mobility faster than that HMW-GS Dy10 and Dy12.

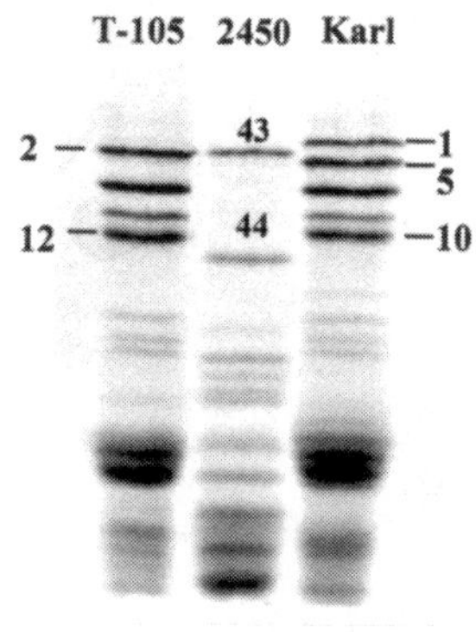

Figure 1. *SDS-PAGE analysis of TAM-105, TA2450 and Karl-92.*

3.2 PCR and sequence analysis

PCR products amplified from genomic DNA were analyzed by agarose gel electrophoresis stained with ethidium bromide (Figure 2). Bands were present at the expected molecular size for the Dx and Dy coding regions, approximately 2.5 Kb and 1.8 Kb. The specificity of the 2.5 Kb band was confirmed by the strong hybridization signal of the DIG labeled probe for HMW glutenin Dx gene.

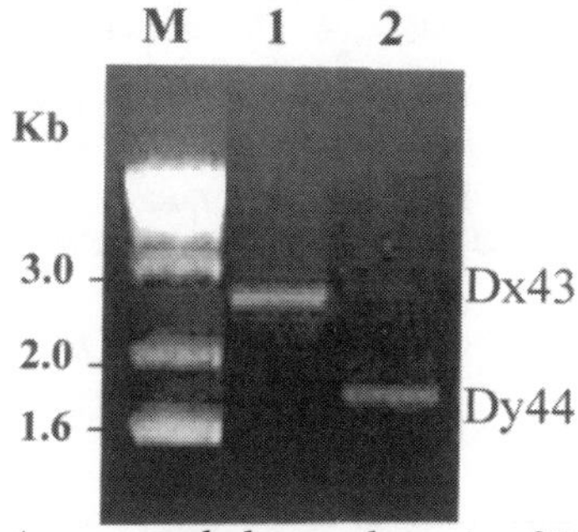

Figure 2 *Agarose gel electrophoresis of PCR products. Lane1, product using primers for amplification of x-type HMW-GS. Lane 2, product using primers designed for amplification of y-type HMW-GS. M= markers.*

The DNA sequence of HMW-GS Dx43 is that of a typical Dx glutenin gene of 2535 base pairs, with a predicted protein of 845 amino acids. Upon comparison to Dx5, Dx43 is slightly larger by 6 amino acids due to an insertion of the hexapeptide repeat PGQGQQ. (Figure 3). Dx43 lacks the 4th cysteine residue present in the N-terminal region of Dx5. The DNA sequence of HMW-GS Dx44 is that of a typical Dy glutenin gene of 1872 base pairs, with a predicted protein of 624 amino acids that displays greatest identity to subunit Dy10 and to a previously described Dy HMW-GS from *T. tauschii*[10].

```
Dx43 MAKRLVLFVAVVVALAALTVAEGEASEQLQCERELQELQERELKACQQVMDQQLRDISPECHPVV  65
Dx5  MAKRLVLFVAVVVALVALTVAEGEASEQLQCERELQELQERELKACQQVMDQQLRDISPECHPVV  65

Dx43 VSPVAGQYEQQIVVPPKGGSFYPGETTPPQQLQQRIFWGIPALLKRYYPSVTSPQQVSYYPGQAS 130
Dx5  VSPVAGQYEQQIVVPPKGGSFYPGETTPPQQLQQRIFWGIPALLKRYYPSVTCPQQVSYYPGQAS 130

Dx43 PGQWQQPEQGQQGYYPTSPQQPGQLQQPAQGQQPGQGQQGQQPGQGQPGYYPTSSQLQPGQLQQP 195
Dx5  PGQWEEPEQGQQGYYPTSPQQPGQLQQPAQGQQPGQGQQGQQPGQGQPGYYPTSSQLQPGQLQQP 195

Dx43 AQGQQGQQPGQGQQGPQRPGQGQQPGQGQQPGQGQQGYYPTSPQQQPGQGQQSGQGQQGQQPGQ 260
Dx5  AQGQQGQQPGQAQQGPQRPGQGQQPGQGQQ------GYYPTSPQQQPGQGQQPGQGQQGQQPGQ 256

Dx43 GQQPGQGQQGQQLGQGQQGYYPTSLQQSGQGQPGYYPTSLQQLGQGRSGYYPTSPQQPGQGQQPG 325
Dx5  GQQPGQGQQGQQLGQGQQGYYPTSLQQSGQGQPGYYPTSLQQLGQGQSGYYPTSPQQPGQGQQPG 319

Dx43 QLQQPAQGQQPGQGQQGQQPGQGQQGQQPGQGQQPGQGQPGYYPTSPQQSGQGQPGYYPTSSQQP 390
Dx5  QLQQPAQGQQPGQGQQGQQPGQGQQGQQPGQGQQPGQGQPGYYPTSPQQSGQGQPGYYPTSSQQP 384

Dx43 TQSQQPGQGQQGQQVGQGQQAQQPGQGQQPGQGQPGYYPTSPQQSGQGQPGYYLTSPQQSGQGQQ 455
Dx5  TQSQQPGQGQQGQQVGQGQQAQQPGQGQQPGQGQPGYYPTSPQQSGQGQPGYYLTSPQQSGQGQQ 449
```

```
Dx43 PGQLQQSAQGQKGQQPGQGQQPGQGQQGQQPGQGQQGQQPGQGQPGYYPTSPQQSGQGQQPGQWQ 520
Dx5  PGQLQQSAQGQKGQQPGQGQQPGQGQQGQQPGQGQQGQQPGQGQPGYYPTSPQQSGQGQQPGQWQ 514

Dx43 QPGQGQPGYYPTSPLQPGQGQPGYDPTSPQQPGQGQQPGQLQQPAQGQQGQQLAQGQQGQQPAQV 585
Dx5  QPGQGQPGYYPTSPLQPGQGQPGYDPTSPQQPGQGQQPGQLQQPAQGQQGQQLAQGQQGQQPAQV 579

Dx43 QQEQQPAQGQQGQQPGQGQQGQQLGQGQQGQQPGQGQQGQQPAQGQQGQQPGQGQQGQQPGQGQQ 650
Dx5  QQGQQPAQGQQGQQLGQGQQGQQPGQGQQGQQPAQGQQGQQPGQGQHGQQPGQGQQGQQPGQGQQ 644

Dx43 PGQGQPWYYPTSPQESGQGQQPGQWQQPGQGQPGYYLTSPLQLGQGQQGYYPTSLQQPGQGQQPG 715
Dx5  PGQGQPWYYPTSPQESGQGQQPGQWQQPGQGQPGYYLTFSVAARTGQQGYYPTSLQQPGQGQQPG 709

Dx43 QWQQSGQGQHGYYPTSPQLSGQGQRPGQWLQPGQGQQGYYPTSPQQSGQGQQLGQWLQPGQGQQG 780
Dx5  QWQQSGQGQHWYYPTSPKLSGQGQRPGQWLQPGQGQQGYYPTSPQQPPQGQQLGQWLQPGQGQQG 774

Dx43 YYPTSLQQTGQGQQSGQGQQGYYSSYHVSVEHQAASLKVAKAQQLAAQLPAMCRLEGGDALSASQ 845
Dx5  YYPTSLQQTGQGQQSGQGQQGYYSSYHVSVEHQAASLKVAKAQQLAAQLPAMCRLEGGDALSASQ 839
```

Figure 3 *Predicted amino acid sequence of Dx43 compared to that of HMW-GS Dx5.*

4 CONCLUSIONS

The coding regions of HMW-GS 43 and 44 from *T. tauschii* accession TA2450 have been PCR amplified, cloned and sequenced. Sequence analysis of subunits 43 and 44 reveals coding regions of 2535 and 1872 base pairs with predicted proteins of 845 and 624 amino acids. The subunits have identity to x-type (43) and y-type (44) HMW-GS encoded at the *Glu-D1* locus. This data along with that of others[10] confirms the use of *T. tauschii* as a source of novel HMW-GS that may be utilized for quality improvement of bread wheat.

References

1. Cox, T.S., W.J. Raupp, D.L. Wilson, B.S. Gill, S. Leath, W.W. Bockus, and L.E. Browder, *Plant Dis.*, 1992, **76**, 1061
2. Lagudah, E.S. and G.M. Halloran, *Theor. Appl. Genet.*, 1988, **75**, 592
3. Cox, T.S., R.G. Sears, R.K. Bequette and T.J. Martin. *Crop Sci.*, 1995, **35**, 913
4. Knackstedt, M.A., Ph.D. thesis, Kansas State University, 1995
5. Lookhart, G.L., K. Hagman and D.D. Kasarda, *Plant Breeding*, 1993, **110**, 48
6. Edwards K, C. Johnstone and C. Thompson, *Nucl. Acids Res.*, 1991, **19**, 1349
7. D'Ovidio, R.D., E. Porceddu and D. Lafiandra, *Theor. Appl. Genet.*, 1994, **88**, 175
8. D'Ovidio, R., S. Masci, and E. Porceddu, *Theor. Appl. Genet.*, 1995, **91**, 189
9. D'Ovidio, R.D. and D. Lafiandra. 1996. pp. 103-106, *In* C.W. Wrigley, (ed.) Gluten '96, Proceedings of the 6th International Gluten Workshop, Sydney, Australia.
10. Mackie, A.M., P.J. Sharp and E.S. Lagudah, *J. Cereal Sci.*, 1996, **24**, 73

Acknowledgements

We thank Dr. Renato D'Ovidio for supplying Glu-Dx PCR primer sequence and the pHMWDx4 clone. This research is supported by Kansas wheat farmers through the Kansas Wheat Commission.

CHARACTERIZATIONS OF LOW MOLECULAR WEIGHT GLUTENIN SUBUNIT GENES IN A JAPANESE SOFT WHEAT CULTIVAR, NORIN 61

T. M. Ikeda*, T. Nagamine, H. Fukuoka and H. Yano

Chugoku National Agricultural Experiment Station, 6-12-1 Nishifukatsu, Fukuyama, Hiroshima 721-8514 JAPAN

*: Corresponding author.

1 INTRODUCTION

Glutenin subunits polymerize by intermolecular disulphide bonds. These bonds play important roles in the rheological properties of wheat flour doughs. It has previously been shown that the allelic variation in high and low molecular weight glutenin subunits is involved in dough properties both in durum and hexaploid wheats[1]. The majority of Japanese wheat cultivars are classified as soft wheats and have been traditionally used for Japanese white salty noodles (Udon). A recent demand has emerged in Japan for breadmaking quality wheat flour. High quality Japanese bread wheats need to be developed, but comprehensive studies to identify specific glutenin subunits and their corresponding genes as related to dough properties have not been carried out in Japanese cultivars. In this study, we attempt to provide new information on the composition of glutenin subunits in Japanese wheat cultivars by cloning and characterizing low molecular weight glutenin subunit (LMW-GS) genes from a standard Japanese soft wheat cultivar, Norin 61.

2 MATERIALS AND METHODS

2.1 Isolation of LMW-GS cDNA clones

Poly (A)+ RNA was prepared from immature seeds harvested two weeks after flowering. A cDNA library was constructed with ZAP-cDNA synthesis kit (STRATAGENE). The cDNA library was screened with a DNA probe specific to LMWGS genes corresponding to the *C*-terminal conserved region. LMW-GS cDNA inserts from individual phage clones were amplified by PCR and sequenced by automated sequencing with an ABI 373A (PE Biosystems).

2.2. Isolation of LMW-GS genomic clones

Total DNA was prepared from seedlings of Norin 61 by the potassium acetate method described by Dellaporta *et al.*[2] followed by phenol/chloroform extraction. PCR reactions were performed in a total volume of 50□l containing 1.5 mM $MgCl_2$, 0.1 mM of each dNTP, 10 pmol of each primer and 1 U *TaKaRa LA-Taq* DNA polymerase (Takara), x1 LA PCR buffer II (Takara) and 100 ng of the total DNA. Reactions were performed according to the following protocol: denaturation at 94°C for 3 min; 35 cycles of 94°C for 30 sec, 55°C for 30 sec, 72° C for 90 sec, followed by an extension at 72°C for 5 min. All PCR products were cloned into pGEM-T and pGEM-T Easy vectors (Promega) and sequenced.

2.3. Sequence analysis

Alignments of the predicted amino acid sequences of LMWGS were assembled using CLUSTAL W Program (ver. 1.7)[3].

3 RESULTS AND DISCUSSION

3.1 Cloning of LMW-GS genes in Norin 61

A LMW-GS specific probe was used to isolate 47 LMW-GS clones from a cDNA library. All clones contained the 3' part of the coding regions and non-translated regions, but none of clones contained the 5' part of the coding regions. To obtain full-length LMW-GS genes, we carried out PCR reactions using total DNA with LMW-GS gene specific primer pairs located in the 5' and 3' non-translated regions based on published LMW-GS gene sequence data and the sequences of the cDNA clones, respectively. We could clone 106 full-length LMW-GS genes from PCR products with different primer combinations. Based on the alignment of conserved regions of the predicted amino acid sequences, these sequences were classified into 12 groups containing 18 different sizes of polypeptides (due to the presence of various deletions/insertions within repeated-sequence and glutamine-rich domains). Among these groups, nine groups share high similarity with published sequences of non-Japanese wheat varieties including LMW-m and LMW-s type LMW-GS, but the remaining three groups seem to be unique for Norin 61. The predicted molecular weights of mature subunits ranged from 22 to 43 kDa, which suggests that these proteins correspond to the B- and C- type glutenin subunits.

3.2 Classification and characterization of LMW-GS sequences in Norin 61

All LMW-GS sequences contain eight cysteine residues, which are conserved among previously published LMW-GS sequences. We classified the LMW-GS amino acid sequences into five types based on the distribution of cysteine residues. The deduced *N*-terminal sequences of these types are listed in Table 1. Type 1 includes sequences that share high similarity to the sequence of 42K LMW-GS (LMW-s type), a major component of glutenin polymer in some good-quality bread wheats[4,5], which has a cysteine in the *N*-terminal repeated sequence domain instead of at position 5 in the *N*-terminal conserved domain. Type 2 also has a cysteine residue in the repeated sequence domain, but this is located much closer to the *N*-terminal end. This type of LMW-GS sequence is unique to Norin 61. The type 2 also has two hydrophobic amino acid clusters interrupting the *N*-terminal repeated sequence domain. These hydrophobic clusters may change the structure of the repeated sequence domain and consequently the rheological properties of dough. Types 3 and 4 have cysteines at position 5 in the *N*-terminal conserved sequence domain, as in most of the nucleotide-based sequences of LMW-GS (LMW-m type). These two types are distinguished from each other by the position of a cysteine residue in a glutamine-rich sequence in the *C*-terminal domain.

In contrast to other types, type 5 contains all eight cysteine residues in the *C*-terminal conserved domain, but no cysteine residue which is likely to be involved in intermolecular disulphide bonds in the *N*-terminal region. Type 5 subunits also lack the *N*-terminal repeated sequence domain between the two cysteine residues that are potentially involved in intermolecular disulphide bonds. Since the repeated domain is predicted to have a positive influence on gluten quality[4], polymerization of this type glutenin lacking the repeated sequence domain as a part of network should have a negative effect on gluten quality. It may also be expected that the cysteine residue relocated from the *N*-terminal domain does not participate in an intermolecular disulphide bond. In such a case, this type of LMW-GS might act as a chain-terminator instead of a chain-extender in gluten polymerization; it would reduce the size of glutenin polymers and result in poor gluten quality.

Based on the analysis of amino acid sequences, we expect type 2 and type 5 subunits to have negative effects on the rheological properties. Based on the number of LMW-GS clones isolated from the cDNA library, type 1 is one of the most abundantly accumulated LMW-GS in immature seeds in Norin 61 (38 % of total clones), followed by type 5 (36 % of total clones). Type 2 should be a minor component of LMW-GSs (4 % of total clones). Type 5 is therefore expected to have a severe effect on the dough strength of Norin 61 in comparison to type 2. Type 5 LMW-GS subunit has so far only been found in an extra strong wheat, Glenlea, and its related cultivars (as band 50). This subunit seems to contribute positively to gluten strength in Glenlea[6]. Although the abundance of type 5 subunit in Glenlea is unknown, the difference in genetic background between Norin 61 (a soft wheat) and Glenlea (an extra strong wheat) may also contribute to the individual dough properties. Further analysis is necessary to clarify the effects of these LMW-GSs in Norin 61.

Table 1 *Deduced N-terminal sequences of Norin 61 LMW-GS*

Type 1	Type2	Type 3	Type 4	Type 5
METSHIPGL	METSRVPGL	MDTSCIPGL	METRCIPGL	ISQQQQQ
METSHIPSL		METSCISGL		
MENSHIPGL		METSCIPGL		
IENSHIPGL				

4 CONCLUSIONS

We cloned Norin 61 LMW-GS genes and classified their predicted amino acid sequences into five types based on the distribution of the cysteine residues. One of these types, which was unique to Norin 61, contains two hydrophobic amino acid clusters interrupting the repeated sequence domain. Another type, which was also found in other cultivars, contained all eight conserved cysteine residues in the *C*-terminal region. Both types may have negative effects on gluten polymerization in Norin 61. The latter type seems to be abundantly expressed in immature seeds. The abundance of this type LMW-GS may weaken the dough strength of Norin 61.

References

1. J.H. Skeritt, *AgBiotech News and Information* 1998, **10**: 247N.
2. S.L. Dellaporta, J. Wood and J.B. Hicks, *Plant Mol. Biol. Rep*. 1983, **1**: 19.
3. J.D. Thompson, D.G. Higgins and T.J. Gibson, *Nucleic Acids Res*. 1994, **22**: 4673.
4. S. Masci, R. D'Ovidio, D. Lafiandra and D.D. Kasarda, *Plant Physiol*. 1998, **118**: 1147.
5. E.J.-L. Lew, D.D. Kuzmicky and D.D. Kasarda, *Cereal Chem*. 1992, **69**: 508.
6. S. Cloutier and O.M. Lukow, Proc. 9th Int. Wheat Genet. Symp. 1998, **3**: 2.

CHARACTERIZATION OF THE *LMW-GS* GENE FAMILY IN DURUM WHEAT

D'Ovidio R.[1], Masci S.[1], Mattei C., Tosi P[2]., Lafiandra D.[1], and Porceddu E.[1]

1. Dipartimento di Agrobiologia e Agrochimica, Università della Tuscia, Via San Camillo De Lellis, 01100 Viterbo, Italy. 2 .Biochemistry and Physiology Department, IACR-Rothamsted, Harpenden, Herts, AL5 2JQ, UK

1 INTRODUCTION

Low molecular weight glutenin subunits (LMW-GSs) are encoded by gene families located at the orthologous *Glu-3* loci, located on the short arm of chromosome group 1. They are part of the endosperm proteins and have a strong influence on technological properties of wheat flour and semolina. Evidence supports the hypothesis that the presence of specific components and the total amount of LMW-GSs are strictly correlated to wheat quality characteristics[1,2]. However, the number and characteristics of the different genes composing the *lmw-gs* gene families at the *Glu-3* loci have not been determined yet. This lack of information has hampered the possibility of establishing a direct correlation between the structure of specific LMW-GSs and their functionality and, in general, the organisation of the *Glu-3* loci.

We are characterising the structure of the *Glu-3* loci through the analysis of the durum wheat cultivar Langdon. Since this cultivar has poor quality characteristics, we have also analysed the biotypes γ-42 and γ-45 of the Italian cultivar Lira, which have contrasting quality properties[3].

2 MATERIALS AND METHODS

2.1 Materials

Several durum wheat cultivars have been used for DNA extraction and two-dimensional gel analysis. Chromosome assignment was performed by using D-genome-chromosome substitution lines of durum wheat cultivar Langdon[4] and intervarietal chromosome 1B substitution line Edmore/Langdon.

2.2 Methods

2.2.1 Two-dimensional electrophoresis (IEF vs. SDS-PAGE and NEpHGE vs. SDS-PAGE). First dimension of IEF vs. SDS-PAGE was performed by the IPGphor™ Isoelectric Focusing System (Amersham Pharmacia Biotech, Uppsala, Sweden), using the

13 cm long strips with the linear pH gradient 3-10, according to the instruction manual. Second dimension was performed on a Hoefer™ SE600 apparatus (T=11, C=2.67). NEpHGE vs. SDS-PAGE was performed according to Holt et al.[5].

2.2.2 DNA extraction. Genomic DNA was isolated from 5 g of leaves from single plants as reported in D'Ovidio et al.[6].

2.2.3 PCR amplification, cloning, and nucleotide sequencing. PCR conditions and cloning were as reported in D'Ovidio et al.[7,8]. Sequencing analysis were performed manually with the Thermo Sequenase™ radiolabelled terminator cycle sequencing kit (Amersham, UK.) and semiautomatically on a ABI PRISM 310 (PE Applied Biosystem).

2.2.4 Southern blotting analysis. Hybridisation experiments were carried out following standard procedures[9]. Probes were prepared by labelling 300 ng of the LMW-GS insert contained into the pLMW21 clone[10] either using digoxigenin or α-^{32}P deoxycytidine. The labelling was performed using the 'Nick Translation Kit' (Boehringer) and following the recommended procedures

3 RESULTS AND DISCUSSION

Two-dimensional electrophoresis indicated a total of 30-35 LMW-GSs in bread wheat and 20-25 in durum wheat, including modified gliadins. The different subunits are within the pI range 6.5-9.5, and the molecular weight range 25.000-45000, and show varying levels of expression.

Our efforts in isolating LMW-GS genes produced a total of four different genes from the cultivar Langdon, two from cultivar Lira biotype 45 and one from biotype 42. Three *lmw-gs* genes from cv. Langdon are located at the *Glu-A3* locus and one at the *Glu*-B3 locus, whereas two genes from cv. Lira, one from biotype 45 and one from biotype 42, are located at the *Glu-B3* locus and are allelic. The remaining clone from cultivar Lira biotype 45 was isolated from a genomic library and has not yet assigned to any specific chromosome.

Comparative analysis showed that these genes differ in the number of repeats within the repetitive domain, in the position of the first and the seventh cysteines, and in the presence/absence of the short *N*-terminal region. In particular, comparative analysis of the deduced *N*-terminal region allowed the recognition of the following three types of LMW-GSs: i) those with a cysteine residue at position 5 and a methionine as first amino acid in the deduced mature protein (i.e. pLDNLMW1A3); ii) those lacking cysteine residues and having a methionine as first amino acid in the deduced mature protein (i.e. pLNDLMW1B); iii) those lacking the *N*-terminal region and having an isoleucine as first amino acid in the deduced mature protein (i.e. pLNDLMW1A1) (Fig. 1). Sequence analysis at the protein level revealed the presence of types i) and ii) only[2,11,12]. Moreover, comparison of a type ii) LMW-GS, deduced from gene sequencing, with the corresponding native protein, indicates that this group of protein may undergo different maturation processes, in comparison to those belonging to type i). In fact, this group of proteins, in addition to having methionine as first amino acid of the mature protein, can also contain LMW-GSs which have a serine as first amino acid of the mature protein[13].

On the basis of the characteristics of the reported genes and of their encoded proteins, it appears that the genes located at the *Glu-A3* locus are more polymorphic with respect to those located at the other loci. In fact, of three genes located at this locus, only one encodes a LMW-GS with the typical structure, the remaining two are inactive or encode a

LMW-GS missing the *N*-terminal region. These genes also contain regions encoding polyglutamine stretches which are characteristic of α-gliadin genes.

In order to obtain more information about the organisation of the *lmw-gs* family within durum wheat, we have performed RFLP analysis and two-dimensional electrophoresis on 14 Italian durum wheat cultivars, seven of which show the poor quality LMW-1 protein pattern, and seven with the good quality LMW-2 protein pattern. RFLP analysis of LMW-GS sequences showed a slightly different pattern between durum wheat type-45 and durum wheat type-42, with the type-45 having a larger number of hybridising fragments or slightly more intense signal. An intervarietal chromosome 1B substitution line of the durum wheat cv. Edmore in cv. Langdon allowed the following assignment of the main hybridising bands: the 7.5 Kb hybridising fragment is located on chromosome 1B, the 6.3 Kb fragment is located on chromosome 1A, whereas the 5.0 Kb fragment contains sequences located on chromosome 1B and probably also on chromosome 1A.

Two dimensional electrophoretic analysis of the fourteen cultivars showed that the main difference between the good (LMW-2) and the poor (LMW-1) quality durum wheats resides in the different amount of LMW-GSs, and that the slowest moving LMW-GS (42K LMW-GS) present in the LMW-2 allelic form contributes most to this difference.

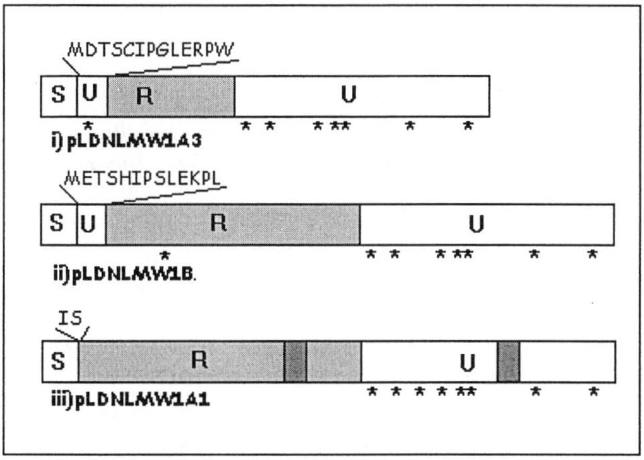

Fig. 1. *Diagrams showing the main characteristics of the three types of LMW-GSs deduced from the nucleotide sequences of genes identified in durum wheat. S, signal peptide; U, unique region; R, repetitive domain; *, cysteine residue. The deduced N-terminal amino acid sequence is reported in the upper part of each scheme.*

4 CONCLUSIONS

The LMW-GSs are encoded by a gene family whose members have a highly conserved *C*-terminal region, several types of *N*-terminal sequences and a very polymorphic repetitive domain.

Our analysis is producing the basic data for understanding the structure, evolution and function of the different members composing the *lmw-gs* gene family, and furnishing specific *lmw-gs* genes for wheat transformation experiments aimed at modifying the technological properties of gluten. The nucleotide sequence data can also be used for developing PCR-based assays for marker-assisted selection

The combined approaches at the protein and DNA levels should give novel information on the expression and maturation processes of this group of protein, and should facilitate the understanding of the relationship between structure and function. Preliminary results on this subject indicate the existence of different maturation processes within the LMW-GSs and point out the importance of the quantitative levels of specific LMW-GSs in the final gluten properties. This latter indication comes from comparison analyses of deduced amino acid sequences of allelic subunits related to contrasting effects on the visco-elastic properties of gluten. These subunits have similar structural characteristics with few differences that do not seem able, by themselves, to account for their different contribution to the final gluten strength[8], but are also present in a different amounts[2].

References

1. J.C. Autran, B. Laignelet, M.H. Morel, 1987, *Biochimie*, **69**, 699
2. S. Masci, E.J.L. Lew, D. Lafiandra, E. Porceddu and D.D. Kasarda, *Cereal Chem.*, 1995, **72**, 100
3. N.E. Pogna, J.C. Autran, F. Mellini, D. Lafiandra and P. Feillet, *J. Cereal Sci.*, 1990, **11**, 15
4. L.R. Joppa and N.D. Williams, *Genome*, 1988, **30**, 222
5. L.M. Holt, R. Astin and P.I. Payne, *Theor. Appl. Genet.*, 1981, **60**, 237
6. R. D'Ovidio, O.A. Tanzarella and E. Porceddu, *J. Genet. & Breed.*, 1992, **46**, 41
7. R. D'Ovidio, M. Simeone, S. Masci, E. Porceddu ,1997, *Theor. Appl. Genet.* **95**: 1119-1126
8. R. D'Ovidio., C. Marchitelli., L. Ercoli Cardelli, E. Porceddu, *Theor. Appl. Genet.*, 1999, **98**, 455
9. J. Sambrook, E.F. Fritsch, T. Maniatis, Molecular Cloning. A laboratory manual, 1989, Cold Spring Harbor Laboratory Press
10. R. D'Ovidio, O.A. Tanzarella, E. Porceddu, *Plant Mol. Biol*, 1992, **18**, 781
11. H.P. Tao and D.D. Kasarda, *J. Exp. Bot.*, 1989, **40**, 1015
12. E.J.L. Lew, D.D. Kuzmicky and D.D. Kasarda, *Cereal Chem.*, 1992, **69**, 508
13. S. Masci, R. D'Ovidio, D. Lafiandra, D.D. Kasarda, *Plant Physiology*, 1998, **118**, 1147.

Acknowledgements

The authors wish to thank Dr L.R. Joppa for providing seeds of D genome chromosome substitution lines of durum wheat cultivar Langdon and intervarietal chromosome 1B substitution line Edmore/Langdon.

The research was supported by the Italian 'Ministero delle Politiche Agricole', National Research Project 'Plant Biotechnology' and by the Italian 'Ministero dell'Università e della Ricerca Scientifica e Tecnologica', grant ex 60% to RD and SM.

WHEAT-GRAIN PROTEOMICS; THE FULL COMPLIMENT OF PROTEINS IN DEVELOPING AND MATURE GRAIN

W.G. Rathmell[1], D.J. Skylas[1-4], F. Békés[1,2] and C.W. Wrigley[1,2]

1. Quality Wheat CRC, North Ryde (Sydney), NSW 1670, Australia. 2. CSIRO Plant Industry, North Ryde (Sydney), NSW 1670, Australia. 3. Australian Proteome Analysis Facility, Macquarie University (Sydney), NSW 2109. 4. University of Sydney, NSW 2006, Australia.

1 INTRODUCTION

For any specific grain sample, the functional properties of gluten in the resulting dough are the result of the interaction of genotype with growth and storage conditions. Knowledge of the gluten-coding genes involved thus provides only half the story that is needed to understand (and thus to predict) gluten function at the molecular level. The 'other half of the story' requires knowledge of environmental factors and their effects on a further set of genes – those that have a regulatory role. Such factors determine the manner of synthesis of the gluten-forming polypeptides during grain development, the manner in which these are (or are not) associated into polymeric structures and the effects of other proteins that may influence these processes.

Study of regulatory genes and of environmental factors has proven to be difficult. A promising approach to understanding such matters is analysis of the full complement of polypeptides in the developing and mature grain, thereby to catalogue the full complement of proteins synthesised.[1,2] This full set of polypeptides has become known as the 'proteome' of the tissue involved, being the expression of the part of the genome that was appropriate to that tissue, to the stage of development, and to the conditions of growth.

A proteomic study involves the extraction, separation and identification of proteins from a specific tissue of an organism. In applying this approach to wheat endosperm, novel extraction buffers and fractionation conditions were developed. A key part of this new technology is the analysis of the components by *N*-terminal sequencing, followed by database searching to indicate the likely identities of the many component proteins. However, even this approach goes only part way to elucidating the molecular aspects of gluten function, since it is restricted to primary structure; knowledge of higher-level structure is also needed, such as disulphide bonding. We have thus supplemented proteomics with a study of the polymerisation of glutenin subunits throughout grain filling, thus to elucidate the sequence and rate of polymerisation of specific polypeptides.

2 METHODS

For the proteome studies, grain of the wheat variety Wyuna was grown under day/night temperatures of 24/18°C, with samples being taken at 17 days post anthesis (DPA) and at maturity (45 DPA). Endosperm (isolated from immature grain and freeze-dried) or flour (milled from mature grain) were extracted with 10% trichloroacetic acid and 0.07% 2-mercaptoethanol in cold acetone to remove components that interfere with solubilisation and fractionation. [3] This material was extracted with solubilisation buffer (7M urea, 2M thiourea, 2mM tributyl-phosphine (TBP), 4% CHAPS, 1% carrier ampholytes, 40mM Tris and 0.001% Orange G dye) by vortexing and sonication. Nucleic acids were digested with endonuclease (Sigma).

The first dimension (isoelectric focusing) was performed on Immobiline® DryStrips (Amersham Pharmacia Biotech, Sweden), either for pH 4-7 or pH 6-11. These resulting strips were used for second-dimension fractionation in large-format SDS-PAGE gels, with a gradient of 8-18%T and 2.5%C piperazine diacrylamide as crosslinker. Analytical SDS-PAGE gels were stained with diamine silver. Preparative SDS gels were electroblotted to PVDF membranes[4,5]. SDS-PAGE gels were scanned using the Molecular Dynamics Personal Densitometer SI. Automated Edman degradation was performed on a Hewlett Packard G1005A Protein Sequencer employing Routine PVDF 3.1 chemistry. PTH amino acids were separated and analysed with an online Hewlett Packard Series II 1090 LC using the PTH_4.M HPLC method. *N*-terminal sequence data were processed using the software tools TagIdent (http://expasy.proteome.org.au/tools/tagident.html) and Fasta3 version 3 at the European Bioinfomatics Institute (EBI) (http://www2.ebi.ac.uk/fasta3) to interrogate SWISS-PROT (release 35) and TrEMBL databases.

To quantify the accumulation of the major protein classes, wheat plants (variety Wyuna) were grown under day/night temperatures of 18/13°C (16-hour days), with samples being taken at four-day intervals from 4 DPA to maturity. Size-exclusion (SE)-HPLC methods[6,7] were used for assessing the main size classes of endosperm proteins. Changes in the size distribution of polymeric proteins were characterised during endosperm development by field-flow fractionation (FFF)[7] and by SE-HPLC-based determination[8] of the amounts of unextractable polymeric protein (UPP). The accumulation of high- and low-molecular-weight (HMW and LMW) glutenin subunits were monitored by reversed phase (RP)-HPLC.[9]

3 RESULTS AND DISCUSSION

About 690 polypeptide spots were resolved in the acidic range (pH 4-7) for immature grain; an additional 610 basic components were resolved in the range pH 6-11 (Figure 1). The results provide new insight into the complex nature of wheat-grain endosperm proteins. It was not possible to attempt *N*-terminal sequencing of all of the components observed. Altogether, 321 proteins were submitted for post-separation characterisation. From this total, 177 (55%) proteins were identified, 55 (17%) proteins were not matched and 89 (28%) proteins did not yield any *N*-terminal sequence data (because of *N*-terminal blockage or insufficient material).

Examples of the types of polypeptides identified in the immature endosperm are shown in Table 1. As expected, many were gluten-forming components (gliadins and polypeptides of glutenin). Nine HMW subunits of glutenin were identified across the top of the pH 4-7 pattern; none were identified in the pH 6-11 range. A total of 80 gliadin components was identified, covering much of the molecular-weight range (36 and 44 in the acidic and basic ranges, respectively). There was a prominent grouping of 14 isoforms of protein disulphide isomerase[10] on the upper acidic side of the pH 4-7 pattern,

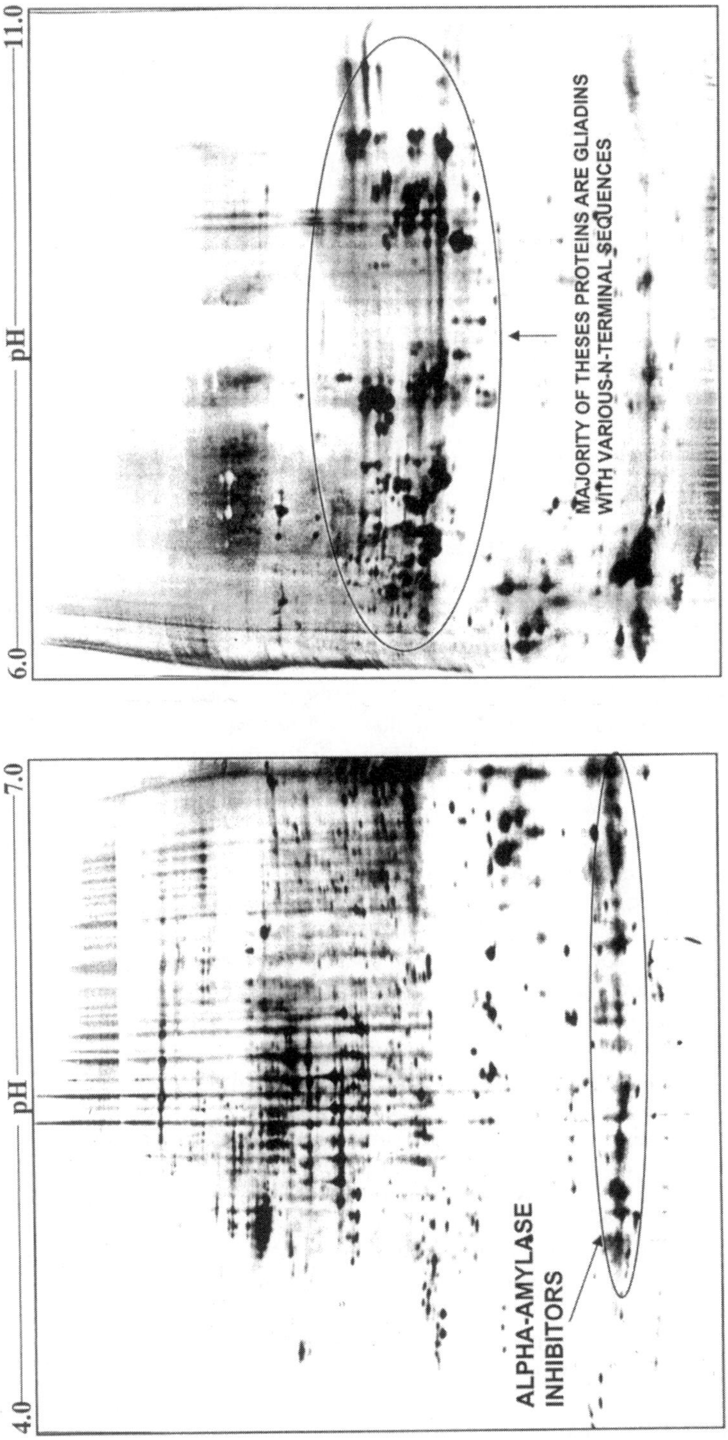

Figure 1. Two-dimensional maps of polypeptides extracted from immature endosperm of wheat (cv. Wyuna) in acidic and basic pH ranges.

presumably important in determining aspects of gluten function related to disulphide-bond formation.[11,12] Across the smaller molecular-weight region of both pH ranges, there was a large number (38) of components identified as having homology to amylase/trypsin inhibitors.

When proteome patterns were compared for immature and mature endosperms, it was evident that there were slightly fewer polypeptides resolved for mature grain (about 650 in the acidic range and 470 basic polypeptides). Presumably there are many polypeptides involved in the synthetic mechanisms of the developing grain that are broken down during ripening. For example, amongst those missing in the mature endosperm were the several of the isoforms of protein disulphide isomerase.

Table 1. *Examples of polypeptides (pH 4-7 range) identified in the proteome analyses. (Spot numbers refer to a paper submitted for publication by Skylas et al.)*

Spot No.	*N*-Terminal sequence	Gene-product	% Identity	Organism
1	KTITLEVE	Ubiquitin (UBQ1 gene)	100% in 8 aa	*T. aestivum*
2	SQAQGSVQ	Alpha/beta-gliadin	100% in 7 aa	*T. aestivum*
9	SGPWSWXD	Alpha-amylase inhibitor 0.28	100% in 6 aa	*T. aestivum*
11	GRLLSPRGK ELHTPQEEFP QQQQFP	Omega-gliadin	46% in 24 aa	*T. monococcum* Einkorn wheat
16-30	VRVPVPQL	Alpha/beta-gliadins	100% in 8 aa	*T. aestivum*

These results illustrate the potential of the proteomics approach to provide detailed information that complements genome studies.[13] The resulting array of polypeptides displays the 'reality' of the phenotype, providing a first indication of the performance of a specific genotype (genome) with respect to a particular combination of tissue and growth environment. As the first products of gene action, the polypeptides[11] are critical to elucidating gene-function relationships, as well as gene-environment interactions[12]. Proteome analysis therefore promises to provide a complementary approach to molecular-marker analyses that have hitherto focused on the DNA level.

The next step of critical importance to elucidating gluten function is the processing of the newly synthesised polypeptides. Proteome studies offer to help at this level by the identification of proteins that may act, for example, as chaperones during processing,[10,11] but more direct studies are also warranted to examine the processing of the polypeptides. This stage is especially important for the glutenin polymers, which are inactive in dough-forming properties as individual polypeptides. To elucidate this aspect of gluten structure, protein composition was studied throughout grain filling.

During the very early stages, SDS-soluble proteins accounted for almost all the material in the 'polymeric' protein class, indicating relatively small glutenin polymers. Differences started to appear after several days (at about 16 DPA), when polymeric proteins grew in size, making them impossible to solubilise without the aid of sonication, resulting in significant proportions of 'unextractable' polymeric protein (UPP). By maturity, the proportion of UPP had risen to nearly 40%. The changes in the size distribution of the UPP was contrasted by FFF analysis, which showed how much smaller was the size distribution for the UPP from immature endosperm. In parallel studies, involving lines deficient in genes for HMW glutenin subunits, the presence of both extractable and unextractable polymeric protein was observed as glutenin polymers formed in the developing endosperm. This result supports earlier reports (Singh and Shepherd[14] and others) that it is possible to have glutenin polymers consisting solely of LMW glutenin subunits.

The amounts of glutenin and gliadin increased steadily during grain development. The glutenin/gliadin ratio was highest at the very early stages of maturity. The relative proportions of the three main protein classes reached a plateau at 20-23 DPA, coinciding with the end of the cellular-division period. The synthesis of HMW subunits commenced slightly before that of the LMW subunits. The different timing of the biosynthesis of HMW and LMW subunits may provide an important indication about the polymeric structure of native glutenin in the mature grain. The relatively higher proportion of HMW subunits, synthesised earlier than the appearance of LMW subunits, suggests that a polymeric structure may form with a backbone of HMW subunits, onto which LMW-subunit branches may be attached. This would not necessarily require a highly sophisticated regulatory mechanism.

The combination of proteome studies with analysis of the glutenin-polymerisation process offers the promise of further elucidation of the molecular basis of gluten function, as it is determined in the developing endosperm, under the combined influence of genotype and environment. New technologies have become available to facilitate these advances, particularly, the proteome approach and the wider range of protein-size distribution that can now be analysed with field-flow fractionation. The new science of proteomics demonstrates the 'reality' of the genome for a specific situation. This will lead to the identification of protein markers that are likely to indicate the 'reality' of gluten quality for actual combinations of genotype and environment, plus the promise of markers (possibly causes) of tolerance mechanisms to stress situations.

References

1. W. P. Blackstock, and M. P. Weir, *Trends in Biotechnol.*, 1999, **17,** 121.
2. H. Thiellement, N. Bahrman, C. Damerval, C. Plomion, M. Rossignol, V. Santoni, D. de Vienne, and M. Zivy, M. *Electrophoresis*, 1999, **20,** 2013.
3. F. Granier, *Electrophoresis* 1988, **9,** 712.
4. J. Kyhse-Andersen, *J. Biochem. Biophys. Methods* 1984, **10** (3-4), 203-209.
5. P. Matsudaira, *J. Biol. Chem.* 1987, **262,** 10035.
6. I. L. Batey, R. B. Gupta, and F. MacRitchie, *Cereal Chem.* 1991, **68,** 207.
7. O. R. Larroque, L. Daqiq, N. Islam, and F. Bekes, in *Cereals '99*, eds J. Panozzo, M. Radcliffe, M.Wootton, and C. W. Wrigley, Royal Australian Chemical Institute, Melbourne, Australia, 2000, (in press).
8. R. B. Gupta, K. Khan, and F. MacRitchie, *J. Cereal Sci.* 1993, **18,** 3.
9. B. A. Marchylo, J. E. Kruger, and D. W. Hatcher, *J. Cereal Sci.*, 1989, **9,** 113.
10. R. S. Boston, P. V. Viitanen, and E. Vierling, E. (1996) *Plant Mol. Biol.* 1996, **32,** 191.
11. P. R. Shewry, *Cereal Foods World*, 1999, **44,** 587.
12. D. Lafiandra, S. Masci, C. Blumenthal, and C. W. Wrigley, *Cereal Foods World*, 1999, **44,** 572.
13. I. Humphery-Smith, S. J. Cordwell, and W. P. Blackstock, *Electrophoresis*, 1997, **18,** 1217.
14. N. K. Singh, and K. W. Shepherd, *Theor. Appl. Genet.*, 1985, **71,** 79.

Gluten Protein Analysis, Purification and Characterization

UNDERSTANDING THE STRUCTURE AND PROPERTIES OF GLUTEN: AN OVERVIEW

Rob J. Hamer [1,2,3] and Ton Van Vliet [1,2]

1. Centre for Protein Technology, Wageningen University, Wageningen, The Netherlands. 2.Wageningen Centre for Food Sciences, Wageningen, The Netherlands. 3.TNO Voeding, Zeist, The Netherlands.

1 INTRODUCTION

It is the progress in our knowledge of the proteins of wheat which has driven the progress in cereal science and technology over the past 10-20 years. It is these proteins that have to a large extent the unique properties that allow the making of a viscoelastic dough with gas holding properties and hence the preparation of bread. At the same time it is the variation in the quantity and composition of these proteins from which the different processing properties and end-use quality of wheat arises.[1] It is clear that understanding the relation between these parameters and end-use quality is crucial to the successful use of wheat. Knowledge of wheat protein structure is the key to this understanding. In this review our current knowledge of wheat protein structure and functionality is discussed.

2 PROBING THE STRUCTURE OF THE GLUTENIN POLYMER

The obvious importance of gluten proteins has led to widespread research and numerous reviews have appeared on this topic.[2,3] Most of these studies are aimed at unravelling the chemical structure of the glutenin polymer. With hindsight these studies can be categorised by the three different approaches used:

- Probing the structure by 'pulling' it apart;
- Understanding the structure by changing its composition;
- Building the structure from molecular data.

2.1 Probing the structure by pulling it apart.

Fractionation studies are among the oldest approaches to study gluten. Osborne's classical work of using different solvents to fractionate wheat proteins still holds today. Since then, scores of sophisticated fractionation studies have followed. In combination with electrophoresis and chromatography, this has resulted in a general consensus on the existence of a polymeric fraction consisting of high and low molecular weight glutenin subunits (HMWGS and LMWGS) and monomeric gliadin proteins. Typically, a proportion of the HMW glutenin polymers remained unextractable in the solvents used.[4]

Only with the use of reduction or sonication,[5,6] could this fraction be solubilized and further studied. The discovery of this unextractable protein polymeric aggregate, called Gelprotein[4] or Glutenin Macropolymer (GMP[7]), can be considered a true landmark in cereal science. It is this fraction that gives rise to some of the unique properties of gluten.[8] A number of different models have been proposed for the glutenin polymer, (Ewart,[9] Graveland[10] and Kasarda[11]), agreeing on the disulphide crosslinked nature of the structure. Further information comes from studies following the changes in the polymeric glutenin upon mixing and resting.[8] It is difficult, however, to interpret these studies in terms of the chemical structure of GMP since the structure is not completely solubilized by mixing.

Fractionation studies have made a major contribution, but at the same time have their limitations. Rearrangements can occur during extraction, as well as co-extraction phenomena and losses of material. Nevertheless, recent reological studies of the GMP fraction demonstrate a clear relation with dough properties and thus provide additional evidence for the importance of this fraction.[12,13]

2.2 Understanding the structure by changing its composition.

2.2.1. Breeding. When Payne first discovered a relation between HMWGS 5+10 and breadmaking quality[14] and developed his scoring system, this knowledge was immediately put to use by breeders to produce wheats carrying the 'baking quality bands'. This did not solve all quality problems, and some of the resulting wheats were considered 'overstrong'. More recently, supported by large EU collaborative projects, excellent materials have been developed for the specific purpose of structure-functionality studies. An overview of wheat lines currently available is given in the chapter by Lafiandra in this volume. Studies with specific wheat lines have clearly corroborated the role of HMW glutenin polymers. Using these materials, the relation could be demonstrated between dough viscoelastic properties and the quantity of HMWGS[15] or GMP.[8] Very recently, Lefebvre *et al*[16] demonstrated that the presence of only HMWGS 5+10 contributes to gluten strength.

2.2.2. Reconstitution experiments. Reconstitution studies present another way of studying structure by changing the composition. Classic studies highlighted the importance of the ratio of glutenin : gliadin in determining dough viscoelastic properties.[2] A breakthrough in this respect was made by Bekes *et al*,[17] who developed a technique, allowing the incorporation of added GS into the existing endogenous glutenin polymeric structures. This technique, in combination with miniature mixing and testing methods[18,19] has been highly successful in demonstrating relations between HMWGS content and size and dough development time. More recently, the system was used with genetically modified HMWGS, indicating the importance of the length of the central repeat region of HMWGS 1Dx5 in relation to stability against breakdown during mixing. Current work in this group focuses on synthetic doughs allowing an even greater flexibility to study the contribution of different GS fractions. In general, these studies lead to an understanding of the relative importance of individual (groups of) proteins rather than understanding of gluten structure.

2.3. Building the structure from its molecules

Recent information on the amino acid sequence of a whole series of glutenin polypeptides allows us in principle to build a 'chemical model'. It is however not possible to use this information and build glutenin polymers in the computer. As recently reviewed,[20] computer modeling can be used for parts of the GS molecule, or at most an individual molecule. Computer modeling studies can help explain possible preferential S-S linkages between subunits. For example, a key question still relates to the x-type y-type alternated chemical backbone structure as postulated by Graveland.[10] Wieser, Kohler and others[21,22,23] working on the exact positions of the various crosslinks, have not yet identified the related disulphide bonded peptides. On the other hand support for x-type-y-type head-to-tail polymers has come from an unexpected source. Shani *et al*[24] reported that a hybrid subunit containing the HMWGS Dy10 *N*-terminus and the HMWGS Dx5 *C*-terminus preferentially formed circular forms. It is expected that these studies, together with the ability to express modified glutenin subunits, will eventually lead to a consensus on the chemical structure of the network.

3 THE NON-COVALENT STRUCTURE OF GLUTEN

The structure of the gluten network is a superimposition of both covalent and non covalent interactions; hydrogen bonds, hydrophobic interactions and electrostatic interactions (salt links, metal ion bridges) are relatively weak but can give great strength due to their numbers. With their glutamine rich repeats,[25] glutenin molecules are well suited to form hydrogen bond stabilized structures.[26] Early 'functional' studies already pointed to the role of hydrogen bonds. In a classic experiment, dough was mixed with deuterated water. Deuterium bridges are stronger than hydrogen bridges and a strengthening effect can be observed.[27] More recent studies on the effect of glycine and arginine (able to promote or disrupt hydrogen bonds respectively) on gluten formation point to the same direction and also highlight the importance of such interactions for the formation of GMP.[28] Clearly, non-covalent interactions contribute to gluten functionality. The disulphide linked glutenin polymer should be regarded as a backbone. Knowledge on the chemical structure of the backbone needs to be integrated with knowledge on additional non-covalent interactions to explain gluten properties.

4 STRUCTURE-FUNCTIONALITY RELATIONSHIPS

Structure-functionality relationships form the rationale for gluten structure research. The knowledge of these relationships allows us to bridge the gap between breeding and end-use characteristics. However, only recently have efforts been made to integrate our current knowledge and design working models allowing these relations to be studied and understood.

4.1. What is functionality?

It is not feasible to directly link composition and end-use quality. This is both due to the many factors affecting this relationship and to the fact that end-use quality in itself is an extremely complex term. More progress has been obtained by focusing on dough properties. These are important both for processing and end-product quality. Moreover, specific dough properties can be identified that allow a definition of functionality in terms that come closer to properties of a structure. For example, from dough rheological properties like loss and storage moduli information is obtained on the number of crosslinks and the concentration and average length of the polymers. Strain hardening properties and extensibility of a dough are considered to be important factors related to gas holding properties.[29] Also, strain hardening properties yield information on the properties of the gluten network. Linking gluten network properties to dough physical properties also allows a more generic approach in defining the various functionalities related to the wide variation in possible end products.

4.2. Models to explain functionality

4.2.1. The entanglement model. MacRitchie[1] were among the first to advocate a more physical approach to glutenin structure. Inspired by polymer theory, he suggested that the glutenin polymer can be best described as an entangled polymer network. In such a network covalent aggregates become joined through physical entanglements. If these junctions live longer than the time required to pull them apart, these cannot be distinguished from covalent junctions. The amount of entanglements is related to the structure, size and concentration of the polymer. MacRitchie uses the entanglement model to qualitatively explain the behaviour of a dough in an Extensigraph. In doing so, MacRitchie arrives at the same conclusions as Weegels *et al.*[8] Weegels found a clear correlation between the GMP fraction and the resistance of dough to extension. MacRitchie[1] increased the proportion of LMW to HMW glutenins and demonstrated increased viscosity of the resulting dough. This can be interpreted by stating that only the highly aggregated polymers present as GMP contribute to elastic properties at large deformation, while the smaller polymers together with monomeric proteins contribute to the viscous behaviour.

The emphasis on glutenin polymer size has prompted research to assess the molecular size distribution of GMP using techniques like multistacking gel electrophoresis,[30] free field flow techniques and light scattering.

4.2.2. The loop and train model. The elastic properties of gluten have intrigued many researchers. Belton *et al*[26] point out the importance of hydrogen bridges between glutenin subunits. In his 'loop and train model' Belton proposes that the interactions between different glutamine-rich domains act to store or release elasticity. The model contains some attractive ideas, but does not take into account the importance of glutenin chemical network structure in inferring elastic properties and the probability of GS-GS loop and train interactions with respect to, for example, GS-gliadin interactions. The model suggests that elasticity arises at a molecular level as is assumed with elastin. This still remains to be proven.

4.3 Towards a comprehensive model

Both 'functional' and 'chemical' models have yet to lead to a consensus and are first steps on the way to resolving structure-functionality relationships. Important questions remain on the formation of the glutenin polymer in the wheat kernel and its significance for final gluten structure. Also, effects of processing, additives etc are not yet dealt with. In the following, a framework is presented that is developed to integrate our current knowledge into a single, comprehensive model: the hyperaggregation model. The hyperaggregation model integrates chemical and physical aspects of gluten structure and focuses on the different scale at which different processes play a role. Three levels are distinguished which are described below (Figure 1).

4.3.1. The molecular level, I. Level one concerns the formation of polymeric glutenin in the wheat kernel. At this level, depending on the genetically determined GS composition and GS synthesis, HMWGS and LMWGS form a covalent polymer (see Figure 1). By definition only covalent bonds are encountered at this level of aggregation. The position of these disulphide bridges will be determined by GS conformation and type (i.e. number and position of reactive sulphydril groups). Aggregation will hence be determined by the presence of individual subunits and their ability to propagate or terminate the network (MacRitchie[1]).

4.3.2. The mesoscopic level, II. At the second level of aggregation glutenin polymers will form larger aggregates by entanglement, stabilised by hydrogen bonding and additional disulphide bridges. The size of such aggregates will be < 100 μm. The quantity and incidence of such aggregates will be largely determined by both the amount and size of level 1 aggregates. For example, long and flexible polymers will aggregate more readily than short and stiff polymers. At this time, it is not known how such polymer properties are related to HMWGS/LMWGS composition and how gliadins play a role. It is expected that this so called mesoscopic aggregation will occur in the wheat kernel and in dough during the first phase of resting.

4.3.3. The macroscopic level, III. Mesoscopic protein particles can aggregate further to a third macroscopic level as further aggregation occurs by the formation of entanglements between protein particles (> 100 μm). Here, by definition, covalent bonds do not play a predominant role in the aggregate formation. Covalent stabilisation, however, can occur in time. The formation of aggregates at this level is thought to be predominantly influenced by process conditions (shear, stress) and other particles interfering with the formation of entanglements (hard fat, arabinoxylans).

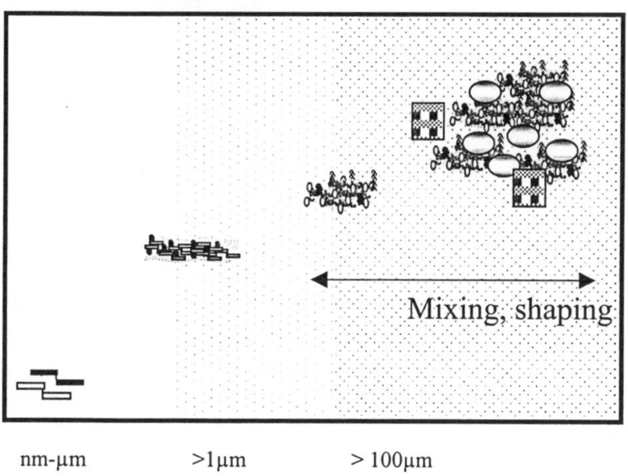

Figure 1: *Hyperaggregation model for glutenin aggregation and gluten formation.*

4.4. Using the hyperaggregation model.

The hyperaggregation model provides a means to link the various aspects of wheat production and processing to structure and functionality. Figure 2 presents a first, qualitative example how the model can be used. On the left three categories are listed: composition, formula and process. Parameters in each of these categories are assessed in terms of their effect on the three levels of aggregation. A link with a specific end-product can be made if we interpret its typical dough properties in terms of the extent of aggregation required. As an example, for biscuits, a non-elastic, non-gasholding dough is required, so parameters should be selected preventing hyperaggregation (low protein content, high gliadin vs GS ratio, hard fats, reducing agent). With bread, a dough is required of certain strength and having good gasholding properties. Therefore, the parameters should be selected to promote hyperaggregation. Although it is still very qualitative, these examples demonstrate the relevance of the model.

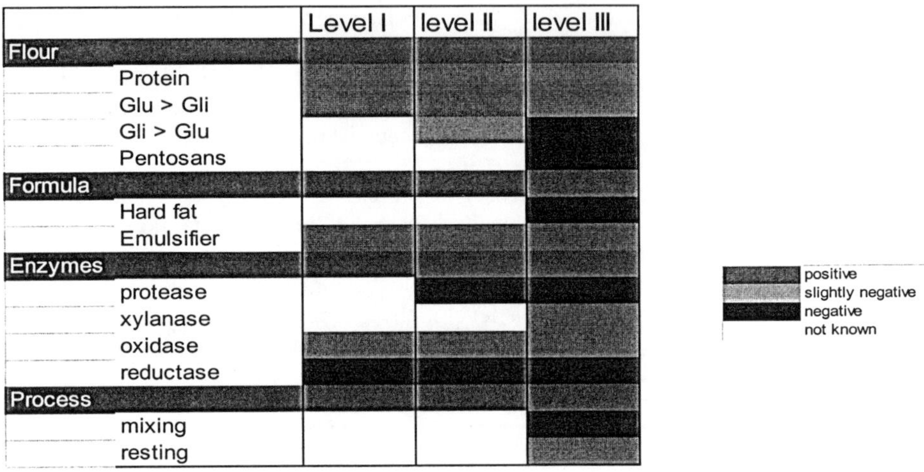

Figure 2: *Scheme listing various parameters and their effect on glutenin aggregation.*

5 CONCLUSIONS

Cereal science has already come a long way in providing knowledge to improve both the production and use of wheats. Recently, various research approaches are converging into a focused effort on understanding structure-functionality relationships. This effort needs to involve both chemical and physical approaches. The qualitative model presented in this overview forms a starting point for more quantitative studies leading to success in understanding how and to what extent gluten composition is reflected in gluten structure and wheat end-use quality.

References

1. F. MacRitchie, *Cereal Foods World* 1999**, 44,** 188.
2. Y. Pomeranz, *Wheat: Chemistry and Technology.,* 3 ed.; Eagan Press: Minneapolis, St Paul, 1989.
3. C.W. Wrigley, J.L. Andrews, F. Békés, P.W .Gras, R. Gupta, F. MacRitchie, J.H. Skerritt, *Interactions: The Keys to Cereal Quality*, R.J. Hamer, R.C. Hoseney, Eds.; Americ. Assoc. Cereal Chemists, Inc.: St. Paul, MN, USA, 1999; Chapter 2.
4. A. Graveland, *Annales de Technologie Agricole* 1980**, 29,** 113.
5. A. Graveland, *Getreide* 1986**, 42,** 35.
6. N.K. Singh, G.R. Donovan, I.L. Batey, F. MacRitchie, *Cereal Chem.* 2000**, 67,** 150.

7. P.L. Weegels, A.M.van de Pijpekamp, A. Graveland, R.J. Hamer, J.D. Schofield, *J. Cereal Sci.* 1997, **23**, 103.
8. P.L. Weegels, R.J. Hamer, J.D. Schofield, *J. Cereal Sci.* 1996, **23**, 1.
9. J.A.D. Ewart, *J. Sci. Food. Agric.* 1979, **30**, 482
10. A. Graveland, P. Bosveld, W.J. Lichtendonk, J.P. Marseille, J.H.E. Moonen, A. Scheepstra, *J. Cereal Sci.* 1985, **3**, 1.
11. D.D. Kasarda, *Wheat is Unique*, Y. Pomeranz, Ed.; Amer. Assoc. Cereal Chem.: St Paul, MN, USA, 1989.
12. P.E. Pritchard, C.J. Brock, *J. Sci. Food Agric.* 1994, **65**, 401.
13. P.L. Weegels, Thesis. King's College London, University of London, London, UK, 1994.
14. P.I. Payne, M.A. Nightingale, A.F. Krattiger, L.M. Holt, *J. Sci. Food Agric.* 1987, **40**, 51.
15. O.M. Lukow, S.A. Forsyth, P.I. Payne, *J. Genetics and Breeding* 1992, **46**, 187.
16. J. Lefebvre, Y. Popineau, G. Deshayes, L. Lavenant, *Cereal Chem.* 2000, **77**, 193.
17. F. Békés, P. Gras, *Cereal Foods World* 1999, **44**, 580.
18. C.R. Rath, P.W. Gras, C.W. Wrigley, C.E. Walker, *Cereal Foods World* 1990, **35**, 572.
19. R. Kieffer, H. Wieser, M.H. Henderson, *J. Cereal Sci.* 1998, **27**
20. D.D. Kasarda, *Cereal Foods World* 1999, **44**, 566
21. S. Muller, W.H. Vensel, D.D. Kasarda, P. Kohler, H. Wieser, *J. Cereal Sci.* 1998, **27**, 109.
22. B. Keck, H.P.E. Kohler, H. Wieser, *Z. Lebensm. Unters. Forsch.* 1995, **200**, 432.
23. P. Kohler, B. Keck-Gassenmeier, H. Wieser, D.D. Kasarda, *Cereal Chem.* 1997, **74**, 154.
24. N. Shani, J.D. Steffen-Campbell, O.D. Anderson, F.C. Greene, G. Galili, *Plant Physiol.* 1992, 433.
25. P.R. Shewry, N.G. Halford, A.S. Tatham, *Oxford Surveys of Plant Molec. Cell Biol.* 1989, **6**, 163.
26. P.S. Belton, *J.Cereal Sci.* 1999, **29**, 103.
27. R. Tkachuk, I. Hlynka, *Cereal Chem.* 1968, **45**, 80.
28. A.-C.A.P.A. Bekkers, W.J. Lichtendonk, and R.J. Hamer, Studying the formation of glutenin polymeric networks; the effect of NEMI, Arg and Gly. (Personal Communication)
29. J.J. Kokelaar, T. Van Vliet. A. Prins, *J. Cereal Sci.* 1996, **24**, 199.
30. R. Gupta, K. Khan, F. MacRitchie, *J. Cereal Sci.* 1993, **18**, 23.

A SMALL SCALE WHEAT PROTEIN FRACTIONATION METHOD USING DUMAS AND KJELDAHL ANALYSIS

O. M. Lukow[1], J. Suchy[1] and B. X. Fu[2]

1. Cereal Research Centre, Agriculture and Agri-Food Canada, 195 Dafoe Rd., Winnipeg, Manitoba, Canada, R3T 2M9. 2. Canadian International Grains Institute, 1000-303 Main St., Winnipeg, Manitoba, Canada, R3C 3G7.

1 INTRODUCTION

Considerable interest has developed in the quantitative fractionation of wheat flour proteins in recognition of the unique role these components play in visco-elastic dough properties. However, the complexity of the existing extraction procedures, and a laborious and hazardous process of nitrogen analysis limits a wide application of wheat flour protein fractionation. Our objective was to develop a fractionation method that will: 1) allow quantification of relatively pure wheat protein solubility classes, 2) decrease the number of nitrogen analyses required per sample, 3) measure the total nitrogen in each solubility group directly from dry residue using the efficient Dumas method and 4) permit the measurement of total nitrogen by the standard Kjeldahl method.

2 MATERIALS AND METHODS

2.1 Plant Materials

Two wheats with weak and extra-strong dough strength properties: Alpha 16 (Canada Prairie Spring wheat class) and Glenlea (Canada Western Extra Strong Red Spring wheat class) were used. Both wheats had an identical high molecular weight-glutenin subunit (HMW-GS) composition of 2*, 7+8, 5+10. Samples were grown in one western Canadian location in 1998 and milled on Ottawa flour mill.

2.2 Protein Fractionation

Flour samples were extracted according to the method outlined in Fig. 1. Three flour samples (100 mg at 14% mb) were concurrently extracted with 7.5% 1-propanol and 0.3M NaI at 25°C (stage 1). In the second stage only residues from flour 2 and 3 were extracted with 50% 1-propanol at 25 °C. Finally, only residue from flour 3 was extracted with 40% 1-propanol and 0.2% DTT at 50 °C (stage 3). Three extractions (30, 30 and 5 min) per stage were performed followed by centrifugation at 1000 x *g*. The dry insoluble residue was analyzed for total nitrogen content using the modified Dumas method (FP-528, LECO Corp.) and the micro-Kjeldahl method (AACC method 46-13, 1995). The nitrogen content of the four flour protein solubility groups[1] were obtained mathematically: monomeric protein (difference between flour and insoluble residue in stage 1), soluble glutenin (difference between insoluble residue in stage 1 and 2), insoluble glutenin (difference between insoluble residue in stage 2 and 3) and residue protein (insoluble residue after stage 3). Ten replicates of protein fractionation were performed. FP-528 instrument was calibrated using analytical-grade reference materials (EDTA, L-lysine-HCl and L-tryptophan) and check sample (soy flour). Unpaired t-test with t-test procedure (SAS Institute Inc. Cary, NC) was used to compare protein content of the solubility fractions derived from the micro-Kjeldahl and modified Dumas methods.

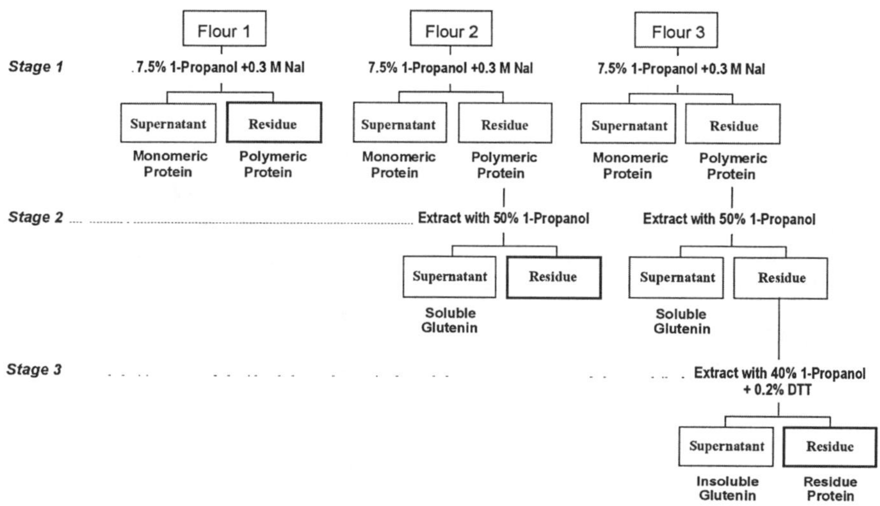

Figure 1 *Wheat protein fractionation scheme.*

3 RESULTS AND DISCUSSION

The suitability of the outlined fractionation procedure for a routine method was evaluated according to objective factors such as (1) reproducibility of dry residue weights and total nitrogen content; (2) degree of separation for different wheat protein solubility groups;(3) cost and time efficiency. Other factors such as complexity of the procedure, labour intensity and safety were also considered.

Low standard deviation of the insoluble residue dry weights (0.4 to 1.9 mg, Table 1) indicated good reproducibility of the proposed extraction procedure. The variation in dry residue weights for Alpha 16 and Glenlea reflected the difference in total protein content (11.9 and 13.5 %, respectively).

Table 1 *Weight of the residue remaining after extraction of 100 mg (14% mb) flour for two wheats (mean ± standard deviation, 10 replications per cultivar).*

Sample	Dry Weight of Residue (mg)		
	(stage 1)	(stage 2)	(stage 3)
Alpha 16	82.7 ± 0.5	75.0 ± 0.4	72.5 ± 1.0
Glenlea	83.1 ± 1.9	75.5 ± 1.8	73.4 ± 1.9

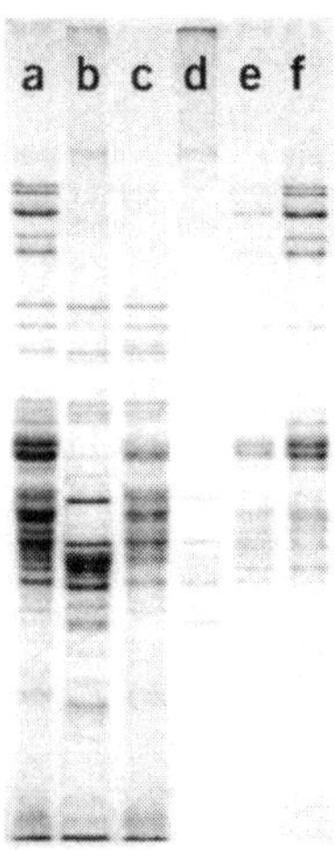

Figure 2. *SDS-PAGE composition of (a) total flour protein reduced, (b) monomeric protein unreduced, (c) monomeric reduced, (d) soluble glutenin unreduced, (e) soluble glutenin reduced, (f) insoluble glutenin for cv. Glenlea.*

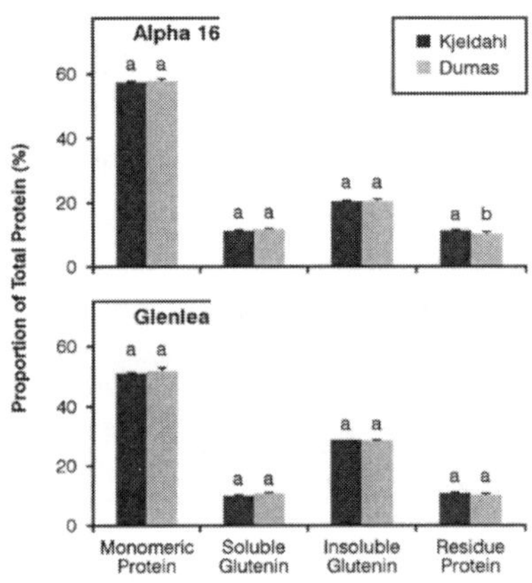

Figure 3. *Distribution of flour nitrogen among four solubility groups for two wheats using Dumas and Kjeldahl total nitrogen analysis.*

SDS-PAGE (Fig. 2) showed that there were no significant amounts of high molecular weight glutenin in the monomeric protein fraction (lane c, reduced) and no significant amounts of monomeric protein (gliadin) in the insoluble glutenin fraction (lane f), indicating relatively pure separation of different protein groups in agreement with previous results[2].

The quantity of the four flour protein solubility groups were identical ($P \leq 0.05$) whether determined by the micro-Kjeldahl or modified Dumas method with exception of residue protein for Alpha 16.(Fig. 3). Nitrogen recoveries from flour and most of the solubility fractions tended to be higher when derived by the modified Dumas method compared to micro-Kjeldahl method and were observed earlier[3]. Some reports[4] suggest that the Kjeldahl method underestimates nitrogen present in the nitrates/nitrite form. Despite an identical HMW-GS composition, a significant difference ($P \leq 0.01$) in the amount of monomeric protein (51 vs 57 %) and insoluble glutenin (28 vs 20 %) was observed for Glenlea and Alpha 16 wheats (respectively), reflecting closely their differences in dough strength. The ability to measure the quantitative variation in the four wheat protein solubility groups may be a useful tool to explain the intercultivar differences in dough rheology.

4 CONCLUSIONS

Advantages of the fractionation method outlined here include:(i) high throughput of up to 20 samples/day, (ii) indefinite storage of processed sample at room temperature, (iii) lower number of nitrogen analyses per sample (lower cost per sample), (iv) small amount of flour required (100 mg) and (v) suitability for nitrogen analysis by the Kjeldahl or modified Dumas methods. The modified Dumas method of nitrogen assay offers a reproducible, rapid, cost efficient, hazard- and waste-free method to characterize major wheat protein solubility groups.

References

1. H.D. Sapirstein and B.X. Fu, *Cereal Chem.* 1998, **75**, 500.
2. B.X. Fu and M.I.P. Kovacs, *J. Cereal Sci.* 1999, **29**,113.
3. P. Williams, D. Sobering and J. Antoniszyn, in *Proceedings of the Wheat Protein Symposium*, ed. D. Fowler, W.E. Geddes, A.M. Johnston and K.R. Preston, University Press, Saskatoon, SK, 1998, p. 37.
4. R.A. Sweeney and P.R. Rexroad, *J. Assoc. Anal. Chem.*, 1987, **70**, 1028.

Acknowledgments

The technical assistance of the Wheat Quality Research Team and Sheila Woods for statistical advice is gratefully acknowledged.

ANALYSIS OF GLUTEN PROTEINS IN GRAIN AND FLOUR BLENDS BY RP-HPLC

O.R. Larroque[1,2], F. Bekes[1,2], C.W. Wrigley[1,2] and W.G. Rathmell [2]

1. CSIRO Plant Industry, P.O.Box 7, North Ryde, NSW 1670, Australia. 2. Quality Wheat CRC Ltd., Locked Bag 1345, North Ryde, NSW 1670, Australia.

1 INTRODUCTION

Flour blends are commonly used to fulfil industry requirements with the final aim being the optimisation of food production. Experiments with mixtures are an important area of research in food processes; thus statistical science has given attention to the matter of mixtures themselves.[1] In that sense, flour blends were chosen as case studies by various research groups.[2,3] The factors considered in a blend can be divided into those subject to control (control factors) and those under no control (noise factors). The proportion of each component in the blend can be considered as a factor under control and can be easily evaluated by analysing, for example, the resulting protein content and comparing it with the estimated values. After thoroughly homogenising the blend, the standard deviation (SD) is expected to be insignificant in statistical terms. Conversely, this linear relationship is not followed in other cases, for example, when considering dough properties.[3,4] In these situations, knowing the exact contribution of each component at the polypeptide level is basic to the prediction of the behaviour of the blend.

Reversed-phase high performance liquid chromatography (RP-HPLC) and gel electrophoresis are commonly used for determining the composition of wheat storage protein present in flour. From a quantitative point of view, the former is preferred over the latter.[5] Despite considerable research effort on RP-HPLC techniques in recent years, it has not been reported for application to the individualisation and quantification of components in complex blends.

Our aim was to adapt the RP-HPLC technique to clearly determine protein components in a blend, at the polypeptide level, both from the qualitative and quantitative points of view. This involved developing methodology to determine the proportions of specific gliadin and glutenin polypeptides, and using this to quantify the proportions of component flours in a blend according to polypeptides specific to each flour in the blend.

2 MATERIALS AND METHODS

2.1 RP-HPLC

The method of Marchylo *et al.*,[6] with some modifications, was used for the qualitative/quantitative analysis of individual subunits of glutenin. After gliadin extraction from 50mg of flour with 70% ethanol (1mL), the residue was extracted with 50% propan-1-ol (1mL). After discarding the supernatant, the pellet was treated with 1 mL of a buffer containing 50% propan-1-ol, 2 M urea, 0.2 M Tris, pH 6.6 to which 1% of DTT was added. The samples were extracted for 1h at 60°C in a water bath. After that, the solubilised proteins were alkylated by adding 10 µL of 4-vinylpyridine during 15 min at 60°C. Supernatants consisted of reduced and alkylated subunits of polymeric proteins, mainly high molecular weight (HMW) and low molecular weight (LMW) glutenin subunits (GS). Following centrifugation and filtration through 0.45 µm PVDF filters, samples were ready to inject. Propiophenone (Sigma, USA), used as internal standard, was incorporated in the extraction buffers at a rate of 5 µl/20 mL.

Protein extracts were analysed by HPLC using a Beckman System Gold HPLC, configured with two 126 Pumps, a 166 Detector and a 507E Autosampler. A Vydac C18 column (Vydac, Hesperia, California) was used throughout the study.

3 RESULTS AND DISCUSSION

Figures 1 and 2 show profiles of gliadins and glutenin polypeptides from a blend. The first peak corresponds to propiophenone, which served as a marker of elution time clear of eluting proteins.

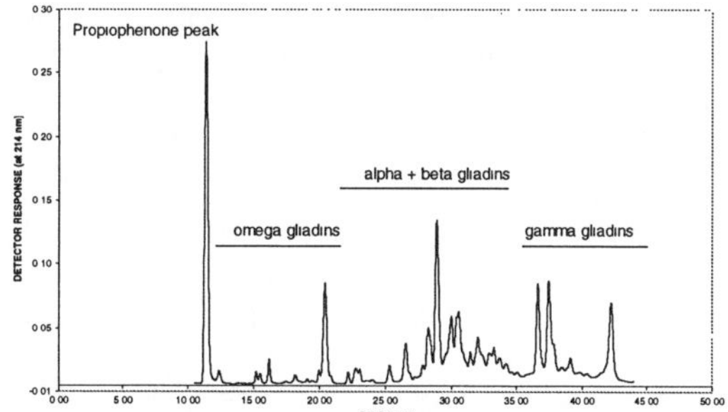

Figure 1: *RP-HPLC profile showing a gliadin extract (obtained by extraction with 70% EtOH) from a blend.*

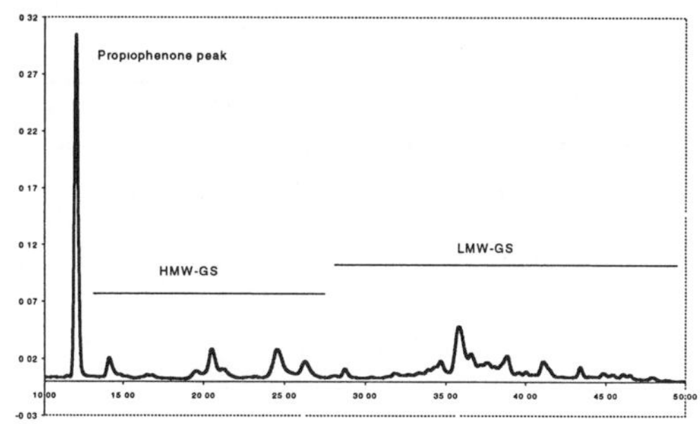

Figure 2: *RP-HPLC profile showing a reduced glutenin extract (obtained by extraction as indicated in the Methods section) from a blend.*

After establishing that the internal standard was suitable for our purpose, the next step was to define areas of the chromatograms where it was possible to identify differences in the quantity of specific components. Figures 3 and 4 show areas of the gliadin and glutenin subunit elution profiles, respectively, where it was possible to identify polypeptides specific to each flour component in the blend.

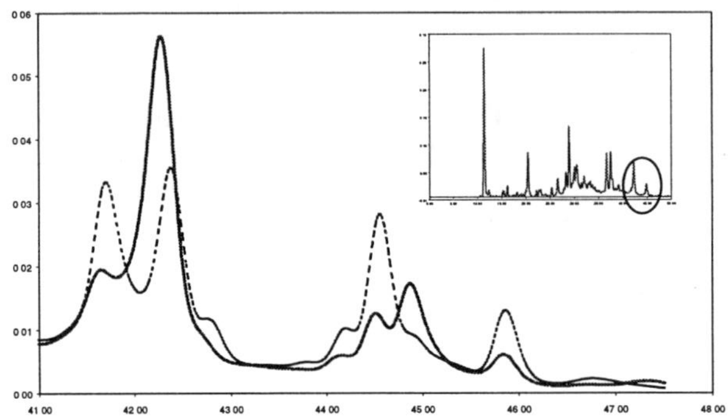

Figure 3: *RP-HPLC profile of some gamma gliadins from two blends (25:75 solid line and 75:25 dotted line) of two flours that differ in the presence/absence of certain gamma gliadins. Inset: Full chromatogram.*

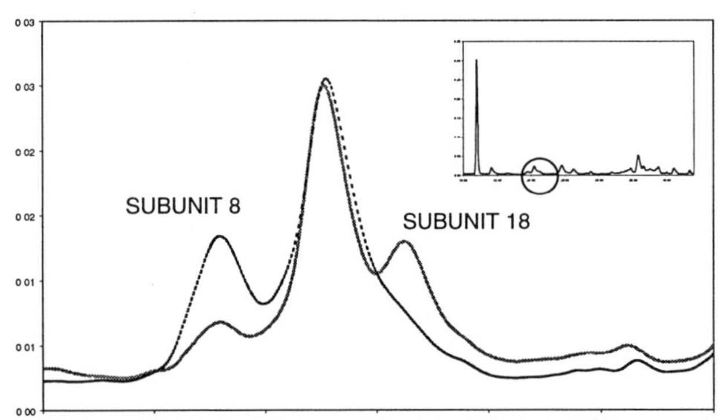

Figure 4: *RP-HPLC profile of glutenin subunits from two blends (25:75 solid line and 75:25 dotted line) of two flours that differ in the presence/absence of subunits 8 and 18. Inset: Full chromatogram.*

4 CONCLUSION

Improved methods have been developed to determine the proportions of flour components in blends. This methodology has been needed particularly in studies of the non-linearity of dough properties resulting from blends made of wheat before milling, compared with flour blends made after milling. The studies showed that the milling step itself caused a degree of departure from linear behaviour due to different proportions of the original grain proteins resulting in the flour blends.

References

1. H. Scheffé, 'Experiments with mixtures', *J. Roy. Stat. Soc.*, 1958, **B, 20**, 344.
2. S. Ghosh and T. Lui, *J. Stat. Plann. Inference* 1999, **78**, 219.
3. T. Næs, F. Bjerke and E.M. Færgestad, *Food Qual. Prefer.* 1999, **10**, 209.
4. L.D. Simmons and K.H. Sutton, in *Cereals '97: Proceedings of the 47th Australian Cereal Chemistry Conference,* ed. A.W. Tarr, A.S. Ross and C.W. Wrigley, RACI, Melbourne, Australia, 1997, 208.
5. F.R. Huebner and J.A. Bietz, *Cereal Chem.*, 1999, **76**, 299.
6. B.A. Marchylo, J.E. Kruger and D.W. Hatcher, *J. Cereal Sci.*, 1989, **9**, 113.

RELIABLE ESTIMATES OF GLIADIN, TOTAL AND UNEXTRACTABLE GLUTENIN POLYMERS AND TOTAL PROTEIN CONTENT, FROM SINGLE SE-HPLC ANALYSIS OF TOTAL WHEAT FLOUR PROTEIN EXTRACT

Marie-Hélène Morel[1] and Christine Bar-L'Helgouac'h[2]

1. INRA-Unité de Technologie des Céréales et des Agropolymères, 2 place Viala, 34060 Montpellier cdx 01- France. 2. Institut des Céréales et des Fourrages, 16, rue Nicolas Fortin, 75013 Paris-France

1 INTRODUCTION

Several contrasting wheat flour dough characteristics are required to process successfully the French "baguette". The dough must exhibit a high extensibility in order to be shaped into elongated dough-pieces and an adequate elasticity to withstand the long proof time. In France particular attention is paid to the wheat flour quality, protein quantity being of secondary interest. Whereas flour dough quality is generally assessed by the Chopin Alveograph, P/L and W indexes, breeders need reliable small-scale screening methods. In this respect, Size-Exclusion High-Performance Chromatography (SE-HPLC) of total wheat flour protein extracted by sonication seems to be a very promising method since the pioneer works of Singh and coworkers[1,2]. In theory, reliable estimate of total protein, gliadin, total glutenin and unextractable glutenin polymer contents might be obtained from routine SE-HPLC analyses from total flour protein extracts. To achieve this goal several requisites are needed: (a) the protein extract must be stable in order to allow SE-HPLC analyses from batch samples; (b) sonication must ensure total protein extraction; (c) sonication must be adjusted in order to limit the depolymerisation of unextractable polymers into the size distribution range of the true SDS-soluble polymeric protein.
The aim of this study was to develop a procedure based on SE-HPLC analysis of total flour protein extracts to estimate contents of total protein, gliadin, SDS-soluble and SDS-insoluble glutenin macropolymers.

2 MATERIALS AND METHODS

A Vibra Cell 72434 sonifier (Bioblock Scientific, Illkirch, France) delivering ultrasonic vibrations at 20 kHz and equipped with a 3 mm diameter tip probe was used. Total protein extract was obtained from 160 mg flour. Prior to sonication (3 min at 30% power setting), flour was dispersed for 80 min at 60 °C by rotation at 60 rpm with 20 mL of 1% sodium dodecyl sulphate (SDS), 0.1M sodium phosphate buffer (pH 6.9). The protein extract was obtained by collecting the supernatant after 30 min centrifugation at 37,000 g at 20 °C in a Beckman centrifuge (model JA-221). The SE-HPLC apparatus was a Waters model (LC Module1 plus) controlled by the Millenium software (Waters) and equipped with a TSK G4000-SW (TosoHaas) size exclusion analytical column (7.5 x 300 mm) and a TSK G3000-SW (TosoHaas) guard column (7.5 x 75 mm). SE-HPLC analysis was performed as described by Dachkevitch and Autran[3].

3 RESULTS AND DISCUSSION

3.1 Stability of flour protein extracts

SE-HPLC analysis of a protein extract obtained by sonication of a flour sample (160 mg) that had been suspended for 80 min at ambient temperature with 20 mL 1% SDS-phosphate buffer showed a marked instability on re-injection. Whereas the total SE-HPLC area remained unchanged, an alteration in the size distribution range of proteins was observed, as a time-dependent drop of fraction F1 (Figure 1). This fraction, which consists of the largest glutenin macropolymers (GMP) eluting at the void volume of the column decreased with an increase in fraction F2, which consists of smaller GMP of M_r ranging from 630,000 to 116,000. The possibility of proteolysis considered and the inhibitory effects of antipain (1μg/mL) and phenylmethanesulfonyl fluoride (PMSF) (1 mM) were studied. Antipain, a competitive inhibitor of papain, a cysteine protease, was shown to be very effective, limiting the extent of fraction F1 depolymerisation, whereas PMSF, a serine protease inhibitor was less effective. Taking into account the study of Wang and Grant[4] on the heat instability of wheat flour proteases above 50°C, our extraction procedure was modified and performed at 60°C in order to obtain stable flour protein extracts.

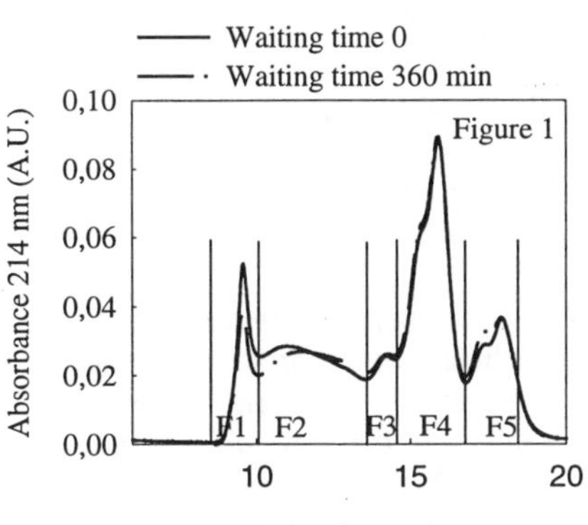

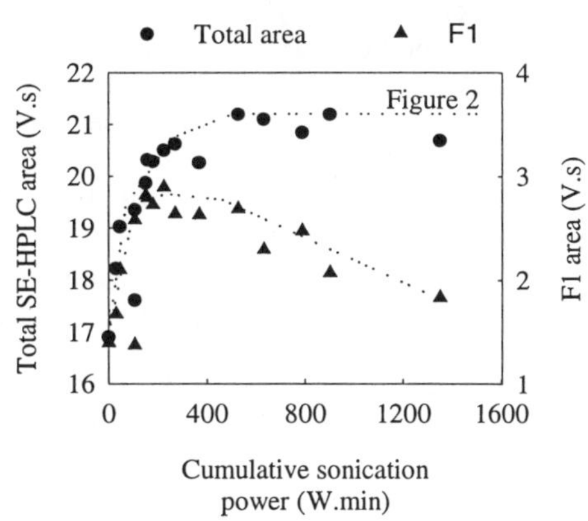

3.2 Extraction of total flour protein by sonication

Flour protein extraction, as judged by the total surface area of the SE-HPLC profile, was found to be related to the cumulative sonication power (Figure 2). No sonication time or power threshold was identified and long sonication time combined with a low power setting equalled short sonication time at high power setting. The increase in protein extractability through sonication treatment was not modified by the position of the sonication probe inside the tube, by the solid-to-solvent ratio (from 4 to 16 g/L) or by the weight of flour (80 or 160 mg). On the other hand, a smaller sample volume (10 mL) increased the sonication efficiency and shifted the protein extractability dependency towards lower values of cumulative sonication power. The greater amount of protein eluted in fractions F1 and F2 accounted for the increase in protein extractability upon sonication. Therefore oversonication was seen as a drop in fraction F1, with an increase in fraction F2, and was observed for cumulative power values exceeding 400 W.sec. For the following studies, sonication was fixed at 630 W.sec in order to limit depolymerisation of SDS-insoluble GMP. The sonication method was applied to a 27 flour sample set showing high variability in protein content from 8.9 to12.6% (mean 10.62% ±1.22, db) and in baking quality score as assessed from the *W* index of the Chopin Alveograph (from 115 to 311 x10^{-4} J; mean 195.4±53.7 x10^{-4} J). A strong relationship was found between the total SE-HPLC area from protein extracts and the flour protein contents (R^2 = 0.97, n = 27), confirming that sonication at 630 W.sec was successful in providing total flour protein extraction.

3.3 Protein estimation from peak area

A reliable estimate of protein content from the SE-HPLC area is possible provided that total protein is eluted from the column. Owing to the huge size of GMP, this is questionable and needs to be validated. Concentrated protein extracts in 1% SDS-phosphate buffer, from bovine serum albumin (BSA), flour, gliadin and gluten were obtained in the laboratory. Flour and gluten samples were sonicated at 630 W.sec in order to extract all proteins. The protein contents of samples were determined by the Kjeldahl method using nitrogen conversion factors of 5.7 and 6.25 for wheat proteins and BSA, respectively. For each protein extract, four series of 10 diluted samples were applied as 20 µL samples onto the SE-HPLC system without filtration. Total SE-HPLC areas were integrated and plotted against the amounts of injected protein. Linear responses were observed and average response factors relating total SE-HPLC area to the amount of injected protein where calculated from the pooled data (n=40) for each extract (Table 1). Factors were not significantly different for gliadin, gluten and flour extracts. According to this finding it was likely that all GMP from gluten and flour extracts was successfully eluted from the column.

A lower response factor was found for BSA. As the peptide bond contributes to absorbance at 214 nm, a greater area is expected for wheat storage protein compared to BSA, owing to the side chain of glutamine. Indeed, the response for flour protein was 1.086 times greater than for BSA; a value comparable to the ratio of the nitrogen conversion factors of BSA and flour protein (6.25/5.7 = 1.096).

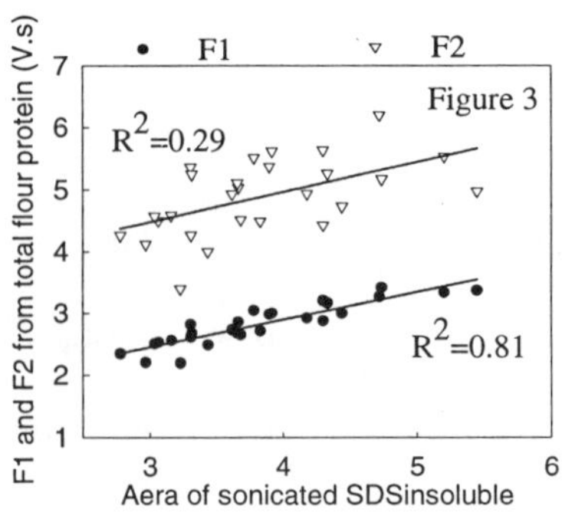

3.4 Size distribution range of sonicated SDS-insoluble GMP

SDS-soluble protein were extracted in SDS-phosphate buffer (160 mg/20 mL), from 24 flour samples, of various protein contents and bread making qualities. The protein residue was suspended with 10 mL of the same buffer and sonicated at 630 W.sec, in order to extract SDS-insoluble GMP. Figure 3 compares, on a similar solid-to-solvent basis, the total SE-HPLC areas from SDS-insoluble GMP extracts with F1 and F2 SE-HPLC areas from the corresponding total flour protein extracts. A very strong correlation ($R^2 = 0.81$, n = 24) is found between the amounts of SDS-insoluble GMP and the area of fraction F1 of the total flour extracts. On average, 74% ± 6% of fraction F1 arises from the insoluble GMP fraction. The use of a carefully adjusted sonication procedure ensures the disruption of insoluble GMP into soluble polymers, large enough to be eluted mainly at the void volume of the column, thus showing a limited overlap with the original SDS-soluble polymers.

Table 1 *Conversion factors between injected protein amounts and total SE-HPLC areas from total flour protein, gliadin, sonicated gluten and bovine serum albumin (BSA) extracts.*

Extract	Concentration range tested (g/L)	Total area for 1 μg of protein injected in 20 μL (V.s)
Total flour protein	0.088 - 2.54	1.37 ±0.024
Gliadin	0.096 - 2.08	1.42 ±0.089
Sonicated gluten	0.020 - 0.50	1.41 ±0.018
BSA	0.129 - 2.48	1.29 ±0.089

4 CONCLUSION

The use of a cumulative sonication power of 630 W.sec allowed extraction of total flour protein from flour samples showing variable protein contents (8.9-12.6%, dmb) and flour dough strengths (*W* from 115 to 311 $x10^{-4}$ J). The optimised sonication procedure limits the depolymerisation of SDS-insoluble GMP, in such a way that the amount of fraction F1 is highly related to the amount of SDS-insoluble GMP in flour. It was verified that flour protein recovery from SE-HPLC column was exhaustive, and a conversion factor allowing the accurate quantification of protein content from SE-HPLC area was calculated. Thus, reliable characterisation of flour protein can be obtained from a single SE-HPLC analysis of total flour protein extract.

References

1. N. K. Singh, G. R. Donovan, I. L. Batey, and F. MacRitchie, *Cereal Chem.*, 1990, **67**, 150.
2. N. K. Singh, G. R. Donovan and F. MacRitchie, *Cereal Chem.*, 1990, **67**, 161.
3. T. Dachkevitch and J. C. Autran. *Cereal Chem.* , 1989, **66**, 448.
4. C.C. Wang and D. R. Grant, *Cereal Chem.*,1969, **46**, 537.

USE OF A ONE-LINE FLUORESCENCE DETECTION TO CHARACTERIZE GLUTENIN FRACTION IN THE SEPARATION TECHNIQUES (SE-HPLC AND RP-HPLC)

T. Aussenac and J.-L. Carceller

Department of Plant Physiology, Ecole Supérieure d'Agriculture de Purpan, 75 voie du T.O.E.C., 31076 Toulouse cedex 03, France. Email : aussenac@esa-purpan.fr.

1 INTRODUCTION

Recent research on the biochemical basis of breadmaking quality of wheat has intensified the need for an accurate and reliable method for separing the polymeric (unreduced or native) glutenin from the monomeric or single polypeptide chain wheat flour proteins (albumins, globulins, and gliadins). The rationale for such a separation is twofold. First, the relative amount of polymeric protein in a flour or a gluten appears to be strongly related to the functionality of the flour or the gluten in breadmaking[1-5]. Second, after reduction of the polymeric glutenins, their subunit composition can be used to predict the breadmaking potential of a wheat[6-8]. Many fractionation procedures have been reported to separate the glutenins from other classes of wheat proteins. Physicochemical approaches have almost invariably been based on the separation of the very large molecular mass polymeric glutenins. Methods have included ultrafiltration[9], gel filtration[10,11], and size-exclusion high-performance liquid chromatography[12-14]. However, with these procedures the quantification and characterization of glutenins was not accurate because of the lack of sufficient polymers-monomers separation. The aim of the present work was to investigate the potential of on-line fluorescence detection to characterize glutenin fraction separated by SE-HPLC and RP-HPLC.

2 MATERIALS AND METHODS

2.1 Materials

The two common wheat cultivars used in this study were Soissons and Thésée possessing *Glu-D1* subunits 5+10 and 2+12, respectively. These cultivars were grown at the experimental farm of ESAP Toulouse, France (1997-1998). For all the biochemical determinations, the grains were freeze-dried, ground in a Janke A10 grinder and kept at -20°C.

2.2 Methods

2.2.1 Glutenin extraction. Two different extraction procedures were used to obtain purified polymeric proteins. In procedure 1, glutenins were isolated by applying propan-1-ol extraction and precipitation according to Fu and Sapirstein[15]. In procedure 2, glutenins were purified by applying NaI/propan-1-ol extraction according to Fu and Kovacs[17].

2.2.2 Glutenin mBBr labelling. To obtain purified and mBBr labelled polymeric proteins, ground kernels were extracted by a 0.3 M sodium iodine (NaI) - 7.5 % (v/v) propan-1-ol solution containing 0.25 mM mBBr. The extraction (1 h at room temperature under continual stirring) was followed by centrifugation (15 900 g, 15 min, 20°C) to obtain a supernatant (monomeric proteins) and a pellet (polymeric proteins). The monomeric and polymeric proteins were then freeze-dried and stored at -20°C. These fractions were used for SE-HPLC and RP-HPLC according to Carceller and Aussenac[17]. Purified glutenins (NaI/propan-1-ol) were characterised by diagonal SDS-PAGE (NR/R) according to Laurière *et al.*[18].

3 RESULTS AND DISCUSSION

Accurate quantification and characterization of glutenins require the isolation of total glutenins in samples without monomeric contamination. Propan-l-ol has been widely used as a solvent for monomeric flour proteins, usually at 50% (v/v) concentration[19].

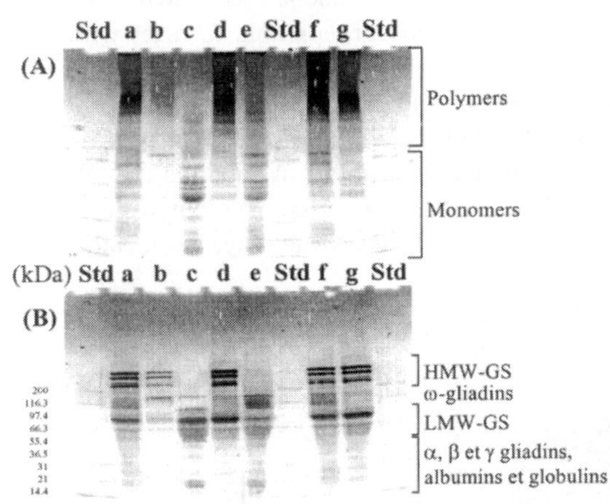

Figure 1 SDS-PAGE of (A) unreduced and (B) reduced proteins. (A) polymers insoluble in 50% propan-l-ol, (b) polymers insoluble in 70% propan-l-ol, © fraction soluble in 70% propan-l-ol, (d) polymeric fraction purified with NaI, (e) monomeric fraction soluble in NaI. (f and g) total glutenin fractions obtrined by propan-l-ol and NaI purification, respectively. *(cv. Soissons-53 DAA).*

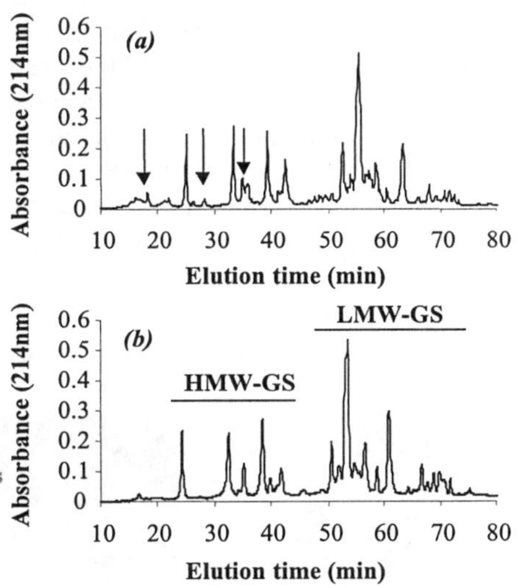

Figure 2 RP-HPLC of reduced and alkylated total glutenin fractions obtained by (a) propan-l-ol and (b) NaI purification. Arrows indicate the contamination by monomeric proteins. *(cv. Soissons-53 DAA).*

However, under those conditions, a significant among of glutenin is also removed along with the monomeric proteins[15]. We have demonstrated that polymeric proteins which were soluble in 50% propan-1-ol can be precipitated by increasing the propan-1-ol concentration to 70% (Figure 1a,b , lanes a and b). This procedure (procedure 1 in Materials and Methods) has been used by Fu and Sapirstein[15] to quantify polymeric glutenins in wheat flour. Even if total glutenins can be recovered by using this procedure, a small amount of monomeric proteins, mainly ω-gliadins and some albumins and globulins, coprecipitate with the polymeric glutenins in 70% propan-l-ol. This coprecipitation can be visualized by both SDS-PAGE (Figure 1a,b, lanes b and f) and by RP-HPLC (Figure 2a).

A recent study[16] suggested that NaI could selectively remove the contaminating ω-gliadins from the glutenins. Fu and Kovacs[16] demonstrated that there is a synergistic effect between NaI and propan-1-ol in removing monomeric proteins. This synergistic effect is probably due to their effects on the interactions between proteins; propan-l-ol would be expected to disrupt mainly hydrophobic interactions, while NaI would interfere with both hydrophobic and electrostatic interactions among proteins. The SDS-PAGE (Figure 1a,b, lanes c and g) and RP-HPLC (Figure 2b) results showed that the NaI/propan-1-ol insoluble fraction (procedure 2 in Materials and Methods) contained mainly polymeric proteins. Thus, the monomeric and polymeric grain proteins were completely separated by single-step extraction without cross-contamination.

The different fractions (polymeric and monomeric) obtained by NaI/propan-1-ol extraction (procedure 2) were characterised by different responses to mBBr. When mBBr was added to the extraction solvant, only the polymers were labelled as shown in Figure 3a,b. Only glutenins were thus characterised by free -SH groups able to react with the mBBr solution, irrespective of the genotype studied. These results were in agreement with the fact that among the monomers (gliadins, albumins and globulins) all cysteine residues took part in intramolecular disulphide bridges. Using RP-HPLC enabled us to confirm the results obtained by SDS-PAGE. As shown in Figure 4, all glutenin subunits (HMW-GS and LMW-GS) were indeed labelled by mBBr. The total polymeric proteins (glutenins) were thus readily identified by fluorescence detection. In the same way, the labelling of

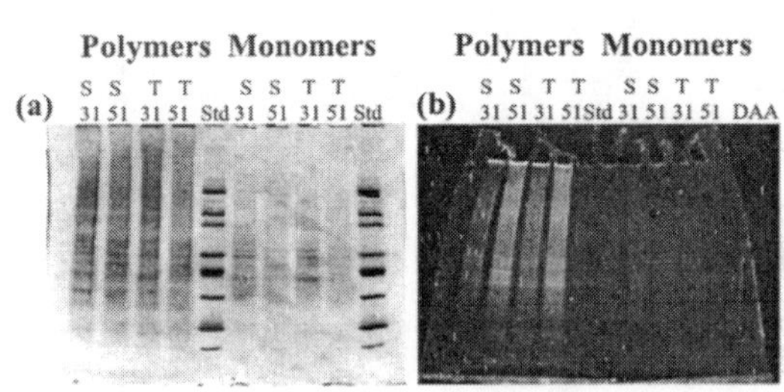

Figure 3 SDS-PAGE of unreduced polymeric and monomeric fractions obtained by NaI purification. (a) Coomassie blue staining, (b) mBBr derivatized proteins visualized under UV light. (S and T) cv. Soissons and Thésée respectively.

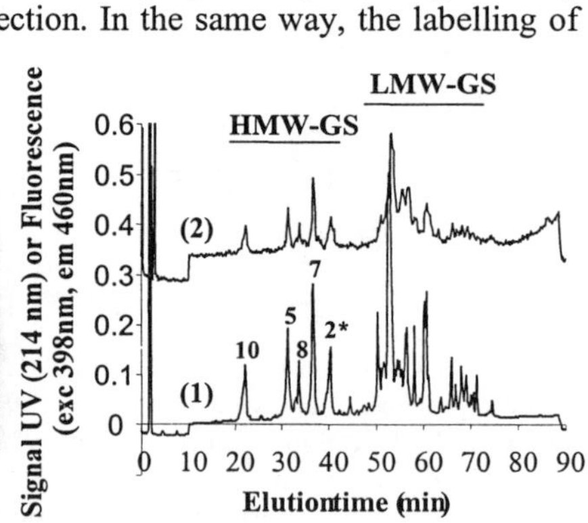

Figure 4 RP-HPLC of reduced, alkylated and mBBr labelled total glutenin fraction obtained by NaI purification. (trace 1) UV detection, (trace 2) Fluorescence detection. (*cv. Soissons-53 DAA*).

free -SH groups by mBBr during the extraction of total proteins by SDS 2%, allows the polymeric fraction to be specifically detected during SE-HPLC.

As shown in Figure 5, the chromatographic pattern obtained from the labelled total proteins (fluorescence detection) is indeed equivalent to the one obtained from the prior purification of polymers by NaI/propan-1-ol. These observations also demonstrate that an important part of the glutenin aggregates is dissociated in the presence of 2% SDS. As shown by the fluoresence detection, certain glutenins are eluted as monomers. These results may be confirmed by diagonal SDS-PAGE. Indeed, NR/R-SDS-PAGE (Figure 6) allowed covalent aggregates (which are linked by interchain disulphide bridges and require a reducing agent for depolymerisation) to be separated from noncovalent glutenin aggregates (which are linked by protein-to-protein interactions, dissociated in presence of SDS). These observations are in total agreement with the results obtained by Laurière *et al.*[18]. The mBBr-SE-HPLC and the NR/R-SDS-PAGE patterns provide clear evidence that some HMW-GS and LMW-GS were not linked by disulphide bridges, and demonstrate that the UV detection in SE-HPLC may not alone allow the rigorous quantification of the glutenin fraction.

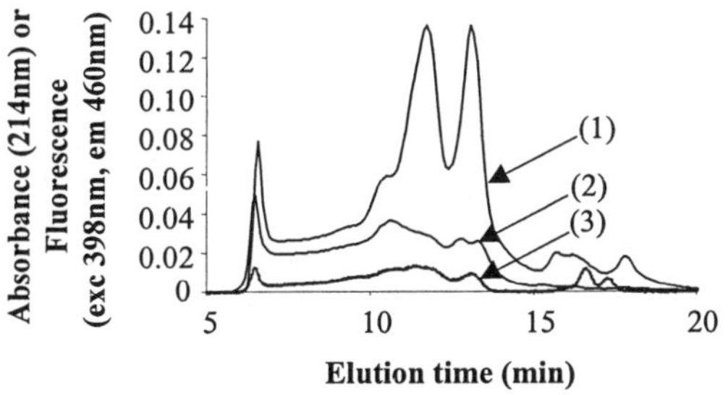

Figure 5 SE-HPLC of (trace 1) total SDS-soluble proteins, (trace 2) total NaI purified glutenins, and (trace 3) total SDS-soluble and mBBr labelled proteins. (*cv. Soissons-53 DAA*).

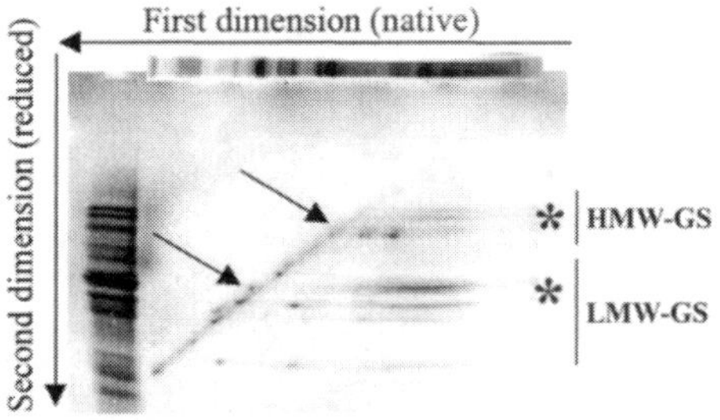

Figure 6 Diagonal SDS-PAGE separation of total glutenin fraction obtained by NaI purification. The arrowheads and stars mark the position of aggregated glutenin subunits and the position of dissociated glutenin subunits respectively. (*cv. Soissons-53 DAA*).

References

1. C. C. Tsen, *Cereal Chem.*, 1967, **44**, 308.
2. K. Tanaka, and W. Bushuk, *Cereal Chem.*, 1973, **50**, 590.
3. J. M. Field, P. R. Shewry, and B. J. Miflin, *J. Sci. Food Agric.*, 1983, **34**, 370.
4. F. MacRitchie, *J. Cereal Sci.*, 1987, **6**, 259.
5. R. B. Gupta, K. Khan, and F. MacRitchie, *J. Cereal Sci.*, 1993, **18**, 23.
6. P. I. Payne, K. G. Corfield, L. M. Holt, and J. A. Blackman, *J. Sci. Food Agric.*, 1981, **32**, 51.
7. P. K. W. Ng, and W. Bushuk, *Cereal Chem.*, 1988, **65**, 408.
8. R. B. Gupta, K. W. Sheperd, and F. MacRitchie, *J. Cereal Sci.*, 1991, **13**, 221.
9. D. R. Goforth, and K. F. Finney, *Cereal Chem.*, 1976, **53**, 608.
10. F. R. Huebner, and J. S. Wall, *Cereal Chem.*, 1976, **53**, 258.
11. R. C. Bottomley, H. F. Kearns, and J. D. Schofield, *J. Sci. Food Agric.*, 1982, **33**, 481.

12. G. Lundh, and F. MacRitchie, *J. Cereal Sci.*, 1989, **10**, 247.
13. T. Dachkevitch, and J. –C. Autran, *Cereal Chem.*, 1989, **66**, 448.
14. I. L. Batey, R. B. Gupta, and F. MacRitchie, *Cereal Chem.*, 1991, **68**, 207.
15. B. X. Fu, and H. D. Sapirstein, *Cereal Chem.*, 1996, **73**, 143.
16. B. X. Fu, and M. I. P. Kovacs, *J. Cereal Sci.*, 1999, **29**, 113.
17. J. –L. Carceller, and T. Aussenac, *Aust. J. Plant Physiol.*, 2000, (submitted).
18. M. Laurière, I. Bouchez, C. Doyen, and L. Eynard, *Electrophoresis*, 1996, **17**, 497.
19. B. A. Marchylo, D. W. Hatcher, and J. E. Kruger, *Cereal Chem.*, 1989, **65**, 28.

EXTRACTABILITY AND SIZE DISTRIBUTION STUDIES ON WHEAT PROTEINS USING FLOW-FIELD FLOW FRACTIONATION

L. Daqiq[1,2,3], O.R. Larroque[1,2], F.L. Stoddard[1,3] and F. Békés[1,2]

[1]Quality Wheat Cooperative Research Centre Ltd., Locked Bag 1345, North Ryde, NSW 1670, Australia. [2]CSIRO Plant Industry, Grain Quality Research Laboratory, PO Box 7, North Ryde, NSW 1670, Australia. [3]Plant Breeding Inst., Woolley Bldg. A20, The University of Sydney, NSW 2006, Australia.

1 INTRODUCTION

The structure-function relationships of wheat polymeric proteins, and hence the quality of the end products made from wheat flour, are affected by the size distribution of the proteins. Firmly establishing the relationship between the size of the protein and its functional properties has been hindered by the difficulty of completely extracting the polymer without alteration of its molecular size or shape. Many solvents have been tested, one of the most effective being 2% SDS in buffer[1]. Sonication improves the efficiency of extraction[2] but can lead to disruption of large polymers[3]. The first part of this study therefore investigated ways to optimise solubilisation of wheat polymeric proteins with the minimum possible changes from their native state. Once the protein has been solubilised, it is necessary to develop a method to evaluate its size and shape. Asymmetrical Flow – Field Flow Fractionation (FFF) has recently been adapted to this purpose[4,5]. This methodology is suitable for separating macromolecules and particles ranging in size from molecular weight of 10,000 to a diameter of 50 μm[6]. The second part of this study involved determining the molecular size of wheat polymeric proteins by using different sonication times.

2 MATERIALS AND METHODS

2.1 Flour samples

Samples of wheat flour of cultivars Cunningham, Vectis, Stiletto and Galahad were obtained from AWB Ltd., Werribee, Victoria. Protein and moisture content of the flours were determined by near-infrared spectroscopy.

2.2 Extraction of proteins

Four extraction procedures were compared (Table 1). There were two replicates and method 4 followed Gupta et al.[7]. In each case, 10 mg of flour was extracted with the first solvent, vortexed for 1 min, shaken for 15 min then centrifuged at 10,000 x *g* for 10 min.

The supernatant, containing extractable proteins (mostly monomeric), was passed through a 0.45 μm filter. The pellet was resuspended in the second solvent with a needle, sonicated in some cases (Table 1), centrifuged and filtered, to give a supernatant containing "unextractable" polymeric proteins. The pellet was resuspended in 0.5 % SDS in 0.05 M sodium phosphate buffer, pH 6.9, with a needle, sonicated for 5 seconds, centrifuged and filtered, to give "remaining" or "residual" polymeric proteins in the supernatant. The effect of varying the sonication time from 5 to 40 seconds in the second extraction step of methods 2 and 4 was further investigated using cvs Cunningham and Vectis.

2.3 Separation techniques

2.3.1 Size-Exclusion HPLC. Aliquots of 20 μL of each extract were injected into a Biosep SEC-4000 column (Phenomenex, Torrance, CA, USA) on a System Gold HPLC (Beckman Instruments Inc., Fullerton, CA, USA) and run for 10 min. The eluent buffer was acetonitrile : water (1:1) with 0.05 % (v/v) trifluoroacetic acid[8]. The areas of the peaks of monomeric and polymeric protein in the HPLC chromatogram were integrated. Extract 1 thus gave peak areas Monomeric 1 (M1) and Polymeric 1 (P1), with the total protein (T1) given by M1+P1. Similarly extract 2 gave M2, P2 and T2 and extract 3, M3, P3 and T3.

2.3.2 Flow – FFF. Aliquots of 20 μL of the second extract were injected into an asymmetrical flow – field flow fractionation system F–1000 (FFFractionation LLC, Salt Lake City, UT, USA) consisting of a theoretical channel of 0.127 mm width, regenerated cellulose membrane of 10,000 cut-off and a Waters 441 detector at 214 nm wavelength. The cross and channel flow rates were 2 mL/min and the sample relaxation time in the channel was 1 min 45 sec. The eluent buffers were 0.05 M sodium phosphate, pH 6.9, with 0.5 % SDS for samples solubilized by SDS and 0.05 M acetic acid containing 0.01 % Brij 35 for samples solubilized by acetic acid. The average retention time was calculated by integrating the area under the curve and determining the time at the midpoint of the area. This time was used as a measure of average molecular size.

2.3.3 SDS – polyacrylamide gel electrophoresis. Freeze-dried fractions were dissolved in an extraction buffer[8] (0.5625 M Tris-HCl, 2% w/v SDS, 15% glycerol, 0.0025% bromophenol blue) and loaded onto 10% polyacrylamide gels for electrophoresis at a constant voltage of 200-250 V for 4 h at 20°C. The gels were stained overnight with Coomassie Brilliant Blue G-250, destained with water and scanned[9].

2.4 Statistical analysis

The proportions of T3 to total protein in all three extracts (T1+T2+T3), P3 to P1+P2+P3 and P3 to T1+T2+T3 were calculated. The proportion of P2 to P1+P2 (equivalent to %UPP [7] when extracted with SDS) was also calculated. Data on retention time and these proportions were subjected to analysis of variance using Genstat 5 v 4.1 for Windows (Rothamsted Experimental Station, UK).

Table 1 *Methods for dissolving wheat proteins; method 4 follows Gupta et al.*[7]

Method	Solvent 1	Solvent 2
1	0.3 M NaI, 7.5 % propan-1-ol	50 % acetic acid, no sonication
2	0.3 M NaI, 7.5 % propan-1-ol	50 % acetic acid, sonication for 10 sec
3	50 % acetic acid	50 % acetic acid, sonication for 10 sec
4	0.5 % SDS in 0.05 M sodium phosphate buffer, pH 6.9	0.5 % SDS in 0.05 M sodium phosphate buffer, pH 6.9, sonication for 10 sec

Table 2 *Mean retention times and amounts of protein extracted following four extraction methods. Pn and Tn, polymeric protein or total protein in extract n, respectively*

Method	Average size (min. retention time)	P2/(P1+P2), %	T3/(T1+T2+T3), %	P3/(P1+P2+P3), %
1	8.45	61.8	14.7	34.1
2	8.29	78.6	2.89	5.14
3	8.82	44.6	2.08	3.76
4	6.69	48.2	2.13	3.85
SE	0.12	1.4	0.32	0.49

3 RESULTS AND DISCUSSION

The analysis of variance of the four measures of protein size and solubility showed that the extraction method was significant but the effect of cultivar and the cultivar x method interaction were not significant. The average size of the protein was largest, and the percentages of the polymeric protein fractions were lowest, with method 3 (Table 2). Method 2 gave the greatest content of polymeric protein in the second extraction. Methods 1 and 2 gave clear separation of monomeric and polymeric proteins[10] but method 1 left a substantial amount of protein in the residue after the second extraction. Method 2 was used further in this study.

Sonication time during the second extraction significantly affected both the amount and the FFF retention time of the extracted protein. Linear regression to zero sonication time gave expected retention times of 10.6 min for acetic acid extracts and 6.7 min for SDS extracts (Figure 1). This provided an estimate of the molecular size of native glutenin. Longer sonication times were associated with longer smears in each SDS-PAGE lane, confirming the decrease of molecular size. Ten seconds sonication extracted as much protein in acetic acid as longer extractions, but 20 seconds was needed in SDS (Figure 2a). The amount of protein remaining after the second extraction to be dissolved in the third also declined with increasing sonication time (Figure 2b). SDS extracted more of the total protein, but acetic acid extracted more of the polymeric protein and the two cultivars responded slightly but significantly differently to the two methods (Table 3).

Table 3 *Residual protein as a proportion of total protein*

Cultivar	Method	P3/(P1+P2+P3), %	T3/(T1+T2+T3), %
Cunningham	2	2.00	3.17
	4	3.77	2.10
Vectis	2	2.06	3.17
	4	2.23	1.25
SE		0.18	0.12

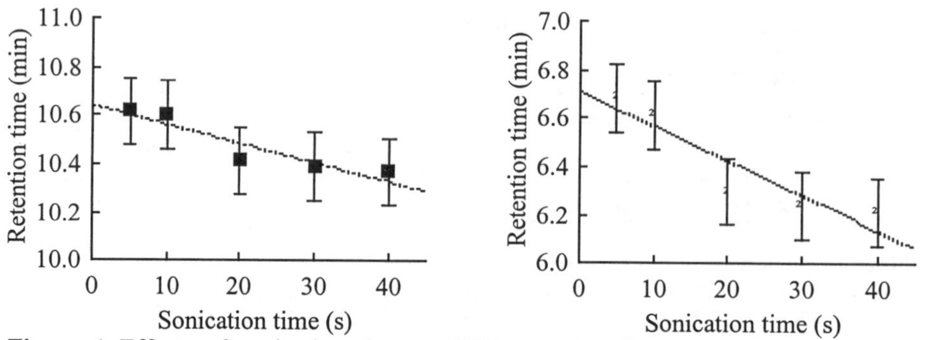

Figure 1 Effects of sonication time on FFF retention time (average molecular size) following extraction in (■) acetic acid or (Δ) SDS. Bars show ±1 standard error.

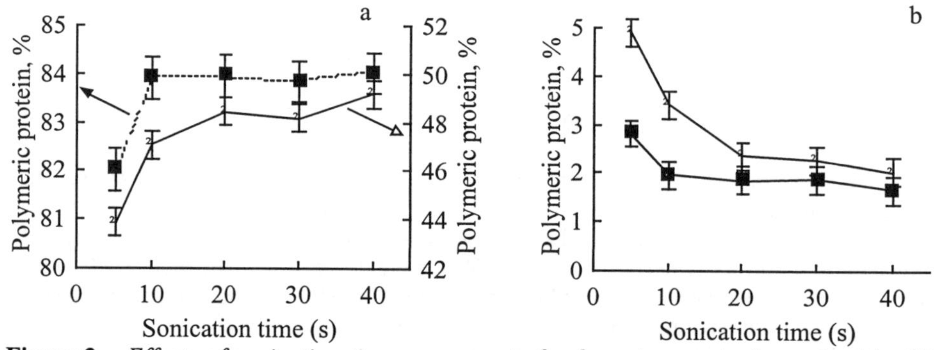

Figure 2 *Effects of sonication time on amount of polymeric protein determined by SE-HPLC in (a) the second extract (P2/(P1+P2), %) and (b) the third extract (P3/(P1+P2+P3), %), following extraction in (■) acetic acid or (Δ) SDS.*

Retention time measurements can be converted to molecular weights by calibrating the FFF with macromolecules of known size. There are few such standards available, but preliminary results indicate that DNA and dextran may be suitable.

4 CONCLUSION

Native size distribution of polymeric protein in wheat endosperm can be estimated using different times of sonication and extrapolating to zero. SDS and acetic acid affect the shape of the polymer differently, with acetic acid giving much larger apparent molecular sizes.

References

1. R. C. Bottomley, H. F. Kearns and J. D. Schofield, *J. Sci. Food Agric.*, 1982, **33**, 481.
2. N. K. Singh, G. R. Donovan, I. L. Batey and F. MacRitchie, *Cereal Chem.*, 1990, **67**, 150.
3. F. MacRitchie, *Cereal Foods World*, 1999, **44**, 188.

4. K. G. Wahlund, M. Gustavsson, F. MacRitchie, T. Nylander and L. Wannerberger, *J. Cereal Sci.*, 1996, **23**, 113.
5. S. Uthayakumaran, M. Southan, O. Larroque and F. Békés, *Proceedings of the 48th Australian Cereal Chemistry Conference*, 1998, 74.
6. J. C. Giddings, *Science*, 1993, **260**, 1456.
7. R. B. Gupta, K. Khan and F. MacRitchie, *J. Cereal Sci.*, 1993, **18**, 23.
8. I. L. Batey, R. B. Gupta and F. MacRitchie, *Cereal Chem.*, 1991, **68**, 207.
9. N. Neuhoff, N. Arold, D. Taube and W. Ehrhardt, *Electrophoresis*, 1988, **9**, 252.
10. B. X. Fu and P. I. M. Kovacs, *J. Cereal Sci.*, 1999, **29**, 113.

DURUM WHEAT GLUTENIN POLYMERS : A STUDY BASED ON EXTRACTABILITY AND SDS-PAGE

Andrea Curioni[1], Nadia D'Incecco[2], Norberto E. Pogna[3], Gabriella Pasini[2], Barbara Simonato[2] and Angelo D. B. Peruffo[1].

1. Dipartimento Scientifico e Tecnologico, Università di Verona, strada Le Grazie, I-35134 Verona (Italy). 2. Dipartimento di Biotecnologie Agrarie, Università di Padova. 3. Istituto sperimentale per la Cerealicoltura, Roma.

1 INTRODUCTION

Glutenin polymers are considered as the key factor in determining the unique properties of wheat flour[1]. The molecular characteristics of their component subunits have been the object of several studies, but direct evidence on the organisation of these subunits within the polymer structure is scant. Most of the proposed models have been based on studies performed on common (bread) wheat, in which the high molecular weight glutenin subunits (HMW-GS) seem to play the most important role in determining breadmaking quality. In contrast, durum wheat pasta-making quality seems to correlate better with the presence of specific low molecular weight glutenin subunits (LMW-GS)[2], indicating different roles played by the two types of subunit in determining the end-uses (bread or pasta) of these wheats. This difference should be related to the glutenin polymer composition and structure. Here we report a study performed on durum wheat semolina with the aim of explaining its suitability to be processed into pasta products on the basis of the structure of its glutenin polymers.

2 MATERIALS AND METHODS

Semolina from the durum wheat cv. Adamello was reduced to flour and extracted in SDS/phosphate buffer and SDS/phosphate buffer with sonication[3].

SE-HPLC was performed as described by Batey *et al.*[4]. Protein peaks to be analysed by SDS-PAGE were collected manually, dried under reduced pressure, and re-dissolved in SDS-PAGE sample buffer with or without 2-mercaptoethanol (2-ME).

SDS-PAGE was carried out in a Mini-protean II cell (Bio Rad) with a total polyacrylamide concentration of 11%. Gels were scanned with a Bio Rad Gel Doc 1000 and analysed with the Molecular Analyst software.

3 RESULTS

Semolina from the Italian durum wheat variety Adamello (HMW-GS: *Glu-B1* x-type 7 and *Glu-B1* y-type 8, LMW-GS 2) was extracted by SDS/phosphate buffer and proteins (extractable fraction, F1, 80% of the original protein) were fractionated by SE-HPLC. The excluded peak (pF1) contained "extractable" glutenin polymers of relatively small molecular size (MS)[3] (not shown). The residue was then sonicated, resulting in the extraction of polymerised protein [3] (fraction 2, F2, 9% of the original protein) (not shown). However, sonication did not give complete protein extraction, since 11% of the original semolina protein remained as a pellet, as assessed by nitrogen analysis. Upon addition of 2-ME to the SDS/phosphate buffer, the protein in this residue was solubilised (fraction 3, F3,), indicating that it consisted of polypeptides linked by disulphide bonds.
Assuming the percentage of polymers in F1 could be calculated as the area of the SE-HPLC excluded peak expressed as a percentage of the area of the whole chromatogram[3], that F2 consisted almost exclusively of polymeric proteins (not shown), and that the insoluble residue proteins (F3) were also polymeric, it could be calculated that polymeric proteins were distributed in proportions of 46% in F1, 24% in F2 and 30% in F3. The polymeric proteins in F1 (pF1), F2 and F3 were then analysed by SDS-PAGE after reduction (Figure 1).

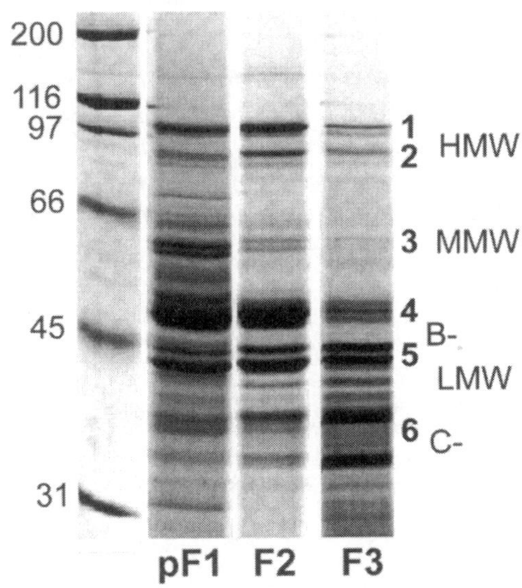

Figure 1 *SDS-PAGE analysis of the reduced polymers of the extractable fraction (pF1) and of the polymer fragments extractable (F2) and unextractable (F3) after sonication of the residue. Numbers on the right-hand side indicate the bands considered for densitometric quantification. HMW, MMW and LMW indicate the groups of high-, medium- and low-molecular weight subunits, respectively.* M_r *in k are indicated on the right side.*

The results obtained indicated that the three fractions comprised the same component polypeptides, although in different relative amounts. These amounts were quantified by densitometric analysis of the SDS-PAGE patterns of each fraction, calculating the areas of the peaks corresponding to HMW-GS 7 (x-type) (band 1) and 8 (y-type) (band 2), the two bands of M_r about 60,000 (MMW) (taken together, bands 3), the major component of the B-group of LMW-GS (band 4), the two minor components of the B-group of LMW-GS (bands 5) and the C-group of LMW-GS (bands 6) (Table 1). The following results were thus obtained:

1. The highest proportion of MMW bands was present in the extractable polymers (pF1). These bands correspond to bound beta-amylases, whose relative proportion is inversely related to the molecular size of the glutenin polymers [6].
2. The highest proportion of HMW-GS was present in the polymer fractions extracted by sonication (F2), and these fractions also had the highest x-type/y-type subunit ratio. Conversely, in the unextractable fractions (F3) this ratio was the lowest.
3. These latter fractions (F3) had the highest proportions of LMW-GS, and a strong variation was noted in the relative amounts of the different LMW-GS components (Table 1). In fact, in the (unextractable) fractions of higher mass the main component of the B-group LMW-GS (band 4) was present in a much lower proportion in comparison to both the extractable polymers (F1) and fragments extracted by sonication (F2), whereas the C-group LMW-GS (bands 6) increased sharply.

Table 1 *Area percentages determined by densitometry of SDS-PAGE analyses of the subunits present in polymers in the extractable fraction (F1) and in the polymer fragments extractable (F2) and unextractable (F3) after sonication of the residue. HMW-GS/LMW-GS (H/L), HMW x-type/HMW y-type (x/y) and B-group LMW-GS/C-group LMW-GS (B/C) ratios are also shown. x and y: x-type and y-type HMW-GS; MMW: medium molecular weight (M_r 60,000) bands; B-L and C-L: B- and C-group LMW-GS. See Figure 1 for. band numbers.*

	x band 1	y band 2	MMW Bands 3	B-L band 4	B-L bands 5	C-L bands 6	H/L	x/y	B /C
F1	8.8	4.5	10.2	38.6	24.0	14.0	0.11	1.9	4.5
F2	13.2	6.0	3.5	35.4	24.4	17.5	0.25	2.1	3.5
F3	6.1	4.6	2.6	15.3	30.5	40.9	0.12	1.4	1.1

4 DISCUSSION

Since it has been shown that sonication causes the fragmentation of the glutenin polymers, thus reducing their mass and allowing their extraction from the flour[5], we assume that both F2 and F3 were comprised of glutenin polymer fragments. It is, in fact, unlikely that the "unextractable" fraction F3 was composed of intact polymers of very high mass, since these latter are the first to be broken down by physical means[7]. Therefore, due to the relationship between mass and solubility, the "extractable" fragments of F2 should have lower masses compared to those of the "unextractable" fragments of F3. The higher mass of these latter fragments should be due to the presence of SS bonds, since complete solubilisation could be obtained after addition of a reducing agent. If the effect of sonication is the rupture of only some of the SS bonds[5] linking the polymer subunits, then it is likely that a linear polymer will be reduced in size quite easily by sonication. This fact allows us to hypothesise that the polymer fragments solubilised by sonication derive essentially from linear polymers or from linear portions of polymers comprising also branched zones that are spatially distinct within the polymer. In contrast, the presence of a relatively high frequency of branching in the polymer is likely to reduce the effect of sonication in reducing the MS, because of the limited number of covalent bonds broken. Therefore, the branched polymer, or the highly cross-linked portions of the polymer, will remain essentially unchanged with respect to its mass and will not be solubilised unless all the SS bonds have been completely split by chemical reduction.

Based on these assumptions, the results reported here would indicate that, in durum wheat semolina, the gluten polymers are not random structures, as previously suggested by others for bread wheat[8]. In particular, two main type of structural organisation would coexist: a more linear structure, in which the x-type HMW-GS and the main component of the B-group LMW-GS are preferentially represented and a more branched one, enriched in y-type HMW-GS and C-group LMW-GS. Indeed, it was suggested that both subunit Bx7 (and probably also the other Bx alleles) and the major component of the B-group of LMW-GS are "linear" subunits, having a molecular structure not allowing the formation of intermolecular branching (for a review see Kasarda[9]). In contrast, the y-type HMW-GS seems to be potentially able to also form branches, due to the presence of an additional cysteine towards the *C*-terminal end of the repetitive domain. On the other hand, from the results presented here it seems that components with the highest tendency to branch are the LMW-GS of type C. Literature data supporting this finding at the molecular level are lacking. However, it was demonstrated that, in bread wheat, the C-group of LMW-GS are more resistant to chemical depolymerisation than the B-group of the LMW-GS, suggesting that they have a higher degree of intermolecular bonding[8].

In conclusion, the data reported here confirm the importance of the LMW-GS in determining the quality characteristics of durum wheat. In fact, the typical tenacity of durum wheat dough and its lack of elasticity can be due to a relatively high degree of reticulation in its glutenin polymers, allowing the production of superior quality pasta.

References

1. F. MacRitchie, *Adv. Food Nutr. Res.*, 1992, **36**, 1
2. J.-C. Autran and G. Galterio, *J. Cereal Sci.*, 1989, **9**, 195
3. R.B. Gupta, K. Khan and F. MacRitchie, *J. Cereal Sci.*, 1993, **18**, 23
4. I. L. Batey, R. B. Gupta and F. MacRitchie, *Cereal Chem.*, 1991, **68**, 207
5. N.K. Singh, G.R. Donovan, I.L. Batey and F. MacRitchie, *Cereal Chem.*, 1990, **67**, 150
6. A. Curioni, N.E. Pogna and A.D.B. Peruffo, in: *Gluten '96,* ed. C.W. Wrigley, Royal Australian Chemical Institute, Melburne, 1996, p. 312
7. F. MacRitchie, *J. Pol. Sci.*, 1975, *Symposium No.* **49**, 85
8. M.P. Lindsay and J.H. Skerritt, *J. Agric. Food Chem.*, 1998, **46**, 3447
9. D.D. Kasarda, *Cereal Foods World*, 1999, **44**, 566

REACTIVITY OF ANTI-PEPTIDE ANTIBODIES WITH PROLAMINS FROM DIFFERENT CEREALS

S. Denery-Papini[1], M. Laurière[2], I. Bouchez[2], B. Boucherie[1], C. Larré[1], Y. Popineau[1].

1. INRA – Unité de Biochimie et Technologie des Protéines – BP 71627 – 44316 Nantes Cedex 03 – France. 2. INRA – Unité de Chimie Biologique – CBAI – 78850 Thiverval Grignon - France.

1 INTRODUCTION

Prolamins of the Triticeae (wheat, rye and barley) and possibly those of oats are responsible for coeliac disease in susceptible individuals. The only treatment of this disease is a diet free of toxic cereal prolamins. Commercial tests for the control of gluten-free food are based on the method developed by Skerritt and Hill[1] and use monoclonal antibodies directed against ω-gliadins that cross-react only weakly with barley prolamins. The objective of our work is to develop an ELISA allowing the detection of prolamins from wheat, barley and rye to the same extent. For this purpose we have studied the reactivity of polyclonal antibodies directed against three gliadin peptides (Table 1) towards prolamins from different cereals.

2 METHODS

We used polyclonal anti-peptide antibodies directed against *N*-terminal and *C*-terminal sequences and against a repetitive motif of $\alpha\beta$-gliadins (Table 1). Total proteins of bread wheat cv. Capitol, durum wheat cv. Brindur, spelt wheat cv. Hercule, einkorn cv. Aubaine Blanche, triticale cv. Dagro, rye cv Petkus, barley cv. Plaisant, oat cv. Gelald, rice cv. Ballila and maize cv. W64A were analysed using one and two-dimensional electrophoreses according to Laurière *et al.*[2]. Protein blotting and immunostaining were performed according to Laurière[3].

Table 1 *Anti-peptide antisera used in this study*

$\alpha\beta$-gliadin peptide	Antiserum	Sequence
N-terminal	Anti-NT-$\alpha\beta$	VRVPVPQGC
C-terminal	Anti-CT-$\alpha\beta$	GFGIFGTN
Repetitive	Anti-R-gliadin	PQQPYPQQPC

3 RESULTS

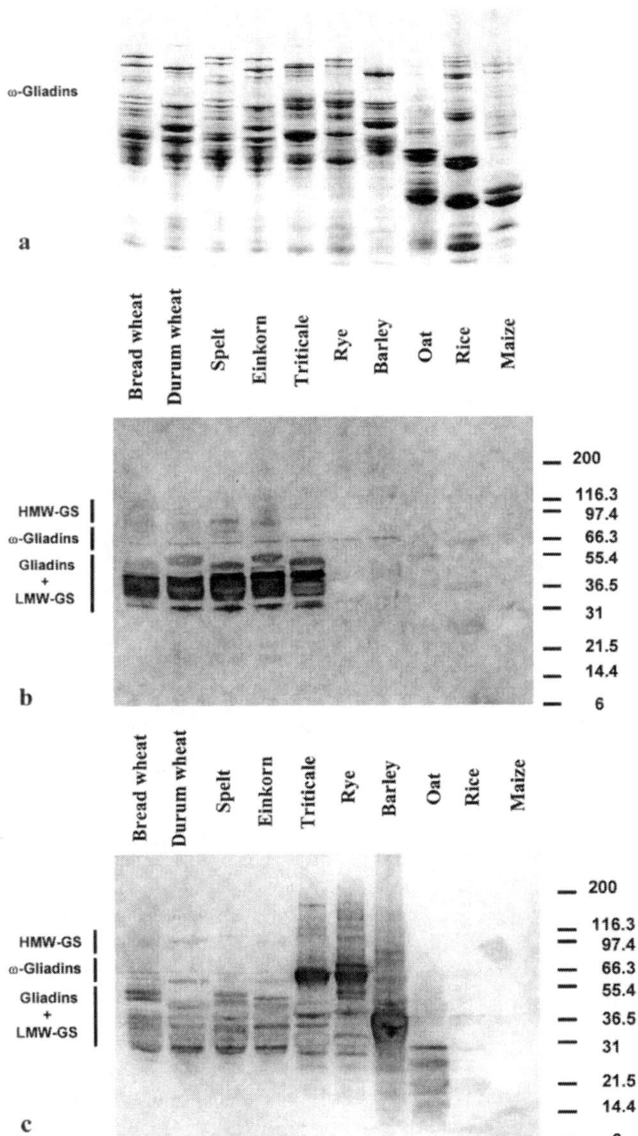

Figure 1 *Gradient SDS-PAGE of total proteins from different cereals a) Gel stained with Coomassie blue; b and c) Immunoblotting analysis using anti-NT-αβ and anti-R-gliadin antibodies, respectively.*

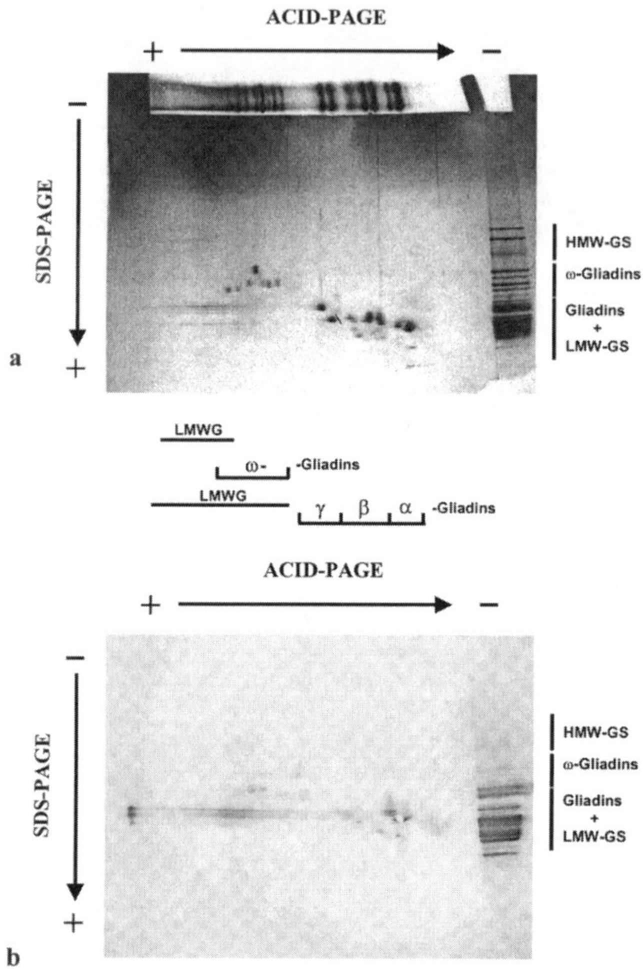

Figure 2 : *Acid-PAGE x SDS-PAGE 2D separation of wheat prolamins a) gel stained with Coomassie blue and b) immunoblotting analysis with anti-R-gliadin antibodies*

3.1 Reactivity of antibodies directed against terminal sequences

Anti-NT-αβ and anti-CT-αβ antibodies react specifically with αβ-gliadins of bread wheat[4]. Figure 1b show that anti-NT-αβ antibodies also cross-reacted with prolamins from all species of the genus *Triticum*, but not with those of other cereal genera. The same reactivity was observed for anti-CT-αβ antibodies (not shown). These results show that αβ-type prolamins are only expressed in wheat species and that the short *N*- or *C*-terminal peptides chosen are very specific for the αβ-gliadin type.

3.2 Reactivity of the antibodies directed against a repetitive peptide

Figure 1c shows that the anti-R-gliadin antiserum had a broader reactivity. It reacted strongly with prolamins from wheat, rye and barley and more weakly with some prolamins from oats. It did not cross-react with proteins from maize or rice. Two-dimensional electrophoresis of wheat prolamins (Figure 2) indicates that the antibodies reacted with some (but not all) ω, γ and αβ-gliadin spots, as well as with some low M_r glutenin subunits. In rye, they reacted strongly with ω and 75k γ-secalins and in barley with B, γ and some C-hordeins. In oats, they detected prolamins of low M_r. The cross-reactivity of anti-R-gliadin antiserum with the different prolamin classes (αβ, γ and ω-gliadins, low M_r glutenin subunits and homologous proteins in barley and rye) can be explained by sequence homologies (Table 2) between the repetitive domains of these proteins and the peptide used for immunisation.

Table 2 *Sequences homologous to the immunogen peptide (PQQPYPQQP) that can be found in wheat, barley and oat prolamins.*

Sequence	Prolamin class
PQQPYPQQP PQQPFPQQP PQQPYPQQQ PQQPYPQPQ PQQPFPQPQ PQQTFPQQP	Some γ-gliadins Some γ-gliadins Some α-gliadins Most α-gliadins Several γ-gliadins Some low M_r GS
PQQPFPQQP QPQPY/FPQQP	B, C and γ-hordeins B and C-hordeins
PFVQQQ/EQPFVQQQQ QPYPEQQEP	Avenins 5-8 Avenins 3, 5-10

4 CONCLUSION

According to the sequence selected for immunisation, anti-peptide antibodies provide various degrees of specificity in the detection of cereals. Antibodies directed against a repetitive peptide of gliadin recognised all the prolamins involved in coeliac disease and could be used for the control of gluten-free food.

References

1. J.H. Skerritt, and A.S. Hill, *J. Agric. Food Chem.,* 1990, **38,** 1771.
2. M. Laurière, I. Bouchez, C. Doyen, L. Eynard, *Electrophoresis* 1996, **17**, 497.
3. M. Laurière, *Anal. Biochem.*, 1993, **212**, 206.
4. S. Denery-Papini, J.P. Briand, L. Quillien, Y. Popineau and M.H.V. Van Regenmortel, *J. Cereal Sci.,* 1994, **20**, 1.

PURIFICATION OF Y-TYPE HMW-GS

Patacchini C., Masci S. and Lafiandra D.

Dipartimento di Agrobiologia e Agrochimica, Università degli Studi della Tuscia, Via S. Camillo de Lellis, 01100 Viterbo, Italy

1 INTRODUCTION

Wheat technological properties are correlated with glutenin polymers that consist mainly of high (HMW-GS) and low molecular weight subunits (LMW-GS), linked together by disulphide bonds.

HMW-GS are object of intensive studies because of their correlation with technological properties of bread wheat flours[1]. Previous studies have indicated a direct correlation between the size and the amount of the glutenin polymers and quality characteristics[2]. It has been found that higher molecular weight glutenin polymers contain a higher proportion of HMW-GS[3,4].

There are two types of high molecular weight subunits, termed x- and y-type subunits, in order of decreasing molecular weight. Whether bread-making quality is mainly influenced by x- or y-subunits or if both play an important role has not been ascertained. The importance of y-subunits, in particular of 1Dy10 subunit, has been stressed by Flavell and collaborators[5]. These authors attributed the good quality properties of wheat flours obtained from cultivars showing the allelic pairs 5+10 compared to those showing the pair 2+12, to the more regular organization of β-spirals in subunit 10, that might determine intrinsic elasticity of glutenin. However, there has not been any direct evidence yet that gluten is intrinsically elastic[1]. The possible differences between subunits 1Dx2 and 1Dx5 are more intuitive. In fact, the observation that subunit 1Dx5 has an extra-cysteine compared to 1Dx2, explains its capability to form higher molecular weight polymers[6,7]. Recently evidence has been reported that supports the hypothesis that both x- and y-subunits might be necessary for an efficient polymerisation process[8].

In order to study the different characteristics of x- and y-type HMW-GS in the polymerisation process and, consequently, in quality, *in vitro* re-oxidation experiments have been performed on combinations of different HMW-GS[9,10] or in micro-mixographic experiments[10]. For this kind of study it is desirable to have the possibility to separate efficiently x-subunits from y-subunits. Moreover, structural studies such as nuclear magnetic resonance (NMR)[11] or circular dicroism[12] may require larger amounts of purified protein than it is possible to obtain by conventional methods. HMW-GS can be separated individually by RP-HPLC[13-15], but most laboratories are equipped with analytical or semi-preparative systems that do not allow the purification of large amounts of protein.

The aim of this paper is to present a procedure that allows the purification, or at least the enrichment, of y-subunits using chemicals and equipment that are present in most biochemical laboratories, and to obtain single y-type HMW-GS from those wheat genotypes that possess only a single allelic pair, such as diploid wheats, some durum wheats, and particular hexaploid genotypes[16].

2 MATERIALS AND METHODS

Flours from the following bread wheat genotypes have been used: Cheyenne (HMW-GS composition 2*, 7+9, 5+10), Chinese Spring (7+8, 2+12), Yecora Rojo (1, 17+18, 5+10), W29323, showing six HMW-GS (21*+21*y, 14+15, 2+12)[17]; from durum wheat cultivars Langdon (6+8), Negridur (7+8), Flavio (7+8), Fenix (1, 7+8), Cosmodur (6+8); from two diploid wheat accessions of *Triticum urartu*, expressing both subunits Ax and Ay, and two *Aegilops squarrosa* accessions, the donor of the D genome of bread wheat, which contains Dx and Dy subunits. Moreover, we have analysed the Langdon disomic substitution line 1D(1B) (LDN1D(1B)), which contains only the allelic pair 1Dx2+1Dy12, and the bread wheat line WRU6979, containing only the allelic pair 1Dx5+1Dy10.

Since this is a methodological paper, details of the methods used are reported in the Results section. The starting material (bulked HMW-GS) was obtained from each genotype analysed by using the precipitation procedure reported in Marchylo et al.[13] performed on 50 mg of flour. Methods of purification included solubilisation of HMW-GS in different ampholyte solutions and pH.

3 RESULTS AND DISCUSSION

After precipitation, proteins were dissolved in a buffer containing different combinations of ampholytes, precipitated and eventually re-dissolved with different solvents. We observed that it was not possible to obtain pure x-subunits, because there was always a significant contamination with y-subunits. The procedure developed is reported in Scheme I.

Figure 1 reports the SDS-PAGE pattern relative to the purification of y-subunits in the material analysed. It is interesting to note that this procedure seems not applicable to the allelic HMW-GS pair 7+8. In fact, it was not possible to obtain any *Glu-B1* coded HMW-GS in a significant amount in the bread wheat cultivar Chinese Spring and in the durum wheat cultivars Flavio, Fenix and Negridur, after treatment with ampholytes. Because it was possible to obtain pure subunit 1By8 from the durum wheat cultivars Langdon and Cosmodur, both possessing the allelic pair 6+8, the different behaviour observed might be due to structural differences between subunits 1By8 belonging to the two allelic pairs, 6+8 and 7+8. Minor contamination of y-subunits with subunit 1Bx17 was observed in cultivar Yecora Rojo, but the amount of such contaminant subunit was variable, depending on the preparation.

Scheme I: *Procedure developed to obtain purified* y-*subunits*

Precipitate HMW-GS from flour (50 mg) according to Marchylo et al[13]

⇓

Dissolve precipitated HMW-GS o/n at RT in 500 μl of 50% (v/v) propan-1-ol containing 2M Urea + 2% (v/v) ampholytes pH 3.5-5 + 2% ampholytes pH 4-6

⇓

Precipitate proteins with 3 volumes of acetone for 10 min at RT. Recover pellet by centrifugation at 14,000 x*g* for 10 min at 20°C.

⇓

Wash pellet with 3 volumes of acetone for 5 min at RT. Recover pellet by centrifugation at 14,000 x*g* for 10 min at 20°C. Dry down the sample.

⇓

Dissolve the sample with 60 μl of 0.1M acetic acid by stirring for at least 2h at RT

⇓

Precipitate proteins with 4 volumes of acetone for 10 min at RT. Recover pellet by centrifugation at 14,000 x*g* for 10 min at 20°C. Dry down the sample

⇓

Dissolve again the sample with 60 μl of 0.1M acetic acid by stirring for at least 2h at RT. Centrifugate at 14,000 x*g* for 5 min at 20°C. Dry down or freeze-dry the supernatant, containing y-subunits

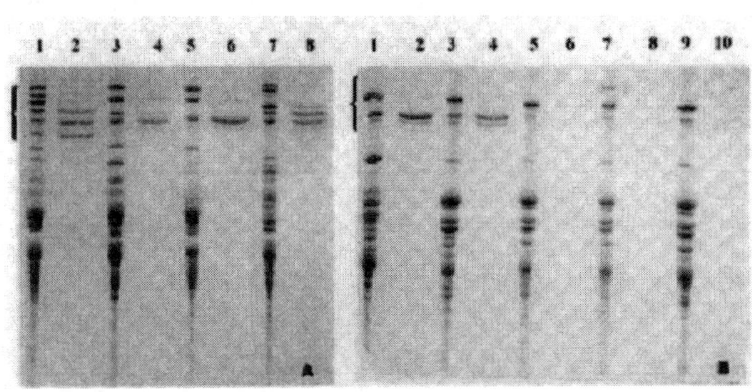

Figure 1: *(A) Hexaploid wheat line W29323 (1-2), cv. Chinese Spring (3-4), cv. Cheyenne (5-6), cv. Yecora Rojo (7-8); (B) durum wheat cultivars Langdon (1-2), Cosmodur (3-4), Flavio (5-6), Fenix (7-8), Negridur (9-10). Odd numbers: total protein patterns; even numbers: purified* y-*type subunits. The HMW-GS compositions of the different genotypes, in order of increasing mobility, are: 1A) 21*x, *2, 14, 15, 12, 21*y; *3A) 2, 7, 8, 12; 5A) 2*+5, 7, 9, 10; 7A) 1, 5, 17, 18, 10; 1B) 6, 8; 3B) 6, 8; 5B) 7, 8; 7B) 1, 7, 8; 9B) 7, 8. Brackets indicate HMW-GS.*

4 CONCLUSIONS

We have presented a preparative procedure aimed at obtaining bulked y-type HMW-GS, or single subunits in genotypes possessing only one HMW-GS for each type, such as particular genotypes, diploid or most durum wheats. Bound ampholytes can be easily removed by washing proteins with 15mM NaCl followed by extensive dialysis[18]. In this way such subunits are readily available to perform structural studies. This allows a better characterisation of such subunit types in relation to their role in the polymerisation processes and to their structural properties.

References

1. P.R. Shewry, N.G. Halford. and A.S. Tatham, *J. Cereal Sci.*, 1992, **15,** 105
2. T. Dachkevitch and J.C. Autran, *Cereal Chem.*, 1989, **66**, 448
3. F.R. Huebner and J.S. Wall, *Cereal Chem.*, 1976, **53**, 258
4. J.M. Field, P.R. Shewry and B.J. Miflin, J. *Sci. Food Agric.*, 1983, **34**, 370
5. R.B. Flavell, A.P. Goldsbrough, L.S. Robert, D Schnick and R.D. Thompson, *Bio/Tech.*, 1989, **7**, 1281
6. D. Lafiandra, R. D'Ovidio, E. Porceddu, B. Margiotta and G. Colaprico, *J. Cereal Sci.*, 1993, **18**, 197
7. R.B. Gupta and F. MacRitchie, *J. Cereal Sci.*, 1994, **19**, 19
8. Y. Shimoni, A.E. Blechl, O.D. Anderson and G. Galili, *J. Biol. Chem.*, 1997, **272**, 15488
9. P. Schropp, H.D. Belitz, W. Seilmeier and H Wieser, Cereal Chem., 1995, **72**, 406
10 D. Candler, C. Szabo, D. Murray and F. Bekes, in *Proceedings of the VI International Gluten Workshop,* 1996, p 133
11. P.S. Belton, I.J. Colquhoun, A. Grant, N. Wellner, J.M. Field, P.R. Shewry and A.S. Tatham, *Int. J. Biol. Macromol.*, 1995, **17**, 74
12. A.S. Tatham, A.F. Drake and P.R. Shewry, *J. Cereal Sci*, 1990, **11**, 189
13. B.A. Marchylo, J.E. Kruger and D.W. Hatcher, *J. Cereal Sci.,* 1989, **9**, 113
14. J.A. Bietz and D.G. Simpson, *J. Chrom.*, 1992, **624**, 53
15. B.N. Margiotta, G. Colaprico, R. D'Ovidio and D. Lafiandra, *J. Cereal Sci.*, 1993, **17**, 221
16. G.T. Lawrence, F. MacRitchie and C.W. Wrigley, *J. Cereal Sci*, 1988,. **7**, 109
17. B. Margiotta, M. Urbano, G. Colaprico, E. Johansson, F. Buonocore, R. D'Ovidio and D. Lafiandra, *J. Cereal Sci.*, 1996, **23**, 203
18. L.S. Rodkey, *J. Chrom.*, 1988, **437**, 147

Acknowledgments:

Research supported by the EU project-FAIR "Eurowheat", "Improving the quality of EU wheats for use in the food industry".

BIOCHEMICAL ANALYSIS OF ALCOHOL SOLUBLE POLYMERIC GLUTENINS, D-SUBUNITS AND OMEGA GLIADINS FROM WHEAT CV. CHINESE SPRING

Tsezi Egorov[1], Tanya Odintsova[2], Alexander Musolyamov[1], Arthur Tatham[3], Peter Shewry[3], Peter Hojrup[4] and Peter Roepstorff[4]

1. Engelhardt Institute of Molecular Biology, Russian Academy of Sciences, Vavilov str. 32, 117984 Moscow, Russian Federation. 2.Vavilov Institute of General Genetics, Russian Academy of Sciences, Gubkin str. 3, 117809 Moscow, Russian Federation. 3. IACR-Long Ashton Research Station, Department of Agricultural Sciences, University of Bristol, Long Ashton, Bristol, BS 18 9AF, UK. 4. Department of Molecular Biology, University of Odense, Campusvej 55, DK-5230 Odense M, Denmark

1 INTRODUCTION

The main storage proteins of bread wheat are the gluten proteins consisting of monomeric gliadins and polymeric glutenins. Of these the glutenins play an important role in dough visco-elastic properties. The glutenins include components considered to be modified gliadins such as the chromosome 1B and 1D-encoded D glutenin subunits.[1] It has been suggested that these may act as glutenin chain terminators.[2] Masci *et al.* isolated two of at least three D-type glutenin subunits, D_1 and D_2, and characterised them by *N*-terminal amino acid sequencing[3] and showed that they contain a single cysteine residue.[4]

The ω-gliadins have been studied less than some other groups of gluten proteins. It has been shown that they contain no cysteine residues[5] with three major *N*-terminal sequences types, called ARQ, KEL, and SRL.[6] No sequences of clones encoding ω-gliadins have so far been reported.

The aim of this work was to study in more detail the D-type glutenin subunits and related ω-gliadins from bread wheat cv. Chinese Spring. One new D_3 glutenin subunit, one ω-type protein and five ω-gliadins were identified by *N*-terminal amino acid sequencing and MALDI-TOF mass spectrometry.

2 MATERIALS AND METHODS

2.1 Preparation and separation of PSG

Flour (500 mg) was extracted with 2.5 ml of 50% propan-l-ol, 2% acetic acid at 37°C for 45 min on a model 5436 Eppendorf Thermomixer. The suspension was centrifuged in a Eppendorf microcentrifuge, and the supernatant was collected (50PS fraction). About 2 ml of the PS50 fraction were applied to the Sephacryl S-400 column (1.6 x 90 cm) equilibrated with 50% ethanol, 0.1% trifluoroacetic acid (TFA) at 30°C and eluted with the same solvent at a flow rate of 16 ml/h. Proteins were detected at 254 nm, and 4 ml-fractions were collected. 0.1 and 0.2 ml aliquots were used for SDS-PAGE analysis before and after reduction. Fractions of interest were dried in a SpeedVac concentrator and analyzed by RP-HPLC after reduction with 2% 2-mercaptoethanol in 0.1 M Tris-HCl buffer, pH 8.0, 5 M guanidinium chloride (Buffer A) at 60°C for 1 h. Protein samples were separated by an acetonitrile gradient from 23% B to 48% B in 120 min on an Aquapore RP-300 column (4.6 x 220 mm) at 50°C and a flow rate of 0.5 ml/h. Proteins were detected at 210 nm (5mm cell).

2.2 Isolation of ω-gliadins and D subunits

2.2.1 Method A:Isolation from propan-l-ol soluble fraction PS50. This method may be used to isolate ω-gliadins, D subunits and low molecular weight (LMW) subunits of glutenin. 100 mg flour was extracted with 50% propan-l-ol after Fu and Sapirstein,[7] with two fractions being collected, namely PS50 and PI50, consisting of gliadins and propan-l-ol soluble glutenins (PSG) and propan-l-ol insoluble glutenins (PIG), respectively. PSG were precipitated by increasing the concentration of propan-l-ol to 70%. The precipitate (PI70 fraction) consisting of PSG and ω-gliadins was partially dried, redissolved in 50% propan-l-ol, 1% DTT (Solvent A) and incubated at 60°C for 1 h. The concentration of propan-l-ol was then increased to 60% in order to precipitate high molecular weight (HMW) subunits of glutenin after Marchylo *et al.*[8] The supernatant (fraction PS60-1) was dried, redissolved in buffer B and subjected to RP-HPLC as described above after incubation at 60°C for 30 min.

2.2.2 Method B: Isolation from propan-l-ol insoluble fraction PI50. The same protocol was used as for method A with the exception that the PI50 fraction was additionally washed with 50% propan-l-ol (5 x 1 ml). As a result, the PS60-2 fraction containing D subunits and LMWs was obtained.

2.2.3 Covalent chromatography. The fraction containing the D_3 subunit and ω-type gliadins was also purified by covalent chromatography on Thiopropyl Sepharose 6B[5].

2.3 Analytical methods

N-Terminal amino acid sequence analysis was carried out on a model 816 protein sequencer (Knauer, Berlin). Mass spectra were obtained on a model Vision 2000 MALDI-TOF mass spectrometer (Thermo Biosystems). Protein samples were dissolved in 50% ethanol containing 2% acetic acid and applied to a target with sinapinic acid as a matrix by the dried-droplet method.

3 RESULTS AND DISCUSSION

We have developed a procedure for the isolation of D-type LMW subunits of glutenin by two procedures based on differential solubility of gluten proteins. Using method A they were isolated from the propan-l-ol soluble polymeric glutenins (PSG) together with ω-gliadins and LMW subunits. Using method B only D subunits and LMW subunits are isolated from propan-l-ol insoluble glutenins (PIG). Enriched preparations of D subunits and ω-gliadins (fraction PS60-1) and of D subunits (fraction PS60-2) obtained by methods A and B, respectively, were further separated by RP-HPLC (Figure 1). Using this purification procedure we have identified by *N*-terminal amino acid sequence, covalent chromatography and mass spectrometry several D subunits and ω-gliadins (Table 1). LMW subunits of glutenin were not studied. In addition to the D_1 and D_2 subunits first isolated by Masci *et. al.*[3] we isolated at least one new subunit (D_3). It is interesting to note that the proportions of all D subunits in fraction PS60-1 are similar. However the amount of subunit D_3 is approx. 3-5 times less in fraction PS60-2 (not shown) than the D_1 and D_2 subunits, which are present approx. in equal amounts.

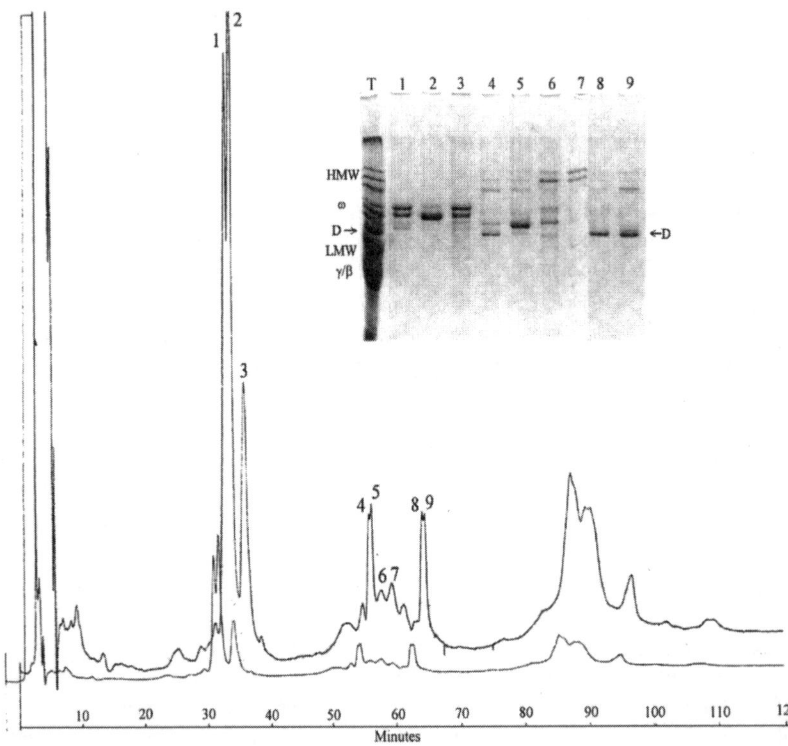

Figure 1 *RP-HPLC of PS60-1 fraction and SDS-PAGE of fraction 1-9*

Table 1 M_r *and* N-*terminal amino acid sequences of ω-gliadins and D subunits*

Peak No	M_r^{x3}	*N*-Terminal sequence	Number of cysteine res.	Type of protein	References
1	50.3, 55.5	SRL	No	ω	This paper
2	50.9	SRL + ARQ	No	ω	"
3	51.5, 56.3	SRL + ARQ	No	ω	"
4	41.7	ARP	1	D_3	"
5	42.8	KEL	?	ω or D	"
6	n.d.			HMW	"
7	n.d.			HMW	"
8	41.8	KEL	1	D_1	Masci *et. al.*[4]
9	40.9	ARQ	1	D_2	"

In addition, five ω-gliadins and one ω-type protein (peak 5) were characterized for the first time by mass spectrometry. The solubility of the ω-type protein (peak 5) differs from that of other ω-gliadins. It does not completely precipitate with 70% alcohol as do the other ω-gliadins and PSG isolated by method A. The presence of cysteine residue(s) in this component is still to be determined. The presence of additional ω-gliadins in peaks 1-6 cannot be excluded. Thus, at least two minor components, with electrophoretic mobilities similar to those of ω-gliadins are observed in peaks 1 and 3 and one in peak 6.

In addition, we investigated the composition of propan-l-ol soluble polymeric glutenins separated by size-exclusion chromatography (not shown). The fractions of interest were analyzed by SDS-PAGE before and after reduction as well as by RP-HPLC of reduced fractions. PSG obtained by flour extraction with 50% propan-l-ol in the presence of 2% acetic acid contained all monomeric gliadins and a small proportion of low molecular weight polymeric glutenins. Comparative analysis of fractions obtained by size-exclusion LC demonstrates a decrease in HMW subunits and an increase in D-type glutenin subunits. Preliminary results show that the proportion of D subunits is higher in PSG than in PIG, providing additional evidence for their role in glutenin chain termination.

References

1. Jackson, L. M. Holt, and P. I. Payne, *Theor. Appl. Genet.* 1983, **66**, 29.
2. S. Masci, E. Porchedu, G. Colaprico, and D. Lafiandra, *Biochem. Genet,* **29**, 403.
3. S. Masci, D. Lafiandra, E. Porcedu, E. J.-L. Lew, H. P. Tao, D. Kasarda, *Cereal Chem.* 1993, **70**, 581.
4. S. Masci, T. A. Egorov, C. Ronchi, D. D. Kuzminsky, D.D. Kasarda, D. Lafiandra, 1999, *J. Cereal Sci.* **29**, 17.
5. T. I. Odintsova, T. A. Egorov, A. A. Sozinov, *Biokhimia*, 1986, **51**, 1124.

6. D. D. Kasarda, J.-C. Autran, E.J.-L. Lew, C. C. Nimmo, and P. R. Shewry, *Biochim. Biophys. Acta,* 1983, **747**, 138.
7. B. X. Fu and H. D. Sapirstein, *Cereal Chem.*, 1996, **73**, 143.
8. B. Marchylo, J. Kruger and D. Hatcher, *J.Cereal Sci.*, 1989, **9,** 113.
9. T. A. Egorov, T. I. Odintsova, A. A. Sozinov, A.A., *Bioorganicheskaya khimiya,* 1986, **12**, 599.

Acknowledgements

This work was supported by grant No. 99-04-48417 from Russian Foundation for Basic Research (T. E.), BBSRC (P. S.) and Danish Biotechnology Programme (P. R.). IACR receives grant-aided support from the Biotechnology and Biological Sciences Research Council of the United Kingdom.

ISOLATION AND CHARACTERIZATION OF THE HMW GLUTENIN SUBUNITS 17 AND 18 AND D GLUTENIN SUBUNITS FROM WHEAT ISOGENIC LINE L88-31

Tanya Odintsova[1], Tsezi Egorov[2], Alexander Musolyamov[2], Arthur Tatham[3], Peter Shewry[3], Peter Hojrup[4] and Peter Roepstorff[4]

1. Vavilov Institute of General Genetics, Russian Academy of Sciences, Gubkin str. 3, 117809 Moscow, Russian Federation. 2. Engelhardt Institute of Molecular Biology, Russian Academy of Sciences, Vavilov str. 32, 117984 Moscow, Russian Federation. 3. IACR-Long Ashton Research Station, Department of Agricultural Sciences, University of Bristol, Long Ashton, Bristol, BS41 9AF, UK. 4. Department of Molecular Biology, University of Odense, Campusvej 55, DK-5230 Odense M, Denmark.

1 INTRODUCTION

Glutenins and gliadins are the major storage proteins of wheat, which play a major role in determining wheat technological properties. Numerous investigations have shown that glutenins have a greater impact on breadmaking quality of wheat than gliadins. The glutenin subunits are divided into two groups, high molecular weight (HMW) subunits and low molecular weight (LMW) subunits. Several lines of evidence have shown that the amounts and composition of the HMW subunits are particularly important in determining dough properties.[1]

The objective of this work was to isolate and characterize HMW subunits from the line L88-31, which contains only two HMW subunits, 17 and 18, which are allelic variants of subunits 7 and 9, respectively, and correlate with good quality characteristics. The gene for HMW subunit 17 has been isolated and sequenced,[2] and the amino acid sequence deduced from the nucleotide sequence. In addition, the D glutenin subunits and the propanol soluble glutenins (PSG) were also studied.

2 MATERIALS AND METHODS

HMW subunits 17 and 18 were isolated from *Triticum aestivum* L. line L88-31. The enriched HMW preparation was obtained according to Marchylo *et al.*[3] by extraction of wheat flour with 50% propan-l-ol and subsequent precipitation of HMW subunits in 60% propan-l-ol. The preparation was separated by RP-HPLC on an Aquapore RP-300 column (C-8, 300 nm, 4.6x220) with an acetonitrile gradient from 23 to 40% B for 1 hour at 50°C and a flow rate of 0.5 ml/min. *N*-terminal sequencing was carried out on a protein/peptide sequencer (Knauer, Berlin) equipped with a model 120A PTH-analyzer (Applied Biosystems). Selective isolation of cysteine-containing peptides was according to Egorov *et al.*[4] PSG was extracted with 50% propan-l-ol containing 2% acetic acid and separated by size-exclusion LC on a Sephacryl S-400 column (1.6 x 90 cm) in 50% ethanol containing 0.1% triflouroacetic acid at 30°C with a flow rate of 16 ml/h. Mass spectra were acquired on a Voyager Elite MALDI mass spectrometer (PerSeptive Biosystems). All samples were prepared using the dried-droplet or sandwich method with sinapinic acid as a matrix.

3 RESULTS AND DISCUSSION

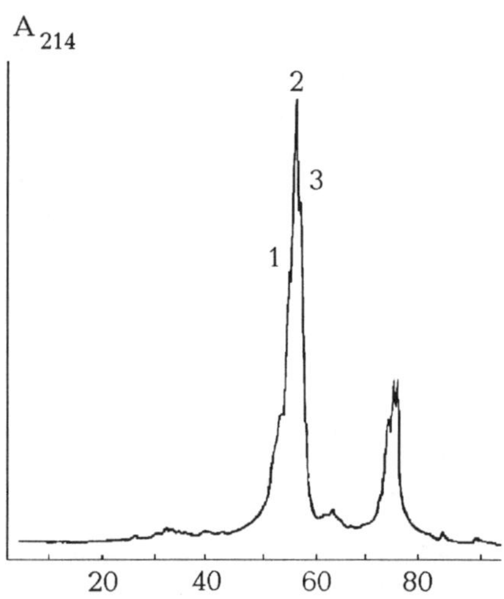

Figure 1 *RP-HPLC of the enriched HMW subunit preparation from wheat isogenic line L88-31.*

As shown in Fig. 1, HMW subunits 17 and 18 are poorly resolved by RP-HPLC. Changing the column type, acetonitrile gradient or alkylation of the subunits with 4-vinylpyridine did not improve the resolution. The subunits were still poorly resolved indicating strong interactions between them. The fractions obtained, designated 1, 2, and 3, were reduced, alkylated with 4-vinylpyridine and subjected to rechromatography on the same column. The purified fractions were characterized by SDS-PAGE, *N*-terminal sequence analysis and mass spectrometry. SDS-PAGE of the fractions showed three poorly resolved bands of different intensity corresponding to HMW subunits and one protein band of approximately 60 kDa. In fraction 1, HMW subunit 18 predominated and in fraction 2 HMW subunit 17, while fraction 3 contained only HMW subunit 17. The *N*-terminal amino acid sequences of these fractions were:

Fraction 1: EGEAS**R**

Fraction 2: EGEAS**G**

Fraction 3: blocked

More data were obtained by mass spectrometry with four proteins being clearly detected in each fraction (63, 71, 75 and 78 kDa) as well as several components of about 20 kDa. The nature of the 71 kDa protein is unknown. The 63 kDa protein probably corresponds to α-amylase, which co-precipitates with HMW subunits. The 78 kDa protein corresponds perfectly in mass to subunit 1Bx17 indicating that there is little or no post-translational modification of this protein. The *N*-terminal sequence is similar to that of subunit 1Bx17 (arginine residue at position 6). *N*-terminal sequencing of the 75 kDa protein showed that it was similar to subunit 1By9 (glycine at position 6), but was larger by 2 kDa. To characterize fraction 3 with a blocked *N*-terminus, we digested it with trypsin and separated the peptides obtained. The sequences of two peptides (peptide 1: DVXPGX, peptide 5: QXAGQXQX) showed that it was homologous to HMW subunit 17, however, with a blocked *N*-terminus.

The number of cysteine residues calculated as the difference between the masses of reduced and alkylated HMW subunits 17 and 18 showed the presence of 3.5 and 5.5 residues, respectively, which is close to the numbers of cysteine residues in subunits 1Bx17 (4 residues) and 1By9 (7 residues).

To determine the primary structure of HMW subunits 17 and 18 in more detail, we isolated cysteine-containing peptides by immobilization on Thiopropyl Sepharose 6B with subsequent RP-HPLC separation. Since it was impossible to isolate both subunits in a pure form on a preparative scale, we used the mixture of subunits. The major peptide fractions were sequenced (see Table 1). Two sequences were obtained for peptide fractions 4 and 7 but they could be clearly assigned to subunits 17 or 18. The results obtained indicate that the amino acid sequences around the cysteine residues in HMW subunits 17 and 18 are conserved and similar to those in subunits 1Bx7 and 1By9.

In addition, PSG was studied by size-exclusion LC on a Sephacryl S-400 column. The chromatographic pattern of PSG of L88-31 is similar to that of Chinese Spring (see paper of T. Egorov *et al.* in this volume). Analysis of fractions obtained by agarose gel electrophotesis[5] showed that they were separated according to size. SDS-PAGE of these fractions under reducing conditions demonstrated the presence of both HMW subunits and LMW subunits as well as D subunits. The mass spectra of chromatographic fractions showed the presence of proteins of 30-39 kDa, 41 kDa and 42 kDa, corresponding to the LMW subunits and D subunits respectively, a 63 kDa protein and HMW subunits (71, 75, and 78 kDa), demonstrating that all of these proteins are the constituents of PSG. A more detailed analysis of D subunits isolated by RP-HPLC showed the presence of a D_2 subunit with a typical *N*-terminal amino acid sequence ARQLN and a new D_3 subunit (ARPLN), which differed by an amino acid residue at position 3. Subunit D_1 (KELOS) was not identified.

Table 1 *The N-terminal amino acid sequences of Cys-containing tryptic peptides*

Fraction number	*N-terminal amino acid sequence*	*HMW homology*
4	^{27}DVSPGXXPI^{35}	1Bx7
	^{13}ELQE^{16}	1By9
6	^{540}QLGQ^{543}	1By9
7	^{747}AQQLAAQ^{753}	1Bx7
	^{661}VQQPAXQ^{667}	1By9
8	^{1}EGEA^{4}	1Bx7, 1By9

References

1. P.R. Shewry, N.G. Halford and A.S. Tatham, *J. Cereal Sci.*, 1992, **15**, 105.
2. P. Reddy and R. Appels, *Theor. Appl. Genet.*, 1993, **85,** 616.
3. B. Marchylo, J. Kruger and D. Hatcher, *J.Cereal Sci.*, 1989, **9,** 113.
4. T. A. Egorov, A. K. Musolyamov, J. S. Andersen and P. Roepstorff P, *Eur. J. Biochem.*, 1996, **224,** 631.

Acknowledgements

This work was supported by a grant No. 99-04-48417 from the Russian Foundation for Basic Research (T.E.), BBSRC (P.S.) and Danish Biotechnology Programme (P.R.). IACR receives grant-aided support from the Biotechnology and Biological Sciences Research Council of the United Kingdom.

VERIFICATION OF THE cDNA DEDUCED SEQUENCES OF GLUTENIN SUBUNITS BY MALDI-MS

S. Foti[1], R. Saletti[1], S.M. Gilbert[2], A. S. Tatham[2] and P. R. Shewry[2]

1. Dipartimento di Scienze Chimiche, Università degli Studi di Catania, Viale A. Doria 6, I-95125 Catania, Italy. 2. IACR-Long Ashton Research Station, Department of Agricultural Sciences, University of Bristol, Long Ashton, Bristol BS41 9AF, UK.

1 INTRODUCTION

The complete amino acid sequences of eight high molecular weight (HMW) glutenin subunits, derived from gene sequencing, are presently available.[1-4] The known sequences show that the M_r of these proteins are between 83,000 and 88,000 for x-type and 67,000 and 74,000 for y-type subunits. The proteins have a conserved structure consisting of a long central repetitive domain (about 640 to 830 amino acids) flanked by shorter *N*-terminal (81-105 residues) and *C*-terminal (42 residues) domains.[5]

Although DNA sequencing is the most efficient way to determine the amino acid sequences of large proteins, and in particular those that contain extensive repeated sequences, it does have some disadvantages. Firstly, it is easy to introduce errors, and a substantial proportion of the sequences in DNA databases are considered to have errors resulting from the sequence analysis or data handling. Secondly, it does not provide any information on post-translational modifications. Direct verification of the gene deduced amino acid sequence is therefore desirable.

In recent years, the development of "soft" desorption/ionisation methods of mass spectrometry (MS), such as electrospray ionisation (ESI) and matrix-assisted laser desorption/ionisation (MALDI), have provided powerful tools for protein characterisation.[6,7] In particular, MALDI-MS was used successfully to determine the molecular weight of several purified HMW subunits,[8] and also for the rapid and sensitive identification of gliadins in food.[9-11] ESI-MS has been used for mapping the disulphide bonds in gliadins[12] but is less suitable than MALDI-MS for gluten proteins which have low contents of charged residues.

We have reported[13] the sequence verification of the reduced and S-pyridylethylated HMW glutenin subunit 1Dx5 and a repetitive M_r 58,000 (58K) peptide (based on residues 102 to 643 of subunit 1Dx5 and expressed in *E. coli*) by combined use of tryptic digestion, high-performance liquid chromatography and MALDI-MS analysis. By this method about 80% of the sequence of 1Dx5 was confirmed, covering residues 1 to 669 with the exception of two short peptides. There was also good agreement with the sequence of the 58K peptide in the region of identity. The results also indicated the absence of substantial levels of post-translational modification. This investigation has now been extended to subunits 1Dx2, 1Dy10 and 1Dy12.

2 EXPERIMENTAL

2.1 Materials

Dithiothreitol (DTT), 4-vinylpyridine (VP), L-1-tosylamide-2-phenylethyl chloromethyl ketone (TPCK) treated trypsin from bovine pancreas, ammonium acetate and calcium chloride were purchased from Sigma (Milan, Italy); trifluoracetic acid (TFA) was obtained from Aldrich (Milan, Italy). High-performance liquid chromatography grade H_2O and CH_3CN were provided by Lab-Scan (Dublin, Ireland).

2.2 Tryptic Cleavage

The reduced and S-pyridylethylated subunits were dissolved to a final concentration of 1 mg mL^{-1} in 20 mM ammonium acetate, pH 8.3, containing 1mM calcium chloride. Trypsin, dissolved in the same buffer, was added to the protein at a molar enzyme/substrate ratio of 1:50 and the solution was incubated at 37 °C for 4 h. The digestion was stopped by cooling in liquid nitrogen and the mixture was immediately freeze-dried.

2.3 HPLC Separation of the Tryptic Digests

The freeze-dried tryptic peptide mixtures were dissolved in aqueous 0.05% (v/v) TFA, filtered on Millipore Ultrafree-MC and loaded onto a reverse-phase Vydac C18 column (0.46 x 25 cm, 300 Å, 5 μ). Tryptic fragments of the x-type subunits were eluted at room temperature from the column with 95% (v/v) of solvent A [H_2O + 0.05% (v/v) TFA] and 5% (v/v) of solvent B [CH_3CN + 0.05% (v/v) TFA] for 5 min and then with a linear gradient of solvent B in A from 5% to 47% (v/v) over 55 min. Fractionation of the tryptic digests of the y-type subunits was achieved using the same isocratic conditions, but with a linear gradient of 5 to 28 % (v/v) solvent B in A over 55 min. Peaks were detected by their absorption at 224 nm, collected manually and freeze-dried.

2.4 Mass Spectrometry

MALDI mass spectra were acquired on a Voyager Elite-DE Time-of-Flight Mass spectrometer (PerSeptive Biosystems Inc., Framingham, USA) equipped with a UV nitrogen laser (337 nm) operated in linear mode. Freeze-dried HPLC fractions were dissolved in 50% (v/v) CH_3CN, 0.1% (v/v) aqueous TFA to provide a final concentration of about 10 pmol μL^{-1}. Sample preparation was according to the dried droplet or the sandwich method,[14] using α-cyano-4-hydroxy cinnamic or sinapinic acid as matrix. Spectra were obtained in positive mode at an acceleration voltage of 20 kV and a delay time of 250 ns. Spectra from about 250 laser shots were averaged to improve the signal-to-noise level. Mass assignment was made using insulin (5,733.6 Da), trypsinogen (23,981.1 Da), lysozyme (14,305.1 Da) and cytochrome *c* (12,360.9 Da) as external standards.

3 RESULTS AND DISCUSSION

Subunit 1Dx2 shows high homology with 1Dx5 (over 98% at the deduced protein sequence level)[15] and therefore most of the tryptic fragments from both subunits are identical. Identification by MALDI-MS of the separated fragments from tryptic digestion of 1Dx2 confirmed the gene deduced sequence from residues 1 to 658, with the exception of fragments T3 (3 residues) and T7 (10 residues), which were not detected, and with the inclusion of an additional Pro residue in position 59 (Table 1). The only peptide detected in the *C*-terminal region was T13, suggesting possible errors or modifications in this part of the sequence. The results closely resemble those from subunit 1Dx5.

MALDI analysis of the tryptic HPLC fractions of subunit 1Dy10 resulted in the identification of almost all the predicted fragments, except for four short peptides (T1, T7, T8 and T21) and for fragment T16 (Table 1). For the latter, the experimentally determined M_r was 40 mass units higher than the calculated average M_r, indicating a possible sequence error in this region. Analogously, the cDNA deduced sequence of subunit 1Dy12, which has high sequence homology with subunit 1Dy10, was verified but the absence of residues Gly455-Gln456 was demonstrated (Table 1).

Table 1. *MALDI-MS identification of the tryptic fragments of subunits 1Dx2, 1Dy10 and 1Dy12*

Fragment	1Dx2		1Dy10		1Dy12	
	M_r^a	Identif.	M_r^a	Identif.	M_r^a	Identif.
T1	1482.6^b	+	619.3^b	-	619.3^b	-
T2	1043.5^b	+	880.4^b	+	880.4^b	+
T3	388.2^b	-	1368.6^b	+	1368.6^b	+
T4	1423.7^b	+	1112.6^b	+	1112.6^b	+
T5	3021.5*	+	1187.6^b	+	1187.6^b	+
T6	1991.1	+	959.4^b	+	959.4^b	+
T7	1156.7^b	-	518.3^b	-	518.3^b	-
T8	9912.3 us T8+9	+	382.2^b	-	382.2^b	-
T9	9756.3	+	915.5^b	+	915.5^b	+
T10	27760.0	+	1187.6^b	+	1187.6^b	+
T11	37001.0 us T11+12 ?	+	4221.6	+	4251.6	-
T12	37001.0 us T11+12 ?	+	2288.4	+	2288.4	+
T13	1504.8^b	+	3118.4	+	3118.4	+
T14	1046.5^b	-	3454.7	+	203.1^b	-
T15			4191.5	+	3865.1	+
T16			14601.3 modified?	+	19128.1	+
T17			8587.0	+	8586.9**	+
T18			1417.7^b	+	14069.7	+
T19			14941.6	+	2364.6	-
T20			1516.8^b	+	1516.8^b	+
T21			1064.4^b	-	1064.4^b	-

^{a} Calculated average or monoisotopic^b mass

*Considering the presence of Pro 59

**Considering the absence of the sequence Gly455-Gln456

Finally, none of the subunits investigated showed substantial levels of glycosylation or other post-translational modifications in the regions covered by the sequences.

References

1. P. R. Shewry, N. G. Halford and A. S. Tatham, *J. Cereal Sci.*, 1992, **15**, 105.
2. W. Seilmeier, H.-D. Belitz and H. Wieser, *Z. Lebensm. Unters. Forsch.*, 1991, **192,** 124.
3. N. G. Halford, J. M. Field, H. Blair, P. Urwin, K. Moore, L. Robert, R. Thompson, R. B. Flavell, A. S. Tatham and P. R. Shewry, *Theor. Appl. Genet.*, 1992, **83**, 373.
4. P. Reddy and R. Appels, *Theor. Appl. Genet.*, 1993, **85**, 616.
5. A. S. Tatham, P. R. Shewry and P. S. Belton, in *Adv. Cereal Sci. Technol.*, ed. Y. Pomeranz, AACC, St Paul, MN, 1990, vol. 10, p. 1.
6. R. D. Smith, J. A. Loo, R. R. Ogorzalek, M. Busman and H. R. Udseth, *Mass Spectrom. Rev.,* 1991, **10**, 359.
7. U. Bahr, M. Karas and F. Hillenkamp, *Fresen. J. Anal. Chem.*, 1994, **348,** 783.
8. D.R. Hickman, P. Roepstorff, P.R. Shewry and A.S. Tatham, *J. Cereal Sci.*, 1995, **22**, 99
9. E. Méndez, E. Camafeita, J.S. Sebastián, Y. Valle, J. Solís, F.J. Mayer-Posner, D. Suckau, C. Marfisi and F. Soriano, *J. Mass Spectrom. Rapid Commun. Mass Spectrom.,*1995, S123.
10. E. Camafeita, P. Alfonso, B. Acevedo and E. Méndez, *J. Mass Spectrom.*, 1997, **32**, 444.
11. E. Camafeita, J. Solís, P. Alfonso, J.A. Lopez, L. Sorell and E. Méndez, *J. Chromatogr. A*, 1998, **823**, 299.
12. T.A. Egorov, A.Kh. Musolyamov, S.F. Barbashov, O. Zolotykh, J. Andersen, P. Roepsdorff and Y. Popineau, in *Proceedings of the International Meeting on Wheat Kernel Proteins, Molecular and Functional Aspects*, ed. S. Martino al Cimino, Viterbo (Italy), 1994, p. 41.
13. S. Foti, G. Maccarrone, R. Saletti, P. Roepstorff, S. Gilbert, A.S. Tatham and P.R. Shewry, *J. Cereal Sci.*, 2000, **31**, 173.
14. M. Kussmann, E. Nordhoff, H. Rahbek-Nielsen, S. Haebel, M. Rossel-Larsen, L. Jakobsen, J. Gobom, E. Mirgorodskays, A. Kroll-Kistensen, L. Palm and P. Roepstorff, *J. Mass Spectrom.*, 1997, **32**, 593.
15. F.C. Greene, O.D. Anderson, R.D. Yip, N.G. Halford, J.M. Malpica Romero and P.R. Shewry, in *Proceedings of the 7th International Wheat Genetics Symposium.*, eds. T.E. Miller and R.M.D. Koebner, IPSR, Cambridge, 1988, p.735.

Acknowledgements

This work was supported by EU FAIR Project CT96-1170 (Improving the Quality of EU Wheats for use in the Food Industry, "EUROWHEAT"). Mass spectra were acquired on instruments of the "Rete di Spettrometria di Massa del CNR".

DEVELOPMENT OF A NOVEL CLONING STRATEGY TO INVESTIGATE THE REPETITIVE DOMAIN OF HMW GLUTENIN SUBUNITS.

K.A. Feeney, N.G. Halford, A.S. Tatham, P.R. Shewry and S.M. Gilbert.

IACR-Long Ashton Research Station, Department of Agricultural Sciences, University of Bristol, Bristol BS41 9AF.

1 INTRODUCTION

There has been considerable interest for some years in the high molecular weight (HMW) subunits of glutenin, principally as a consequence of the work of Payne and others[1,2] who correlated differences in allelic composition with breadmaking performance. Despite detailed studies of the HMW glutenin subunits, little is known about the molecular basis for the differences in quality associated with the various alleles. For example, although 1Dx5 is associated with good quality and 1Dx2 is associated with poor quality,[3] they are more than 98% similar at the protein sequence level.[4]

A key feature of the HMW glutenin subunits is the quantitatively dominant central repetitive domain. It has been proposed that the repeat motifs in this domain contribute to the formation of a β-spiral supersecondary structure[5], evidence from hydrodynamic studies, scanning tunnelling microscopy and CD spectroscopy, together with structure prediction data, support this hypothesis.[5-9] It has also been suggested that some quality differences are attributable to variation in this central repetitive domain.[10] However, detailed analysis of the structural and functional properties of the repetitive domain are limited by the presence of flanking non-repetitive sequences in the whole HMW subunit protein. In order to overcome this we have recently described the expression of an M_r 58,000 peptide derived from the repetitive domain of subunit 1Dx5.[8] In the present paper we extend this work by describing a novel cloning strategy which has allowed us to express and purify "perfect repeat" peptides of varying length. This will allow detailed studies of the relationships between repeat domain length, peptide motif sequence and functional properties to be explored.

2 MATERIALS AND METHODS

2.1 Expression and purification of the M_r 58,000 and perfect repeat peptides

A 1591 bp *Hin*dIII/*Nco*I fragment from the *Glu-1Dx5* gene, encoding an M_r 58,000 peptide corresponding to residues Ser[102] to Thr[643], had previously been cloned into a pET17b expression vector.[11] This construct was transferred to BLR(DE3)pLysS , a *rec*A$^-$

host strain derived from BL21(DE3)pLysS. It was anticipated that this would improve yields of the expressed peptide as a consequence of the increased stability of tandem repeats in this strain. Cells were grown in batch culture in 2YT medium and induced with IPTG according to established procedures prior to harvest. The expressed peptide was extracted with 70% ($^v/_v$) ethanol at 60°C and purified to homogeneity using CMC ion-exchange chromatography and gel-permeation.[12] Expression and extraction of the perfect repeat peptides were done in a similar manner. However, precipitation of these peptides was achieved with acetone and final purification accomplished using HPLC.

3 RESULTS AND DISCUSSION

A novel cloning strategy has been developed to provide a range of perfect repeat peptides of various sizes (Figure 1). Pairs of complementary oligonucleotides were heat-denatured and annealed together to form sets of double stranded DNA sequences called 'Linker' and 'Repeat' sequences. These DNA sequences were specifically ligated together to form the genes encoding the perfect repeat peptides. Although different linkers were produced, attention was focused on the one shown in Figure 1. The Cys residues either side of the internal *Pst*I site were included to allow investigation of the effects of inter-chain disulphide bond formation on peptide structure.

The DNA sequences have been designed so that the insertion of a 'Repeat' sequence into the internal *Pst*I site of the 'Linker' (or another 'Repeat' sequence) does not regenerate the *Pst*I site at either end of the sequence. Although the ends of the 'Repeat' sequence are compatible with the "sticky ends" generated by *Pst*I digestion, the sequences produced at each end of the 'Repeat' sequence after insertion are CTGCAA and TTGCAG and not the CTGCAG sequence which is recognised by *Pst*I. This means that the internal *Pst*I in the newly inserted sequence is unique, allowing it to be used as the site

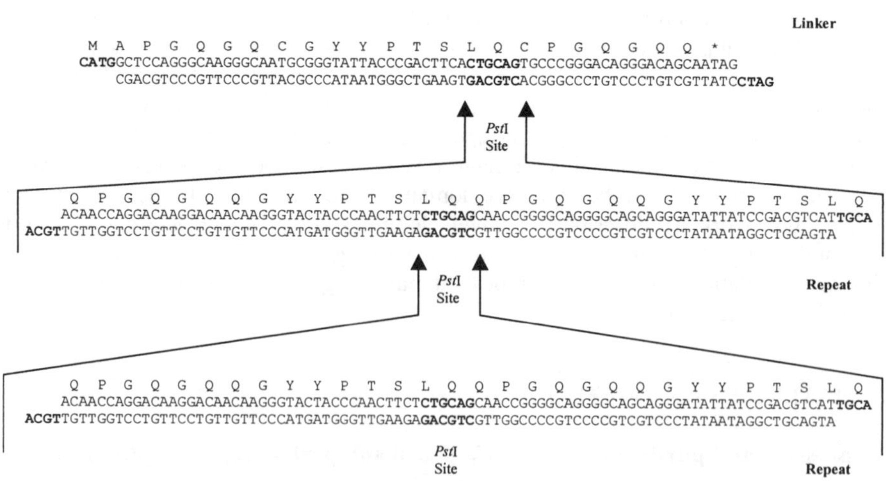

Figure 1, *Insertion of a new 'Repeat' sequence into the PstI cut site of a previously ligated 'Repeat' sequence.*

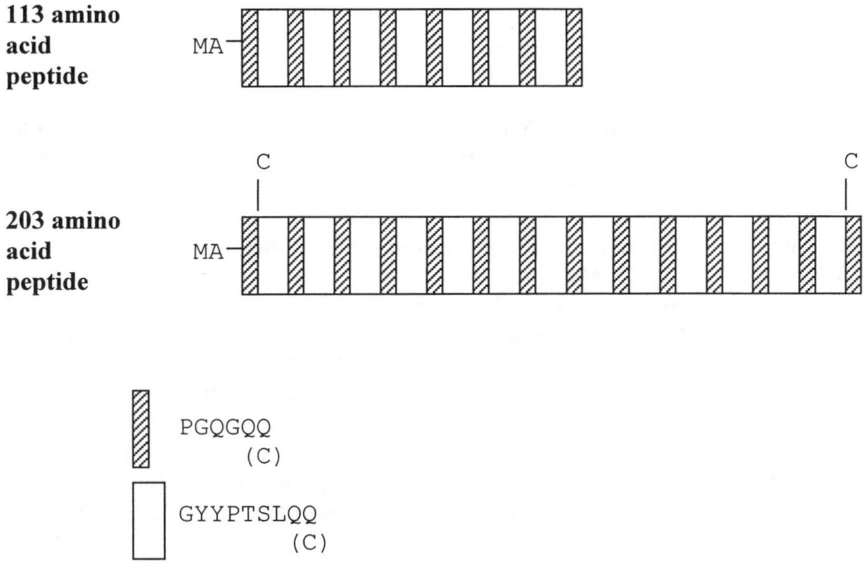

Figure 2, *Diagram of the repetitive nature of the positioning of the hexapeptide and nonapeptide motifs of the peptides*

for insertion for another 'Repeat' sequence. This process can be repeated, each time adding another block of 'Repeat' sequences, to create a family of genes that encode peptides with the same amino acid sequence motifs but varying in size (Figure 2).

Two perfect repeat peptides were purified for characterisation. The first peptide contains 113 amino acids, based on a 'Linker' plus three inserted 'Repeat' sequences. This peptide has eight hexapeptide and seven nonapeptide motif sequences in total. The second peptide contains 203 amino acids, based on a 'Linker' plus six inserted 'Repeat' sequences and has fourteen hexapeptide and thirteen nonapeptide motif sequences in total.

Initial characterisation of the perfect repeat peptides using far-UV CD spectroscopy showed spectra similar to those reported previously for random coil[13] when the peptides were in aqueous solution, with some structuring evident when dissolved in TFE (data not shown). Similar results were observed with the M_r 58,000 peptide, although in this case more extensive investigations at low temperature revealed an isodichroic point indicating that there were two structures in equilibrium with each other. At room temperature, the signals from these structures (polyproline II and type I/III β-turns) effectively cancelled each other to result in an apparently random coil signature.[8] Temperature studies of the largest perfect repeat peptide suggest that a similar equilibrium occurred with the peptide (data not shown). The perfect repeat peptides produced using the cloning strategy described above, together with the 'wild type' M_r 58,000 peptide, will allow us to determine how the 3D structures, interactions and biochemical properties of the peptides are affected by length and by differences in peptide repeat motif.

References

1. P.I. Payne, P.A. Harris, C.N. Law, L.M. Holt and J.A. Blackman, *Ann. Technol. Agric.,* 1980, **29**, 309.
2. P.I. Payne, *Annu. Rev. Plant Physiol.*, 1987, **38**, 141.
3. P.I. Payne, K.G. Corfield, L.M. Holt and J.A. Blackman, *J. Sci. Food Agric.,* 1981, **32,** 51.
4. F.C. Greene, O.D. Anderson, R.D. Yip, N.G. Halford, J-M. Malpica Romero and P.R. Shewry, in *Proceedings of the 7th International Wheat Genetics Symposium*, eds. T.E. Miller and R.M.D. Koebner, IPSR, Cambridge, 1988, p.735.
5. J.M. Field, A.S. Tatham and P.R. Shewry, *Biochem. J.,* 1987, **247**, 215.
6. A.S. Tatham, B.J. Miflin and P.R. Shewry, *Cereal Chem.,* 1985, **62**, 405.
7. M.J. Miles, H.J. Carr, T.C. McMaster, K.J. I'Anson, P.S. Belton, V.J. Morris, J.M. Field, P.R. Shewry and A.S. Tatham, *Proc. Natl. Acad. Sci. USA,* 1991, **88**, 68.
8. S.M. Gilbert, N. Wellner, P.S. Belton, J.A. Greenfield, G. Siligardi, P.R. Shewry and A.S. Tatham, *Biochim. Biophys. Acta., ,*2000, **1479**, 135-146.
9. O. Parchment, A.S. Tatham, P.R. Shewry and D.J. Osguthorpe. 2000. Submitted.
10. A.P. Goldsborough, N.J. Bulleid, R.B. Freedman and R.B. Flavell, *Biochem. J.,* 1989, **263**, 837.
11. F. Buonocore, L. Bertini, C. Ronchi, F. Bekes, C. Caporale, D. Lafiandra, P. Gras, A.S. Tatham, J.A. Greenfield, N.G. Halford and P.R. Shewry, *J. Cereal Sci.,* 1998, **27**, 209.
12. F. Castelli, S.M. Gilbert, S. Caruso, D.E. Maccarone and S. Fisichella, *Thermochim. Acta.* In Press.
13. N. Greenfield and G.D. Fasman, *Biochemistry*, 1969, **8**, 4108.

Acknowledgements

IACR receives grant-aided support from the Biotechnology and Biological Sciences Research Council of the United Kingdom. This work was supported in part by EU FAIR project CT96-1170 (Improving the Quality of EU Wheats for use in the Food Industry, 'EUROWHEAT').

MOLECULAR STRUCTURES AND INTERACTIONS OF REPETITIVE PEPTIDES BASED ON HMW SUBUNIT 1Dx5

N. Wellner[1], S. Gilbert[2], K. Feeney[2], A.S. Tatham[2], P.R. Shewry[2] and P.S. Belton[1]

1. Institute of Food Research, Norwich Research Park, Colney, Norwich NR4 7UA, UK.
2. IACR-Long Ashton Research Station, Department of Agricultural Sciences, University of Bristol, Long Ashton, Bristol BS41 9AF, UK

1 INTRODUCTION

A characteristic feature of high molecular weight (HMW) subunits, as with many gluten proteins, is their large repetitive central domain. This consists of tandem and interspersed repeats based on two or three consensus motifs: a nonapeptide (Gly.Tyr.Tyr.Pro.Thr.Ser.Pro/Leu.Gln.Gln), a hexapeptide (Pro.Gly.Gln.Gly.Gln.Gln) and, in the x-type subunits only, a tripeptide (Gly.Gln.Gln).

Secondary structure prediction and spectroscopic studies of purified HMW subunits and short synthetic peptides based on the consensus repeat motifs have indicated that in solution the repetitive domain forms a loose spiral consisting of regularly repeated β-reverse turns[1-3]. In the solid state numerous non-covalent intermolecular interactions exist. We have used Fourier-transform infrared spectroscopy to study perfect repeat peptides with different lengths in order to understand how this affects the molecular structures.

2 MATERIALS AND METHODS

Samples:

Synthetic and expressed peptides based on the consensus repeat, with 21, 45, 110 and 203 amino acid residues length, as well as a 58 kDa peptide based on the repetitive domain of 1Dx5, were prepared as described by Feeney *et. al.*,in this volume.

All samples were stored in a desiccator over P_2O_5 at ambient temperature. For the experiments, the freeze-dried powders were mixed in distilled water to a nominal concentration of 10 mg/ml (6mg/ml for 2.3 kDa peptide). Only the 2.3 and 5 kDa peptides gave clear solutions in H_2O.

Sample (Mw)	Sequence	Preparation
2.3 kDa	PGQGQQGYYPTSLQQPGQGQQ	Synthesised
5 kDa	[PGQGQQGYYPTSLQQ] x 3	Synthesised
12 kDa	contains 3 x P45 repeat	Expressed
22.3 kDa	contains 6 x P45 repeat	Expressed
58 kDa	1Dx5 central domain	Expressed

FT-IR measurements:

Spectra were recorded on a Bio-Rad FTS 6000 Fourier-transform spectrometer equipped with a HgCdTe detector. The samples were injected in a MicroCircle ATR cell with a ZnSe crystal and a spectrum of the solution measured. Then the cell was drained slowly, the remaining protein dried onto the surface of the ATR crystal, and the spectrum of the dry protein was recorded. Rehydration was achieved by flushing the cell with moist air - bubbled through saturated NaCl solution (76% r.h.) or water (100% r.h.) until the absorption of the sample had become constant. 256 scans at 2 cm^{-1} resolution were averaged for all spectra. The empty cell was used as background. Water spectra were recorded for solvent subtraction. The amide band region of the spectra were Fourier-deconvoluted (enhancement factor = 2.0, halfwidth = 16 cm^{-1}) .

3 RESULTS

Figure 1 shows the spectra of the peptides in the dry state. The amide I band had a maximum around 1655 cm^{-1} which could not be resolved by Fourier-deconvolution. This peak resulted most likely from a mixture of unordered structures and β-turns as well as from glutamine side chain contributions. Shoulders at 1630 an 1694 cm^{-1} indicated β-sheet structures. The spectra of all peptides were very similar in spite of the different lengths, indicating that in the dry state the 'local' secondary structure was comparable. The only notable difference in the 58 kDa peptide spectrum was the lower intensity of the tyrosine ring band at 1516 cm^{-1}.

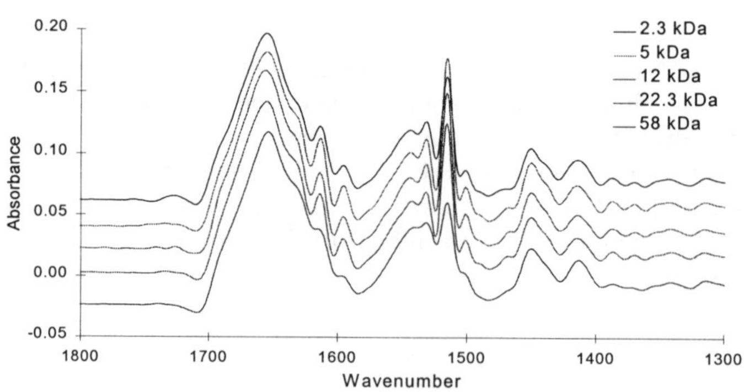

Figure 1: *Fourier-deconvoluted FT-IR spectra of the dry peptides*

Moderate hydration caused limited structure changes in all samples. There was a marked decrease of the amide I band component at 1694 cm^{-1} due to non-hydrogen bonded peptide carbonyl groups. The overall intensity of the 1654 cm^{-1} band maximum decreased and there was an increase in the 1640-1620 cm^{-1} region attributed to β-sheet. At 1669 cm^{-1} a distinct shoulder became visible which indicated β-turn structure. The amide II band maximum shifted towards higher wavelengths in line with an increase in protein backbone hydration. To compare the hydration behaviour the normalised intensities at 1666, 1654, 1630 and 1616 cm^{-1} were plotted for each sample. These plots (Figure 2) show that the initial change on hydration was comparable in all samples. However, a different structure change occurred going from the air-hydrated solid (100% r.h.) to the aqueous system. The β-sheet content which initially went up decreased and β-turns became more prevalent. This second step of hydration was very noticeable in the water-soluble 2.3 and 5 kDa peptides, but less so in the insoluble longer peptides.

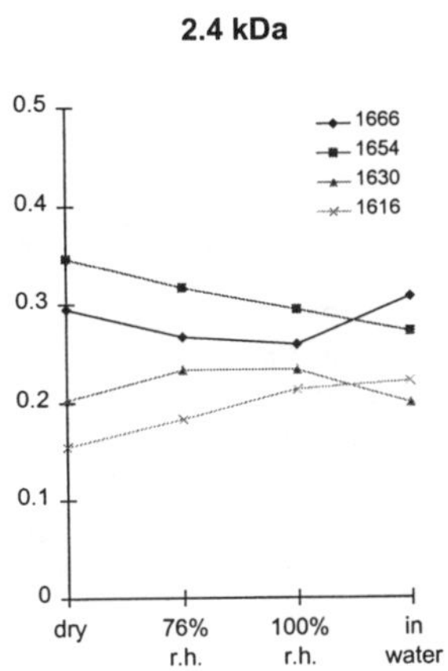

2a

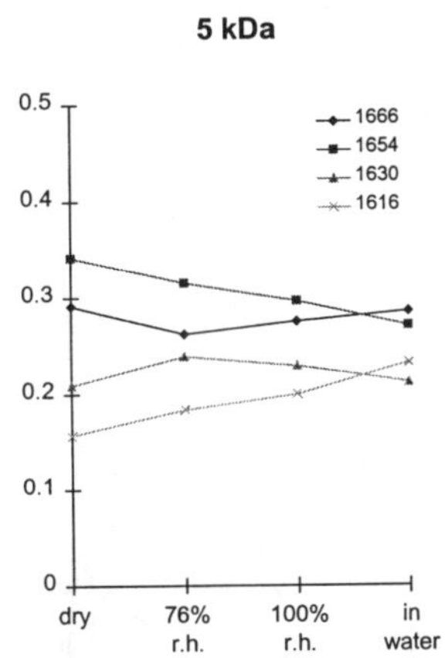

2b

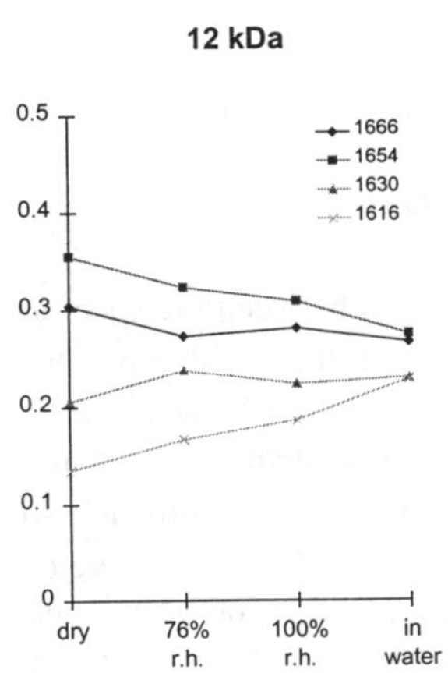

2c

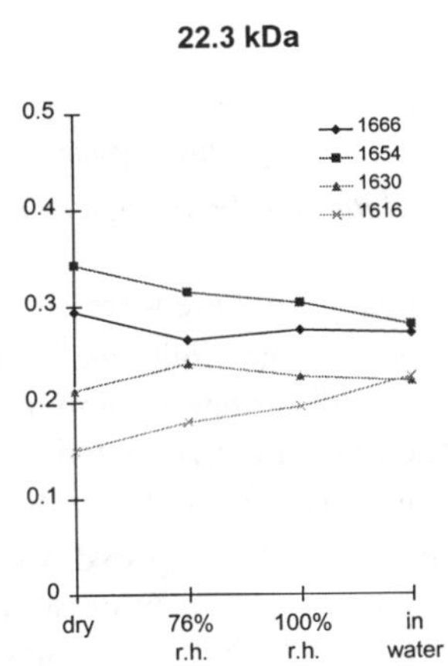

2d

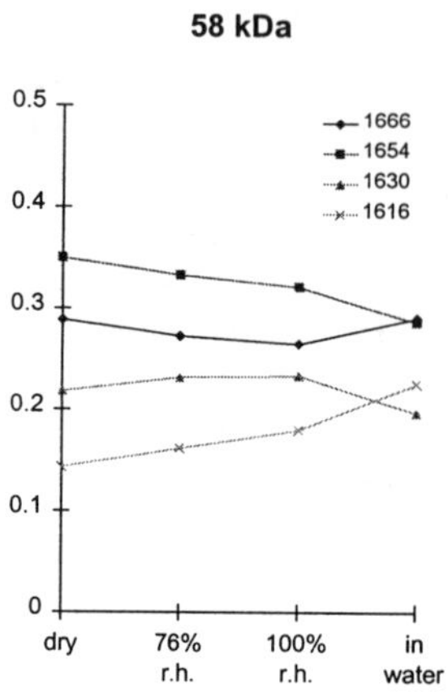

2e

Figure 2 *Changes in the amide I band during hydration*

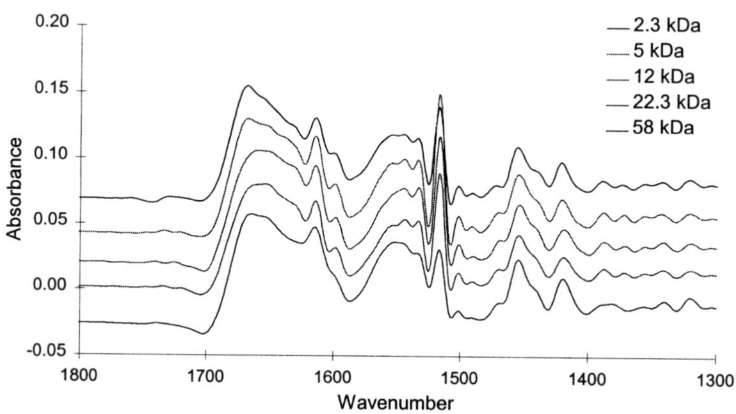

Figure 3 *Fourier-deconvoluted FT-IR spectra of the peptides in water*

Due to their different hydration there were clear differences between the spectra of the samples in water (Figure 3). The 1669 cm^{-1} (β-turns) and the 1614 cm^{-1} (hydrated extended chain) peaks were the strongest components of the amide I band of the 2.3 kDa peptide in aqueous solution. Both the 1653 and 1630 cm^{-1} components were smaller than in the hydrated solid state. This indicated that the peptide adopted a β-turn-rich solution conformation in accordance with the proposed β-spiral structure of the repetitive domain. The solution spectrum of the 5 kDa peptide was similar, but the 1668 cm^{-1} peak was slightly smaller and the 1630 cm^{-1} component somewhat higher, which may indicate some aggregates in the solution. In the 12 and 22.3 kDa peptides the 1652 cm^{-1} and 1630 cm^{-1} components progressively increased and the 1665 / 1614 cm^{-1} bands decreased.

Interestingly, this behaviour could not be extrapolated to the 58 kDa repeat peptide which had a lower content of β-sheet and more β-turns in water than the much smaller 22.3 kDa peptide. The last step of solubilisation of the 58 kD peptide was more comparable with the shorter repeat peptides, indicating that the sequence imperfections countered the effect of longer chain length, making intermolecular aggregation less favourable than in the perfect repeat structure.

4 CONCLUSIONS

It can be concluded that

i Secondary structures are variable.

ii Intermolecular interactions in the solid state (dominated by hydrogen bonding of glutamine side chains, eventually backbone) create unordered and β-sheet structures.).

iii Hydration gives rise to rigid and solvated domains. The equilibrium depends on the environment (T, co-solvents).

iv The repetitive domain structure adopts a β-turn rich structure in solution.

v Solubility and structure change on hydration depend on the molecular weight and possibly also the sequence. This indicates polymer-like aggregation behaviour via a large number of hydrogen bonds.

References

1. Tatham, A.S., Shewry, P.R. and Miflin, B.J. (1984) *FEBS Lett.* **177**, 205
2. Tatham, A.S., Drake, A.F. and Shewry, P.R. (1990) *J. Cereal Sci.* **11**, 189
3. Belton, P.S., Colquhoun, I.J., Field, J.M., Grant, A., Shewry, P.R., Tatham, A.S. and Wellner, N. (1995) *Int. J. Biol. Macromol.* **17**, 74

Acknowledgements

IACR and IFR receive grant-aided support from the Biotechnology and Biological Sciences Research Council of the United Kingdom. This work was supported by European Union FAIR grant CT96-1170: Improving the Quality of EU Wheats for use in the Food Industry, 'EUROWHEAT'.

CHARACTERISATION AND CHROMOSOMAL LOCALISATION OF C-TYPE LMW-GS

Rovelli L.[1], Masci S.[1], Kasarda D.D.[2], Vensel W.H.[2] and Lafiandra D.[1]

1. Dipartimento di Agrobiologia e Agrochimica, Università degli Studi della Tuscia, Via S. Camillo de Lellis, 01100 Viterbo, Italy. 2. U.S. Department of Agriculture, Agriculture Research Service, Western Regional Research Center, 800 Buchanan St., Albany, CA, 94710, U.S.A.

1 INTRODUCTION

The glutenin fraction, composed mainly of high (HMW-GS) and low (LMW-GS) molecular weight subunits, plays the major role in determining gluten viscoelastic properties. Although most abundant, LMW-GS have still not been characterised in detail, because of their large number and difficulties in their purification. LMW-GS are classically divided into B, C and D groups, on the basis of molecular weights and isoelectric points[1]. The B group consists mainly of typical LMW-GS (Ser-type and Met-type)[2], which are also present in minor amount in the C group, that is in contrast made up mainly of α- and γ-gliadin type LMW-GS[2]; the D group of LMW-GS corresponds to ω-gliadin type subunits[3]. The presence of gliadin-like subunits in glutenin preparations is apparently due to differences in the number and/or position of cysteine residues in the proteins incorporated into glutenin relative to their equivalent gliadins. These differences enables such proteins to be incorporated into the glutenin polymer.

Among the three LMW-GS groups, C subunits are the least characterised, because of their low level of expression compared to B subunits. In order to study C subunits in detail, we have improved an existing protocol[4], based on differential precipitation of glutenin subunits with increasing concentration of propan-1-ol, that has allowed specific isolation of C-subunits. In order to confirm the effectiveness of the method, N-terminal sequencing has been performed on combined fractions of B and C subunits and the polypeptide composition of the two groups has been compared. Chromosomal localisation of C subunits has been determined and the presence of polymorphism assessed in different durum wheat cultivars.

2 MATERIALS AND METHODS

The bread wheat cultivar Chinese Spring (CS), together with its nullisomic-tetrasomic lines involving chromosomes 1 and 6, has been used to determine the chromosomal localisation of C subunits. To confirm this localisation, CS ditelosomic lines, intervarietal substitution lines and genotypes with nulls at *Gli-A2* and/or *Gli-D2* have also been used.

In order to analyse allelic variation in C subunits, the following durum wheat cultivars have been used: Creso, Duilio, Langdon, Lira, Neodur, Ofanto, Simeto and Svevo.

The procedure used is based mainly on that reported by Verbruggen et al[4]. A further precipitation step with 85% propan-1-ol was added. This latter precipitation, performed overnight, allowed us to obtain a fraction containing exclusively C subunits.

The B and C subunit groups were separated by one-dimensional SDS-PAGE (with T=12 and C=1.28) and by a two-dimensional method (A-PAGE vs. SDS-PAGE) described by Morel[5], with some modifications.

In order to determine the percentage of each sequence type, fractions corresponding to B and C subunits were further purified by RP-HPLC (gradient: 29-43% acetonitrile/water containing 0.05% TFA, in 50 minutes), the various peaks obtained for each fraction combined, and the bulked material analysed by N-terminal amino acid sequencing that was performed with a Procise Model 492 sequencer (PE-Applied Biosystems).

3 RESULTS AND DISCUSSION

The protocol devised resulted in an enrichment in C subunits that allowed detection of even minor polymorphic forms. Figure 1 shows the effectiveness of the procedure and the variation present in C subunits among the durum wheat varieties assayed.

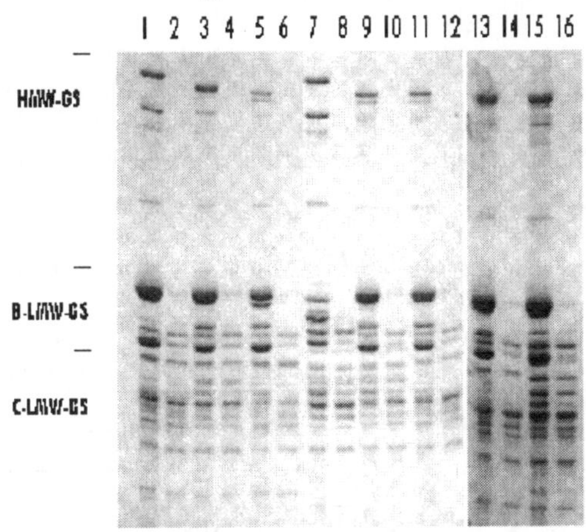

Figure 1: *SDS-PAGE of total glutenin subunits (odd numbers) and corresponding C subunits (even numbers) of the following durum wheat cultivars: (1-2) Creso, (3-4) Duilio, (5-6) Lira, (7-8) Langdon, (9-10) Neodur, (11-12) Ofanto, (13-14) Simeto, (15-16) Svevo. The cultivars here analysed show variation in C subunits.*

Further heterogeneity in C subunits was assessed by two-dimensional electrophoresis. In Figure 2, the two-dimensional patterns of C subunits obtained from the bread wheat cultivar Cheyenne and the durum wheat cultivar Creso are compared.

The use of appropriate genetic stocks allowed us to assign the resolved subunits to the short arms of the homeologous groups 1 and 6 chromosomes. Analysis of null-types (either at the *Gli-A2* or *Gli-D2* loci, or both) showed consistency between the absence of particular gliadin components and C components, suggesting a tight linkage between loci coding for gliadins and those coding for C-type LMW-GS. The availability of such a procedure, combined with two-dimensional analysis, and the existence of polymorphism, will make it possible to establish genetic linkages between loci coding for C components and gliadins.

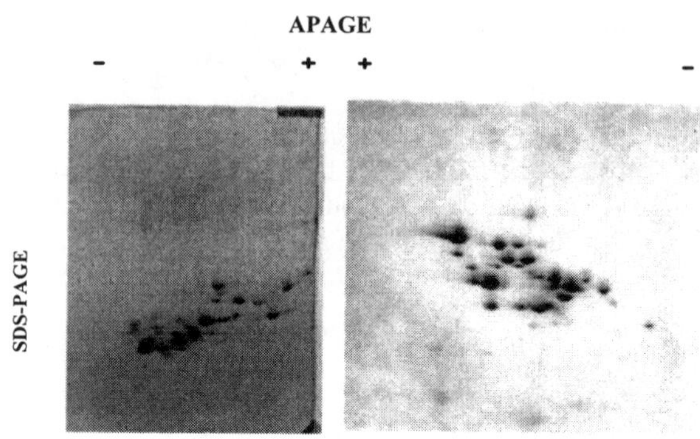

Figure 2: *Two dimensional electrophoresis (A-PAGE vs. SDS-PAGE) of C subunits belonging to cv. Cheyenne (bread wheat, left side) and cv. Creso (durum wheat, right side)*

N-Terminal amino acid analysis of polypeptides present in RP-HPLC purified bulks of B and C subunits showed that, as expected, B subunits mostly have typical Ser- or Met-type LMW-GS sequences (76%), whereas C subunits have gliadin-like sequences almost exclusively (95%). The percentages attributed to each sequence type are reported in Table 1.

Table 1: *Percentages of N-terminal sequences present in B and C subunits*

	Sequence type	Percentage
B subunits	α	9
	γ	15
	LMW-GS Met-type	24
	LMW-GS Ser-type	52
C subunits	α	40
	γ	55
	LMW-GS Met-type	5
	LMW-GS Ser-type	0

4 CONCLUSIONS

The results reported here show that C subunits are encoded by genes located on the short arms of chromosomes 1 and 6. N-terminal amino acid sequencing confirmed that B subunits are mainly composed of typical LMW-GS with Ser- and Met-type N-terminal sequences, whereas C subunits are almost entirely composed of either α– or γ-gliadin type sequences. These observations, together with the parallelism between the presence/absence of particular gliadin component and the presence/absence of some C subunits, suggest a tight linkage between genes coding for gliadin subunits and those coding for C subunits.

The presence of gliadin-like components among LMW-GS can be explained by mutations that introduce or remove cysteine codons and these changes enable them to be incorporated into the glutenin polymer. The presence of such a process has already been demonstrated in α- and γ-gliadin genes, and in ω-components[6-8]. If only one cysteine residue is available for intermolecular disulphide bond formation in such subunits, they would act as chain terminators and would decrease the molecular size distribution of the glutenin polymer[9]; where more than one cysteine is formed, as has been recently demonstrated in a ω-secalin gene from rye[10], they might instead act as chain extenders.

References

1. E.A. Jackson, L.M. Holt and P.I. Payne, *Theor. Appl. Genet.*, 1983, **66**, 29
2. D.D. Kasarda, H.P. Tao, P.K. Evans, A.E. Adalsteins and S.W. Yuen, *J. Exp. Bot.*, 1988, **39**, 899
3. S. Masci, D. Lafiandra, E. Porceddu, E.J.-L. Lew, H.P. Tao and D.D. Kasarda, *Cereal Chem.*, 1993, **70**, 581
4. I.M. Verbruggen, W.S. Veraverbeke, A. Vandamme and J.A. Delcour, *J. Cereal Sci.*, 1998, **28**, 25
5. M.H. Morel, *Cereal Chem.*, 1994, **71**, 238
6. O.D. Anderson and F.C. Greene, *Theor. Appl. Genet.*, 1997, **95**, 59
7. R. D'Ovidio, M. Simeone, S. Masci, E. Porceddu and D.D. Kasarda, *Cereal Chem.*, 1996, **72**, 443
8. S. Masci, T.A. Egorov, C. Ronchi, D.D. Kuzmicky, D. Lafiandra and D.D. Kasarda, *J. Cereal Sci.*, **29**, 17
9. H.P. Tao and D.D. Kasarda, *J. Exp. Bot.*, 1989, **40**, 1015
10. B.C. Clarke and R. Appels, *Pl. Syst. Evol.,* 1999, **214**, 1

Acknowledgments

Research supported by the Italian Ministero dell'Università e della Ricerca Scientifica e Tecnologica (M.U.R.S.T.), National Research Project "Studio delle proteine dei cereali e loro relazioni con aspetti tecnologici e nutrizionali".

CHARACTERIZATION OF A MONOCLONAL ANTIBODY THAT RECOGNISES A SPECIFIC GROUP OF LMW SUBUNITS OF GLUTENIN

S. Hey[1], J. Napier[1], C. Mills[2], G. Brett[2], S. Hook[3], A.S. Tatham[1], R. Fido[1] and P.R. Shewry[1]

1. IACR-Long Ashton Research Station, Department of Agricultural Sciences, University of Bristol, Long Ashton, Bristol BS41 9AF, UK. 2. Institute of Food Research, Norwich Laboratory, Colney Lane, Norwich NR4 7UA, UK. 3. RHM Technology Limited, The Lord Rank Centre, Lincoln Road, High Wycombe, Bucks HP12 3QR, UK.

1 INTRODUCTION

Monoclonal antibodies are an important tool to identify specific sequences and structural features in cereal proteins. They can also be used as the basis of test kits to identify and quantitize components which influence the processing properties of cereals or aspects of consumer acceptability (for example, the presence of gluten in foods for consumption by those with coeliac disease). The LMW subunits of glutenin are a particularly appropriate target for analysis using monoclonal antibodies, as they are extremely difficult to purify in sufficient quantities for detailed characterization. Consequently, with the exception of pioneering work by Kasarda and co-workers,[1] most of our knowledge of this group of proteins comes from sequences of cDNA and genomic clones.[2]

The monoclonal antibody IFRN0067 was raised against a total glutenin fraction from wheat cv. Avalon.[3] It reacts on western blotting with a small number of LMW subunits, indicating that it recognises unique rather than repetitive sequences (and contrasting with many other monoclonals).[4] Preliminary analyses also showed a correlation between reaction with IFRN0067 and loaf score in twenty-one flour samples obtained from a French mill.[3] We therefore constructed a cDNA expression library from developing wheat endosperms in order to identify and characterize clones encoding proteins that reacted with IFRN0067.

2 CONSTRUCTION OF A cDNA EXPRESSION LIBRARY AND ISOLATION OF A CLONE BY SCREENING WITH IFRN0067

Endosperms were dissected by hand from heads of wheat cv. Chinese Spring at stages between 14 and 24 days after anthesis. Equal weights of endosperms from different stages were combined for extraction of total RNA. Several RNA preparations were combined for preparation of poly A^+ mRNA. The cDNA expression library was produced by Clontech (USA) by cloning cDNA into the *Eco*RI site of the phage vector λ gt11 to give an estimated 2.0×10^6 independent clones and an amplified library titre of 4.375×10^{10} plaque forming units/ml. The mean insert size was estimated as 1.5kb and the range 0.3 to 3kb. Screening with IFRN0067 with alkaline phosphatase-conjugated goat anit-mouse secondary antibody identified a single reactive plaque which was purified and the 1kb insert sub-cloned into a plasmid vector.

3 IDENTIFICATION OF THE IFRN0067 EPITOPE

The protein encoded by the cDNA comprises 264 amino acids with an M_r of about 30,700. Four distinct regions can be recognised (Fig. 1). Residues 13 form a unique *N*-terminal sequence and residues 14 to 85 a repetitive domain with high homology with the LMW subunit sequences. A similar level of homology is shown by residues 86 to 222 which are non-repetitive but residues 223-269 show no homology and may be derived from a gene rearrangement which has occurred either *in vivo* or, perhaps more likely, during the cloning.

In vitro transcription and translation of a sub-clone corresponding to residues 1-222 followed by immunoprecipitation with IFRN0067 confirmed the presence of the epitope within this part of the protein.

M E T S R V P G L E K P W Q Q Q P L P P Q Q Q P S F S Q Q Q L P
↑
P F S Q Q Q S P F S Q Q Q Q I V L Q Q Q P P F L Q Q Q Q P S L P Q

Q P P F S Q Q Q Q Q L V L P Q Q Q I P F V H P S I L Q Q L N P C K
↑
V F L Q Q Q C S P V A M P Q S L A R S Q M L Q Q S S C H V M Q Q

Q C C Q Q L P Q I P Q Q S R Y E A I R A I I Y S I I L Q E Q Q Q V Q

G S I Q T P Q Q Q P Q Q L G Q C V S Q P Q Q Q S Q Q R L G Q Q P

Q Q Q Q L A Q G T F L Q P H Q I A Q L E V M T S I A **S N H T N L**
↑
N N H I H N N N H M G V V L Q A S M A N E M K T C N T T R M

D H R C L V N E C S M *

Figure 1 *The amino acid sequence of the protein encoded by the cDNA clone. the divergent N-terminal and C-terminal sequences (residues 1-13 and 223-264, respectively), are in bold and the repetitive and non-repetitive domains (residues 14-85 and 86-222, respectively) indicated by arrows.*

Comparison of residues 1-222 with the sequences of other LMW subunits shows that the major region of divergence is unique *N*-terminal sequence. Thus residues 1-13 of the IFRN0067-reactive protein (METSRVPGLEKPW) are similar to the *N*-terminal sequences determined by direct analysis of several LMWm subunits[1] but no identical sequences over the whole length were identified.

A peptide corresponding to residues 1-113 was, therefore, synthesised and shown to react with IFRN0067 on western blotting. This confirms the presence of an epitope for

IFRN0067 in this region but does not rule out the presence of further reactive sequences in residues 14-222.

4 REACTION OF IFRN0076 WITH WHEAT CULTIVARS

Western blots of IFRN0067 with gluten protein fractions from eleven cultivars of wheat are shown in Fig. 2. All showed reactions with a group of bands of M_r about 45,000 and with at least two bands of lower M_r, but two allelic patterns can be recognised which differ in the intensity of reaction and the mobilities of the lower M_r bands. These can be defined as alleles 2 (Hereward, Cadenza, Axona) and 1 (all other cultivars) (Fig. 2).

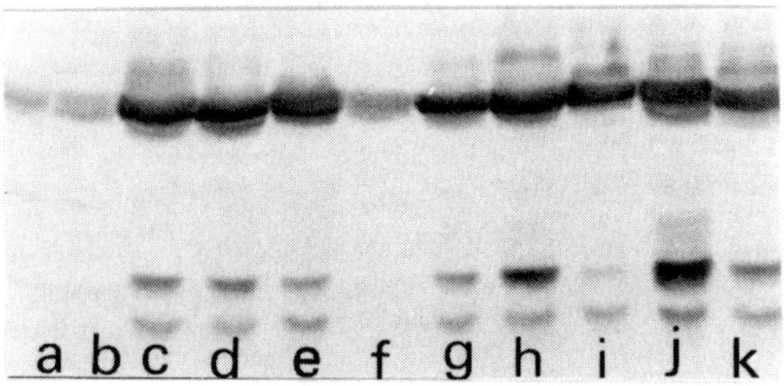

Figure 2 *Western blotting of IFRN0067 with gluten protein fractions from eleven bread wheat cultivars: a, Hereward (allele 1); b, Cadenze (1); c, Mercia (2); d, Soissons (2); e, Riband (2); f, Axona (1); g, Spark (2); h, Avalon (2); i, Brigadier (2); j, Hussar (2) and k, Magellan (2).*

Quantitative analysis of 66 flour samples derived from the cultivars in Fig. 2 showed no relationship between IFRN0067 binding and loaf volume (Fig. 3). However, flours of cultivars with allele 1 tended to give lower absolute values, which is consistent with their weaker reactions on western blotting.

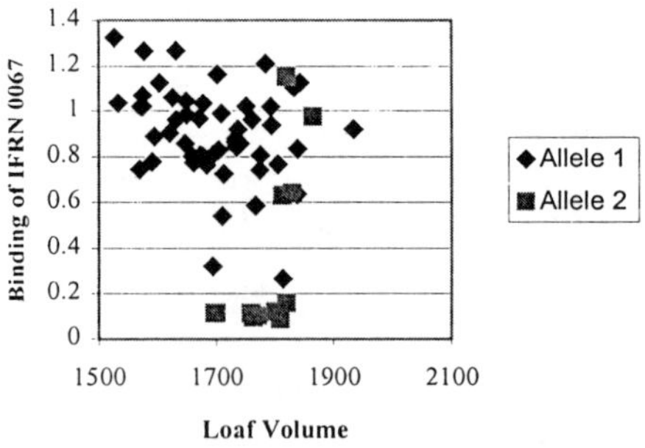

Figure 3 *The relationship between binding of IFRN0067 and loaf volume for 66 flour samples of the eleven cultivars shown in Fig. 2. Cultivars with alleles 1 and 2 are shown as squares and diamonds, respectively.*

Our failure to demonstrate a relationship between the binding of IFRN0067 and loaf volume contrasts with the results of Brett *et. al.* (1993).[3] We do not know the reason for this but can speculate that the French flours used in the previous studies were mixtures of two or more varieties in which alleles 1 and 2 were fortuitously associated with poor quality and good quality (e.g. cv. Soissons), respectively.

References

1. E. J.-L. Lew, D. D. Kuzmicky and D. D. Kasarda, *Cereal Chem.*, 1993, **69**, 508.
2. P. R. Shewry and A. S. Tatham, *J. Cereal Sci.*, 1997, **25**, 207.
3. G. M. Brett, E. N. C. Mills, A. S. Tatham, R. J. Fido, P. R. Shewry and M. R. A. Morgan, *Theor. Appl. Genet.*, 1993, **86**, 442.
4. G. M. Brett, E. N. C. Mills, B. J. Goodfellow, R. J. Fido, A. S. Tatham, P. R. Shewry and M. R. A. Morgan, *J. Cereal Sci.*, 1999, **29**, 117.

Acknowledgements

IACR and IFR receive grant-aided support from the Biotechnology and Biological Sciences Research Council of the United Kingdom. Work described in this paper was supported by a DTI Agro-Food LINK grant.

TEMPERATURE INDUCED CHANGES IN PROLAMIN CONFORMATION

E.N.C. Mills[1], G.M. Brett[1], M.RA Morgan[1], A.S. Tatham[2], P.R.Shewry[2].

1. Institute of Food Research, Norwich Research Park, Colney, Norwich, NR4 7UA.
2. IACR-Long Ashton Research Station, Long Ashton, Bristol, BS41 9AF.

1 INTRODUCTION

Wheat prolamins are characterised by unusual amino acid compositions, being rich in glutamyl and prolyl residues (up to 70 mol%), which results from the presence of repeated sequences which can account for over 50% of the total protein[1]. In comparison with many globular proteins, prolamins have a high degree of molecular mobility when hydrated[2]. In addition, heating results initially in solubility rather than denaturation. One of the consequences of this behaviour is that wheat gluten appears to have no significant protein-related co-operative transition on heating, implying that there is no transition from a 'folded' to a 'denatured' state, as is the case for globular proteins. However, CD and fourier transformed IR spectroscopy of the protein and of a synthetic peptide based on the consensus octapeptide repeat motif, have shown more subtle conformational transitions between poly-L-proline II-like and βI/III reverse turn structures, the latter predominating at higher temperatures and in solvents of low dielectric constant[3]. This report describes three anti-prolamin Mabs which have been used to study temperature-related alterations in prolamin structure.

2 MATERIALS AND METHODS

2.1 Prolamin and peptide preparations

Total glutenins (cv Avalon), and barley C hordein were prepared as reported previously[4]. A linear unblocked peptide based on the consensus octapeptide repeat motif of C hordein with *N*- and *C*-terminal glycine residues (Gly.Gln.Pro.Gln.Gln.Pro.Phe.Pro.Gln.Gly) was synthesised by the Microchemical Facility, Institute of Animal Physiology, Babraham, U.K. The peptide was conjugated to bovine serum albumin (BSA) using 1-ethyl-3 (3-dimethylaminopropyl) carbodiimide, as previously described[4].

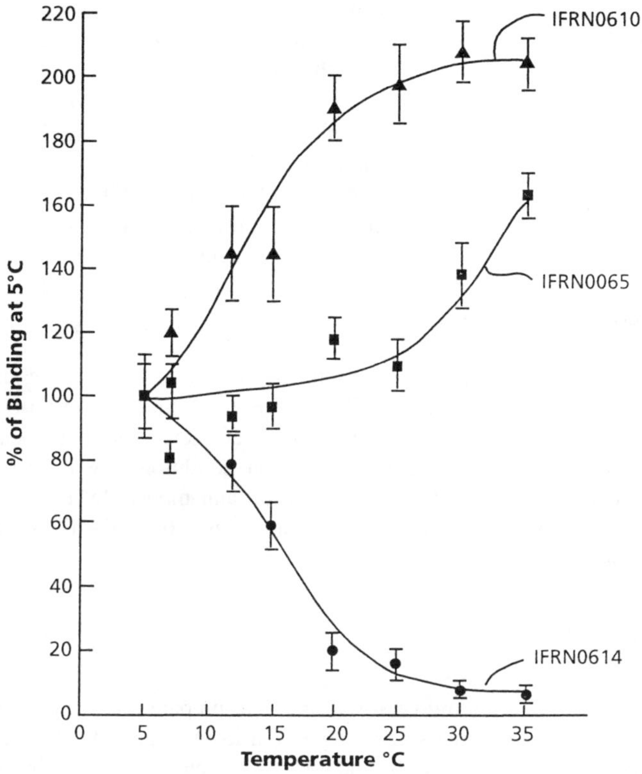

Figure 1. *Effect of increasing incubation temperature on the binding of three anti-prolamin monoclonal antibodies to total glutenins from wheat (cv Avalon)*
Results are expressed as a percentage of the binding of each Mab determined by direct ELISA at 5°C.

forces which predominate at lower temperatures being compensated for by an increase in van der Waals attractions as temperature increases[6]. Temperature also affects the kinetics of antibody reactions, the off-rate increasing about ten-fold and the on-rate increasing two-fold, resulting in a loss of antibody affinity at higher temperatures[7]. In an ELISA for human chorionic gonadotrophin, Mab binding was reduced between 4°C and 37 °C, but there was no dramatic loss of binding until temperatures greater than 60 °C were reached, when the antibody would have become denatured[8]. Such effects do not completely account for our observations and imply that conformational changes in epitope structure also play a role.

Thus, the ability of Mabs to recognise prolamins at low, but not higher, temperatures can be only partly explained by thermodynamic and kinetic considerations of antibody binding

Table 1. *Anti-prolamin monoclonal antibody specificities.* (based on data from Brett *et. al.*)[4,5]

Monoclonal Antibody Preparation	Specificity
IFRN 0061 (IgGM) Wheat, IFRN 0614 (IgM)	Binds predominantly to S-poor types of barley and rye. Binds to the repeat motif PQQPFPQQ.
IFRN 0065 (IgG_1)	Recognises the motif QPFP which is present in α, β, γ, ω-gliadins and LMW. subunits of glutenin.
IFRN 0610 (IgG_1)	Recognises the motif QQSF/Y which is present in α, β, γ, ω-gliadins and LMW subunits of glutenin.

2.2 Monoclonal antibody preparations and immunoassays

A number of Mabs specific for motifs present in the repetitive domains of different prolamin types were selected (see Table 1). They were used in the form of culture supernatant and antibody binding was determined by enzyme-linked immunosorbent assay (ELSIA) procedures performed under different incubation conditions.[4] Briefly, a direct ELISA format was employed using anti-mouse IgG, or anti-mouse IgM, horseradish peroxidase (Sigma Chemical Co., Poole, UK) and a substrate based on 3,3',4,4' - tetramethyl benzidine (Vetoquinol, Bicester, UK).

3 RESULTS

The effect of temperature on Mab binding to a total glutenin preparation (cv Avalon) was investigated using enzyme-linked immunosorbent assay (ELISA) (Figure 1). Two types of binding were observed. Mabs IFRN 0610 and 0614, which recognise the epitope Pro.Gln.Gln.Pro.Phe.Pro.Gln.Gln, bound well at 4°C but not at all at 37°C (data only shown for IFRN 0614). The same effect was also observed towards C hordein and a synthetic peptide (Gly.Gln.Pro.Gln.Gln.Pro.Phe.Pro.Gln.Gly) corresponding to the repeat motif of the S-poor prolamins conjugated to BSA (data not shown). In contrast, IFRN 0610, which recognises the core epitope Gln.Gln.Ser.Phe/Tyr, showed an increase in binding to total glutenins of around 100% whilst the binding of IFRN 0065, which recognises the sequence Gln.Pro.Phe.Pro, remained unchanged until reaching temperatures above 25°. Similar results were obtained for all the Mabs using a variety of prolamin preparations including α-, γ-, and ω-gliadins and LMW subunits of glutenin from wheat, C hordein from barley, and ω-secalins from rye.

4 DISCUSSION

Antibody-binding reactions are generally exothermic in nature, with a ΔG° of around -10 kcal/mol. ΔG° remains constant over a wide range of temperatures, the loss of electrostatic reactions and it seems likely that alterations in prolamin conformation between 4°C and 37°C also contribute to the loss of binding. Both IFRN0061 and 0614 recognise the same repeat motif, Pro-Gln-Gln-Pro-Phe-Pro-Gln-Gln.

The temperature dependent binding indicates that they may only recognise the motif when present in the poly-L-proline II type structure which is favoured at lower temeratures[3], but not in the β-reverse turn conformation present at higher temperatures. The observation that other Mabs bound more strongly at higher temperatures can only be the result of endothermic alterations in prolamin conformation, as has been observed for antibody recognition of the D determinant on Rho positive erythrocytes[9]. Thus, the Pro-Gln-Gln-Ser-Phe/Tyr or Gln-Pro-Phe-Pro motifs appear to be recognised more strongly by Mabs IFRN 0610 and IFRN 0065, respectively, when in the β-reverse turn conformation than in the poly-L-proline II conformation.

These changes in antibody binding occur both in solution and with protein adsorbed to the surfaces of microtitration plates but it remains to be shown whether such conformational transitions also occur in dough. The availability of such Mabs will allow the presence of poly-L-proline II type structures and β-reverse turn conformations to be identified in complex systems, such as doughs, using immunolabelling techniques.

References

1. P.R. Shewry, A.S. Tatham, *Biochem. J.,* 1990, **267,** 1.
2. P.S. Belton, A.M. Gil, A.S. Tatham, *J. Chem. Soc. Faraday Trans.,* 1994, **90,** 1099.
3. A.S. Tatham, A.F Drake, P.R. Shewry, *Biochem. J.,* 1989, **259,** 471.
4. G.M. Brett, E.N.C., Mills, S. Parmar, A.S. Tatham, P.R. Shewry, M.R.A. Morgan, *Cereal Sci.,* 1990, **12,** 245.
5. G.M. Brett, E.N.C., Mills, B.J. Goodfellow, R.J. Fido, A.S. Tatham, P.R. Shewry, M.R.A. Morgan, *J. Cereal Sci.*, 1999, **29,** 117.
6. C.J. van Oss and D.R. Absolom, in *The Antigens* vol VI ed. M. Sela Academic Press, New York, 1982, p337.
7. D.W. Mason, A.F. Williams, *Biochem. J.,* 1980, **187,** 1.
8. R.H.J. van der Linden, L.G.J. Frenken, B. de Geus, M.M. Harmsen, R.C. Ruuls, W. Stok, L. de Ron, S. Wilson, P. Davis, C.T. Verrips, *Biochim. Biophys. Acta* 1999, **1431,** 37.
9. F.A. Green, *Immunol. Commun.,* 1982, **11,** 25.

CHARACTERISATION OF ω-GLIADINS FROM DIFFERENT WHEAT SPECIES

H. Wieser[1], W. Seilmeier[1], I. Valdez[2] and E. Mendez[2]

1. German Research Institute of Food Chemistry, Garching, Germany. 2. Centro Nacional de Biotecnologia, Cantoblanco, Madrid, Spain.

1 INTRODUCTION

Extensive studies on ω-gliadins have been performed only with common (bread) wheat. Accordingly, ω-gliadins are minor components of gluten proteins and are characterised by high proportions of glutamine and proline. In contrast to other gluten protein types, complete amino acid sequences have not been determined up to now, and molecular masses have been derived from SDS-PAGE mobility ranging from 55,000 – 79,000[1]. With respect to differences in amino acid compositions and molecular masses, ω-gliadins of common wheat were classified into the ω5- and ω1,2-types[2].

ω-Gliadins from other cultivated wheat species have been less investigated. Therefore, ω-gliadins from representatives of hexaploid spelt, tetraploid durum wheat, tetraploid emmer and diploid einkorn were isolated by RP-HPLC and characterised by amino acid compositions, N-terminal amino acid sequences and actual molecular masses. In the case of hexaploid common wheat, three different classes (winter wheat, spring wheat, wheat rye hybrid) were compared.

2 MATERIAL AND METHODS

2.1 Materials

Kernels of winter wheat (cv. Rektor), spring wheat (cv. CWRS), wheat rye hybrid (cv. Herzog), spelt (cv. Schwabenkorn), durum wheat (cv. Biodur), emmer (unknown cultivar) and einkorn (unknown cultivar) were milled to white flour using a laboratory mill. The flours were dafatted with light petroleum (40 – 60°C boiling range).

2.2 Methods

2.2.1 Extraction. Defatted flours (1 g) were extracted stepwise with 2 x 10 ml of NaCl (0.4 mol/L)/KHNa PO_4 (0.067 mol/L, pH 7.6) and with 2 x 10 ml of 60 % (v/v) ethanol (= gliadins).

2.2.2 RP-HPLC. Aliquots (= 500 μl) of the combined ethanol extracts were filtered through a 0.45 mm-membrane and separated on a C_8 silica gel column (4.6 x 240 mm, 50°C)[3]. The gradient of the elution solvents A (0.1 % trifluoracetic acid) and B (99.9 % acetronitrile, 0.1 % trifluoroacetic acid) was linear from 25 % B (0 min) to 37 % (60 min). The flow rate was 1.0 ml/min and the dection wave-length was 210 nm. The eluates corresponding to the peaks of the chromatograms were collected and dried by means of a vacuum centrifuge.

2.2.3 Protein analysis. Amino acid compositions were determined after hydrolysis with HCl using an amino acid analyser LC 3000 (Biotronic). N-terminal amino acid sequences (5 – 10 cycles) were analysed with a protein sequencer Procise (PE Biosystems). MALDI-TOF mass spectrometry was performed with a Reflex II spectrometer (Bruker).

3 RESULTS AND DISCUSSION

3.1 RP-HPLC

Gliadins were isolated from the flours of three different classes of common wheat, and of spelt, durum wheat, emmer and einkorn by extraction with 60 % (v/v) aq. ethanol, after albumins and globulins had been removed. Preparative separation of ω-gliadins was achieved by RP-HPLC of the gliadin extracts on C_8 silica gel using an optimised elution gradient. The ω-gliadin patterns obtained revealed typical differences amongst wheats; they demonstrated a marked variability in numbers, elution times and quantities of single components. Six to nine protein fractions from each wheat were collected and characterised by amino acid analysis and by determination of N-terminal amino acid sequences and molecular masses (Table 1).

Table 1 *Characterisation of ω-gliadins from different wheat species*

	Amino acid compositions (mol-%)	
Gln	ω5	ω1,2
Pro	51 – 57	39 – 45
Phe	18 – 21	22 – 31
	9 – 10	6 – 8
	N-Terminal amino acid sequences	
	ω5	ω1,2
	SRQLSP	KELQSP
	SRLLSP	ARQLNP
	SMELQR	RQLNPS
	Molecular masses	
	ω5	ω1,2
	44,000 – 55,000	34,000 – 44,000

3.2 Amino acid compositions

All ω-gliadins analysed had significantly higher proportions of glutamine, proline and phenylalanine compared with other gluten protein types. These three amino acids accounted for 70 – 86 % of the total composition. Typical differences in amino acid compositions allowed a clear differentiation into ω5- and ω1,2-types. ω5-Gliadins, the most hydrophilic components upon RP-HPLC (R_t = 19 – 30 min), had 52 – 57 mol-% Gln, 18 – 21 mol-% Pro and 9 – 10 mol-% Phe. ω1,2-Gliadins were eluted after ω5-gliadins (R_t = 38 – 50 min); they had 39 – 45 mol-% Gln, 22 – 31 mol-% Pro and 6 – 8 mol-% Phe. Typical for both ω-types were the low values for Cys (0.0 – 0.4 mol-%), Met (0.0 – 0.3 mol-%) and Lys (0.3 – 0.8 mol-%). With respect to wheat species, principal differences in the compositions could not be observed except that emmer and einkorn did not contain any ω1,2-gliadin.

3.3 N-terminal amino acid sequences

Though many modifications of single residues were present in the N-terminal sequences of ω-gliadins, they could be classified into a few basic types. SRLLSP or SRQLSP were typical for ω5-gliadins of all wheats except emmer (SMELQT). ω1,2-gliadins were characterised by the sequences KELQSP and ARQLNP. ω-Secalins which had been introduced by 1B/1R chromosome translocation into wheat Herzog and appeared in the HPLC chromatogram in the elution area of ω1,2-gliadins, had the N-terminal sequences RQLNPS.

3.4 Molecular masses

The masses of the isolated ω-gliadins were determined by MALDI-TOF analysis. The results revealed a range of 44,000 – 55,000 for most proteins of the ω5-type and a range of 34,000 – 44,000 for the ω1,2-type. Thus, actual masses were by far lower than those derived from SDS-PAGE mobility[1]. Significant differences between wheat species could not be detected.

4 CONCLUSIONS

The present study demonstrates that RP-HPLC is an efficient method for the characterisation and preparation of ω-gliadins from different wheat species. Amino acid compositions of all ω-gliadins analysed reveal significantly higher values for Gln, Pro and Phe compared with the other gluten proteins. Differences in the proportions of these three amino acids allow a clear differentiation into the ω5- and ω1,2-type. Emmer and einkorn have only ω5-, but not ω1,2-gliadins. The N-terminal amino acid sequences occur in three basic variants in ω5-gliadin and in three basic variants in ω1,2-gliadins. The actual masses of ω5- and ω1,2-gliadins determined by MALDI-TOF mass spectrometry are much lower than those derived from SDS-PAGE mobility.

References

1. I. Krause, U. Müller and H.-D. Belitz, *Z. Lebensm. Unters. Forsch.*, 1988, **186**, 398.

2. D.D. Kasarda, J.-C. Autan, E.J.-L. Lew, C.C. Nimmo and P.R. Shewry, *Biochim. Biophys. Acta*, 1983, **747**, 138.
3. H. Wieser, S. Antes and W. Seilmeier, *Cereal Chem.*, 1998, **75**, 644.

IDENTIFICATION OF WHEAT VARIETIES USING MATRIX-ASSISTED LASER DESORPTION/IONIZATION TIME-OF FLIGHT MASS SPECTROMETRY

W. Ens[1], K R. Preston[2], M. Znamirowski[1], R. G. Dworschak[1], K. G. Standing[1], and V. J. Mellish[2]

1. Department of Physics and Astronomy, University of Manitoba, Winnipeg, MB, Canada R3T 2N2. 2. Grain Research Laboratory, Canadian Grain Commission, 1404-303 Main Street, Winnipeg, MB, Canada R3C 3G8.

1 INTRODUCTION

Variety identification by gliadin protein fingerprinting is being widely used in conjunction with visual analysis for the grading and/or classification of commercial wheat samples. The most common procedure for fingerprinting is polyacrylamide gel electrophoresis (PAGE) under acidic conditions[1]. Sufficient bands (about 40) are normally available to accurately identify most varieties, although closely related varieties may present problems. Equipment for PAGE is relatively inexpensive and throughput in batch mode can be quite high. However, the technique is slow (normally > 1hr) and can not be automated. Other techniques that have been used for wheat variety identification based on gliadin protein patterns include RP-HPLC[2] and capillary electrophoresis[3].

Matrix-assisted laser desorption/ionization time-of-flight mass spectrometry (MALDI-TOF MS) is being widely used to characterize purified, partially purified and complex mixtures of proteins with masses up to several hundred kDa or more[4]. Recent studies in our laboratory[5] have demonstrated its ability to characterize mass profiles of crude or partially purified extracts of wheat gliadins, low molecular weight (LMW) glutenin subunits and high molecular weight (HMW) glutenin subunits. HMW glutenin subunits show relatively simple spectra while complex spectra are obtained for gliadins and LMW glutenin subunits. Using delayed extraction to improve resolution, gliadin and low MW glutenin subunits spectra showed a large number of well resolved peaks in the 30-40 kDa range. In the present study, we report on the ability of MALDI-TOF MS to identify sixteen varieties representing five different Canadian wheat classes.

2 MATERIALS AND METHODS

2.1 Wheat Samples

Reference samples of sixteen varieties representing five wheat classes were obtained from a collection maintained at the Grain Research Laboratory. The five wheat classes included Canada Western Red Spring (CWRS), Canada Western Amber Durum (CWAD), Canada Western Extra Strong (CWES), Canada Prairie Spring Red (CPSR) and Canada Prairie Spring White (CPSW). Six varieties representing four wheat classes grown at eight high grade sites were also obtained from the 1996 Saskatchewan wheat variety trials to assess the impact of environment on mass spectra patterns. All samples were subjected to acid PAGE[1] to confirm variety purity before further testing.

2.2 Extraction and Preparation of Samples for MS

Gliadins were extracted from ground grain with 600 μL of 70% ethanol at room temperature for 1 hour in 1.5 mL micro-centrifuge tubes with mixing every 10 min using a vortex mixer. The supernatant was retained for analysis after centrifugation at 8800 *g* for 10 minutes.

A saturated matrix solution of sinapinic acid (3,5-dimethoxy-4-hydroxycinnamic acid) in aqueous 50% acetonitrile containing 0.1% trifluoroacetic acid was prepared and mixed 10:1 with gliadin extract. An aliquot (3 μL) was placed on a metal target probe and dried with a hot air blower to form a deposit about 2 mm in diameter. Approximately four samples were applied to the probe plus an external calibrant (myoglobin). Myoglobin was also added to samples as an internal calibrant to improve mass determination accuracy. All sample deposits were washed with water to remove contaminants and redried prior to analysis.

2.3 MALDI-TOF MS

Positive ion spectra were obtained on a custom-built MALDI TOF instrument[6] in the linear mode with delayed extraction as described previously[5]. A two-grid delayed extraction system was employed using a 25 kV d.c. accelerating potential on the probe and first grid with a pulse of 3 kV applied to the probe 1.2 μs after the laser pulse (VST 337ND, 2-3 Hz). Approximately 150-200 shots were accumulated per spectrum. Digital files of m/z versus intensity spectra were converted to compact grey-scale or false-colour plots to facilitate comparison among varieties.

3 RESULTS AND DISCUSSION

3.1 Variety and class identification

Improved resolution is evident from the spectra for the sixteen varieties representing five Canadian wheat classes (Figure 1). This can be attributed primarily to the use of delayed extraction which gives substantial improvements in mass resolution in this mass range[5]. In some cases, more than 100 resolved peaks were evident from spectra representing individual hexaploid varieties.

Differences are clearly evident in patterns between wheat classes and between varieties within a class. Varieties within each wheat class showed a number of strong easily identifiable bands, which were common to the class and were lacking in the other wheat classes. These common bands can be summarized as follows:

CWAD	34280, 38350	CPSR, CPSW	35539, 39310
CWES	34980	CWRS	33952, 35200

These common bands are probably a reflection of the common genetic background within Canadian wheat classes[7]. These results suggest the possibility of identifying Canadian wheat varieties by class by assessing the presence or absence of a few strong bands.

Within each class, most varieties could be identified by the presence or absence of a few strong to medium intensity bands. In a few cases, analysis of less prominent band patterns were required for variety discrimination. For example, for the most widely and least genetically variable wheat class, CWRS, the six varieties could be differentiated as follows:

CDC Teal, Leader	peak at 32258 absent in all other varieties, peak at 31296 for Leader, absent in CDC Teal
Columbus	distinctive peak at 38913
Roblin	distinctive peak at 33151, no peak at 34642 (present in all other varieties)
Laura, Katepwa	very similar patterns, Laura shows peak at 30104 absent in Katepwa

A similar approach was used which allowed differentiation of varieties within the other four wheat classes.

3.2 Environmental effects

Little change was evident in patterns obtained for two CWAD wheat varieties (Kyle, AC Melita), one CPSR wheat variety (AC Crystal), one CPSW wheat variety (AC Karma) and two CWRS varieties (Katepwa and CDC Teal) grown at six Saskatchewan sites (data not shown). The only exception was at one site for each of the CWRS varieties where some less intense peaks seemed to be fainter or absent. However, changes in these bands did not impede the ability to identify varieties since the patterns of more intense peaks did not change. Overall, these results suggest that environment does not generally have a major impact on MS spectra, consistent with electrophoretic and HPLC studies[8,9].

4 CONCLUSIONS

The potential of matrix-assisted laser desorption/ionization time-of-flight mass spectrometry (MALDI-TOF-MS) for wheat variety identification was assessed using alcohol soluble protein (gliadin) extracts from sixteen Canadian wheat varieties representing four wheat classes. Using washing and delayed extraction to optimize resolution, reproducible mass spectra with a large number of resolved peaks in the 30-40,000 Da range were obtained. Distinct spectra were evident for all sixteen varieties. Varieties within each class exhibited a number of characteristic strong bands absent from the other wheat classes. Environment had little impact on mass spectra. These results suggest that MALDI-TOF MS shows promise as a method for identifying varieties and/or classes of wheat.

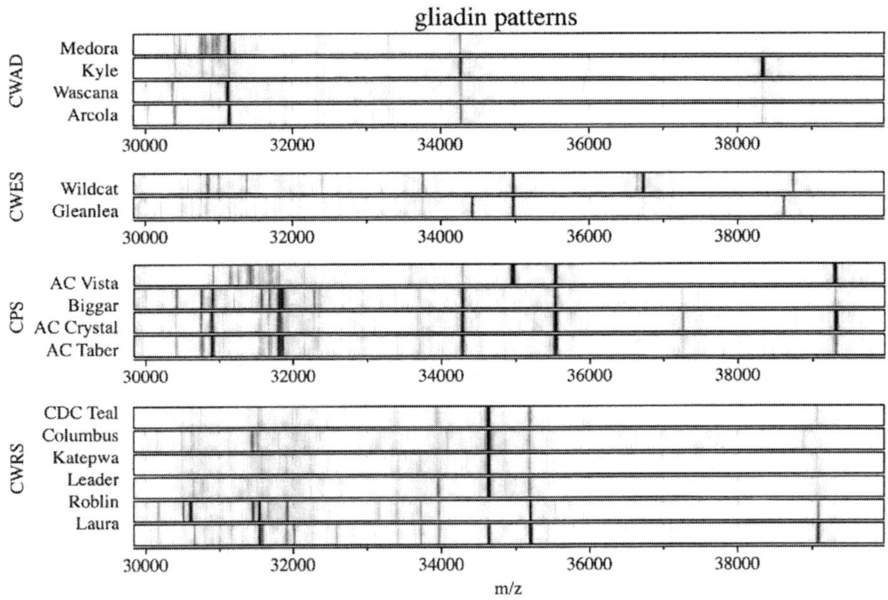

Figure 1 *MALDI-TOF MS gray scale plots of wheat variety spectra*

References

1. R. Tkachuk and J. Mellish, *Ann. Technol., agric.*, 1980, **29**, 207.
2. T. Burnouf and J. A. Bietz, *Seed Sci. & Technol.*, 1987, **15**, 79.
3. G. Lookhart and S. A. Bean, *Cereal Chem.*, 1995, **72**, 42.
4. F. Hillenkamp, M. Karas, R. C. Beavis and B. T. Chait, *Anal. Chem.*, 1991, **63**, 1193A.
5. R. G. Dworschak, W. Ens, K. G. Standing, K. R. Preston, B. A. Marchylo, M. J. Nightingale, S. G. Stevenson and D. W. Hatcher, *J. Mass Spectrom.*, 1998, **33**, 429.
6. X. Tang, R. C. Beavis, W. Ens, F. Lafortune, B. Schueler and K. G. Standing, *Int. J. Mass Spectrom. Ion Processes*, 1988, **85**, 43.
7. K. R. Preston, B. C. Morgan and K. H. Tipples, *Can. Inst. Food Sci. Technol. J.*, 1988, **5**, 520.
8. R. R. Zillman, and W. Bushuk, *Can. J. Plant Sci.*, 1979, **59**, 281.
9. F. R. Huebner, and J. A. Bietz, *Cereal Chem.*, 1988, **65**, 362.

Disulphide Bonds and Redox Reactions

QUANTITATIVE DETERMINATION AND LOCALISATION OF THIOL GROUPS IN WHEAT FLOUR

S. Antes and H. Wieser

Deutsche Forschungsanstalt für Lebensmittelchemie and Kurt-Hess-Institut für Mehl- und Eiweißforschung, Lichtenbergstrasse 4, D-85748 Garching, Germany

1 INTRODUCTION

Disulphide bonds play a key role in determining the structure and properties of wheat gluten proteins. The well-known effects of oxidising or reducing agents on the rheological properties of dough and gluten are undoubtedly due to changes in the thiol/disulphide structure of gluten proteins. About 95 % of total cysteines in wheat flour are present in the disulphide (SS) form. Most α- und γ-type gliadins have only intramolecular disulphide bonds located in the *C*-terminal domains[1,2]. LMW and HMW subunits of glutenin form both intra- and intermolecular disulphide bonds and occur in an aggregated state. About 5 % of total cysteines in flour are present in the thiol (SH) form[2]. Only small amounts of free SH groups (≈ 0.5 %) are present in low-molecular-weight compounds, mainly in glutathione and cysteine, most being present in flour proteins (≈ 4.5 %). The aim of the present work was, firstly, to optimise the classical method of Ellman[3] for the quantitative determination of thiol groups in wheat flour. Secondly, to determine the distribution of thiol groups on the Osborne fractions and their location in gluten protein using a fluorescent reagent as a marker.

2 MATERIALS AND METHODS

2.1 Determination of free thiol groups

For the extraction of thiol-containing compounds from wheat flour (cv. Rektor), four extraction solvents (S1-S4) were compared differing in pH values and compositions (Table 1). After adding Ellman's reagent, the suspension was allowed to stand for 30 min at room temperature. Subsequently, the suspension was centrifuged and the coloured supernatant was measured at 412 nm against a corresponding solvent and flour blanks, respectively. The amount of thiols was calculated from a calibration curve for glutathione.

2.2 Rheological studies

Kernels of cv. Rektor (REK) were milled under air (REK-O_2) or under nitrogen (REK-N_2), and the flours were stored for two weeks under these conditions. Microscale extensigrams of gluten and dough were prepared following the method of Kieffer *et al*[4].

2.3 Localisation of free thiol groups

Flour proteins were extracted stepwise with water, with a salt solution, with 60 % ethanol and with an SDS containing solvent. The fractions obtained were mixed with the fluorescent reagent DACM (=N-(7-Dimethylamino-4-methyl-2-oxo-3-chromenyl) maleimide)[5,6]. Proteins soluble in the SDS solvent and treated with DACM were separated after reduction of disulphide bonds by RP-HPLC[7] using a fluorescence detector. Fluorescent proteins were collected and analysed for their N-terminal amino acid sequences. Fluorescent peptides obtained by thermolytic digestion of proteins were isolated by RP-HPLC, analysed for amino acid sequences and assigned to known sequences of gluten proteins.

3 RESULTS AND DISCUSSION

3.1 Determination of free thiol groups

Flour of the wheat variety Rektor (REK) was suspended in the respective extraction solvent (Table 1) and was stirred under nitrogen. After adding Ellman's reagent, the suspension was centrifuged and the coloured supernatant was measured at 412 nm against a corresponding solvent blank. The values obtained ranged from 1.05 to 1.78 µmol/g flour (Table 1). The lowest amounts were achieved with S2 (water), the highest amounts with S4 (urea containing solvent). Because of the slight yellow colour of the supernatants, which appeared before addition of Ellman's reagent and might disturb thiol analysis, the experiments were repeated measuring against a blank which contained the same amount of flour and the same solvent as the sample, but with no Ellman's reagent (flour blank). The results listed in Table 1 demonstrate that about 18-35% less thiol groups were measured using flour blanks. The highest values were obtained with SDS (S3) or urea containing extraction solvents. Enzymic digestion of flour proteins with thermolysin did not significantly improve the accessibility of thiol groups. For routine analysis, extraction of flour with solvent S3 and measurement against a blank derived from the same flour is recommended. The optimised Ellman method was applied to the comparison of flours milled and stored under nitrogen (REK-N_2) in comparison with those milled and stored under air (REK-O_2). The amounts of free thiols were significantly higher in REK-N_2 (1.22 µmol/g) than in REK-O_2 (0.64 µmol/g) after a storage time of two weeks.

3.2 Rheological studies

Increased amounts of free thiol groups might cause differences in the rheological properties of dough and gluten. Therefore, extension tests were performed on dough and gluten, which were produced from the flours REK-O_2 and REK-N_2 under air and nitrogen, respectively. Regarding the extensigrams, both gluten and dough which were prepared under nitrogen showed lower maximum resistance and greater extensibility than gluten and dough prepared under air. Dough prepared under air from flour milled under nitrogen (REK-N_2/O_2) had a lower maximum resistance than dough prepared under air (REK-O_2). Compared with dough prepared under nitrogen (REK-N_2), REK-N_2/O_2 showed a greater maximum resistance. Concerning the extensibility, differences between REK-O_2 and REK-N_2/O_2 but not between REK-N_2 and REK-N_2/O_2 could be found. Regarding gluten which was prepared under air from flour milled under nitrogen, only differences concerning the maximum resistance and extensibility could be observed between REK-N_2 and REK-N_2/O_2.

Table 1 *Amounts of accessible thiol groups measured with Ellman's reagent using different blanks (μmol SH/g flour)*

Solvent[a]	pH	Solvent blank	Flour blank	Difference
S1	2.0	1.23	0.88	0.35 (≡ 28 %)
S2	5.8	1.05	0.83	0.22 (≡ 21 %)
S3	7.0	1.48	1.21	0.27 (≡ 18 %)
S4	8.0	1.78	1.16	0.62 (≡ 35 %)

[a] S1: trifluoroacetic acid (TFA; 0.1 %)
S2: destilled water
S3: SDS (1.5 %)/Tris-HCl (62.5 mmol/l)
S4: urea (6 mol/l)/triethylamine (0.5 %)/NaCl (0.1 mol/l)/Tris (0.05 mol/l)

3.3 Localisation of free thiol groups in wheat flour proteins

Flour REK-O_2 was fractionated into water-soluble albumins and low-molecular-weight thiol compounds, salt-soluble globulins, alcohol-soluble gliadins and SDS-soluble glutenins. Free thiol groups were labelled adding DACM, and fluorescence was measured against a blank. The highest amounts of free thiol groups could be detected within the SDS-soluble glutenins (60 % of total recovered fluorescence), followed by water-soluble compounds (31 %) and globulins (9 %). The gliadin fraction was free of labelled thiol groups. The sum of the determined thiol groups (1.65 μmol/g) was somewhat higher than the content found with Ellman's reagent (1.21 μmol/g).

In order to determine the position of free thiol groups in flour proteins, the SDS-soluble fraction labelled with DACM was reduced with dithioerythritol and separated by RP-HPLC. The measurement of fluorescence during elution indicated that major fluorescent components were located within the elution area of LMW subunits of glutenin. N-terminal sequencing of the fluorescent proteins allowed the assignment of the sequences to known sequences of gluten proteins (Table 2). One protein (peak 8) corresponded to the s-type of LMW subunits and the others (peaks 10 and 11) to α- and γ-type gliadins.

The position of the cysteine residues was determined by partial hydrolysis of the fluorescent proteins with thermolysin and sequencing of the fluorescent peptides. The LMW subunit contained C^x as a binding site for DACM, and thus C^x is present as a free thiol in flour (nomenclature of cysteine residues according to Köhler *et al.*[9]). The γ-type

Table 2 *N-terminal sequences of fluorescent proteins*

Peak	Sequence[a]	Protein type [8]
Peak 8a	SHIPGLERPSQQQPLPPQQTLXXHH	LMW-s
Peak 8b	SHIPGLERPSQQQPLPPXQXXL	LMW-s
Peak 10	VRVPQLQXQN	α-gliadin
Peak 11	NMQVDPXYQVQXPQQ	γ-gliadin

[a] Single-letter nomenclature for amino acids; X: not identified

gliadin included two free thiol groups (C^b and C^z) and the α-type gliadin included one free cysteine residue (C^z).

4 CONCLUSION

The results of the quantitative determination of free thiol groups in wheat flour with Ellman`s reagent are strongly influenced by extraction procedure and solvent. Extraction with an SDS containing solvent and measurement against a flour blank appear to be an appropriate method. Flours milled and stored under nitrogen had higher thiol contents than flours milled and stored under air. Dough and gluten prepared under nitrogen gave lower maximum resistance and higher extensibility than those prepared under air. The addition of a fluorescence reagent (DACM) to Osborne fractions and the isolation of fluorescent proteins enabled the identification of protein-bound thiol groups. The results demonstrated that DACM was specifically bound to one cysteine residue of a LMW subunit, to two cysteine residues of a γ-type gliadin and to one cysteine residue of an α-type gliadin. Thus, these protein-bound thiol groups present in wheat flour might be involved in redox reactions during dough mixing.

References

1. P.R. Shewry and A.S. Tatham, *J Cereal Sci*, 1997, **25**, 207.
2. W. Grosch and H. Wieser, *J Cereal Sci*, 1999, **29**, 1.
3. G.L. Ellman, *Arch Biochem Biophys*, 1959, **82**, 70.
4. R. Kieffer, F. Garnreiter, H.-D. Belitz, *Z Lebensm Unters Forsch*, 1981, **172**, 193.
5. K. Shimada and K. Mitamura, *J Chrom B*, 1994, **659**, 227.
6. B. Kågedal and M. Källberg, *J Chrom*, 1982, **229**, 409.
7. H. Wieser, S. Antes and W. Seilmeier, *Cereal Chem*, 1998, **75**, 644.
8. E. J-L. Lew, D.D. Kuzmicky, D.D. Kasarda, *Cereal Chem,* 1992, **69,** 508.
9. Köhler P, Belitz H-D and Wieser H, *Z Lebensm Unters Forsch*, 1993, 196, 239.

Acknowledgements

Funding for the work, as part of the EU COST programme (contract FAIR-CT97-3010) is gratefully acknowledged.

GLUTEN DISULPHIDE REDUCTION USING DTT AND TCEP

N. Guerrieri, E. Sironi and P. Cerletti

Università degli Studi di Milano, Dipartimento di Scienze Molecolari Agroalimentari
Via Celoria 2- 20133 Milano, Italy

1 INTRODUCTION

Gluten behaviour is fundamental in determining the success of breadmaking. Gluten is formed by the association of gliadins and glutenins, its behaviour depends on the interaction of gluten protein molecules among themselves. Much has been done for a better understanding of the molecular structure of gluten and its constituent proteins[1], but comparatively little attention has been paid to the correlation between the surface properties of the assembly and the redox behaviour. This subject is of major interest because of the easer disulphide interchange and the great importance of protein and other molecular interactions within flour, such interactions determining the role of gluten in structuring and kneading a bread crumb i.e. the success of breadmaking.

The present work investigates the sulphydryl-disulphide structure of the gluten polymer by performing reduction on solubilised gluten proteins. A study was made of how dithiothreitol (DTT) and Tris-(2-carboxyethyl)-phosphine hydrocloride (TCEP) reduction, in neutral or acid medium, affected gluten structure. The effect was determined by SDS-PAGE separation of polypeptides and by fluorescence of 8-aniline-1-naphtalene sulphonate (ANS) when it interacts with the hydrophobic areas of gluten (extrinsic fluorescence).

2 MATERIALS AND METHODS

2.1 Materials

Durum wheat (*Triticum durum*) from the Italian cultivar Capeiti, of good baking quality (BQ) was supplied by the Istituto Nazionale della Nutrizione (I.N.N.), Rome.

Gluten was prepared from remilled semolina, by the AACC standard method of hand washing n°38-10, using 30 min resting time. It was frozen in liquid nitrogen, liophylised, ground to 60 mesh, dry-stored at 4 °C. The gluten contained 78 % proteins (d.w.). All chemicals were of analytical grade.

2.2 Methods

2.2.1 Protein determination and separation. The total protein of the gluten sample was determined by a Carlo Erba NA 1500 automatic nitrogen analyser, with atropine standard; the conversion factor was 5.7. Soluble proteins were quantified spectrophotometrically as described by Eynard *et. al.*[2]. Gluten proteins were extracted either in 0.05 N Tris/HCl buffer pH 7.5, containing 2% SDS (sodium dodecyl sulphate) or in acetic acid: 5 mL solvent were normally added to 64 mg frozen dried gluten and the extraction was carried out for 1 h at room temperature under magnetic stirring. The ratio of gluten to acetic acid was modified as necessary in fluorescence experiments. The yield of extracted protein under such conditions approached 100%.

2.2.2 Electrophoresis. SDS-PAGE was carried out according to Laemmli[3] on 12% polyacrylamide gels: the sample containing denaturing solution was not boiled so as not to favour any reduction if DTT and TCEP were present, and was applied to the gel as such. Protein markers were the Sigma standard mixture of proteins. The gel was stained with Coomassie Brilliant Blue R-250. The images were acquired using a scanner, Studioscan IIsi, AGFA, interfaced with a personal computer and quantification was done with Cream 1D software (Kem-en Tec, Copenhagen Denmark).

2.2.3 DTT quantification after dialysis. DTT was measured using the Ellman method[4], reactivity to DTNB, modified as suggested by S. Antes, Cereal Research Unit, Garching, Bavaria (D) in a personal comunication: about 100 μL dialysed DTT sample in acetic acid were added with 200 μL 50% 0.05 M 1-propanol/phosphate buffer, pH 8.0 and 2 mL 0.5 M phosphate buffer pH 7.0, 10 mM DTNB. After 20 min the yellow samples were measured at 412 nm against the buffer.

2.2.4 Fluorescence measurements. The ANS binding to the proteins was evaluated by titrating the AAE (3 mL, 0.3 mg protein/mL) with 1-10 μL of 1 mM or 10 mM ANS in 0.05 N acetic acid and measuring the fluorescence developed, λ excitation 402 nm, λ emission 480 nm, slit width 2.5 nm. Operations were at 25° C, with magnetic stirring. The best fit to the titration values, as described in a previous paper[5], turned out to be a curve which, using Peakfit software (Jandel Scientific, Erkrath, Germany) could be deconvoluted into two components corresponding to reactivity at high or low ANS concentrations. The components were linearized by the Lineweaver Burk plot (low ANS concentration) and the Hill equation (high ANS concentration)[6]. This allowed the determination of maximal fluorescence for the low- and the high- affinity sites in gluten, of the ANS concentration producing it and of the dissociation costants (Kd) of the ANS-gluten complex at high and low affinity sites. The reproducibility of the fluorescence measurements was ascertained using an ovalene standard in metacrylate.

3 RESULTS AND DISCUSSION

3.1 Reduction studies: separation of polypeptides

The DTT reduction process (25°C, 1 h) in the alkaline medium (0.05 M Tris/HCl pH 7.5, 2 % SDS) was rapid and was maintained throughout the experiment. The released polypeptides of the DTT treated gluten were analysed by SDS-PAGE.

Analysis of the sample (Figure 1) in Tris buffer both with DTT and after DTT removal by dialysis against the SDS containing buffer, revealed the same separation pattern (not shown). No change in polypeptide composition was observed with 0.1 mM DTT (not shown). Modifications based on disulphide bond reduction started at 1 mM DTT in the sample (Figure 1): the HMW aggregates, either immobile or with minimal movement, started decreasing, the major change occurring between 2 and 5mM DTT with complete disappearance above 10 mM DTT. The decrease was accompanied by changes in the 145 kDa band which appeared above 2 mM DTT, and vanished above 15 mM DTT.

A major reduction product was a polypeptide of about 40 kDa in the α and γ type gliadin region (Figure 1); other polypeptides that accumulated from the beginning of reduction were at 61 kDa, probably D glutenin[7] and 103 kDa. At 31 kDa a polypeptide of native gluten became more evident during the reduction while one at 45 kDa became apparent with 5 mM DTT. Some of the newly formed components disappeared when the DTT increased, confirming their disulphide sustained oligomer nature: this is the case for the bands that became apparent with 1 mM DTT at 37kDa, 71 kDa and 138 kDa. These bands first increased but then disappeared above 10 mM DTT.

The sample treated with 25 mM DTT behaved like the completely reduced sample (142 mM 2-mercaptoethanol) and a nitrogen atmosphere during operation (results not shown) had little effect on the above mentioned results.

In order to check the molecular behaviour under the acid conditions required for fluorescence measurement, gluten, solubilized by acetic acid (the AAE), was subjected to reducing conditions: the gluten molecules remained dispersed in the solvent, even though no SDS was present. These conditions led to few changes in gluten polypeptides: only a slight reduction in the HMW oligomers, and an increase of the intensity of the 40 kDa and 71 kDa polypeptides at high DTT (Figure 1). The reduced polypeptides became evident when urea is added before DTT: above 25 mM DTT the gluten appeared to be completely reduced. The results indicate that without urea the DTT reduction of the disulphide bonds in the acid medium was minimal, whereas in the presence of 5M urea the pattern (Figure 1) was the same as that in alkaline SDS medium.

Another reductant (TCEP) with good reducing capacity over a large pH range (1.5-8.5[8]) was used to reduce the solubilised gluten in the acid medium. The released polypeptides analysed by SDS-PAGE revealed the some polypeptides pattern as DTT in the alkaline medium, a nearly completely reduction occurs at 25mM TCEP, while 50 and 100mM TCEP differ only for the 51 and 117 kDa polypeptides (results not shown).

3.2 Fluorescence measurement

The intrinsic fluorescence of tryptophan residues was also affected by low concentrations of TCEP and progressively decreased in intensity whereas the effect on tyrosine was less. The positions of the maxima were not modified. This behaviour indicates surface modifications related to reduction.

The extrinsic fluorescence of the ANS-gluten complex did not change with 1mM TCEP, but decreased when TCEP was 25mM or higher. The titration curves were deconvoluted into two components which correspond to high and low affinity sites for ANS. Kd was calculated from the deconvoluted curves of the high affinity sites. The gluten affinity (1/Kd) for ANS decreased on reduction up to 25mM TCEP and then remained nearly constant.

4 CONCLUSIONS

The gluten surface was changed by reduction with TCEP in acid medium. The

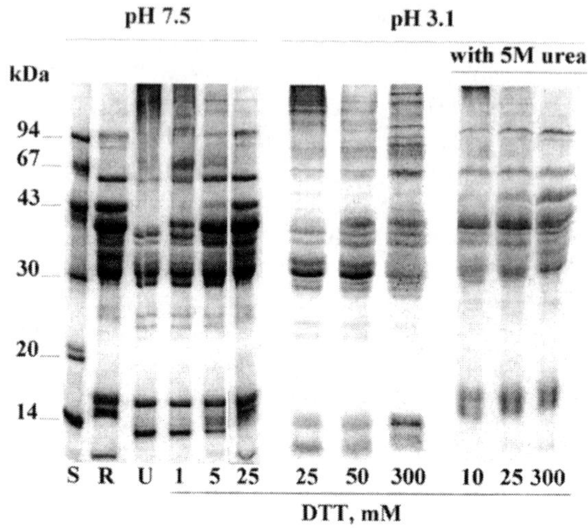

S: standard MW
R: completely reduced gluten (with β-mercaptoethanol)
U: unreduced gluten

decrease in the intrinsic and extrinsic fluorescence indicates that the reduction affected the hydrophobic areas of the gluten structure, which probably decreased. 25mM is an important step for the reductive transition: most of the disulphides connecting different polypeptides have been reduced and fluorescence parameters undergo significative variation.

In our opinion this indicates that hydrophobic sites and areas of gluten complex are sustained by disulphides which are not easily available to reductant or with high conformational energy. The HMW associations are disulphide dependent but probably do not affect hydrophobic surface reactivity.

References

1. P.R. Shewry and A.S. Tatham, *J. Cereal Sci.*, 1997, **25**, 207.
2. L. Eynard, N. Guerrieri, and P. Cerletti, *Cereal Chem.*, 1994, **71**, 434.
3. U.K. Lammli, *Nature*, 1970, **227**, 680.
4. G.L. Ellman, *Arch. Biochem. Biophys.*, 1959, **82**, 70.
5. N. Guerrieri, E. Alberti, V. Lavelli, P. Cerletti, *Cereal Chem.*, 1996, **37**, 368.
6. C.R. Cantor and P.R. Shimmel, 'Biophysical Chemistry. III. The behaviour of biological macromolecules', W.H. Freeman ed., San Francisco, 1980
7. S. Masci, D. Lafiandra, E. Porceddu, E.J.L. Lew, H.P. Tao, D:D. Kasarda, *Cereal Chem.*, 1993, **70**, 581.
8. J.C. Han and G.Y. Han, *Anal. Biochem.*, 1994, **220**, 5.

Acknowledgements

Funding for the work, as part of the EU FAIR Programme (Contract CT97-3010) is gratefully acknowledged.

MODEL STUDIES ON THE REACTION PARAMETERS GOVERNING THE FORMATION OF DISULPHIDE BONDS IN LMW-TYPE PEPTIDES BY DISULPHIDE ISOMERASE (DSI)

N. Bauer and P. Schieberle

Deutsche Forschungsanstalt für Lebensmittelchemie, Lichtenbergstrasse 4,
85748 Garching, Germany

1 INTRODUCTION

It is generally accepted that the formation of disulphide bonds is chiefly responsible for the characteristic visco-elastic properties of wheat doughs. The major part of the cysteine residues in the wheat proteins is, however, fixed by intramolecular bonding. Very few, but important cysteine residues present, in particular, in the LMW- and HMW-subunits of glutenin, are proposed to form the large protein network needed for bread crumb formation. It is, however, unclear whether this "polymerisation" occurs randomly or is strongly directed when the flour is mixed in the presence of water.

Disulphide isomerase (DSI) offers the possibility to form such disulphide bonds in a directed way. The enzyme, a member of the oxidoreductase family, is well known in biochemistry as a catalyst in formation and reshuffling of disulphide bonds during biosynthesis of proteins[1-4]. However, in the literature no data are available on the selectivity of DSI for cysteine residues in gluten proteins. The purpose of the following investigation was to study the effectiveness of the DSI reaction by using synthesised peptides with amino acid sequences identical to those of the LMW-subunits of glutenin.

2 MATERIALS AND METHODS

2.1 Materials

2.1.1 Enzyme. DSI (EC 5.3.4.1) was purchased from Fluka Chemie, Switzerland.

2.1.2 Peptide Synthesis. The cysteine containing peptides LGQCV, FSQQQPCS and QQQPIQQQPQPFPQQPPCSQQ were made by Merrifield solid phase peptide synthesis. After manual cleavage of the solid phase with trifluoroacetic acid, the peptides were precipitated and freeze dried. To obtain disulphide peptides for reshuffling experiments, LGQCV was oxidised with potassium bromate and purified by RP-HPLC.

2.2 Methods

Experiments using substoichiometric amounts of DSI were carried out at pH 5, 7 and 9, respectively. Equimolar solutions in Tris/HCl buffer of either FSQQQPCS and LGQCV (oxidation) or FSQQQPCS and $(LGQCV)_2$ (reshuffling) were mixed and incubated for the times given in the tables after adding DSI in an enzyme to substrate ratio of 1 : 10, based on the amount of thiol groups present.

Experiments using catalytic amounts of DSI were carried out at pH 4 in the presence of 5 µmol/mL oxidised glutathione. Solutions of either LGQCV or QQQPIQQQPQPFPQQPPCSQQ were incubated in the time range of 5 minutes to 80 hours after adding DSI in an enzyme to substrate ratio of 1:2000, based on the amount of thiol groups present.

All reactions were stopped by adding hydrochloric acid to a final pH below 3.

Reaction products were isolated by means of RP-HPLC and identified by means of Edman degradation, ESI-MS and/or MALDI-TOF-MS.

To all experiments the synthetic dipeptide FA was added as an internal standard. Reaction temperature was 25 °C.

3 RESULTS

The oxidation and reshuffling of the pentapeptide LGQCV or the octapeptide FSQQQPCS was shown to be much accelerated when substoichiometric amounts of DSI were used (Figure 1). In reshuffling experiments, however, an accelerated reaction (Figure 2) only occurred at pH 7 and 9. Without enzyme (pH 5, 7 and 9) or with DSI at pH 5 no reshuffling was detectable (Figure 2).

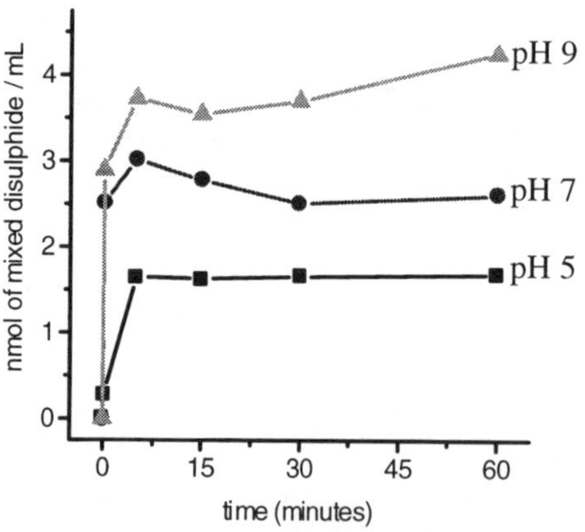

Figure 1: Oxidation of LGQCV and FSQQQPCS by substoichiometric amounts of DSI at pH 5, 7 and 9

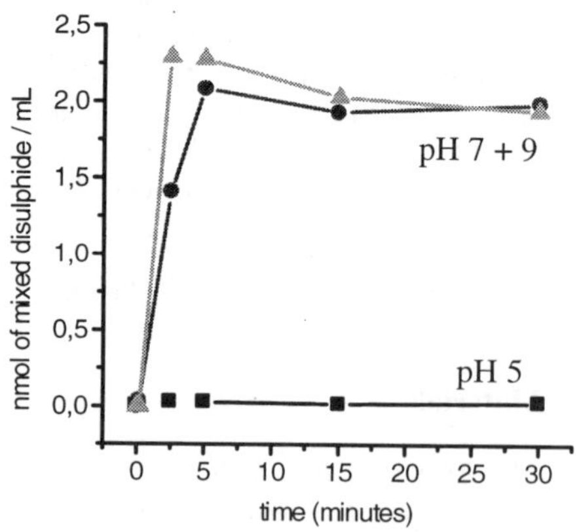

Figure 2: Reshuffling of $(LGQCV)_2$ and FSQQQPCS by substoichiometric amounts of DSI at pH 5, 7 and 9

The use of catalytic amounts of DSI also led to a significant acceleration in the oxidation of a pentapeptide. Because oxidised glutathione had to be present in the reaction mixture, a small amount of mixed disulphide with glutathione was always detectable (Figure 3), even without enzyme addition (data not shown).

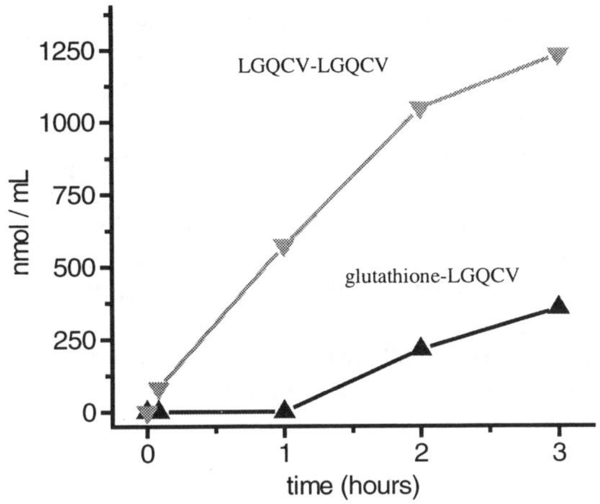

Figure 3: Oxidation of LGQCV by catalytic amounts of DSI at pH 4

When larger peptides (MW > 2500) were studied, the DSI activity against these peptides was drastically decreased. The polypeptide QQQPIQQQPQPFPQQPPCSQQ was oxidised very slowly (Figure 4) whereas the mixed disulphide with glutathione became predominant. From a quantitative point of view it is interesting to note that a large amount of the polypeptide did not react at all.

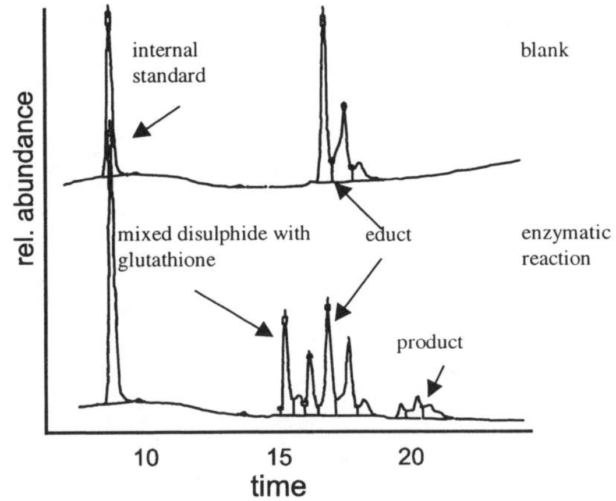

Figure 4: RP-HPLC of the reaction products formed after oxidation of QQQPIQQQPQPFPQQPPCSQQ in the presence of glutathione by catalytic amounts of DSI at pH 4

4 DISCUSSION

The active site of DSI (in the active state) consists of the two cysteine residues ^{30}Cys-Pro-His-Cys^{33} linked to a disulphide bond. Thus, DSI itself is a disulphide reagent. When using DSI as an oxidation catalyst, the disulphide bond in the enzyme is reduced, and intermolecular disulphide bonds between peptides are formed. Substoichiometric amounts of DSI were shown to be active up to 10 minutes under the reaction conditions used. However, without a reoxidising agent present DSI lost its oxidising power, thus the oxidation of peptides stopped.

When catalytic amounts of DSI were studied, oxidised glutathione had to be added to reoxidise the enzyme. The enzyme was now active all the time but a side reaction became predominant: the formation of mixed disulphides with glutathione.

In summary, the results have shown that DSI in substoichiometric or catalytic amounts has a strong oxidising power versus small gluten-type peptides. However, the more the molecular weight of the peptides increased, the more the activity of DSI against these peptides decreased. This suggests that DSI may not influence the formation of disulphide bonds in gluten proteins when used as a flour improver.

References

1. J.C.A. Bardwell, K. McGovern and J. Beckwith, *Cell* 1995, **67**, 581.
2. M. Wunderlich, A. Otto, R. Seckler and R. Glockshuber, *Biochemistry* 1993, **32**, 12251.
3. M. Wunderlich, R. Jaenicke and R. Glockshuber, *J. Mol. Biol.,* 1993, **233**, 559.
4. U. Grauschopf, J.R. Winther, P. Korber, T. Zander, P. Dallinger and J.C.A. Bardwell, *Cell* 1995, **83**, 947.

OXIDATION OF HIGH AND LOW MOLECULAR WEIGHT GLUTENIN SUBUNITS ISOLATED FROM WHEAT

W.S. Veraverbeke[1], O.R. Larroque[2], F. Békés[2] and J.A. Delcour[1]

1. Laboratory of Food Chemistry, Katholieke Universiteit Leuven, Kardinaal Mercierlaan 92, B-3001 Heverlee, Belgium. 2. Grain Quality Research Laboratory, CSIRO Plant Industry, P.O. Box 7, North Ryde, NSW 1670, Australia.

1 INTRODUCTION

Wheat glutenin is a heterogeneous mixture of polymers of high molecular weight (HMW-GS) and low molecular weight-glutenin subunits (LMW-GS) linked by interchain disulphide (SS) bonds. Several studies have demonstrated that differences in the molecular weight (MW) distribution of glutenin are responsible for breadmaking quality differences[1,2]. Furthermore, it was also shown that glutenin subunit composition relates to breadmaking quality. From the above, it may be concluded that at least one way in which the glutenin subunit composition affects breadmaking quality is by its potential to form large polymers. In this study, isolated glutenin subunits were oxidatively polymerised to further investigate this hypothesis.

2 MATERIALS AND METHODS

2.1 Oxidation of glutenin subunits

HMW-GS and LMW-GS, isolated from flour of wheat cultivar Minaret essentially as described previously[3], were oxidised both separately and in a 1:2 (w/w) mixture with different inorganic oxidants (KIO_3, $KBrO_3$ or H_2O_2). Oxidations were performed at room temperature in 1 % (w/v) protein solutions at pH 3.0. A study of the effects of oxidation as a function of time with addition of oxidant in a single step ('single step' oxidation) was compared with a procedure in which the level of oxidant was increased gradually over time ('stepwise' oxidation). In the following, one 'unit' of oxidant is defined as the amount of oxidant that is theoretically needed to oxidise 1 % of the free sulphydryl (SH) present in the isolated HMW-GS.

2.2 Evaluation of the effects of oxidation

2.2.1 Decrease of free SH. The level of free SH during oxidation was measured spectrophotometrically after reaction of the free SH with Ellman's reagent.

2.2.2 Polymer formation. Polymer formation was monitored with three different techniques for measuring polymer MW distribution. Typical MW distribution profiles obtained with the different techniques are shown in Figures 1-3. With size-exclusion HPLC (Figure 1), the disappearance of monomers, the formation of polymers and shifts in the size distribution of these polymers were simultaneously monitored by measuring the changes in the relative proportions of three MW classes [(1) monomers with MW < 90k, (2) low MW glutenin polymers with MW between 90k and 300k and (3) high MW glutenin polymers with MW > 300k]. With multi-layer SDS-PAGE (Figure 2), MW distribution was monitored by measuring the relative proportions of different MW fractions, represented by the different layers of the gel. Multi-layer SDS-PAGE is characterised by a higher upper MW limit for analysis (> 800k) than size-exclusion HPLC (670k). Flow-field flow fractionation (flow-FFF) was only recently introduced in cereal science[2]. However, it offers enormous potential for the MW determination of the very large glutenin polymers since it is not characterised by an upper MW limit for analysis. Using calibration with protein MW standards, an average MW was calculated from the flow-FFF fractograms (Figure 3).

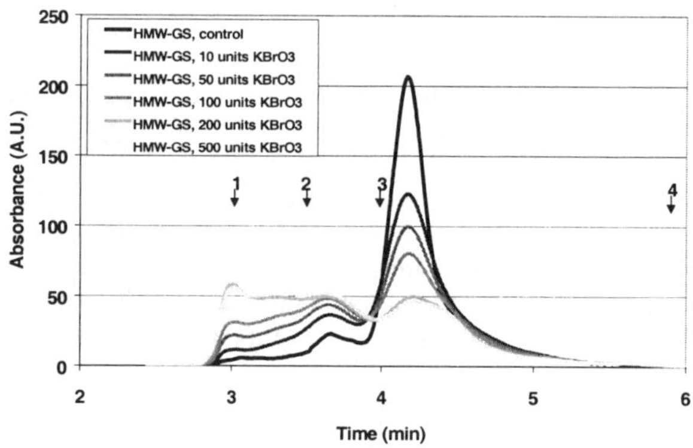

Figure 1 *Size-exclusion HPLC profiles of oxidation products of a 'stepwise' oxidation of HMW-GS with $KBrO_3$ [MW markers: (1) 669k, (2) 135k, (3) 67k and (4) 1.4k are indicated by arrows]*

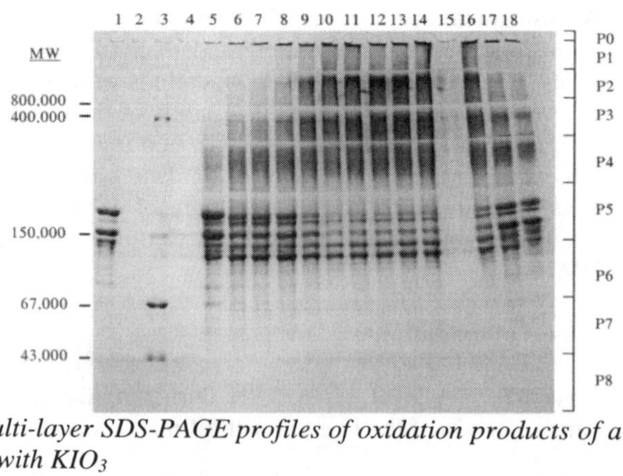

Figure 2 *Multi-layer SDS-PAGE profiles of oxidation products of a 'stepwise' oxidation of HMW-GS with KIO_3*

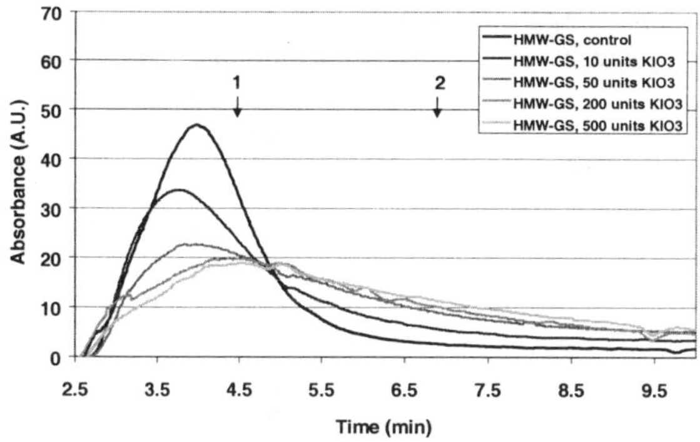

Figure 3 *Flow-field flow fractograms of oxidation products of a 'stepwise' oxidation of HMW-GS with KIO_3 [MW markers (1) 67k and (2) 669k are indicated by arrows]*

2.2.3 Polymerisation behaviour of different glutenin subunits. Since the B and C groups of LMW-GS and the individual HMW-GS in the mixture of HMW-GS were resolved by multi-layer SDS-PAGE, the extent of incorporation of these different glutenin subunits into polymers could be compared.

3 RESULTS AND DISCUSSION

3.1 Oxidation of HMW-GS

The different oxidants differed in their ability to produce large size polymers upon oxidation of HMW-GS and could be ranked in order of increasing efficiency as $KBrO_3$ < KIO_3 < H_2O_2. However, under all conditions significant levels of HMW-GS remained monomeric (see for example Table 1). Interestingly, high concentrations of KIO_3 negatively affected polymerisation in 'single step' oxidations. Differences were observed in the polymerisation behaviour of the different HMW-GS in the mixture. As shown in Table 1, the monomers remaining during 'stepwise oxidation' of HMW-GS became enriched in y-type HMW-GS and 1Dx5. Interestingly, the unpolymerised y-type HMW-GS displayed higher mobilities on SDS-PAGE than their completely reduced forms and unpolymerised HMW-GS 1Dx5 displayed three different mobility forms, suggesting that formation of intrachain SS prevented incorporation of these subunits into polymers.

3.2 Oxidation of LMW-GS

The different oxidants could be ranked in order of increasing efficiency to polymerise LMW-GS as $KBrO_3$ < KIO_3, H_2O_2. Compared with HMW-GS, lower levels of LMW-GS remained monomeric. Multi-layer SDS-PAGE showed that the B group of LMW-GS were incorporated to a larger extent into polymers than the C group.

3.3 Oxidation of a 1:2 (w/w) mixture of HMW-GS and LMW-GS

When a mixture of HMW-GS and LMW-GS was oxidised, no effect was observed on polymer size that would be associated with a possible co-polymerisation of HMW-GS and LMW-GS.

Table 1 *Free SH content, MW distribution and HMW-GS composition of remaining monomers of oxidation products of a 'stepwise' oxidation of HMW-GS with KIO_3*

		Level of KIO_3 (units)								
	Starting material	0	10	50	100	150	200	300	400	500
Free SH content (μmol SH/g protein)	50	45	40	26	9	3	3	3	4	5
Multi-layer SDS-PAGE size fractions (%)										
P0	1	2	3	6	6	8	7	5	5	6
P1	0	0	1	2	5	9	8	6	7	10
P2	0	0	3	7	17	22	22	24	24	20
P3	0	7	10	13	19	20	19	21	22	19
P4	13	26	25	24	22	21	22	22	21	20
Monomer	86	65	58	48	31	21	22	22	21	24
Size exclusion-HPLC size fractions (%)										
HMW polymer	2		7		31	39		39		37
LMW polymer	12		25		33	32		32		33
Monomer	86		68		36	28		29		31
FFF average size (kDa)		91	126	145	252	255	260	307	313	332
HMW-GS composition of monomers (%)										
1Ax1	26	17	19	18	13	10	8	9	9	8
1Dx5	10	15	14	12	14	14	15	14	14	13
1Bx7	29	21	21	22	21	18	18	17	17	18
1By9/1Dy10	35	47	46	48	52	58	59	60	60	61

5 CONCLUSIONS

The MW distribution of the polymers formed by oxidation of HMW-GS and/or LMW-GS depended on the oxidant. However, none of the oxidation conditions resulted in complete polymerisation. Possible explanations are (1) oxidation of SH to a higher oxidation stage than SS and (2) 'alternative' SS bond formation. Finally, differences in polymerisation efficiency were observed for structurally different glutenin subunits: LMW-GS, x-type HMW-GS and B-type LMW-GS polymerised more efficiently than HMW-GS, y-type HMW-GS and C-type LMW-GS, respectively.

References

1. W.S. Veraverbeke and J.A. Delcour, *Crit. Rev. Food Sci. Nutr.* (in press)
2. M. Southan and F. MacRitchie, *Cereal Chem.*, 1999, **76**, 827
3. I.M. Verbruggen, W.S. Veraverbeke, A. Vandamme and J.A. Delcour, *J. Cereal Sci.*, 1998, **28**, 25

Acknowledgements

W.S. Veraverbeke wishes to acknowledge the Fund for Scientific Research - Flanders (Belgium) (F.W.O.) for his position as Postdoctoral Fellow.

INFLUENCE OF THE REDOX STATUS OF GLUTEN PROTEIN SH GROUPS ON HEAT-INDUCED CHANGES IN GLUTEN PROPERTIES

S. H. Mardikar[1] and J. D. Schofield[1]

1. The University of Reading, Department of Food Science and Technology, Whiteknights, P.O. Box: 226, Reading RG6 6AP, UK.

1 INTRODUCTION

It is important to understand the heat-setting properties of gluten for a number of reasons; for example, the functionality of gluten as a baking additive decreases with high temperature treatment[1], necessitating the use of a low temperature drying during the commercial manufacture of gluten. The functional damage that occurs to wheat caused by grain drying at excessive temperatures may also involve gluten proteins. Heat-setting is also important in the development of a suitable gluten-moulding technology, especially to use gluten as a bio-degradable structural material[2]. The rheological properties of gluten change on heating the sample and these are believed to be mainly due to changes in the degree of cross-linking. The total free SH content serves as an index for changes in cross-linking, as heat-induced oxidation of SH groups results in the formation of disulphide (S-S) bonds. The redox status of gluten can be altered by direct chemical blocking with alkylating agents such as iodoacetamide and results in a reduction in the total SH content, which thereby may influence the heat-induced changes in gluten properties.

In this study we have focused on the total free SH groups to reflect the heat induced changes in disulphide cross-linking and relate them to changes in the rheological properties of the gluten. By reducing the level of free SH groups by alkylation, we have proved further evidence that the transition from a viscous behaviour to a more elastic behaviour of heat treated gluten is due to formation of additional disulphide linkages.

2 MATERIALS AND METHODS

2.1 Materials

Flours of the two English wheat varieties namely: Hereward, a strong bread making wheat and Riband, a weak biscuit making variety, were obtained from Chorleywood Campden Food Research Association (CCFRA), Campden, UK.

2.2 Methods

2.2.1 Preparation of gluten samples. Gluten was washed out from dough as follows: 50 g flour was mixed with 30 ml distilled water (or 30 ml iodoacetamide solution in case of iodoacetamide treated glutens) for 3 minutes using a Minorpin™ dough mixer. The dough was hand-washed for 5 minutes in 1 litre of distilled water gently kneading the dough ball under water. The insoluble gluten was further washed two times in 1 litre of distilled water, kneading 3 minutes each time, until the milky starch stopped leaching into the wash water. Wet gluten samples were used for heat treatment and measurement of rheological properties, whereas the samples were freeze dried for SH content determinations. To prevent the oxidation of SH groups in gluten, wet samples were kept under water at all times and dry samples under a nitrogen cover.

2.2.2 Heat treatment of gluten samples. Wet gluten samples kept overnight under water were heated in a water bath at room temperature (20°C), 45, 60, 70, 85 and 100°C for 45 minutes.

2.2.3 Determination of rheological properties. a) *Small deformation rheology* Mechanical spectra of the gluten samples were obtained by carrying out oscillation frequency sweeps between 0.001 and 10 using a StressTech™ rheometer (StressTech AB, Lund, Sweden) in a parallel-plate conformation (plate size: P20). b) *Large deformation - bubble inflation* Bubble inflation experiments to measure the strain hardening[3] of gluten samples were done using the Dobraszczyk-Roberts™ bubble inflation system (Stable Microsystems, Guilford, UK)

2.2.4 Determination of SH groups. Total SH content in the gluten samples was determined using the Ellman reaction, following a recently standardised protocol (S. H. Mardikar and J. D. Schofield; unpublished results).

3 RESULTS AND DISCUSSION

3.1 Blockage of free SH groups using iodoacetamide

Incorporation of iodoacetamide at varying concentrations in the dough-mixing water brings about partial blockage of SH groups in the gluten samples washed out thereafter. There was a decrease in the free SH with increasing concentrations of iodoacetamide until about 25mM concentration, as shown in Figure 1. Gluten networks shredded into fibres with increasing iodoacetamide concentrations making it difficult to wash out the gluten, hence an iodoacetamide concentration of 5mM was used in subsequent studies.

3.2 Effect of heat-treatment on the total free SH content

There was a progressive decrease in the total free SH content as shown in Figure 2. The sudden increase in SH content at around 70°C is difficult to explain, but was a consistent

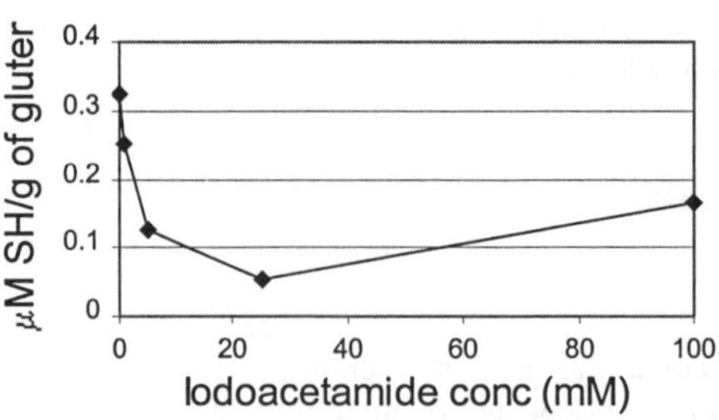

Figure 1 *Partial blockage of free SH at varying iodoactamide concentrations in Hereward gluten*

observation. It occurred at the temperature previously observed to be that at which polymerisation of gluten began to occur and may be related to the beginning of changes in polymer properties.

Figure 2 *The total free SH content of heat-treated glutens*

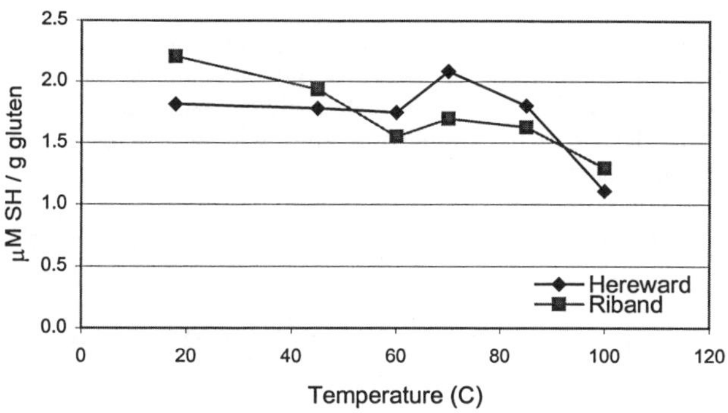

3.2 Changes in the rheological properties of heat-treated glutens

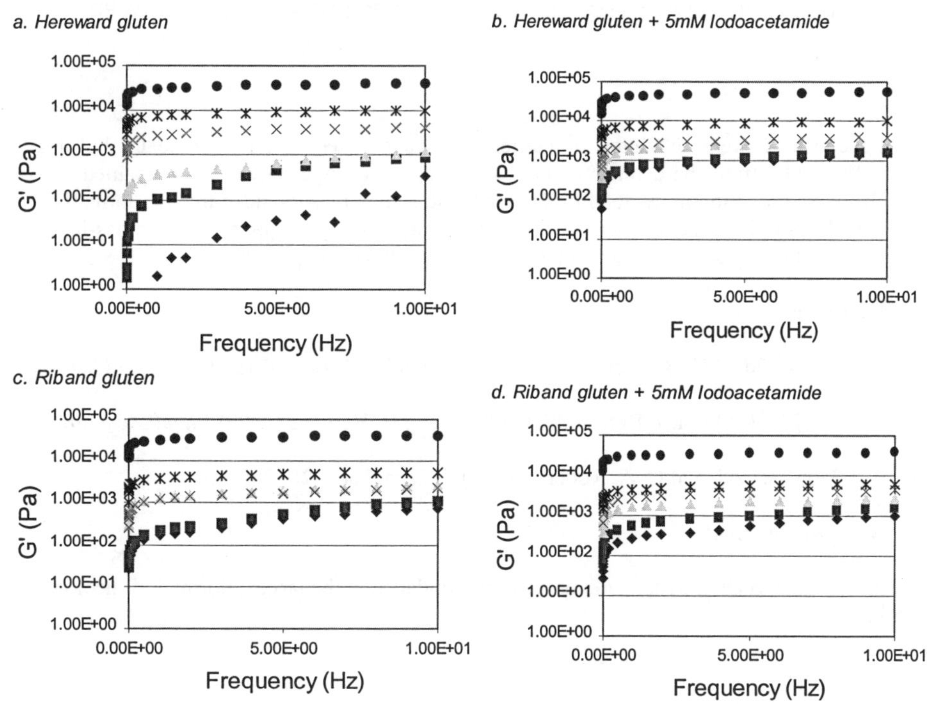

Figure 3 Changes in the storage modulus (G') of glutens heated 20, 45, 60, 70 85 and 100 °C (bottom to top in each graph)

3.2.1 Small deformation rheology. Oscillatory frequency sweep tests were used to examine changes in the rheological properties of gluten. The elastic character of samples increased with increases in heat-treatment temperature as indicated by increasing G'values. (Figure 3)

Similar trends were observed for both untreated and iodoacetamide treated gluten samples, although as expected the changes were less pronounced in the latter case.

3.2.2 Large deformation rheology The bubble inflation test is a large deformation rheological test and involves biaxial extension of the gluten samples, measuring the strain hardening of the sample. Results with heat-treated gluten samples (Table 4) show an increasing resistance to deformation (strain hardening) with an increase in temperature, indicating increasing cross-linking.

Table 4 *Strain hardening indices of heat-treated samples of control and iodoacetamide-treated gluten*

Temperature (°C)	Hereward	Hereward + 5mM iodoacetamide
20	---	2.409
45	2.053	2.452
60	2.45	2.677
70	2.819	---

STRAIN HARDENING = SLOPE LOG (STRESS) / LOG (STRAIN) (POWER LAW FIT)

4 CONCLUSIONS

Heat induced changes in gluten properties were manifested as an increase in the elastic character and strain hardening index. Since these changes were accompanied by a simultaneous reduction in the total free SH content, it may be concluded that the formation of additional disulphide bonds occurred which brought about an increase in cross-linking, altering the rheological properties of the heat-treated gluten samples.

References

1. J. D. Schofield, R.C. Bottomley, M. F. Timms and M. R. Booth (1983) *J. Cereal Sci.,* **1,** 241.
2. A. Redl, M. H. Morel, J. Bonicel, S. Guilbert. and B. Vergnes (1999) *Rheol. Acta,* **38,** 311.
3. B. J. Dobraszczyk and C. A. Roberts (1994) *J. Cereal Sci.*, **20,** 265.

Acknowledgements

Thanks are due to Dr. Bogdan Dobraszczyk for guidance in the large deformation rheology work.

EFFECTS OF OXIDOREDUCTASE ENZYMES ON GLUTEN RHEOLOGY

Clare V. Skinner[1], Amalia A. Tsiami[1], Gitte Budolfsen[2] and J. David Schofield[1]

1: University of Reading, Department of Food Science and Technology, RG6 6AP, UK
2: Novo Nordisk A/S, Novo Allé, 2880 Bagsværd, Denmark

1 INTRODUCTION

For some time oxidoreductase enzymes have been regarded as potential bromate replacers for use in breadmaking, and investigations have been made into finding an enzyme that can elicit functional changes in gluten similar to those thought to lead to increased loaf volume when bromate is added[1]. Bromate is thought to produce it's effects by causing an increase in the size of polymers in the glutenin fraction of gluten. Thus to assess the effect of an enzymic redox additive it would be useful to see if similar polymerisation were occurring.

Small deformation rheology can be used to indicate whether wheat gluten contains large polymers, as these will contribute significantly to the elastic properties of the gluten network.[2] The use of an additive that resulted in larger polymers would result in increased elastic properties in the gluten, and this would suggest the possibility of improving bread loaf volume.

The effects of glucose oxidase (GOX), a broad acting carbohydrate oxidase from *Microdochium nivale* (MCO), cellobiose dehydrogenase (CBDH) and some chemical oxidants on gluten rheology have been assessed after two dough resting times. Small deformation tests (strain sweep and stress relaxation) have been used to assess the effect of these additives on polymer size.

2 MATERIALS AND METHODS

2.1 Materials

Dough was made using Hereward flour and two untreated flours from Meneba Meel BV (Rotterdam, The Netherlands), one unnamed and referred to as Meneba flour and the other called Pelikaan. Most chemicals were obtained from Merck BDH (Poole, UK). Other dough ingredients were cane sugar, sodium chloride, 'Fermipan' instant yeast, deionised water and oxidant solution. The oxidant systems used were 50ppm potassium bromate, 50ppm ascorbic acid, 150U glucose oxidase (GOX),

500U GOX, 200U *Microdochium nivale* carbohydrate oxidase (MCO) and 25ppm cellobiose dehydrogenase (CBDH).

2.2 Methods

2.2.1 Dough preparation. Doughs were made with the different oxidant systems. They were mixed using the appropriate water absorption on a 10g pin mixer for 3.5 min and immediately washed to isolate the gluten. The process was repeated allowing the dough to rest for 45 min at 37°C prior to the gluten extraction.

2.2.2 Gluten extraction. Gluten was extracted from the dough by washing with 2% NaCl solution for 7 min in a Glutomatic machine. The resultant gluten was centrifuged in a Perten centrifuge for gluten preparation.

2.2.3 Rheological tests. The gluten was loaded onto a Bohlin VOR rheometer between rough plate-plate geometry of 30mm diameter with 1mm gap. The strain sweeps were run at a frequency of 1Hz from a strain of 0.00022 to a strain of 0.22. The stress relaxation tests were run with a strain rise time of 2 s and a strain of 0.0399. Both types of test were run at 32°C.

3 RESULTS AND DISSCUSSION

3.1 Strain sweeps

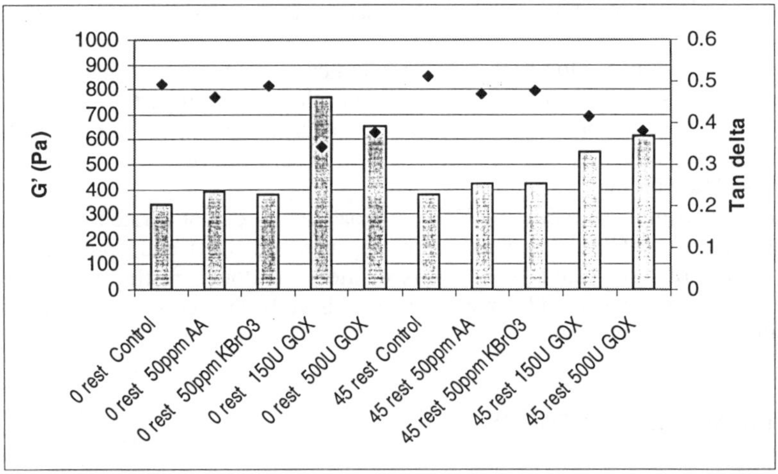

Figure 1 *Effect of standard redox additives on G' (cross hatched bars) and tan δ (closed diamonds)*

Strain sweeps carried out on glutens from doughs made with Meneba flour and treated with various redox systems gave data for G' and tan δ showing a small decrease in tan δ between the gluten from the control dough and those made with the chemical additives. But GOX produced a more substantial increase in G' and a decrease in tan δ (the data for rested doughs is more reliable when treated with GOX in all results). This indicates that overall GOX is acting to increase the size of the glutenin molecules, or the number of linkages per molecule resulting in increased elasticity.

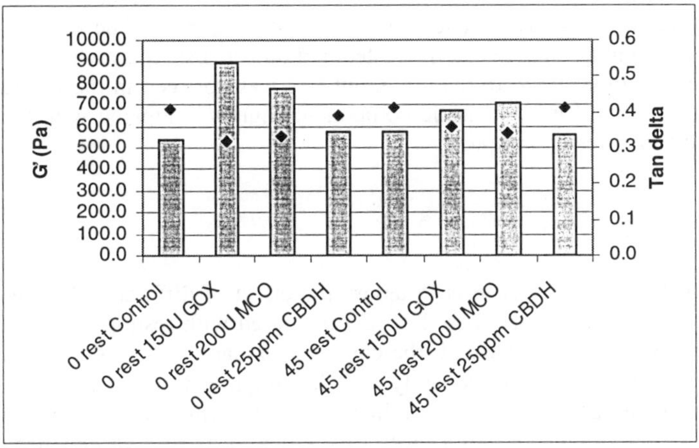

Figure 2 *Effect of enzymic redox additives on G' (cross hatched bars) and tan δ (closed diamonds)*

Strain sweeps carried out on glutens from doughs made with the untreated commercial flour, Pelikaan (Meneba), and treated with various enzymic redox systems showed a similar trend to Figure 1; that is GOX showed an increase in G' and decrease in tan δ when compared with the control. MCO, which acts on a wider range of carbohydrate substrates than GOX, had a similar effect in increasing overall elasticity again indicating that the size or number of linkages per molecule in the glutenin has been increased. The CBDH however, appeared to have little effect compared with the control.

3.2 Stress relaxation

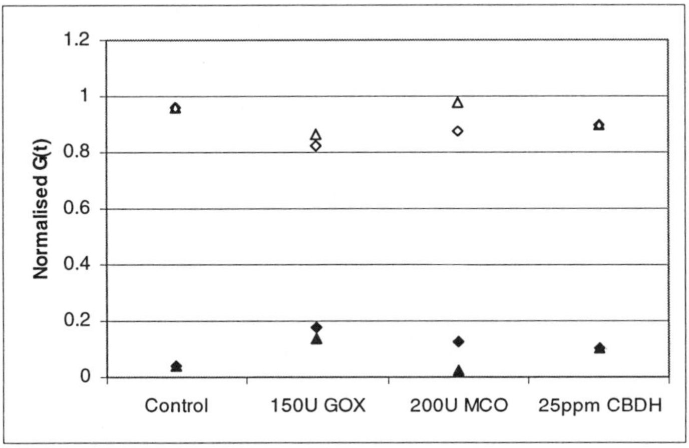

Figure 3 *Effect of enzymic redox additives on the elastic (closed symbols) and viscous (open symbols) components for unrested samples* (◇◆) *and samples rested for 45 min* (Δ▲).

Relaxation tests separate the viscous and elastic components. Normalised data were used. No difference was seen between the rested and unrested control samples. It can be seen that the addition of GOX resulted in an increase in elastic component (and a decrease in viscous component) but this was slightly enhanced in the unrested samples. The addition of MCO gave a greater differentiation between the rested and unrested samples where the unrested samples showed a similar trend to the GOX but this was not seen in the rested samples. CBDH appeared to increase the elastic component slightly compared with the control in both the rested and unrested samples.

4 CONCLUSIONS

The use of small deformation rheological tests to assess the effect of different additives on gluten appeared to show that elasticity (based on G') increased with resting and with the addition of GOX and MCO more than with the chemical oxidants. With GOX and MCO viscosity (G") did not increase proportionately to G' leading to a drop in tan δ. The stress relaxation tests gave similar results to the strain sweep tests. Glucose oxidase and *M. nivale* carbohydrate oxidase appeared from these rheological tests to give enhanced polymerisation in gluten. Cellobiose dehydrogenase had a less pronounced effect.

References

1. S.P. Kaufman and O. Fennema, *Cereal Chem.*, 1987, **64**,172.
2. A.A. Tsiami, A. Bot and W.G.M. Agterof, *J. Cereal Sci*, 1997, **26**, 279.

Acknowlegements

Funding for this research was through a CASE Studentship to C. Skinner provided by MAFF and Novo Nordisk A/S.

GLUTATHIONE: ITS EFFECT ON GLUTEN AND FLOUR FUNCTIONALITY

S.S.J. Bollecker, W. Li and J.D. Schofield

The University of Reading, Department of Food Science and Technology, Whiteknights, PO Box 226, Reading RG6 6AP, UK.

1 INTRODUCTION

The tripeptide glutathione (γ-glutamylcysteinylglycine) occurs in the wheat grain and flour in several forms: free reduced (GSH), free oxidised (GSSG) and protein-bound (PSSG)[1,2,3]. It has long been thought to play a key role in the reduction-oxidation reactions taking place in the flour and doughs, thus influencing the final quality of baked products. It has been shown, for example, that adding GSH or GSSG to doughs results in a loss of the elastic properties of the gluten[4,5].

The aim of the present work was to assess the variation in the levels of the different forms of glutathione occuring in wheat grain/flour, and the significance of such a variation in relation to end-product quality. The benefits of finding a relationship between glutathione levels with flour quality attributes would be wide-ranging; for example, the processors would be able to control raw material intake better, and the plant breeders could provide new material with suitably tailored redox properties.

2 METHODOLOGY

The methodology used to measure the various forms of glutathione was optimised from previously published work[1,2], and is detailed below. The steps requiring improvement were as follows:

- the concentration of iodoacetic acid (IAA) was increased to correspond to a ratio of SH/IAA of about 1/1000 to enable full alkylation of the free thiols. This prevented the GSH standard from appearing as three peaks (alkylated dinitrophenylated GSH, unalkylated GSH and GSSG) in the HPLC chromatograms.
- phenanthroline was introduced as an oxygen scavenger to prevent oxidation of GSH to GSSG.
- the solvent used to perform the extraction of the protein-bound species was changed from MOPS/DTT to Tris/DTT, pH 9, which had an improved capacity to buffer the medium at an alkaline pH value (pH 8.3) and facilitate the action of the DTT. The amounts of protein-bound glutathione increased dramatically after this alteration was made.

It should also be noted that a temperature of 40°C must be maintained for the dinitrophenylation reaction to be carried to completion, especially in the case of GSSG.

1. Extraction of the free compounds (from 120 to 130 mg of flour)
➔ Extract in 1.3 mL of 5% (v/v) perchloric acid (PCA) for 1 h, on ice and under nitrogen.
➔ Centrifuge (20,000 g, 15 min, 4°C).
➔ Take a 1 mL aliquot of the supernatant and mix with 0.1 mL of a 50mM phenanthroline solution in ethanol.

2. Extraction of the protein-bound compounds
➔ Extract the pellet resulting from the above step with 1 mL of a 0.2M Tris, 0.025M DTT solution, pH 9.
➔ Incubate at 40°C for 1 h.
➔ Reprecipitate by addition of 0.1 mL of 70% (v/v) PCA, leave on ice for 5 min, and centrifuge (20,000g, 15 min, 4°C).
➔Take a 1 mL aliquot and mix with phenanthroline.

3. Alkylation of the free SH groups
➔ Add 0.1 mL of 0.1M iodoacetic acid to the aliquots.
➔ Neutralise the solution by adding 105 mg of sodium bicarbonate, and incubate for 1 h at room temperature in the dark.

4. Dinitrophenylation
➔ Take a 0.75 mL aliquot of the alkylated samples and mix with the same amount of a 3% solution of 1-fluoro-2,4-dinitrobenzene in ethanol.
➔ Allow to react for 4 h at 40°C in the dark.
➔ Centrifuge (8,000g, 10 min, room temperature), filter (0.2μm) and HPLC.

Furthermore, the methodology also enabled the quantitative determination of the metabolic intermediates of glutathione (γ-Glu-Cys and Cys-Gly) and cysteine in their free and protein-bound forms (Figure 1).

Figure 1: *HPLC chromatogram of the dinitrophenylated derivatives of glutathione and of its metabolic intermediates (free forms, extracted from flour).*

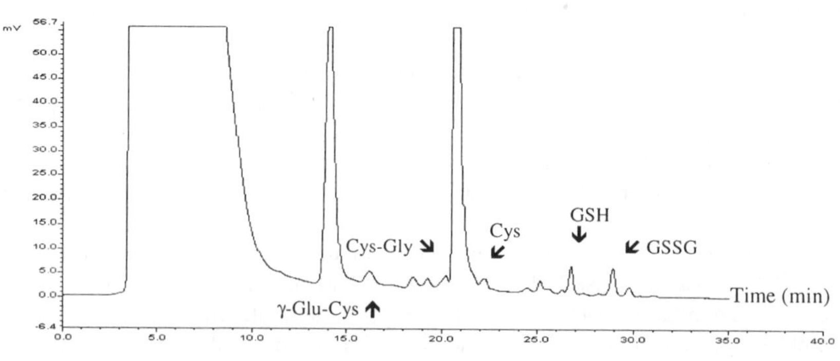

3 RESULTS

The methodology described above was used to measure the levels of the various forms of glutathione and of its derivatives in a set of 36 wheat grain samples, which was kindly provided by Monsanto Cambridge. These wheats were grown in the UK and were of commercial importance, as they formed part of a pre-national list trial set; they exhibited a range of baking performance from very poor to excellent. The grains were milled on a Brabender Quadrumat Junior experimental mill as white flours and analysed immediately in duplicate.

The free GSH, GSSG and PSSG values ranged from 62 to 144 nmoles/g flour, 25 to 100 nmoles/g flour and 66 to 145 nmoles/g flour, respectively. The PSSG values are in agreement with those reported previously[2]. The values for the free glutathione, although in the same order of magnitude, are higher than those presented earlier[1]. A summary of the values for the other compounds is presented in Table 1 below.

Table 1 Average values (± standard deviation) for the glutathione compounds and derivatives in flour

Values in nmole/g flour	Free compounds				
	γ-Glu-Cys	Cys-Gly	Cys	GSH	GSSG
	34 ± 9	26 ± 5	31 ± 9	110 ± 21	53 ± 24
	Protein-bound compounds				
	γ-Glu-Cys	Cys-Gly	Cys	GSH	GSSG
	31 ± 8	13 ± 2	57 ± 20	90 ± 17	N/A

Linear and multiple regressions were used in order to determine correlations between the amounts of the various glutathione compounds with a number of flour quality attributes, including not only baking performance, but also protein content and SDS sedimentation test values. The compounds were also grouped in classes (i.e. total glutathione, total protein-bound compounds) to see if any further links with flour quality could be found.

The results of the linear regression calculations (R^2) are summarized in Table 2.

These results show that no correlation was found between glutathione levels and flour attributes. The use of multiple regression, after selection of the subsets containing parameters showing the best relationships, did not succeed in improving the correlation.

4 CONCLUSIONS

This set of data clearly shows that natural levels of glutathione and of its derivatives in flour cannot be used as predictors of quality. This is rather surprising, considering that they can show a wide variation, as in the case of GSSG.

This finding contrasts with the fact that added glutathione, in either reduced or oxidised form, has a drastic affect on dough and gluten rheology[4]. The analysis of the metabolic intermediates of glutathione and cysteine in this sample set showed that the levels of these components were also not related to quality attributes.

Table 2 *Correlation coefficients (R^2) between glutathione, its derivatives, and flour attributes*

Compound	Bread Volume	Bread Quality	Protein Content	SDS Sedimentation
Free reduced glutathione (GSH)	0.0174	0.0012	0.0007	0.1434
Free oxidised glutathione (GSSG)	0.1047	0.0192	0.1451	0.0567
Total free glutathione (GSH + 2 GSSG)	0.0761	0.0254	0.1661	0.0057
Protein-bound glutathione (PSSG)	0.1009	0.0155	0.0512	0.0179
Total glutathione (GSH, 2 GSSG, PSSG)	0.1020	0.0277	0.1585	0.0103
Total free reduced compounds (GSH, Cys, γ-Glu-Cys, Cys-Gly)	0.0260	0.1284	0.0003	0.0308
Total free compounds (GSH, 2 GSSG, Cys, γ-Glu-Cys, Cys-Gly)	0.1695	0.0912	0.1344	0.0207
Total protein-bound compounds (GSH, Cys, γ-Glu-Cys, Cys-Gly)	0.1947	0.2047	0.1061	0.0041
Free cysteine (Cys)	0.0806	0.1664	0.0003	0.0083
Protein-bound cysteine (PCys)	0.0021	0.0869	0.0027	0.0295
Total cysteine (Cys + PCys)	0.0166	0.1250	0.0019	0.0246

References

1. J.D. Schofield and X. Chen, *J. Cereal Sci.*, 1995, **21**, 127
2. X. Chen and J.D. Schofield, *J. Agric. Food Chem.*, 1995, **43**, 2362
3. R. Sarwin, C. Walther, G. Laskawy, B. Butz and W. Grosch, *Z. Lebensm. Unters. Forsch.*, 1992, **195**, 27
4. W. Li, A.A. Tsiami and J.D. Schofield, in *Gluten 2000* Royal Society of Chemistry (these proceedings)
5. R. Kieffer, J. Kim, C. Walther, G. Laskawy and W. Grosch, *J. Cereal Sci.*, 1990, **11**, 143

Acknowledgments

The authors gratefully acknowledge funding from the LINK Agro Food Quality Programme. They also wish to thank Dr D. Every for his helpful suggestions regarding the glutathione methodology and C. Humphrey and K. Salzedo for their technical assistance.

REDOX REACTIONS DURING DOUGH MIXING AND DOUGH RESTING: EFFECT OF REDUCED AND OXIDISED GLUTATHIONE AND L-ASCORBIC ACID ON RHEOLOGICAL PROPERTIES OF GLUTEN

W.L. Li, A.A. Tsiami and J.D. Schofield

The University of Reading, Department of Food Science and Technology, P.O. Box 226, Reading RG6 6 AP, UK

1 INTRODUCTION

Both reduced and oxidised glutathione (GSH and GSSG, respectively) have been reported to have a weakening effect on dough by cleaving interchain disulphide (SS) bonds between glutenin polymers during dough mixing[1,2]. It has also been demonstrated that the functional properties of gluten proteins are determined by their rheological properties[3,4].

The aim of this study was to determine the significance of GSH and GSSG in relation to the rheological properties of gluten and the influence of redox reactions involving L-ascorbic acid (L-AA) during dough mixing and resting.

2. MATERALS AND METHODS

2.1 Materials

Flour of the U.K. wheat cultivar Hereward was used; it contained 30.3 nmol/g of GSH, 18.8 nmol/g of GSSG and 138.7 nmol/g of PSSG.

2.2 Methods

2.2.1 Preparation of gluten for rheological analysis The *cv.* Hereward flour was mixed with solutions of the additives in a Mixograph for 3.5 min. Gluten was then washed out using a Glutomatic machine either immediately after mixing the dough or after 45-min resting in a proving oven. The gluten rheological properties were then determined using a Stress Tech constant stress, small deformation, oscillatory rheometer.

2.2.2 Stress sweep tests. A cone-plate geometry (C40 4) was used and the stress was varied from 1.0 to 200Pa in linear steps. The frequency was 1Hz and the temperature was 25°C

2.2.3 Time temperature superposition test (TTST). A plate-plate geometry (P20ETC) was used, with a gap setting of 2.0 mm. The strain was in 0.02, the temperatures were at 30°, 20°, 10° and 1°C, and the frequency range was 0.005 – 10 Hz.

3. RESULTS AND DISCUSSION

3.1 Effect of adding GSH or GSSG at different levels on gluten rheology

Stress sweep tests were used in order to determine how gluten rheological properties were affected by addition of a wide range of adding GSH or GSSG levels to the dough from which the gluten was washed out.

The results for gluten washed out from dough immediately after mixing (Figure1a) showed that the dynamic storage modulus, G′, a measure of the dough's elasticity, decreased with increasing levels of addition of both GSH and GSSG, indicating the gluten was weakened due to addition of the two compounds. The weakening effect of GSH was more pronounced than that of GSSG.

The structure breaking stress of gluten was also reduced by addition of GSH or GSSG to the dough. Comparison of the breaking stress of gluten with addition of GSH and GSSG to the dough showed a clear difference in the weakening effects of the two compounds (Figure1b). Over the range of addition levels from 12.5 to 150nmol/g of flour, the breaking stresses of the glutens with addition of GSH were considerably lower than those with the corresponding levels of addition of GSSG. The gluten with addition of GSH was therefore weaker than that with addition of GSSG.

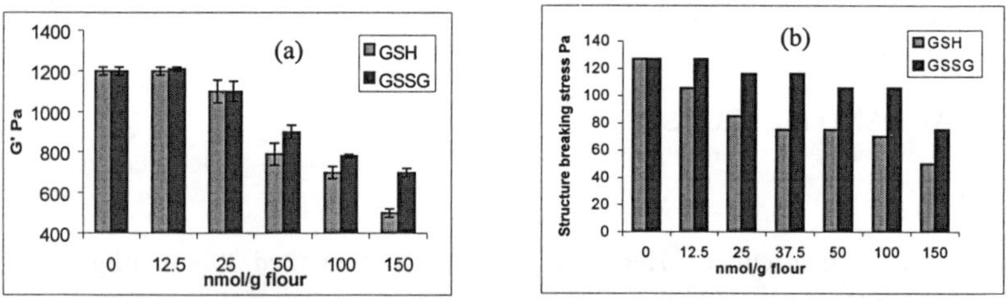

Figure 1 *Effect of GSH and GSSG on the rheology of gluten as determined in stress sweep tests (a) G′ of gluten; (b) structure breaking stress.*

3.2 Effect of GSH, GSSG and L-AA on the rheological properties of gluten from freshly-mixed and rested doughs as determined by time-temperature superposition tests(TTST)

The TTST spectra were qualitatively similar for the glutens washed out from the dough with different additives immediately after mixing and after resting (Figure 2). Over the 0.05 – 10 Hz frequency domain, the G′ values showed a slightly ascending plateau. At all frequencies the G′ values were higher than the G″ values. This type of viscoelastic behaviour is typical of a network structure, which is characteristic of the boundary region of the plateau zone and the transition zone. The height of the plateau can be taken as an indication of the degree of cross-linking of the gluten network, whereas the slope of the G″ curve at the highest frequency is related to the proportion of low molecular weight (LMW) proteins in the gluten network[4].

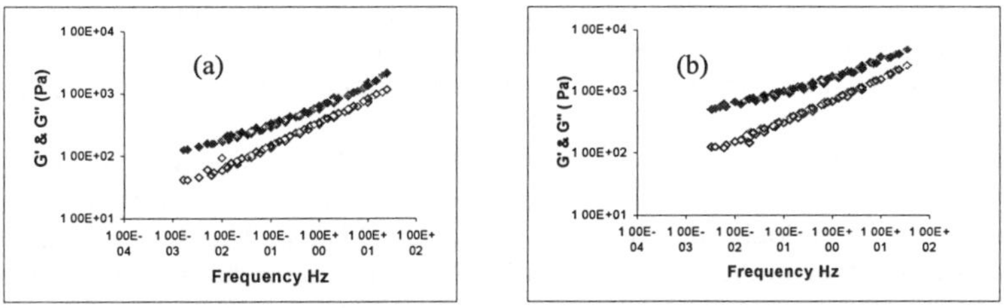

Figure 2 *TTST spectra of gluten washed out from dough containing L-ascorbic acid (L-AA 100 ppm). Gluten washed out (a) immediately after dough mixing and (b) after 45 min rest.*

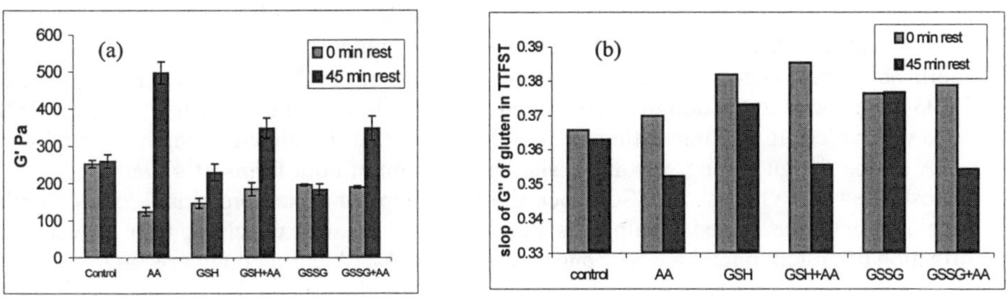

Figure 3 *Effect of GSH, GSSG and L-AA on gluten rheology as determined by TTST. (a) G′ at 0.05 Hz; (b) slope of the G″ curve*

It can be seen (Figure 3) that both forms of glutathione affected the gluten rheological properties both during dough mixing and also during dough resting. The decrease in G′ and the increase in the slope of the G″ curve that occurred with addition of both GSH and GSSG indicated that the interchain SS bonds in glutenin polymers were cleaved. This would have resulted in an increased proportion of low molecular weight (LMW) proteins in the gluten network, which in turn weakened the gluten during dough mixing. The gluten weakening effect of GSH was largely reversed during dough resting, but this reversal did not occur in the case of GSSG addition.

Unexpectedly, addition of L-AA alone also decreased G′ and slightly increased the slope of the G″ curve for the gluten washed out from the dough immediately after mixing, indicating that the gluten structure was weakened by L-AA addition to the dough. This observation is reminiscent of results reported many years ago by Mauseth and Johnston for continuously mixed doughs[5]. In that work addition of L-AA was observed to lower the work input requirement for continuously mixed doughs, implying that L-AA had a reducing effect under the oxygen limiting conditions that occur in the closed mixing chamber of a continuous dough mixer. Galliard has also found that even doughs mixed in open mixing devices can become anaerobic quickly during mixing presumably because of the low rate of transport of oxygen through the dough matrix[6]. Our results imply, therefore, that L-AA has a mainly reducing effect during dough mixing when oxygen is likely to be limiting. Most rheological analyses would fail to detect this effect because prolonged resting times are usually used between mixing and analysis in order to allow stresses built up in the dough during mixing to relax.

In contrast to the effect observed for gluten washed out from dough immediately after mixing, L-AA markedly increased the G′ value and decreased the slope of the G″ curve for gluten washed out from dough that had been rested for 45 min. This indicates that L-AA caused a marked increase in the cross-linking of the gluten proteins during dough resting and a decrease in the proportion of LMW proteins in the gluten network. It appears from these results that oxidation of sulphydryl (SH) groups in gluten proteins to produce interchain SS bonds resulting from addition of L-AA occurs mainly during the dough resting stage, which would be consistent with the view that L-AA act as a long-acting oxidant[7].

When L-AA was added to the dough together with GSH, it slightly increased the G′ value and the slope of the G″ curve compared with the gluten that contained only GSH. No significant change occurred in the G′ value or the slope of the G″ curve when both L-AA and GSSG were added to the dough as compared with the gluten from the dough to which GSSG alone was added. It was interesting to note, however, that the gluten strengthening effect of L-AA during dough resting was diminished by addition of both forms of glutathione. It may be postulated that GSH and GSSG block the SH groups of gluten proteins through forming protein glutathione mixed disulphides during dough mixing, which in turn prevents the formation of protein interchain SS bonds promoted by L-AA during dough resting[1].

4 CONCLUSION

Addition of GSH or GSSG to dough weakened the gluten washed out from those doughs immediately after mixing. There was a loss of elasticity in the gluten, which increased as the level of addition of GSH or GSSG increased. The weakening effect of GSH was more pronounced than that of GSSG.

Both forms of glutathione were able to influence gluten rheological properties during dough resting. The loss of cross-linking and increase in the LMW fraction of gluten proteins resulting from the addition of GSH during mixing was partly restored after dough resting. In contrast, the gluten with addition of GSSG showed no increase in elasticity.

Addition of L-AA alone showed a significant weakening effect on the gluten washed from the dough immediately after mixing, suggesting a predominantly reducing effect of L-AA during mixing. However, after resting, L-AA had a marked strengthening effect on the gluten network caused by increasing the cross-linking of gluten proteins and decreasing the LMW fraction of the gluten, suggesting that the oxidising effect takes place predominantly during dough resting. But addition of GSH or GSSG diminished the strengthening effect of L-AA during the resting period.

The observations reported here suggest that glutathione does have functionally significant effects during dough mixing and resting.

References

1 X. Chen and J.D. Schofield, *Z. Lebensm. Unters. Forsch.*, 1996, **203**, 255.
2 W. Dong and R.C. Hoseney, *Cereal Chem.*, 1995, **72**, 58.
3 A.M. Janssen and T. von Vliet, *J. Cereal Sci.*, 1996, **25**, 19.
4. A.A. Tsiami, A. Bot, W.G.M. Agterof and R.D. Groot, *J. Cereal Sci.*, 1997, 26, 15.
5 R.E. Mausethand W.R.Johnston, *Cereal Sci. Today*, 1967, **12**, 390.
6. T. Galliard, In: *Chemistry and Physics of Baking*, Eds J.M.V. Blanshard, P.J. Frazier and T. Galliard, Royal Society of Chemistry, London, 1986, 199.
7. C.S. Fitchett and P.J. Frazier, In: *Chemistry and Physics of Baking*, Eds J.M.V. Blanshard, P.J. Frazier and T. Galliard, Royal Society of Chemistry, London, 1986, 179.

Acknowledgements

This research was funded by the Biotechnology and Biological Sciences Research Council (BBSRC), the Ministry of Agriculture, Fisheries and Food (MAFF) and several industrial companies under a LINK Agro Food Quality Programme project as well as by an Overseas Research Studentship from the Committee of Vice-Chancellors and Principals of U.K. Universities.

REDOX REACTIONS IN DOUGH: EFFECTS ON MOLECULAR WEIGHT OF GLUTENIN POLYMERS AS DETERMINED BY FLOW FFF AND MALLS

A. A. Tsiami[1], D. Every[2] and J. D. Schofield[1]

1. The University of Reading, Department of Food Science and Technology, Whiteknights, P.O. BOX 226, Reading RG6 6AP, UK. and 2. Crop and Food Research, Private Bag 4704, Christchurch, New Zealand.

1 INTRODUCTION

Ascorbic acid (AA), dehydroascorbic acid (DHAA) and glutathione (GSH) can alter the rheological properties of dough and can influence baking performance[1,2]. There is little evidence of the effect of redox reactions on molecular weight distribution of the glutenin polymers. The aim of this study is to use a new technique that could enable the determination of changes in the molecular weight (M_w) distribution of the gluten polymer. It has been shown by Flow Field-Flow Fractionation (Flow-FFF) and multi-angle laser light scattering (MALLS) that gluten from the strong cultivar, Hereward, has higher M_w glutenin polymers than the weak cultivar, Riband, and that this difference in M_w is related to the rheological properties of the respective glutens[3,4].

2 MATERIALS AND METHODS

2.1 Materials

The cultivar Otane (New Zealand) was used. Doughs were prepared as described by Every *et al.*[5].

2.2 Methods

2.2.1 Sample preparation. The doughs were freeze-dried, ground and defatted. The glutens were extracted and separated into six fractions by sequential centrifugation and fractional precipitation with sodium chloride as described by Tsiami *et al.*[6]. Each fraction was dialysed against 0.05 M acetic acid before being used in the Flow-FFF/MALLS measurements.

2.2.2 Dough preparation. The dough was prepared as described by Every *et al.*[5] (mixed for 1 min slow mixing and 3 min fast mixing). DHAA was added 35 s before the end of the mixing, at a level of 568 nmol/g flour. AA and the GSH were added at the start of mixing, at levels of 568 nmol/g flour and 100 nmol/g flour, respectively.

2.2.3 Flow Field-Flow Fractionation. The technique has been described in literature[7-9]. The Flow-FFF system used, was equipped with a model F-1000 channel frit inlet-frit outlet FFF unit from FFF Inc. (Salt Lake City, Utah). The frit outlet facility only was used. The unit was fitted with a regenerated cellulose semi-permeable membrane with a molecular weight cut-off of 10^4. A Waters pump was used to maintain a carrier flow rate of 1 mL/min. A Pharmacia LKB-P500 syringe pump in a closed circuit drove the cross flow. The solvent was an aqueous solution of 0.05 M acetic acid containing 0.002% (w/v) FL-70 (a detergent).

2.2.4 Multi Angle Laser Light Scattering. The MALLS equipment used here was a DAWN (Wyatt Technology Corporation) fitted with a K5 refraction cell and a laser with a wavelength of 632.5 nm. The RI detector was an Optilab DSP Interferometric Refractometer (Wyatt Technology Corporation) operating at a wavelength of 633 nm and the UV detector was from Pharmacia operating at 220 nm.

3 RESULTS AND DISCUSSION

3.1 Effect of the redox phenomena during dough mixing

The effect of mixing on the gluten proteins was observed by comparing the analogous fractions obtained from the flour and the freeze-dried dough. The data are presented in Table 1. Fraction R2 of the flour had two peaks, one of low M_w ($8x10^7$) and the other of high M_w ($2x10^8$). Comparing the analogous fraction obtained from the dough, a lower M_w distribution was found. The R2 fraction from dough also had two peaks, low M_w ($2x10^7$) and high M_w ($1x10^8$). A similar pattern was observed for the rest of the fractions (Table 1), except for fraction R7 (gliadin), in which no change occurred. The M_w distribution of the gluten polymer decreased dramatically with mixing. This is the first time that a decrease in M_w of the glutenin polymers with mixing has been demonstrated directly in terms of absolute M_w.

Addition of DHAA increased the M_w distribution of the glutenin polymers in all the fractions (Table 1). The HMW fraction R2 had two peaks, the lower of the two had a M_w of $3x10^7$ at a cumulative weight fraction (CWF) of 0.5 and the high peak had a M_w of $2x10^8$ at a CWF of 0.5, which was significantly higher than the control. This effect was also observed for fraction R3. Fractions R4-R7 appeared to be monodisperse for the control dough, but when DHAA was added, all the fractions were polydisperse with HMW polymers.

Unexpectedly, addition of AA reduced the M_ws of the HMW polymer fractions (R2 and R3) (Table 1). Fractions R4 and R5 had higher M_ws than the control dough. Both fractions were polydisperse with two distinct peaks of low and high M_w. The increase in M_w was probably due to fragments produced by breakdown of the polymers in fractions (R2, R3 and R1, the insoluble acetic acid fraction). Fractions R6 and R7 were not changed in terms of M_w.

Also unexpectedly, addition of GSH (a reducing agent) resulted in an apparent increase in the M_w distribution of the fractions compared with the analogous fractions of the control. This observation was unexpected as it is known that GSH breaks S-S linkages. In the fractionation procedure, a starch fraction (R1) is obtained first, which contains some insoluble gluten protein. The amount of insoluble protein in fraction R1 is shown in Figure 1. This shows that the amount of insoluble protein in the case of GSH addition was 15 mg/g flour whereas for the control it was 38 mg/g flour. Thus when adding GSH, a higher percentage of soluble protein was obtained than without GSH. The fractionation is based on the separation of the protein by centrifugation. The insoluble fraction therefore had a higher molecular weight distribution than the soluble protein fraction. It is probable that GSH reduced the insoluble glutenin polymer (the ultra-high

M_w glutenin protein) to soluble glutenin polymers of lower M_w, but still relatively high in comparison with the M_w of fractions R2-R4 from control doughs.

Table 1 *Molecular weight averages according to the peaks that appear in each fraction. The cumulative weight fraction of each peak is shown in parenthesis.*

	Flour		Control dough		DHAA dough		AA dough		GSH dough	
	LMW	HMW	LMW	HMW	LMW	HMW	LMW	HMW	LMW	HMW
R2	8 10^7	2 10^8	2 10^7	1 10^8	3 10^7	2 10^8	4 10^6	2 10^7	1 10^8	7 10^8
	(0.5)	*(0.5)*	*(0.4)*	*(0.6)*	*(0.5)*	*(0.5)*	*(0.5)*	*(0.5)*	*(0.3)*	*(0.7)*
R3	4 10^7	7 10^7	1 10^7	6 10^7	2 10^7	2 10^8	3 10^5	6 10^7	1 10^8	2 10^8
	(0.4)	*(0.6)*	*(0.6)*	*(0.4)*	*(0.4)*	*(0.6)*	*(0.5)*	*(0.5)*	*(0.3)*	*(0.7)*
R4	1 10^6	5 10^6	3 10^4		2 10^5	1 10^8	4 10^4	3 10^5	5 10^5	5 10^7
	(0.5)	*(0.5)*	*(0.9)*		*(0.4)*	*(0.6)*	*(0.6)*	*(0.4)*	*(0.4)*	*(0.6)*
R5	1 10^5	6 10^5	3 10^4		2 10^5	2 10^7	2 10^4	1 10^7	8 10^4	
	(0.4)	*(0.6)*	*(0.8)*		*(0.4)*	*(0.4)*	*(0.5)*	*(0.5)*	*(0.7)*	
R6	4 10^4	1 10^5	3 10^4		4 10^5	1 10^7	3 10^4		3 10^4	
	(0.5)	*(0.3)(0.2>)*	*(0.8)*		*(0.4)*	*(0.4)*	*(0.8)*		*(0.9)*	
R7	3 10^4		3 10^4		3 10^5	2 10^6	3 10^4		3 10^4	
			(0.9)		*(0.6)*	*(0.4)*	*(0.8)*		*(0.9)*	

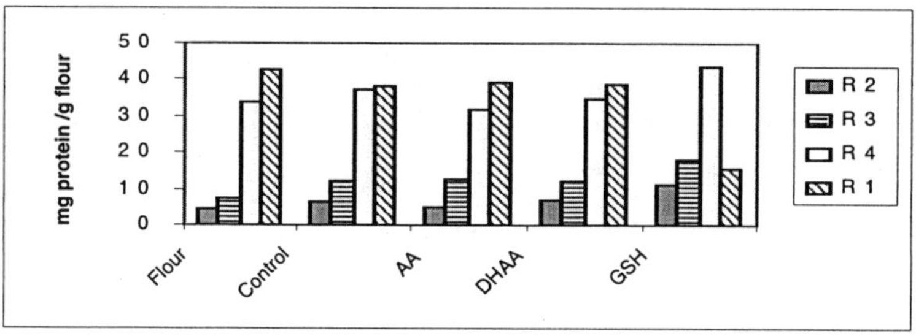

Figure 1. *Yields of gluten fractions (R1-R4), the data were normalised to constant protein content.*

3.2 The effect of dough resting on MW distribution of gluten

Doughs that were mixed under various redox conditions were rested for 55 min before being frozen and freeze-dried. The fractionation procedure was performed, and the fractions were analysed fresh by Flow-FFF and MALLS. There was a clear increase in the M_w distribution of the three HMW fractions R2, R3 and R4 after resting. No change in M_w was observed for the remaining fractions (R5-R7).

There were no significant changes in the M_w distribution with DHAA after resting (Table 2). In contrast, in the gluten with AA after resting, a significant increase in the M_w distribution was observed for all HMW fractions (R2-R4) (Table 2). There was an increase in the M_w of the fraction R5 but only for the high M_w polymers ($1x10^8$; CWF 0.4). The M_w distribution of fractions R6 and R7 were unchanged.

For GSH-treated dough after resting the high M_w polymer fractions had higher M_ws than those for the unrested GSH-treated dough (Table 2). The last two fractions R6 and R7 did not change in size.

4 CONCLUSIONS

Flow-FFF in tandem with MALLS is a powerful technique, which is able to show how treatment of dough with different redox additives can affect the M_w distribution of the gluten proteins. Thus it was possible to determine the absolute M_w before and after mixing of the dough. Mixing for 4 min reduced the M_w distribution of the soluble glutenin polymers. Addition of AA reduced the M_w distributions of the glutenin polymers compared with the control dough. This observation was unexpected since AA is considered to be a dough strengthening agent. But the observation was in line with rheological data obtained recently for another cultivar (Hereward) by Li *et al.*[10]. On the other hand, an increase in the M_w distribution was observed after addition of GSH. In this case, all the fractions showed an increase in M_w but the unextractable fraction was reduced by a significant amount, which indicated a lower amount of the ultra high M_w polymer. DHAA increased the M_w distribution of the glutenin polymers when compared with the control. The M_w distribution of the gluten polymers increased with resting in all cases except for addition of DHAA.

Table 2 *Molecular weight averages according the peaks that appear on each fraction. The cumulative mass of each peak is shown in parenthesis.*

	Control (55 min rest)		DHAA (55 min rest)		AA (55 min rest)		GSH (55 min rest)	
	LMW	HMW	LMW	HMW	LMW	HMW	LMW	HMW
R2	**4 10^7**	**2 10^8**	**5 10^7**	**1 10^8**	**3 10^7**	**2 10^8**	**1 10^8**	**1 10^9**
	(0.4)	*(0.6)*	*(0.5)*	*(0.5)*	*(0.4)*	*(0.6)*	*(0.5)*	*(0.5)*
R3	**2 10^7**	**1 10^8**	**5 10^7**	**1 10^8**	**3 10^7**	**1 10^8**	**1 10^8**	**2 10^8**
	(0.4)	*(0.6)*	*(0.5)*	*(0.5)*	*(0.4)*	*(0.6)*	*(0.7)*	*(0.3)*
R4	**1 10^4**	**5 10^6**	**1 10^5**	**1 10^8**	**1 10^7**	**7 10^7**	**1 10^7**	**4 10^7**
	(0.4)	*(0.6)*	*(0.6)*	*(0.4)*	*(0.3)*	*(0.7)*	*(0.6)*	*(0.4)*
R5	**4 10^4**		**1 10^5**	**3 10^7**	**1 10^5**	**1 10^8**	**7 10^6**	**1 10^6**
	(0.8)		*(0.5)*	*(0.5)*	*(0.6)*	*(0.4)*	*(0.5)*	*(0.5)*
R6	**3 10^4**		**3 10^4**		**3 10^4**		**3 10^4**	
	(0.9)		*(0.9)*		*(0.9)*		*(0.9)*	
R7	**3 10^4**		**3 10^4**		**3 10^4**		**3 10^4**	
	(0.9)		*(0.9)*		*(0.9)*		*(0.9)*	

References

1. S.P. Kaufman, R.C. Hoseney and O. Fennema, *Cereal Chem.* 1986, **31**, 820.
2. D. Every, L. Simmons, K.H. Sutton and M. Ross, *J. Cereal Sc.* 1999, **30**, 147.
3. A.A. Tsiami, C. Stathopoulos and J.D. Schofield, '8th Intl Symp. on FFF' Paris, 1999.
4. A.A. Tsiami and J.D. Schofield, *1999 AACC Annual meeting*, Seattle, 1999, 289.
5. D. Every, L. Simmons, M. Ross, P.E Wilson, J.D.Schofield, S.S.J. Bollecker and B. dobraszczyk, 'Gluten 2000', Roy. Soc. Chem., Cambridge, 2000 *(in press)*.
6. A.A. Tsiami, A. Bot, W.G.M. Agterof and R.D. Groot, *J. Cereal Sc.*, 1997, 26, 15.
7. J.C. Giddings, F.J. Yang, M.N. Myers, *Anal. Chem.* 1976, **48**, 1126.
8. K.D. Caldwell, *Anal. Chem.*, 1988, **60**, 959A.
9. J.C. Giddings, *Science*, 1993, **260**, 1456.
10. W. Li, A.A. Tsiami and J.D. Schofield, 'Gluten 2000', Roy. Soc. Chem., Cambridge, 2000 *(in press)*.

Acknowledgements

Funding for the work, as part of an EU FAIR Programme project FAIR CT97-3010, is gratefully acknowledged.

BACTERIAL EXPRESSION, *IN VITRO* POLYMERISATION AND POLYMER TESTS IN A MODEL DOUGH SYSTEM .

C. Dowd,[1,3] H. Beasley,[2,3] and F. Bekes [2,3]

1 CSIRO Plant Industry, Canberra, ACT, 2601 2. CSIRO Plant Industry, North Ryde, NSW, 1670. 3. Wheat Quality CRC Limited, NSW, 1670.

1 INTRODUCTION

A Model Dough system has been developed in which a dough is built up from the minimum number of pure and defined components[1]. This is achieved by fractionating a base flour into starch, water-soluble and gluten fractions, the gluten is further separated into glutenin and gliadin fractions. Reconstitution studies of these fractions revealed their relative importance in dough development[1]. It has been established that the glutenin fraction is essential to dough development. However, glutenin polymer fractions isolated from flour are impure. Therefore, *in vitro* polymerisation is being developed in order to produce pure glutenin polymers from glutenin subunits. Once synthesised these polymers are added as the sole glutenin fraction to a model dough system. This system has the advantage over base flour methods of eliminating background effects. In order to perform the polymerisation reactions large quantities of starting material are required, specifically, large quantities of pure HMW-glutenin subunits. Bacterial expression has been examined for the production of large quantities of HMW-GS Dx2, Dx5, Dy10 and Dy12. The expression and purification studies of these proteins, *in vitro* polymerisation methods and studies of artificial polymers in a model dough system are discussed in the following paper.

2 MATERIALS AND METHODS

2.1 Materials

Cloned genes encoding HMW-GS Dx2, Dx5, Dy12 and Dy10, were kindly provided by Olin Anderson, USDA, Albany, USA. All were in pET-3a expression vectors and were freshly transformed into *E. coli*. BL21-pLysS cells

2.2. Methods

2.2.1 Bacterial Expression. Fortified LB media, supplemented with 200μg/ml ampicillin, was innoculated and grown at 37°C. At an $O.D._{600}$ of 0.6-1.0, cells were induced by the addition of IPTG(1mM). Cells were grown overnight and harvested by centrifugation at 6000 rpm for 10minutes. Cells were lysed according to a variation on a previous method[2] using sonication in buffer A (50mM Tris-HCl, 50mM NaCl, 1mM EDTA, pH 7.50). Lysates were centrifuged at 18,000 rpm for 20 minutes to separate supernatant from pellet fractions. The pellet was then washed in buffer B (buffer A plus 0.1% Triton X-100 and 10mM EDTA). After centrifugation the pellet fraction was washed and then resuspended in distilled water. All solutions contained PMSF and DTT.

2.2.2 Purification. Alcohol extraction was performed with 50% isopropanol/100mM DTT, followed by sonication and incubation at 65°C for 30 minutes. After centrifugation the supernatant is dialysed against dilute acetic acid, freeze dried and stored at -20°C.

2.2.3 In vitro polymerisation. Subunits (from flour or bacteria) are first dialysed against dilute acetic acid followed by water. *In vitro* polymerisation of subunits (3mg/ml), was performed in water-HCl, pH6.0 with 50μM KIO_3 as oxidising agent, the reaction is allowed to proceed overnight and polymers are then dialysed in dilute acetic acid (Beasley *et al*, in preparation). Efficiency was evaluated by densitometry of multistacking gels.

2.2.4 Model dough experiments. Polymers were then added as the sole glutenin component in a model dough. Two gram mixographs were performed using commercial starch and homogenous bulk water solubles and gliadin fractions. Protein content, glutenin/gliadin ratio and water absorption were all constant.

3 RESULTS AND DISCUSSION

3.1 Bacterial expression of HMW-glutenin subunits

The HMW-GS Dx2, Dx5, Dy10 and Dy12 were expressed in small-scale (200ml) and large-scale (2L) shaker flasks. Large-scale (2L) expression was also performed in a Biostat fermentor. Time courses revealed different patterns of expression. Dx2 and Dx5 consistently had higher expression levels than Dy10 and Dy12. SDS-PAGE of the expressed proteins revealed the correct molecular weight. The expression level changed from about 10-15% of total protein (small-scale) to about 1-5% of total protein in large-scale. This lower expression has been observed with HMW-GS[3]. However, the 2L Biostat fermentor results in a 10-fold higher yield of bacterial cells (40-50g) compared to flasks (4-5g) and an OD_{600} of 20-30 in comparison to an OD_{600} of 3-5 for flasks. We now estimate from a 2L fermentation, 50mg of pure material can be prepared in the case of Dx2 and Dx5 and about 20mg for Dy10 and Dy12.

Fractionation of bacterial cell pellets into soluble cytoplasmic protein (supernatant) and insoluble inclusion body (I.B.) (pellet) fractions proved worthwhile. It was found that Dx2, Dx5 are almost exclusively expressed as inclusion bodies and are found in the pellet fraction. Figure 1A shows Dx5 as an example of this localisation. Dy10 and Dy12 appeared initially in the supernatant as soluble protein, probably due to low expression. We found by increasing expression levels by changing the culture media from fortified LB to ZY we could obtain the y-type proteins in the pellet fraction as I.B., which greatly facilitates their purification, Dy10 is shown as an example in Figure 1B.

Alcohol extraction was used to purify the HMW-GS. While this was efficient for small-scale preparations, several factors needed to be optimised when attempting to purify proteins on a large-scale. An important factor was the starting material. (Fig 1A). Alcohol extraction from I.B. preparations only, results in pure protein. Fifty percent isopropanol was superior to seventy percent ethanol. Sonication after adding solvent is essential to extraction efficiency. An important factor is thorough washing and resuspension of the I.B. pellet in water/100mM DTT. Dx2 and Dx5 have been successfully purified using the above method. A lot of variation was seen in purity between large-scale and small-scale. Variation in purity for Dy10 and Dy12 is possibly because lower expression levels result in lower concentration and possibly increase the labile nature of these proteins.

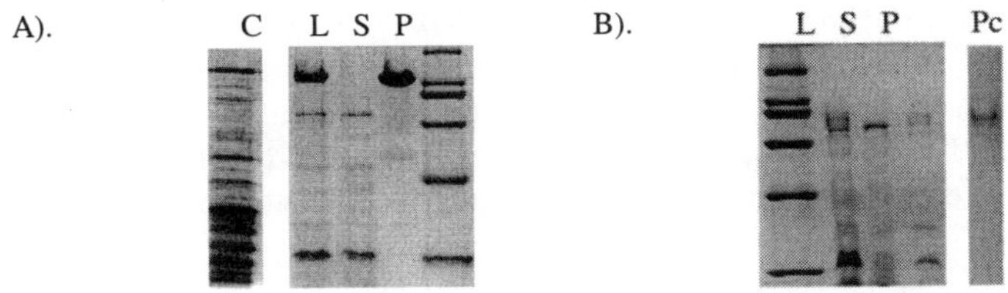

Figure 1. *Localisation of Dx5 (A) and Dy10 (B). Alcohol extractions of cells (C), lysate (L), supernatant (S), pellet (P) fractions and a concentrated pellet fraction (Pc) washed extensively with water is shown for Dy10.*

3.2 *In vitro* polymerisation and model dough experiments

Optimisation of the polymerisation process has been done using Hartog HMW-GS due to the large quantity required for a single reaction (200mg). Optimum parameters for polymerisation efficiency included high concentration of subunits, final pH of 6.0 and low concentration of oxidising agent. Functionality of polymers once formed was dependent on dialysis of both subunits and polymers in dilute acetic acid.

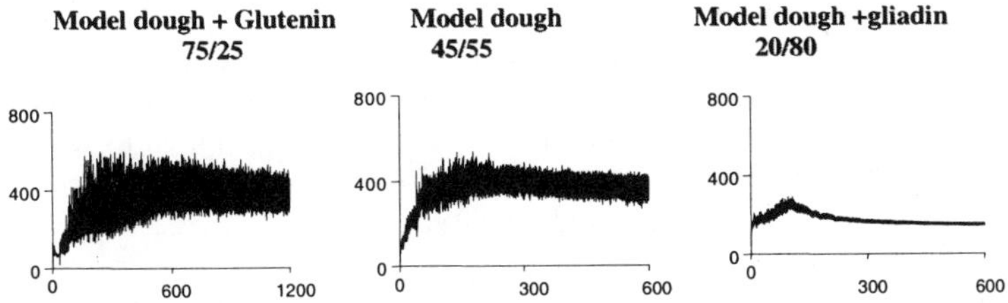

Figure 2. *The effects of altering Glu/Gli ratio. Base flour and model dough mixographs are compared.*

The Model Dough system has been used to examine artificial polymers. It has also been used to examine the effect of altering Glutenin/Gliadin ratio, HMW/LMW ratio and preliminary results have been obtained on the effect of bacterially expressed proteins on dough parameters, specifically Dx5 and Dy10. Model Dough results have been compared to similar experiments performed with base flours, similar trends were found, the effect of glutenin/gliadin ratio in base flour versus model dough is shown in Figure 2.

The advantage of the Model Dough system is that more extreme parameters can be examined and *in vitro* polymers added as the sole glutenin, eliminating background effects. Bacterial protein, recently becoming available, has been used in preliminary polymerisation experiments and tested in model doughs. It appears firstly that the expressed proteins can form polymers indicating that they are functional and that they have an effect on dough development even in small quantities. We have copolymerised bacterial expressed Dx5 (10mg, 5%) with Hartog HMW-GS and examined this polymer in model doughs, the results were very similar to the effect of incorporating (10mg) Dx5 from flour. We also performed copolymerisation of expressed Dx5 and Dy10 (10mg) separately with just LMW-GS, these results (Figure 3) indicated that the proteins in the polymer had similar effects whether Dx5 or Dy10 were incorporated into a base flour.

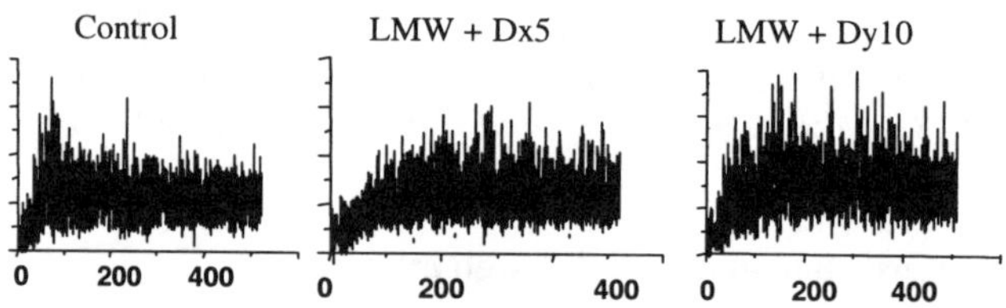

	Control	LMW + Dx5	LMW + Dy10
MT	**100**	**160**	**132**
PR	**219**	**187**	**188**
RBD	**18**	**9**	**11**

Figure 3. *Model dough mixographs using artificial polymers from in vitro co-polymerisation of Hartog LMW-GS and bacterially expressed Dx5 and Dy10.*

4 CONCLUSIONS

Recent optimisation of large-scale expression and development of large-scale purification methods for HMW-GS, means it is now possible to purify large quantities of Dx2, Dx5. Dy10 and Dy12 are becoming available but will require some more optimisation in order to increase their expression levels. In future experiments these proteins can be used with the optimised large-scale *in vitro* polymerisation methods in order to examine the relative importance of x and y-type proteins (singly or in combination) to polymer structure and function. We have the unique advantages of having pure polymers which we can tailor make and test and a dough assay system which allows us the flexibility and specificity to ask important questions about polymer/subunit structure and function in relation to dough quality.

References

1. Blanchard, C. Proc. RACI Cereal Chem. Conf., Cairns,1998, p17-21.

2. Lullien-Pellerin, V.,Gavalda, S., Joudrier, P. and Gautier, M-F. *Prot.Exp.Pur*, 1994, **5**, 218

3. Murayama, N., Ichise, K., Katsube, T., Kishimoto, T., Kawase, S-I., Matsumura, Y., Takeuchi, Y., Sawada, T. and Utsumi, S. *Eur. J. Biochem.* 1998, **255**, 739.

Acknowledgements

This work was funded by the Wheat Quality CRC Limited, NSW, 1670.

IN VITRO POLYMERISATION OF SULPHITE-TREATED GLUTEN PROTEINS IN RELATION WITH THIOL OXIDATION.

Marie-Hélène Morel, Valerie Micard and Stéphane Guilbert

INRA-Unité de Technologie des Céréales et des Agropolymères, 2 place Viala, 34060 Montpellier cdx 01- France.

1 INTRODUCTION

Several studies have attempted to improve the mechanical properties of gluten protein based films by using chemical cross-linkers or by applying thermal treatments. Among forces which stabilise the structure of protein films, disulphide bonds would be highly relevant in determining the properties of wheat gluten proteins. Gennadios et al.[1] showed that addition of sodium sulphite in gluten film-forming solution strengthened gluten films. They supposed that the thiol groups liberated during sulfitolysis would be converted back into disulphides during film drying resulting in the reinforcement of gluten network structure.
In this work, we investigated the effect of sulphite on the size distribution range and on the thiol/disulphide balance of proteins from gluten films. The films were analysed during their preparation from a film forming solution and during their storage. Effects of storage were studied under various temperatures and relative humidities.

2 MATERIALS AND METHODS

Vital gluten was prepared from cultivar Soissons by the Institut Technique des Céréales et des Fourrages (Boigneville, France). Protein content (72.98±0.5%, db) (N x 5.7) was determined in triplicate by the Dumas method (NA 2000, Fisons Instruments, France).

2.1 Film preparation and storage. The film-forming solution was adapted from Gontard[2] et al. except it was supplemented with sodium sulphite (2 mg/g gluten) and incubated for 2 h before being adjusted to pH 4 by acetic acid. The films were dried for 10 h in a ventilated oven at 25 °C and then stored over saturated salt solutions or Silicagel, at 25 °C and 50 °C, for durations ranging from 24 h to 656 h. Saturated salt solutions included NaCl and $MgCl_2$. Before biochemical analyses, films were conditioned for 20 h in a room thermostated at 20 °C and 60% relative humidity (RH).

2.1 Biochemical analysis. Contents of thiol and disulphide groups were performed according to Morel and Bonicel[3] using Ellman's reagent. Triplicate measurement gave an average content of 162.8 ± 4.5 $\mu moles{\cdot}g^{-1}$ of thiol equivalent for the vital gluten. Once corrected for its thiol content, the disulphide content of gluten was calculated as 81.18 ± 0.51 $\mu moles{\cdot}g^{-1}$. SDS-soluble and -insoluble gluten proteins from films were extracted and analysed by SE-HPLC according to Redl[4] et al. From the earliest to the latest fraction eluted from the column, we distinguished, fractions F1 and F2 that include glutenin macro-polymers (GMP) and fractions F3, F4 and F5 that include monomeric protein. Calibration of the column allowed us to estimate the median M_rs of the distribution range of fractions F2: 267,200; F3: 98,035; F4: 33,750 and F5: 8,325. For fraction F1, which was partly eluted at the void volume of the column (7,000,000 according to the manufacturer) we have taken an arbitrary value of 2,500,000 for median M_r. These M_rs values were used to estimate the fraction contents in terms of moles instead of percent of total proteins.

3 RESULTS AND DISCUSSION

3.1 Effect of sulphitolysis on wheat gluten protein.

SE-HPLC analyses of samples, taken throughout the preparation of the film-forming solution, were carried out to follow the effect of sulphitolysis on gluten proteins. Results in Table 1 shows that sulphitolysis increases the extractability of gluten protein in SDS-buffer. The sulphitolysis reaction appeared very efficient and rapid since the effect was noticeable after 5 min reaction. The SDS-insoluble glutenin macropolymers (GMP) fell from about 25% for gluten to 1% for the film-forming solution. The change coincided with a marked increase in fractions F3 and F4, whose molecular masses (M_r) ranged from 145,000 to 17,000. Sulphitolysis promoted extensive depolymerisation of SDS-insoluble GMP, leading to the release of protein monomers such as low molecular weight ($M_r \cong 45{,}000$) and high molecular weight ($M_r \cong 90{,}000$) glutenin subunits.

Drying of film-forming solution resulted in a slight decrease of F4 to the benefit of soluble GMP whereas the content of insoluble protein remained almost unchanged. These results indicated that disulphide bond formation was prevented during film drying. Bisulphite (HSO_3^-), which is in equilibrium with sulphite at pH 4 ($SO_3^{-2} + H^+ \leftrightarrow HSO_3^-$, pKa 6.8), loses some SO_2 and is gradually oxidized to sulphate, upon exposure to air. Sulphinic acid, in equilibrium with bisulphite ion ($HSO_3^- + H^+ \leftrightarrow H_2SO_3$, pKa 1.81) is unstable and decomposes into SO_2 and H_2O. These reactions would occur during the drying of the acidic film-forming solution, leading to the disappearance of all sulphite species through oxidation or decomposition reactions.

Table 1: *Changes in protein solubility and size distribution range from native gluten to freshly dried film.*

	Native gluten	Film forming solution[a]		Gluten film[b]
		5 min	2 h	
% SDS-soluble protein	75.33	98.94	98.61	99.47
%F1	6.20	7.21	4.74	7.32
%F2	13.19	16.33	14.61	17.58
%F3	7.34	11.29	10.80	12.49
%F4	38.08	51.11	54.18	49.98
%F5	10.52	13.00	14.29	12.11

[a] The gluten film forming solution was analysed 5 min and 2 h after sulphite addition.
[b] Freshly dried gluten film.

From native gluten to freshly dried gluten film, the protein thiol content increased by about 8 μmoles·g^{-1}. This would mean that sulphitolysis insured the breakdown of 8 μmoles·g^{-1} of disulphide bonds. Compared with native gluten, freshly dried gluten film showed more soluble proteins (F3 + F4 + F5), which in terms of moles accounted for 5.83 μmoles·g^{-1}. SDS-soluble GMP (F1 + F2) increased by 0.16 μmoles·g^{-1} from gluten to gluten film. Because of the huge size of SDS-insoluble GMP, the release of soluble counterparts could require the breakdown of several inter-chain disulphide bonds. In spite of this uncertainty, we could estimate that the changes observed from gluten to gluten film implied the breakdown of at least 6 μmoles·g^{-1} (5.83 + 0.16 μmoles·g^{1}) of inter-chain disulphide bonds. A maximum of 2 μmoles·g^{-1} of intra-chain bonds would also have been cleaved. Based on the quantitative distribution of the different gluten protein types, Grosch and Wieser[5] calculated that approximately 10% of gluten disulphide bonds are involved in inter-chain bonds. In that respect, inter-chain disulphide bonds appeared to be much more sensitive to sulphitolysis than the intra-chain bonds as already shown for several soluble proteins.

3.2 Change in thiol groups during film storage

Thiol contents of gluten films were measured at various intervals during storage at different relative humidities (0, 33 and 75%) and temperatures (25 and 50°C). Temperature was shown to increase the oxidation rate at all RH levels. The rate of thiol oxidation was estimated from the time ($t_{20\%}$) allowing a drop of 20% of the initial thiol content of freshly dried gluten film (8.9 μmoles·g^{-1}). To establish the effect of gluten plasticization on oxidation rate, we plotted $1/t_{20\%}$ values as a function of the difference between storage temperature (T) and the glass transition temperature of gluten (Tg) (figure 1). Tg values of the gluten films studied were estimated from the data published by Gontard and Ring[7]. Figure 1 shows that the oxidation rate began to increase at – 40 °C below Tg and then rose notably above Tg. This indicates that thiol oxidation occurs even though gluten films are stored within their glassy state and that only short range mobility is allowed for protein side chains. Above Tg, segmental movements of protein chains increase abruptly and the probability of molecular collisions may rise, leading to a sudden increase in the thiol groups oxidation rate. Gontard[2] et al. have shown that the oxygen permeability of gluten films increases exponentially as gluten proteins undergo the glass to rubbery transition. Thus, the abrupt increase of thiol oxidation rate above Tg might follow the increase in oxygen concentration in the vicinity of proteins, whereas the increase in segmental mobility would be less essential.

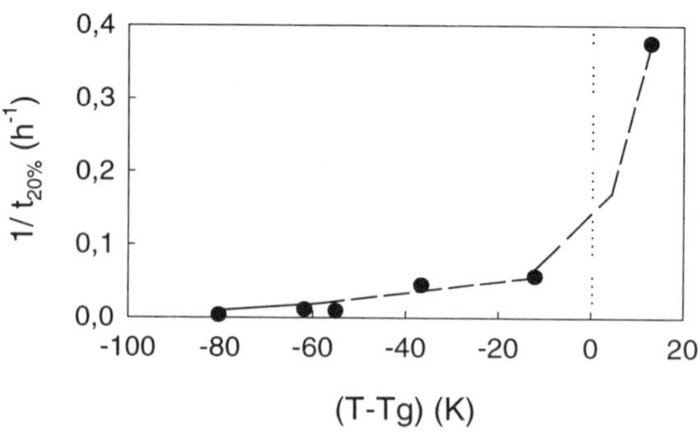

Figure 1 : *Variation in the rate of thiol oxidation according to the glass-to-rubbery state of gluten proteins estimated from (T-Tg).*

3.3 Relationship between thiol groups content and protein size distribution range

The formation of inter-chain disulphide bonds during gluten film ageing was estimated by following changes in the molecular weight distribution of the SDS-soluble proteins. As a general trend, solubility of gluten protein in SDS decrease as thiol groups were oxidised. We observed a continuous drop in F4 and then in F1 and F2 fractions while fractions F3 and F5 remained almost unchanged. When oxidation reached completion, a network structure different from that of gluten was obtained. Gluten films comprised more insoluble proteins (up to 41.8% instead of 25 % for native gluten), and less monomeric proteins (27.6% of F4 instead of 38 %) and soluble GPM (1.21% of F1 and 10.8% of F2 instead of 6.2% and 13.2%, respectively, for gluten). Compared to native gluten, the fully oxidized gluten film had about 2 μmoles$\cdot g^{-1}$ less SDS-soluble proteins. We previously suggested that sulphitolysis led to the cleavage of 8 μmoles$\cdot g^{-1}$ of disulphide bonds, among which 2 μmoles$\cdot g^{-1}$ were intra-chain. These bonds would have been converted into inter-chain bonds, resulting in the insolubilisation of 2 μmoles$\cdot g^{-1}$ more of the soluble proteins in the fully oxidized gluten film than for native gluten.

4 CONCLUSION

At the level used in our study sulphite anions contributed mainly to the breakdown of inter-chain bonds of glutenin macropolymers. During storage, thiol oxidation occurred and was coupled with a large decrease in protein solubility. We showed that thiol oxidation did not allow rebuilding of the initial structure of gluten. Loss in protein solubility was observed whereas translational mobility was limited owing to the storage conditions (close to or below the protein glassy state). So it is likely that specific protein interactions are set up during the drying of the film-forming solution. Those interactions would bring thiol and S-sulphonate groups of proteins in close contact, so that in the presence of some oxidising agent, covalent coupling via disulphide bonds can occur. Thus, the oxygen permeability would be the parameter that governs thiol oxidation of gluten proteins whereas the molecular mobility and in particular segmental or translational mobility of protein chains would not be requisite.

References

1. A. Gennadios, C. L. Weller, and R. F. Testin, *Trans. ASAE* 1993, **36**, 465.
2. N. Gontard, R. Thibault, B. Cuq and S. Guilbert, *J. Agric. Food Chem.* 1996, **44**, 1064.
3. M.-H. Morel and J. Bonicel, in *Gluten '96, Proceedings of the Sixth International Gluten Workshop*; Wrigley, C.W., Ed.; Royal Australian Chemical Institute : Melbourne, Australia, 1996; pp 257.
4. A. Redl, A., M.-H. Morel, J. Bonicel, B. Vergnes and S. Guilbert, *Cereal Chem.,* 1999, **76**, 361.
5. W. Grosch and H. Wieser, *J. Cereal Sci.,* 1999, **29**, 1.
6. N. Gontard and S. Ring, *J. Agric. Food Chem.,* 1996, **44**, 3474.

MODIFICATION OF CHAIN TERMINATION AND CHAIN EXTENSION PROPERTIES BY ALTERING THE DENSITY OF CYSTEINE RESIDUES IN A MODEL MOLECULE: EFFECTS ON DOUGH QUALITY

L. Tamás[1,3], F. Békés[2], P.W. Gras[2], M.K. Morell[1] and R. Appels[1]

1. CSIRO Plant Industry, Canberra, GPO Box 1600, ACT 2601, Australia. 2. CSIRO Plant Industry, North Ryde, Grain Quality Research Laboratory, NSW Australia. 3. present address: Department of Plant Physiology, Loránd Eötvös University, Budapest, GPO Box 330, H-1445, Hungary

1 INTRODUCTION

Not only are the central repetitive region features, such as length and sequence of repetitive motifs of glutenin molecules, interesting in determining dough properties, but so are the number and distribution of cysteine residues. The formation of intra- and inter-chain disulphide bonds needs to be studied to find out more about their effects on gluten matrix quality. A major problem in exploring the relationships between the structure and properties of storage proteins lies in their complexity. To overcome this problem, we have developed a model molecule and a protein expression system, based on *E.coli*. This model molecule has characteristics similar to HMW-GS molecules. The molecule is called Analogue Glutenin (ANG) and has unique, short N- and C-terminal domains, flanking a long central repetitive region. A series of peptides, based on the ANG protein molecule, have been expressed with different numbers of cysteine residues.

In this investigation, functional studies on purified proteins were carried out using pilot scale testers and micro-baking technology. Elasticity and extensibility data on seven engineered polypeptides, different in density of cysteine residues, are presented and compared. We demonstrate that changes in the number, positioning and distance between two cysteine residues in the storage protein molecule strongly influence the structure of the gluten matrix of dough. On the basis of measured data we conclude that a combination of protein engineering and small-scale functional studies is a powerful tool for exploring structure/function relationships in gluten proteins. Pilot scale studies can be used to design special storage protein molecules for special dough quality requirements.

2 MATERIALS AND METHODS

2.1 Cloning of ANG molecules with altered numbers of cysteine residues

The method to manipulate the gene of the S-poor barley storage protein (C hordein) to replace single amino acids by cysteine residues has already been published[1]. Four recombinant plasmids containing the sequences encoding ANG wild type (**WT**), ANGCys7 (**1N**), ANGCys236 (**1C**), ANGCys7Cys236 (**1N1C**) proteins were available from earlier studies[1]. The new plasmids to express ANGCys7Cys230Cys236 (**1N2C**), ANGCys7Cys13Cys236 (**2N1C**), ANGCys7Cys13Cys230Cys236 (**2N2C**) and ANGCys13Cys230 (**1'N1'C**) polypeptides were also constructed by the use of the PCR amplification technique. Amplified DNA was cloned, using the pGEM-T Vector System I (Promega). Genes were excised from positive, sequenced clones and subcloned into pET-11d expression vector.

2.2 Expression and purification of proteins

Proteins were expressed in *E.coli* strain AD494(DE), grown in 5 litres of 2YT medium supplemented with 100 mg/l of ampicillin. Extraction and purification were carried out as described elsewhere[2] with modification. Ethanol (70 % (v/v)) extraction was followed by dialysis of the protein solution against 10 mM of acetic acid. The pellet was collected after centrifugation and dissolved in 0.1 M acetic acid and freeze-dried.

2.3 Functional studies

2.3.1 Mixing. Tests were conducted with a prototype 2g Mixograph using a modification of the standard method[3]. Mixing parameters were determined using a previously reported computer program[4]. Parameters determined were mixing time (MT) and break down in resistance (BDR). Reversible reduction/oxidation procedure for incorporation of 5 mg of added, purified protein into the glutenin matrix, was carried out according to Békés[5].

2.3.2 Extensibility. A small-scale extension tester was used, providing results for resistance in maximum (Rmax) and extensibility (Ext), which are closely related to those from the Brabender Extensograph. Sample preparation and handling method were as published earlier[6].

2.3.3 Baking. Test baking was carried out using a recently developed procedure, employing 2.4g dough prepared in the 2g Mixograph[7].

All tests were performed three times. The least significant differences (LSD) were calculated by Student test.

3 RESULTS AND DISCUSSION

3.1 Proteins

The gene selected for this work encodes a C hordein storage protein from barley homologous to omega-gliadins of wheat. It is an ideal polypeptide for the purpose of this study because it has no cysteine residues. The long, highly conserved repetitive primary structure results in a conserved supersecondary structure, which is able to undergo deformation/reformation under stress/relaxation. This model polypeptide with added cysteine residues is able to be incorporated into the gluten matrix of dough in the same way as glutenin subunit proteins are[1]. Because they have similar functional characteristics as glutenins these modified molecules are called analogue glutenin (ANG) peptides.

Genes for mature proteins (molecular weight 28 kDa) have 723 nucleotides, including a 669 bp long central repetitive domain. Codons for cysteine residues were introduced into the non-repetitive N- and C-terminal domains, substituting the 7th serine, 13th

proline, 230th proline and 236th threonine, depending on the construct. The signal peptide sequence was replaced with the initiation (ATG) codon and cloned into pET11d vector in order to express the protein in *E. coli*. Ten different ANG protein molecules were expressed, extracted and purified in amounts of about 100 mg each.

3.2 Functional properties

3.2.1 Mixing. Data of mixing time (MT) and breakdown in resistance (BDR) of dough samples were measured after the incorporation of ANG protein molecules. Results only of mixing time are shown in Figure 1. ANG proteins with one or two cysteine residues at both ends (1N1C, 1'N1'C, 2N2C) increased MT and decreased BDR, indicating greater strength. These effects are similar to the incorporation of HMW glutenin subunit (HMW-GS) 1Bx7, although of lower magnitude. These molecules behave as chain extender proteins and so the size distribution of the polymeric protein in gluten macropolymer was increased. When the functional property of odd numbers of cysteine residue containing ANG proteins (1N, 1C, 2N1C and 1N2C) were studied, MT of the dough was significantly reduced, compared to the control dough sample. It is interesting to note that while mixing time was reduced greatly, stability was not affected to the same extent. BDR values increased to a lesser extent than expected (data not shown). These four molecules stop the extension of the polymer, reducing the size of it in the gluten matrix. This type of molecule is called *chain terminator*.

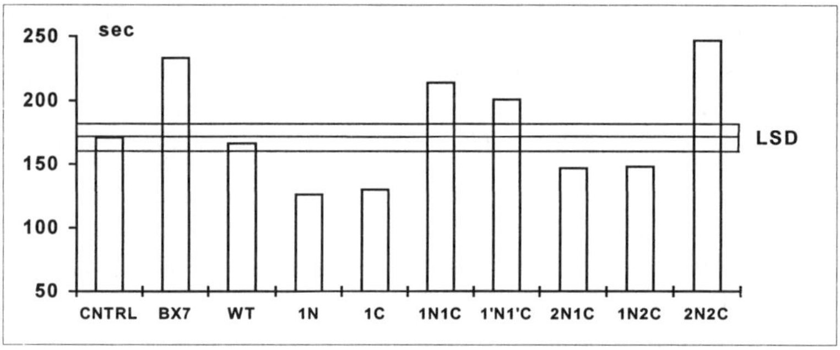

Figure 1, *Comparison of mixing time results of ANG protein samples with altered number of cysteine residues.*

3.2.2 Extensibility. Extensibility was measured in a small scale Extensograph on approximately 1.3g of dough samples. Extensibility and Resistance in maximum (Rmax) values were measured (data not shown). Polypeptides (HMW-GS 1Bx7, 1N1C, 1'N1'C, 2N2C), which can enlarge the size of polymeric protein fraction, decreased the extensibility. HMW-GS 1Bx7 and the 2N2C peptide had the strongest positive effect on extensibility. On the other hand, ANGs containing odd numbers of cysteine residue increased the extensibility. There were small differences in the extent of this increase. Rmax values changed in the opposite way compared to extensibility. When proteins able to increase the size of the polymer were incorporated into the dough Rmax values were increased. On the other hand, chain terminator molecules reduced Rmax. In the case of the wild type molecule, the maximum value of the resistance dropped dramatically, resulting in very weak dough.

3.3.3 Baking. Loaf height (LH) of the small breads was measured and the results are shown in Figure 2. Polypeptides increasing the size distribution increased the LH value as well. The effect of ANG proteins with odd numbers of cysteine residue was the opposite. The height and consequently the volume of the bread were reduced with the added extra molecule, possibly due to the formation of shorter gluten polymers. Protein without cysteine (WT) did not alter the height of the test bread.

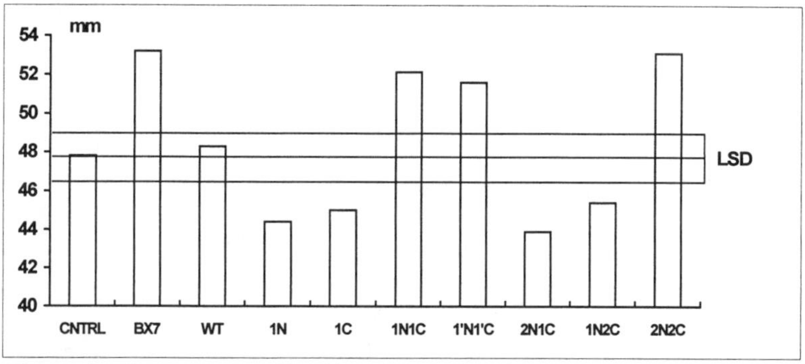

Figure 2, *Comparison of loaf height results of ANG protein samples with altered number of cysteine residues.*

4 CONCLUSIONS

The results of this study provide evidence that gluten structure and dough characteristics are influenced by the number and distribution of cysteine residues within storage proteins.

References

1. L. Tamás, F. Békés, J. Greenfield, A. S. Tatham, P. W. Gras, P. R. Shewry and R. Appels, *J. of Cereal Chem.,* 1998, **27**, 15.
2. L. Tamás, F. Békés, M. K. Morell and R. Appels, in: 'Cereals '97', ed. A. W. Tarr, A. S. Ross and C. W. Wrigley, Melbourne, 1997, p. 202.
3. C. R. Rath, P. W. Gras, C. W. Wrigley and C. E. Walker, *Cereal Foods World,* 1990, **35**, 572.
4. P. W. Gras, G. E. Hibberd and C. E. Walker, *Cereal Foods World,* 1990. **35**, 568.
5. F. Békés, O. D. Anderson, P. W. Gras, R. B. Gupta, A. Tam, C. W. Wrigley and R. Appels, in: 'Imp. of Cereal Quality by Genetic Engineering' ed. R. Henry, 1994, p. 97.
6. P. W. Gras, F. W. Ellison and F. Bekes, in: 'Proc. Int. Wheat Quality Conf.', ed. J. I. Steele and O. K. Chung, Manhattan, Kansas, 1997, p. 161.
7. S. Uthayakumaran, F. Békés and P. W. Gras, in: 'Cereals '97', ed. A. W. Tarr, A. S. Ross and C. W. Wrigley, Melbourne, 1997, p. 212.

Acknowledgements

The authors are grateful to Goodman Fielder Ingredients (NSW Australia) and to Group Limagrain for financial support.

EFFECTS OF TWO PHYSIOLOGICAL REDOX SYSTEMS ON WHEAT PROTEINS

F. Jarraud, K. Kobrehel

Unité de Biochimie et de Biologie Moléculaire des Céréales, I.N.R.A, 2, place Viala, 34 060 Montpellier, France

1 INTRODUCTION

In the complex mechanism of dough formation, the role of storage proteins (gliadins and glutenins) is generally recognized[1]. The role of S-S bonds and -SH groups of the proteins and their redox changes are also known as being of particular importance regarding the viscoelastic characteristics of a dough[2]. In wheat, two endogenous enzyme systems are present which can modify the redox state of proteins, the NADP-dependent thioredoxin system (NTS) and the gluthathione system (NGS). The first one is composed of NADPH, of thioredoxin reductase and of thioredoxin, while the second of NADPH, of glutathione reductase and of glutathione. NTS is a dithiol specific and NGS is a monothiol reducing system.

The results reported here illustrate the specific action of these two systems on wheat gluten protein fractions. The effects of the two enzyme systems were compared to the effects of two chemical reducing agents, 2-mercaptoethanol (2-ME) and dithiotreithol (DTT). Both chemical reducing agents are widely used in the study of cereal proteins, the first one being a monothiol and the second a dithiol.

2 MATERIALS AND METHODS

2.1 Plant materials

Bread wheat (*Triticum aestivum* L.) cultivars were grown on experimental fields in the south of France. Flour samples were obtained by using a Brabrender Quadrumat Junior pilot mill.

2.2 Protein extraction

Proteins were extracted by using a modified Osborne fractionation method. Albumins and globulins were solubilized in 0,5 M NaCl, then, after a washing step in order to eliminate residual NaCl from the pellet, gliadins were extracted with 70% ethanol. In order to avoid inhibitory effects and protein reduction, glutenins were solubilized in distilled water in the presence of sodium myristate[3-4], both the chemical reductants and the enzyme redox systems are functioning in protein solutions obtained under these conditions.

2.3 Protein reductions

The different protein fractions were reduced with the different reducing agents and with the reducing enzyme systems under the conditions previously described[5].

2.4 Fluorescent labelling

The technique of monobromobimane (mBBr) labelling was used as previously described[6]. The reaction of mBBr is specific for free (accessible) sulphydryl groups. In the protein, the bromine atom of the mBBr replaces the proton of the accessible sulphydryl groups, releasing HBr. The bimane, linked with the protein, becomes fluorescent, allowing the visualization of under UV light at 365 nm .Every labelling operation is carried out with an excess of mBBr which was afterwards derivatized by the addition of 2-mercaptoethanol.

2. 5 Electrophoretic analyses

SDS-PAGE were performed in 16/18 cm Hoefer systems by applying 40 mA/gel. Gels were fixed with 12% TCA before washes with 40% ethanol/10% acetic acid. Photographs were taken under UV light at 365 nm with Polaroid type 555 films. After UV detection gels were silver stained.

3 RESULTS

3.1 Effect of different reducing agents

Protein fractions (albumins, globulins, gliadins and glutenins), obtained by sequential extraction, were reduced by the two enzyme systems (NTS and GS) and by the two chemical reducing agents. The free -SH groups were then labelled with mBBr. The proteins became fluorescent when mBBr could fix free sulphydryl groups. With this method, the intensity of the fluorescence is proportional to the degree of reduction. Non-reduced controls did not show any significant fluorescence (Figures 1 and 2), indicating that the extracted proteins were oxidized.

Dithiol reducing systems showed much stronger effect on wheat proteins than monothiols. The weak reducing effect of monothiols, compared to dithiols, suggest that most of the wheat proteins (including both metabolic and storage proteins) are dithiol specific. Thus, NTS and DTT reduced all the main protein fractions in each protein family.

However, differences were also observed between the action of NTS and DTT. Within each protein fraction, the storage protein components were more strongly reduced by NTS than by DTT, the opposite observation could be made for a few minor proteins in the albumin-globulin fraction.

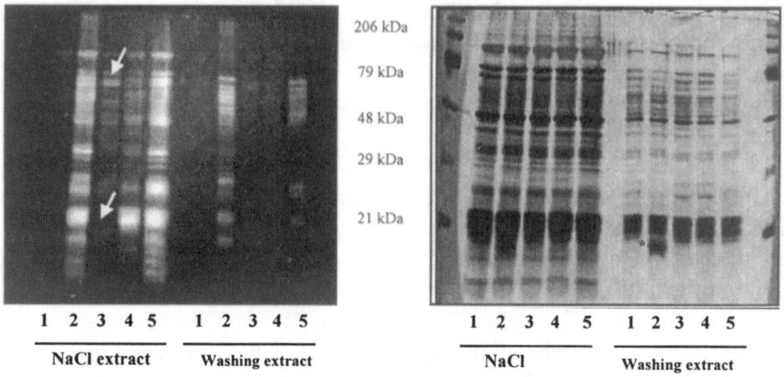

1: non reduced control
2: Thioredoxin reduced
3: Glutathione reduced
* Thioredoxin
4: 2-Mercaptoethanol reduced
5: DTT reduced
MW: Molecular Weight Markers

Figure 1 *SDS PAGE obtained after different reduction of the albumin-globulin fraction (A: mBBr labelling; B: silver stained gel)*

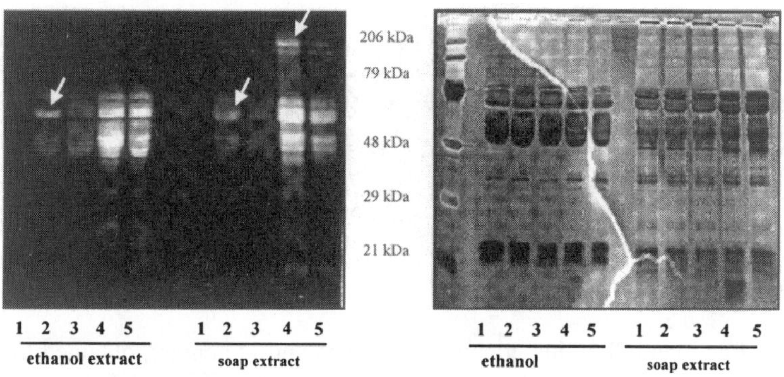

1: non reduced control
2: Glutathione reduced
3: 2-Mercaptoethanol reduced
* Thioredoxin
4: Thioredoxin reduced
5: DTT reduced
MW: Molecular Weight Markers

Figure 2 *SDS PAGE obtained after different reduction of the storage protein fraction (A: mBBr labelling; B: silver stained gel)*

Under the electrophoretic conditions used to detect reduced proteins, the highest number of protein components (protein bands) that were targeted specifically by thioredoxin was found in the albumin globulin fraction. Differences between NTS and DTT were also observed for gliadins and for both low and high molecular weight glutenins. The stronger effect of NTS on HMW glutenins, and in general on storage proteins, compared to DTT, may be particularly interesting in relation to the potential involvement of thioredoxin in wheat processing.

Differences between monothiols were found in their specificity in targeting wheat proteins. Thus, low molecular weight sulphur-rich proteins (MW between around 10 and

15 KDa) and higher MW globulins were specifically targeted by 2-ME and NGS, respectively.

The detection of reduced proteins, using either Coomassie blue or silver staining, showed that the reducing conditions and the degree of reduction of the proteins with the different reducing agents had effects, but a relatively small ones, on the electrophoretic patterns of the proteins (on the mobility and the intensity of the bands), suggesting relatively small conformational modifications.

3.2 Reoxidation of reduced proteins

Proteins reduced with different reducing systems were reoxidized under different conditions (by addition of H_2O_2, by changing the pH etc.). In general, the aggregates generated in these conditions could not enter the gel, or, as illustrated (Figure 3), very few bands were detectable. Methods are being developed to determine the degree of involvement of the different types of covalent bonds or noncovalent interactions leading to the formation of these aggregates.

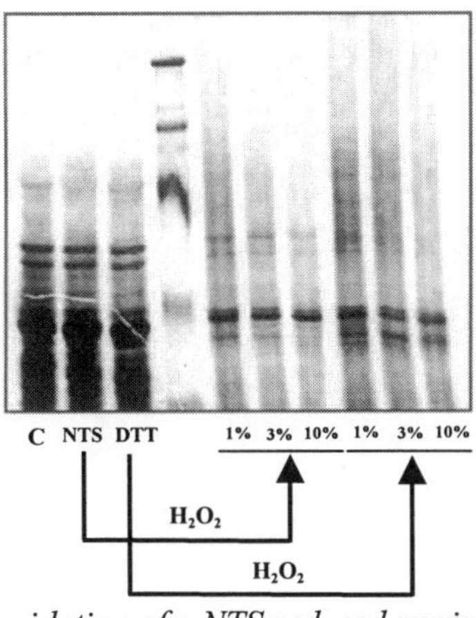

Figure 3 *Reoxidation of a NTS reduced myristate protein extract*

It is of interest, as shown in Figure 3, that the ω-gliadins were among the major protein components that did not participate in the formation of protein aggregates after the reoxidation of reduced proteins. This is consistent with the fact that ω-gliadins are known to contain no disulphide bonds. These results suggested that the free -SH groups, formed through reduction by the different reducing systems, were essential in forming aggregates. This would also imply the potential role of endogenous reducing systems in wheat processing, especially in dough formation and, consequently, in the quality of manufactured wheat products.

4. CONCLUSIONS

The disulphide bonds of most wheat proteins (including both metabolic and storage proteins) are dithiol specific. Thioredoxin, the reducing agent of the endogenous NTS system, seemed to be the most efficient in reducing SS bonds of wheat proteins, especially gliadins, low and high molecular weight glutenins. The reoxidised reduced proteins tended to form large aggregates, for which the presence of free -SH groups seems to be essential.

References

1. B.J Miflin, J.M. Field and P.R. Shewry, in *Seed Proteins*, eds J. Daussant, J. Mosse and J. Vaughan, Academic Press, London, 1983, p 255.
2. C.A. Stear, ed in *Handbook of breadmaking technology*, Elsevier Applied Science, London, 1990, pp 848.
3. K. Kobrehel and W. Bushuk, *Cereal Chem*, 1977, **54**, 833.
4. Z. Hamauzu, K. Khan, W. Bushuk, *Cereal. Chem.*, 1979, **56**, 513.
5. P.Gobin, 'Etats d'oxydoréduction des protéines chez le blé au cours de la maturation du grain et au sein du réseau protéique', thesis, 1995, pp 121.
6. K. Kobrehel, B.C. Yee, B.B. Buchanan, *Plant. Physiol.*, 1992, **99**, 919.

Aknowledgements

This work is supported by the FAIR CEREDOX programm (DG VI, DG XII, DG XIV).

INVOLVEMENT OF REDOX REACTIONS IN THE FUNCTIONAL CHANGES THAT OCCUR IN WHEAT GRAIN DURING POST-HARVEST STORAGE

G. Mann[1], P. Greenwell[2], S.S.J. Bollecker[1], A.A. Tsiami[1] and J.D. Schofield[1]

[1]The University of Reading, Department of Food Science and Technology, PO Box 226, Whiteknights, Reading RG6 6AP. UK. and [2]Campden and Chorleywood Food Research Association, Chipping Campden, Gloucestershire GL55 6LD. UK.

1 INTRODUCTION

It is widely believed in the milling and baking industries that wheat is unsuitable for milling and baking until stored for several weeks post harvest. This improvement in functionality during the first few weeks of post harvest storage is known as the 'new crop phenomenon'. Newly harvested and milled wheat flour tends to produce 'immature doughs' that have poor mixing tolerance, poor gas retention capabilities and yield undesirable loaf characteristics.

Chen and Schofield[1] have shown that the baking performance of wheat flour improved with storage. This was related to falling levels of free reduced (GSH), free oxidised (GSSG) and protein-bound (PSSG) glutathione in the freshly milled flour. It is thus reasonable to hypothesise that redox reactions may be involved in the new crop phenomenon. In this context, glutathione would act as a reducing agent, which would be capable of cleaving the interchain disulphide (SS) bonds linking the subunits in glutenin polymers. This, in turn, would affect the rheological and bread making properties adversely.

The objective of this study was to assess the new crop phenomenon and to explore the mechanisms involved.

2 MATERIALS AND METHODS

2.1 Materials

Flours: 1999 harvest, UK soft wheat grain *cv.* Consort was stored at 20°C and milled into straight-grade flour using a Buhler mill laboratory (model MLU-202). The biochemical and rheological properties of the freshly milled flour ('grain') and the same flour stored for one week ('flour') were determined.

2.2 Methods

2.2.1 Determination of total sulphydryl (SH) groups. A modified version of Ellman's method was used[2].

2.2.2 Reduced (GSH), oxidised (GSSG) and protein bound (PSSG) glutathione determination. A modified version of the HPLC method developed by Schofield and Chen[3] and Chen and Schofield[4] was used.

2.2.3 Gluten rheology. Small deformation oscillatory constant stress rheometry was performed on freshly extracted gluten at 0.5% strain, between 0.01-10 Hz, and at 25°C.

3 EXPERIMENTAL AND RESULTS

3.1 Determination of total SH groups

The total SH group content decreased slightly during the first two weeks of grain storage (Figure 1). A more significant reduction (26%, $P < 0.001$) was observed after six weeks of post harvest storage. The level of total SH groups was reduced significantly during the one week of flour storage for grain samples stored for 0 ($P < 0.001$), 1 ($P < 0.001$) and 2 ($P < 0.001$) weeks. However, the level of total SH groups showed a slight increase during the one week of flour storage for grain samples stored for 6 weeks (Figure 1).

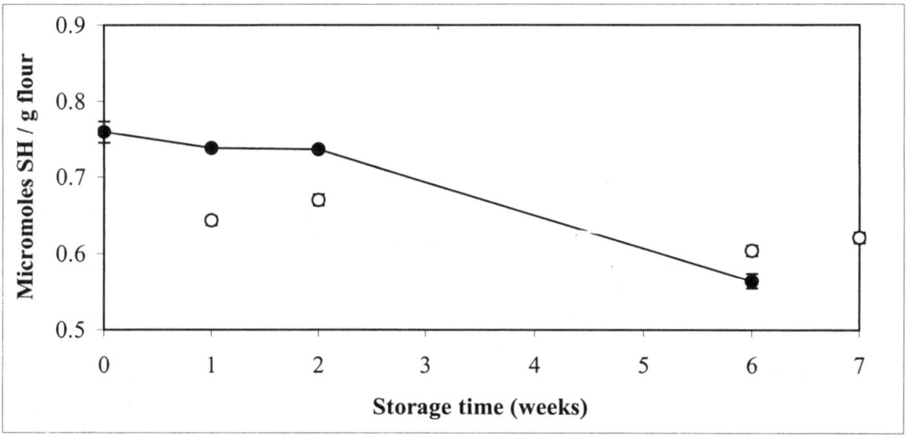

Figure 1 *Changes in total SH groups during post harvest storage of cv. Consort wheat grain (●) and flour (○).*

3.2 Reduced (GSH), oxidised (GSSG) and protein-bound (PSSG) glutathione determination

The content of GSH increased somewhat in grain during the first week of post harvest storage of grain (from 167 to 214 nmol /g flour, $P < 0.005$). It then remained constant to the second week. Then followed a significant decrease of 49% by week six (from 231 to 118 nmol/g flour, $P < 0.001$). The content of GSH also increased during the week of flour storage for grain samples that had not been stored ($P < 0.001$). It remained constant during the week of flour storage for grain samples stored for 1 week. Then followed a significant decrease of 58% (from 231 to 97 nmol/g flour, $P < 0.001$) during the week of flour storage for grain samples stored for 2 weeks. However, the decrease in GSH content during the week of flour storage for grain samples stored for 6 weeks was not significant. The decrease in GSH content was accompanied by an increase in the GSSG content. The GSSG content of grain showed a steady increase during storage (from 38.4 to 88.9 nmol /g flour for stored grain, $P < 0.001$). There were no significant changes in the contents of GSSG during the week of flour storage for grain samples stored for 0, 1, 2 and 6 weeks. The PSSG content decreased after the first week of grain storage, remained constant up to week two and then increased at week six. A similar pattern was also observed for the stored flours. The increase at week six was more prominent for the stored grain than for the stored flour (Figure 3).

The fall in GSH can be accounted for both by oxidation to GSSG and formation of protein-glutathione mixed disulphides since glutathione levels in those two pools were increased (Figure 2).

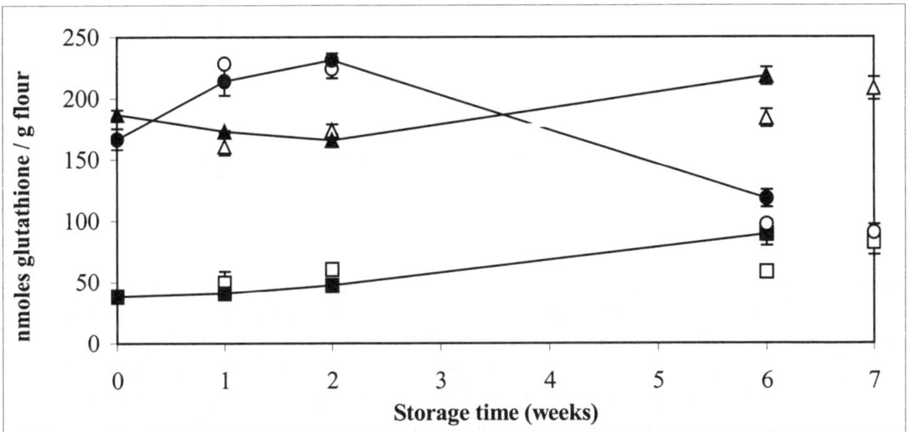

Figure 2 *Changes in GSH (●, ○), GSSG (■, □) and PSSG (▲, △) glutathione levels during post harvest storage of cv. Consort wheat grain (closed symbols) and flour (open symbols).*

3.3 Gluten Rheology

The elastic modulus (G') for gluten increased significantly during grain storage ($P < 0.001$). Increases in G' were also observed during the week of flour storage for grain samples stored for 0, 1 and 2 weeks. The G' values for glutens from stored flours were generally higher than those for the corresponding grain samples. The gluten tan δ values decreased significantly ($P < 0.001$) during grain storage (Figure 3). Tan δ values also decreased during the week of flour storage for grain samples stored for 0, 1 and 2 weeks. Increased G' and decreased tan δ values suggest that the gluten from wheat grain became more elastic during post harvest storage and that flour storage lead to further increases in elasticity. Furthermore the G' and tan δ values correlated well with SH groups, GSH, GSSG and PSSG levels obtained during the post harvest storage of both wheat grain and flour.

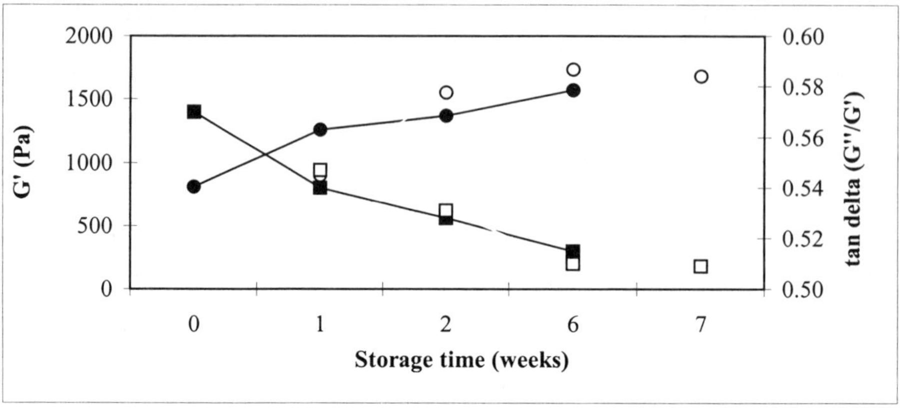

Figure 3 *Effects of post harvest storage on the G' (●, ○) and tan δ values (■, □) of cv. Consort wheat grain (closed symbols) and flour (open symbols) gluten proteins.*

4 CONCLUSIONS

The levels of total SH groups decreased significantly in both post harvest stored grain and stored flour. The levels of GSH showed an initial increase during grain storage then a marked decrease, whereas GSSG showed a slow increase and PSSG first decreased slightly then increased. The increased elastic modulus (G') and reduced tan δ values suggested that both grain and flour gluten proteins became more elastic during post harvest grain and flour storage. Significant correlations occurred between the elastic modulus (G') and tan δ values and the levels of free SH groups and GSH, GSSG and PSSG.

References

1. X. Chen and J.D. Schofield, *Cereal Chem.*, 1996, **73**, 1
2. P. Greenwell, personal communication, 1998
3. J.D. Schofield and X. Chen, *J. Cereal Sci.*, 1995, **21,** 127
4. X. Chen and J.D. Schofield, *J. Agric. Food Chem.*, 1995, **43**, 2362

Acknowledgements

This research was supported through provision of a research studentship funded by MAFF and CCFRA and through an EU funded project FAIR CT 97-3010.

Improvers and Enzymic Modification

STUDY OF THE EFFECT OF DATEM

P. Köhler

Deutsche Forschungsanstalt für Lebensmittelchemie and Kurt-Hess-Institut für Mehl- und Eiweißforschung, Lichtenbergstraße 4, D-85748 Garching, Germany

1 INTRODUCTION

The anionic oil in water emulsifier DATEM improves the handling properties of wheat doughs and increases the volume of bread. As DATEM is produced by the reaction of mono- and diacetyltartaric acid with monoacylglycerols or mixtures of mono- and diacylglycerols derived from edible fats[1] commercial DATEM is a complex mixture of components. The improver effect of the commercial product is well characterised, however, up to now no information is available about the influence of the acyl residue on the effect of DATEM and nothing is known about the effects of individual components because the amounts of material for baking tests and rheological measurements under standard conditions (1000 g of flour) were too low in the past[2-5]. By means of micro-scale methods on the basis of 10 g of flour[6-8] it is now possible to determine the effect of fractions or individual components of DATEM on the rheology and on the baking performance because only minor amounts of material are necessary in comparison to the standard methods.

Aim of the study presented here was (1) the determination of the influence of fatty acid chain length on the improver effect of DATEM, (2) fractionation of DATEM in order to obtain individual components with high baking activity and the determination of their structures, (3) the characterisation of the major components with respect to rheology and baking and (4) the optimisation of DATEM synthesis to produce DATEMs with high amounts of active components.

2 MATERIALS AND METHODS

2.1 Materials

The German wheat variety 'Kraka' from the 1992 harvest was supplied by Petersen, A. S., Lundsgaard, Germany. Eight DATEM samples were obtained from two producers.

2.2 Methods

2.2.1 Synthesis of DATEM. For the synthesis of 1-monoacylglycerols 2,2'-dimethyl-4

hydroxymethyl-1,3-dioxolane (solketal) was acylated with fatty acids 6:0 – 22:0. 18:1 and 18:2 and the ketal was hydrolysed with hydrochloric acid. 1-monoacylglycerols were used for the synthesis of DATEM according to Jacobsberg et al.[9].

2.2.2 Fractionation of DATEM. A commercial DATEM sample was fractionated by gel permeation chromatography on Sephadex LH-20, fractions 3 and 5 of the six fractions were separated by HPLC on LiChrospher 100DIOL into 20 fractions, respectively, individual components were isolated and their structures were determined by mass spectrometry and NMR spectroscopy.

2.2.3 Characterisation of the major DATEM components. The major components P5-8-1, P3-10-1 and P5-12-1 were synthesised and characterised by micro-rheological tests[6,7], a micro-scale (10 g of flour) and a normal-scale (300 g of flour) baking test.

2.2.4 Optimisation of DATEM synthesis. DATEM was synthesised in a 1g-scale by modifying the parameters of a standard procedure[9] systematically. Synthesised samples were separated by analytical HPLC. The amounts of the individual components P5-8-1, P3-10-1 and P5-12-1 were determined by calibration with standard solutions.

3 RESULTS AND DISCUSSION

3.1 Influence of the Acyl Residue on the Activity of DATEM

DATEMs with fatty acids of chain lengths 6:0 - 20:0, 18.1 and 18:2 were synthesised. The activity of synthesised DATEMs and commercial DATEM products was studied by means of rheological methods and a micro scale baking test with 10 g of flour. Variation of the acyl residue (Figure 1) showed that stearic acid (18:0) had the best effect on the baking activity of DATEM (loaf volume increased by 62 %). DATEMs containing unsaturated fatty acids (18:1, 18:2) or DATEMs produced from diacylglycerols instead of monoacylglycerols showed a slight increase of the loaf volumes. A slight effect of DATEM on the rheology of dough was observed. However, much greater was the effect on the gluten isolated from doughs prepared with DATEM. The resistance of gluten to extension was increased after the addition of increasing amounts of DATEM (1 – 5 g/kg of flour). Within the series of DATEMs derived from the homologous series of monoacylglycerols the product based on glycerol monostearate (18:0) showed a maximum increase of the gluten resistance.

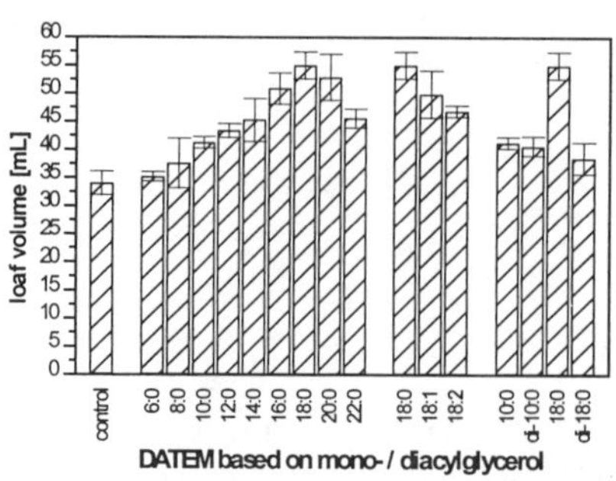

Figure 1 *Micro-scale baking test with synthesised DATEM samples. Influence of the acyl residue on the loaf volume*

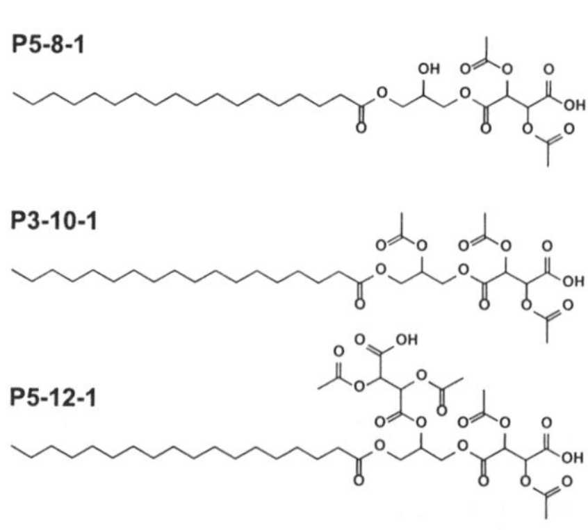

Figure 2 *Major components of DATEM*

3.2 Identification of Major DATEM Components

In order to answer the question which component of DATEM is most effective in baking, a commercial DATEM sample was fractionated by a combination of gel permeation chromatography and high-performance liquid chromatography. The activities of fractions and individual components were determined by the micro-scale methods described above. Three active compounds were isolated which were the major components of DATEM. Their structures were determined by mass spectrometry and NMR spectroscopy (Figure 2). The major component of DATEM (P5-8-1; 35.4 % by weight) was a glycerol molecule esterified with stearic acid and diacetyltartaric acid and a free hydroxyl group at the secondary C-atom. In the second component (P3-10-1; 12.1 % of DATEM) this hydroxyl group was acetylated, whereas in the third compound (P5-12-1; 6.5 % of DATEM) it was esterified with an additional diacetylartaric acid residue. The activity of DATEM was based on the sum of the three major components.

3.3 Characterisation of Major DATEM Components

The major components of DATEM were synthesised and characterised by micro-scale methods and, additionally, by a normal-scale baking test with 300 g of flour. Both the micro-scale and the normal-scale baking test were in good accordance and showed that DATEM components with one carboxyl group exhibited better baking performance than compounds with two carboxyl groups (Figure 3). The best results in baking were obtained with a concentration of 2 g of emulsifier/kg of flour in contrast to 3 g/kg with a commercial DATEM sample, because commercial samples may contain up to 40 % of inactive components. The DATEM component with two carboxyl groups had the lowest baking activity, but it was most effective in dough and gluten rheology. This discrepancy between rheology and baking indicates that for DATEM different mechanisms of action have to be present in the dough phase and during baking.

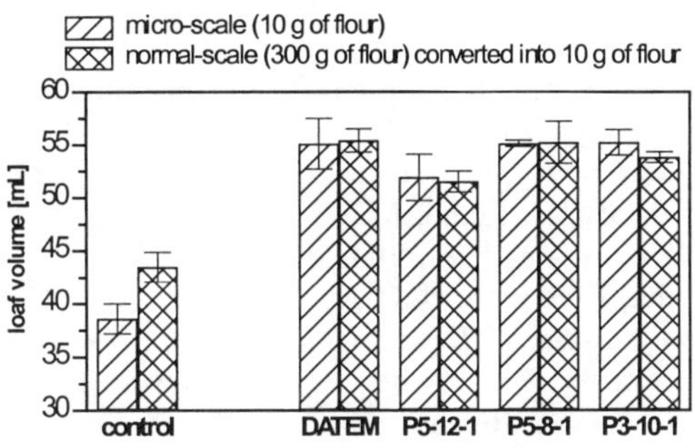

Figure 3 *Baking tests with the major DATEM components P5-8-1, P3-10-1 and P5-12-1*

3.4 Optimisation of DATEM Synthesis

DATEM synthesis was optimised to produce samples with high amounts of active components P5-8-1, P3-10-1 and P5-12-1. Low temperature (100°C) had a positive effect on the formation of active components, especially after the addition of sodium acetate as a catalyst that increased the yield of active components up to 88 %. The highest yield of the major components was obtained by using tetrahydrofuran (THF) as a solvent and pyridine as a catalyst. Under these conditions more than 90 % of the DATEM sample were components with high baking activity (Table 1).

Table 1 *Major DATEM components [%] in relation to the conditions during synthesis*

Procedure	P3-10-1 [%]	P5-8-1 [%]	P5-12-1 [%]	total [%]
Standard	7,5	44,7	6,1	58,3
Commercial product	12,1	35,4	6,5	54,0
100°C	5,4	53,3	2,4	61,1
100°C, NaAc	3,9	64,3	20,0	88,2
200°C	0,0	0,0	0,0	0,0
NaAc, THF, 22 °C	4,5	49,7	14,0	68,2
Pyridine, THF, 22 °C	3,6	69,3	21,1	94,0

4 CONCLUSIONS

The activity of DATEM is caused by only a few components. Commercial samples contain 35 - 60 % of these components. The effect of DATEM depends on the length of the acyl residue and is optimal for stearic acid. The most active components of DATEM consist of a glycerol molecule esterified with a fatty acid and a diacetyl tartaric acid. One hydroxyl group remains free or is esterified with acetic acid or diacetyl tartaric acid. Chemical synthesis of DATEM components makes it possible to characterise their rheological effect, their effect on the baking performance, and gives insight into the mechanism of action. By systematic modification of a standard procedure DATEM synthesis can be optimised to give products with high amounts of active components. The amount of additive for the production of bread can be reduced by using the optimised products. The micro-scale baking test (10 g of flour) is in good accordance with the corresponding normal-scale method (300 g of flour)

References

1. W.F Adams and G. Schuster. 'Emulgatoren für Lebensmittel', Springer Verlag, Berlin, Heidelberg, New York, Tokyo, 1985, p 114.
2. A. Seher and J. Janssen, *Fette Seifen Anstrichmittel,* 1970, **72**, 773.
3. G. Sudraud, J.M. Custard, C. Retho, M. Caude, C. Rosset, R. Hagemann, D. Gaudin and H. Virelizier, *J. Chromatogr.*, 1981, **204**, 397.
4. J.M. Custard, C. Retho, F. Blanchard, G. Sudraud, M. Caude, C. Rosset, R. Hagemann, D. Gaudin and H. Virelizier, *Falsif. Expert. Chim. Toxicol.*, 1982, **75**, 563.
5. N.O. Carr and P.J. Frazier, P. J. 'Wheat Structure, Biochemistry and Functionality', Symposium April 10 – 14, 1995, Reading, U.K.
6. R. Kieffer, F. Garnreiter and H.-D. Belitz, *Z. Lebensm. Unters. Forsch.*, 1981, **172,** 193.
7. R. Kieffer, J.J. Kim and H.-D. Belitz, *Z. Lebensm. Unters. Forsch.*, 1981, **172,** 190.
8. R. Kieffer, H.-D. Belitz, M. Zweier, R. Ipfelkofer and G. Fischbeck, *Z. Lebensm. Unters. Forsch.*, 1993, **191**, 134.
9. F.R. Jacobsberg, S.L. Worman and N.W.R. Daniels, *J. Sci. Fd. Agric.*, 1976, **27**, 1064.

Acknowledgements

This research project was supported by FEI (Forschungskreis der Ernährungsindustrie, e.V., Bonn), the AiF and the Ministry of Economics. Project No. 10634N.

MECHANISM OF THE ASCORBIC ACID IMPROVER EFFECT ON BAKING

D. Every[1], L. Simmons[1], M. Ross[1], P.E. Wilson[1], J.D. Schofield[2], S.S.J. Bollecker[2] and B. Dobraszczyk[2]

1. Crop & Food Research, Private Bag 4704, Christchurch, New Zealand. 2. Department of Food Science and Technology, University of Reading RG6 6AP, UK

1 INTRODUCTION

There are two main hypotheses for the mechanism of the ascorbic acid (AA) improver effect on dough and bread. Both hypotheses require that AA is oxidised to dehydroascorbic acid (DHA) by ascorbate oxidase (AOX) and metal ions. Hypothesis-1[1] proposes that glutathione (GSH) is rapidly oxidized at the early stages of dough mixing by a DHA:GSH oxidoreductase (GSH dehydrogenase; EC 1.8.5.1) catalysed reaction with DHA, and therefore is not available to cleave the disulphide bonds in glutenin that would weaken dough.

Scheme 1. L-DHA + 2GSH $\xrightarrow{\text{DHA:GSH Oxidoreductase}}$ L-AA + GSSG

Hypothesis-2[2] proposes that glutenin thiols (P^{SH}), produced by reductants and SH/SS interchange reactions during mixing, are oxidatively cross linked by DHA to interprotein disulphides ($P^{SS}P$), predominantly during the proof stage of dough, and thus strengthening dough. It is also proposed that a Thiol Disulphide Oxidoreductase (TDOR) in flour may catalyse this reaction. This paper tests these hypotheses by treating doughs at different times of mixing and proofing with various redox agents, and analysing dough for changes in AA, DHA, GSH, GSSG, protein-GSH mixed disulphides of protein fractions, rheology and baking properties.

Scheme 2

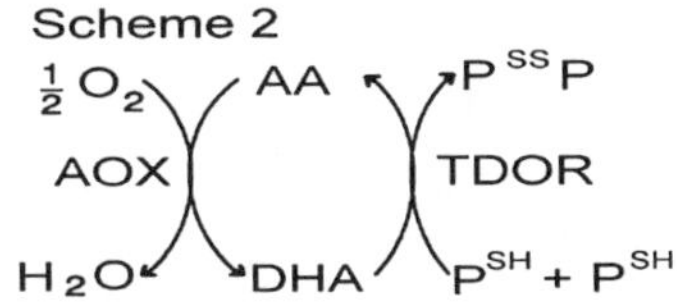

2 MATERIALS AND METHODS

Dough was treated at different times of mixing and proofing with either AA, DHA or GSH (see Figures 1-3). Dough was mixed in either a 1 kg MDD mixer (Morton double Z-blade mixer) or a 10 g MDD mixer (Crop & Food Research, NZ) using optimum work input and water addition. Dough samples were taken at different stages of mixing and proofing, frozen in liquid nitrogen, freeze dried, and stored at -20°C. Bread was made from dough mixed in the 10 g MDD mixer. Dough samples taken immediately after mixing on the Morton mixer were rheologically tested by the Dobraszczyk and Roberts dough inflation system and dough extensibility method (Kieffer rig) using a TA.XT2i Texture Analyser. Dough samples taken after mixing on the 10 g MMD mixer were measured for strength by a compressive stress relaxation test. Freeze dried dough samples from the Morton mixer were analysed directly for GSH, GSSG and PSSG content by HPLC methods.[3,4] Albumin, globulin, gliadin, acetic acid soluble and insoluble glutenin were extracted from the freeze dried samples by a modified Osborne fractionation method[4] and analysed for PSSG content Freeze dried dough samples from the 10 g mixer were analysed for AA and DHA as described by Every.[5] Freeze dried samples of reduced gluten and glutenin were prepared using sodium borohydride[6] or dithiothreitol. DHA in pH 6.2 buffer was added to reduced gluten or glutenin powder (1.3:1 v/w) and mixed with a glass rod in a test tube. The reaction was stopped with 5% perchloric acid and analysed for AA and DHA.[5] Formation of glutenin disulphides was determined by SDS-PAGE.

3 RESULTS AND DISCUSSION

Figure 1 shows that in control dough GSH decreased by 66% during the 1 min of slow mixing, then by the end of fast mixing GSH has decreased to 9% of the initial level. GSSG levels increased during the first min of mixing, then decreased slowly to initial GSSG levels at the end of mixing. GSH and GSSG levels hardly changed during 55 min of proof. In contrast, when AA was added to dough at the start of mixing, GSH decreased by 95% during the first min of mixing, undoubtably by Scheme 1, and by 99% at the end of mixing. The dough inflation and extensibility results (Table 1) show that AA has increased the strength of dough. The dough rheology result together with the rapid removal of GSH by AA, are

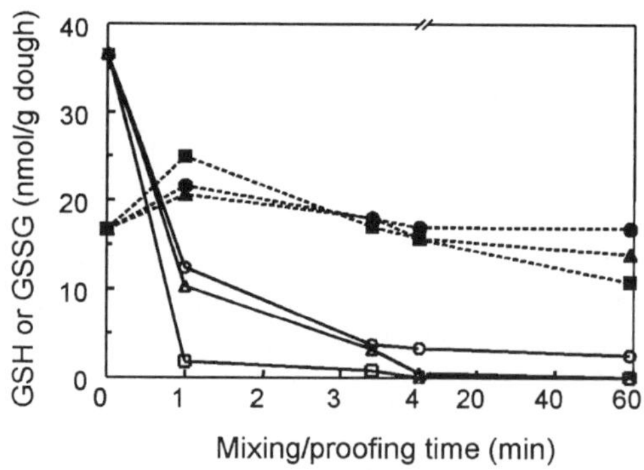

Figure 1 *Contents of GSH (open symbols) and GSSG (solid symbols) in dough at various times of mixing (4 min total mixing) and proofing (55 min total proofing). Dough was treated with no additive (O, ●), or 568 nmol AA/g flour at the start of mixing (□, ■), or 568 nmol DHA/g flour for the final 0.6 min of mixing (Δ, ▲).*

consistent with Hypothesis-1. However, Fig. 1 and Table 1 show that when DHA was added to dough near the end of mixing, after 90% of GSH had disappeared and done its hypothetical damage, the dough was still strengthened to a similar extent as with AA addition at the start of mixing. This result is inconsistent with Hypothesis-1, but consistent with Hypothesis-2.

Table 2 shows that the addition of AA or DHA to dough had little effect on albumin levels, but increased the ratio of insoluble glutenin to soluble glutenin. Addition of GSH to dough decreased the ratio. Addition of AA or DHA together with GSH increased the ratio again (data not shown). These results are consistent with the hypothesis that GSH breaks down very high molecular weight acetic acid-insoluble glutenin to lower molecular weight acetic acid-soluble glutenin. It appears that AA/DHA either prevents this break down or converts acetic acid-soluble glutenin to acetic acid-insoluble glutenin, or a combination of both. This partly supports Hypothesis-1, but is totally consistent with Hypothesis-2. Other work, however, suggests that AA does not prevent break down of glutenin during mixing.

Table 1 *Effect of AA, DHA and GSH on dough extensibility and dough inflation*

	Extensibility measurements		Bubble inflation measurements			
Redox agents added	Maximum peak force (g)	Extensibility (mm)	Max pressure (mm H_20)	Max extensibility (mm)	Failure strain	Viscosity index (m power)
None	24.1	91	38.9	203.2	2.602	2.015
AA	27.5	50.6	119.2	61	1.737	2.546
DHAA	27.1	53.2	101.7	122.2	2.223	2.259
GSH	21.8	96.2	40	225.7	2.692	2.034

[a]AA (568 nmol/g flour) or GSH (100 nmol/g flour) were added at start of mixing. DHA (568 nmol/g flour) was added for the final 35 sec of mixing. Total time of mixing = 4 min.

Table 2 *Protein (mg/g flour or dough) and protein-glutathione mixed disulphide (values in parenthesis are nmol PSSG/g protein) contents of Osborne fractions obtained from flour and dough treated with AA, DHA or GSH as described in Table 1.*

Sample	Albumin	Soluble glutenin[a] = S	Insoluble glutenin[a] = I	Protein Ratio I:S
Flour	18.4 (285)	21.3 (203)	60.3 (377)	2.8
Mixed dough control	15.8 (228)	35.1 (213)	47.2 (564)	1.3
Proofed dough control	15.8 (313)	34.6 (211)	48.1 (509)	1.4
Mixed dough + AA	15.7 (412)	35.3 (194)	57.8 (532)	1.6
Proofed dough + AA	15.5 (463)	32.9 (226)	60.7 (473)	1.8
Mixed dough + DHA	15.7 (269)	34.7 (209)	58.6 (467)	1.7
Proofed dough + DHA	15.7 (381)	32.0 (242)	56.8 (508)	1.8
Mixed dough + GSH	15.8 (974)	42.4 (414)	53.8 (446)	1.3
Proofed dough + GSH	16.5 (943)	44.0 (337)	49.0 (483)	1.1

[a] Soluble or insoluble in 0.1 M acetic acid.

Table 2 shows that the reaction of free GSH/GSSG with albumin, soluble glutenin and insoluble glutenin is similar in control dough and dough with delayed DHA addition, but differs from dough treated with AA at the start of mixing by having less GSH/GSSG reacting with albumin. Since AA and delayed DHA treated doughs have the same improver effect, the above results suggest that the pattern of GSH/GSSG reaction with proteins is not important for the improver effect. When extra GSH was added to dough (100 nmol/g flour), the GSH combined with albumin and soluble glutenin, but not with insoluble glutenin. It is not clear how this relates to the observation that addition of extra GSH to dough actually enhances the AA improver response (Figure 2). This result is inconsistent with Hypothesis-1, but may be explained by Hypothesis-2 as follows. The extra thiols produced by GSH cleavage of disulphide bonds in glutenin during mixing may be oxidized by DHA during proofing to disulphides that are in optimal configuration for dough strength - the SH/SS interchange reaction thus being enhanced. A TDOR may assist this reaction to account for the L-AA stereo isomer specificity of the improver effect.

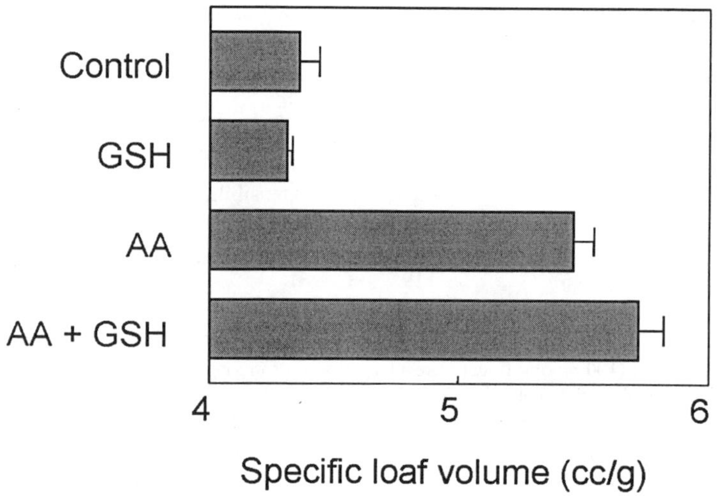

Figure 2
Volume of 125 g MDD bread made from dough treated with 100 nmol GSH/g flour, or with 568 nmol DHA/g flour, or with 568 nmol AA and 100 nmol GSH/g flour, or with no AA or GSH (Control). Mean and standard deviation of three replicates are shown.

The 10 g mixer experiments, with AA or DHA additions at different times of mixing and proofing gave similar rheological results (compressive stress relaxation tests - data not shown) to the Morton mixer experiments, and the dough strengthening effect of AA/DHA translated into improved bread (Figure 3). Thus, even if GSH has done its hypothetical damage up to the end of intermediate proof, it seems that DHA still repairs that damage.

Support for Hypothesis-2 comes from Figures 4 & 5, which indicate that DHA oxidizes protein thiols with formation of disulphides and AA. Figure 4 shows that when DHA was added at the end of mixing, at least 70 nmol DHA/ g flour was reduced to AA within 3 min, and another 12 nmol DHA was oxidized to AA after 13 min. In this system at the end of mixing, only about 3 nmol GSH/g flour was available to reduce DHA. It seems likely, therefore, that protein thiols reduced DHA. The levels of DHA and AA during proof were very similar, whether AA was added at the start of mixing or DHA was added at the end of mixing. Figure 5 shows that glutenin thiols can indeed reduce DHA to AA. SDS-PAGE showed that reduced glutenin subunits were cross linked by oxidation with DHA to SDS-insoluble glutenin (data not shown).

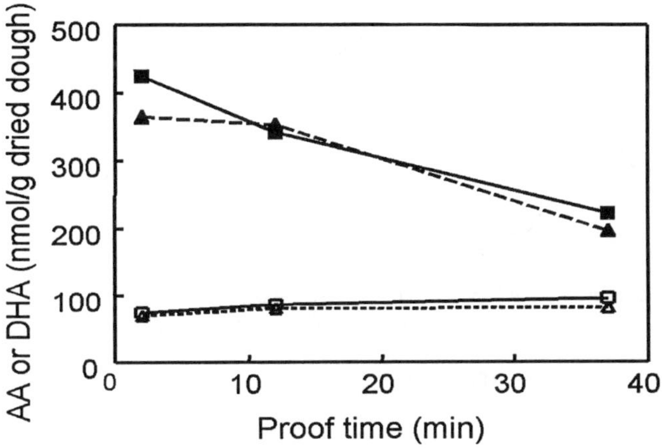

Figure 3
Contents of AA (open symbols) and DHA (solid symbols) in dough at various times of proofing. Dough was treated with 568 nmol AA/g flour at the start of mixing in a 10 g mixer (□, ■), or with 568 nmol DHA/g flour at 20 sec before the end of mixing (Δ, ▲).

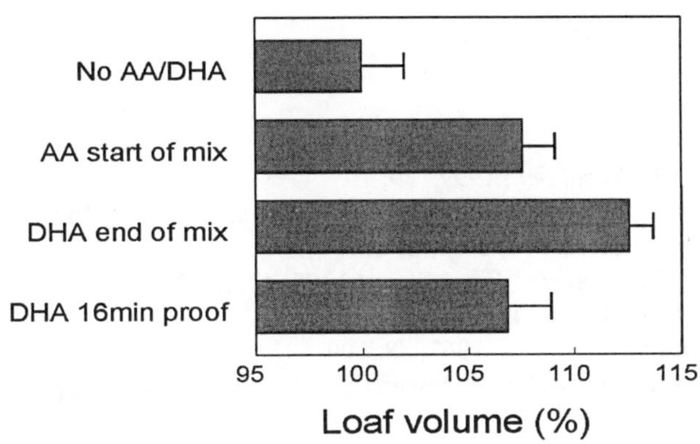

Figure 4
Volumes of bread made from dough treated with 568 nmol AA/g flour at the start of mixing in a 10 g mixer, or with 568 nmol DHA/g flour at 20 sec before the end of mixing, or with 568 nmol DHA/g flour at the end of intermediate proof (16 min), or with no AA or DHA (Control). Mean and standard deviation of four replicates are shown.

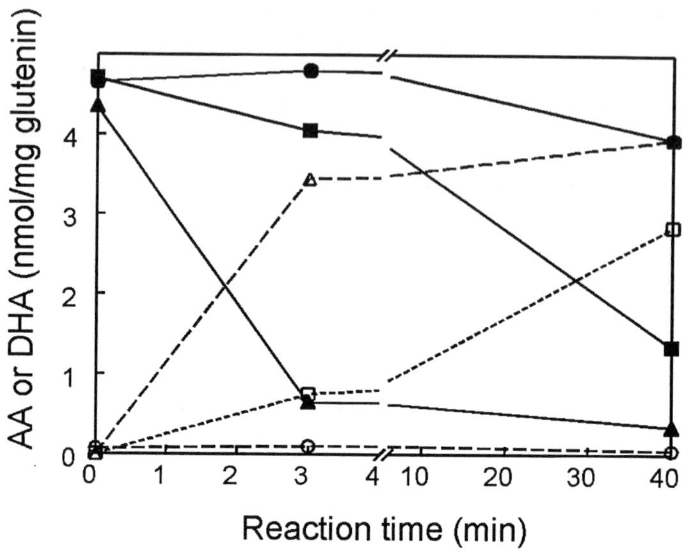

Figure 5
Contents of AA (open symbols) and DHA (solid symbols) in hand mixed doughs made of non-reduced glutenin (○, ●), or DTT-reduced glutenin (△, ▲), or sodium borohydride-reduced glutenin (□, ■) containing 4.8 nmol DHA/mg glutenin.

4 CONCLUSION

The results are most consistent with Hypothesis-2.

References

1. W. Grosch and H. Wieser, *J. Cereal Sci.*, 1999, **29**, 1.
2. D. Every, L. Simmons, K. H. Sutton and M. Ross, *J. Cereal Sci.*, 1999, **30**, 147.
3. J.D. Schofield and X. Chen, *J. Cereal Sci., 1995,* **21** 127.
4. X. Chen and J. D. Schofield, *J. Agric. Food Chem.*, 1995, **43**, 2362.
5. D. Every, *Analytical Biochem.*, 1996, **242**, 234.
6. R. Frater and F. J. R. Hird, *Biochem. J.*, 1963, **88**, 100.
7. X. Chen, *Glutathione in wheat*, PhD Thesis, University of London, 1994, 177pp.

DEGRADATION OF WHEAT AND RYE STORAGE PROTEINS BY RYE PROTEOLYTIC ENZYMES

K. Brijs, I. Trogh and J.A. Delcour

Laboratory of Food Chemistry, Katholieke Universiteit Leuven, Kardinaal Mercierlaan 92, B-3001 Heverlee, Belgium

1 INTRODUCTION

Endoproteolytic, exoproteolytic, carboxypeptidase, aminopeptidase and N-α-benzoyl-arginine-*p*-nitroanilide hydrolysing activities were detected in 0.05 M sodium acetate buffer extracts of ungerminated[1] and germinated rye. During rye grain germination, the proteolytic activity increases strikingly due to the synthesis and secretion of endoproteases. Activities in rye germinated for 3 days are about 5.0 times higher than in ungerminated rye. The four proteinase classes (serine-, cysteine-, metallo- and aspartic-proteinases) are present in rye germinated for 3 days, but the majority of these are cysteine-proteinases. Aspartic proteinases are clearly the most abundant class in ungerminated rye. Pepstatin A, an inhibitor of aspartic proteases, reduced ca. 88% and 75% respectively of the hemoglobin and azocasein hydrolysing activities of the proteases present in ungerminated rye.

The aim of this study was to evaluate the effects of some concentrated enzyme fractions on rye and wheat storage proteins.

2 MATERIALS AND METHODS

2.1 Materials

Rye cultivar Humbolt (AVEVE, Landen, Belgium) was milled at a moisture level of 14.5% on a MLU-202 laboratory mill (Bühler, Uzwl, Switzerland) according to Approved Method 26-31 (AACC, 1995) to yield eight streams: bran, shorts and six flour fractions (B1, B2, B3, C1, C2, C3).

Humbolt was steeped to 45% moisture at 16 °C with several air rests and germinated in the dark, with slow rotation at 16 °C for 3 days. Aliquots were removed prior to steeping and at the end of the process and were frozen at -20 °C until used.

Commercial wheat gluten was from NV Amylum (Belgium) and secalins were extracted as described by Shewry et al.[2]

2.2 Enzyme fractions

An aspartic proteinase and a cysteine proteinase fraction were isolated from the ungerminated bran fraction and from the germinated flour fraction respectively, both by concentration via ammonium sulphate (AS) precipitation and further purification with column chromatography. For aspartic proteinases, affinity chromatography was used and for the isolation of the cysteine proteinases a combination of ion exchange chromatography and gelfiltration.

2.3 Evaluation of the effect of rye proteases on secalins and gluten

Gluten proteins and secalins were suspended in 0.2 M sodium acetate buffer, pH 4.0. In the case of the gluten proteins, the mixture was boiled for 10 min to inactivate the gluten-associated proteolytic enzymes[3]. Then, enzyme solution was added to the substrate and the mixtures were incubated with continuous stirring for different periods at 40 °C. Activity was measured in two different ways. The increase of free α-amino nitrogen as a function of time was assayed with trinitrobenzenesulfonic acid reagent. Digested proteins were characterised by SDS-PAGE.

3 RESULTS AND DISCUSSION

3.1 Increase of free α-amino nitrogen as a function of time

When adding the enzyme fractions to gluten and secalins, we noticed an increase in the free α-amino nitrogen content as a result of hydrolysis of the substrates (Figure 1). All enzyme fractions, and especially the aspartic proteinases, had clearly more affinity for the gluten than for the secalins.

By making use of inhibitors it was clear that under the experimental conditions, both substrates were hydrolysed by aspartic (—) and cysteine (---) proteinases, as the activity could be totally inhibited by pepstatin A and E-64.

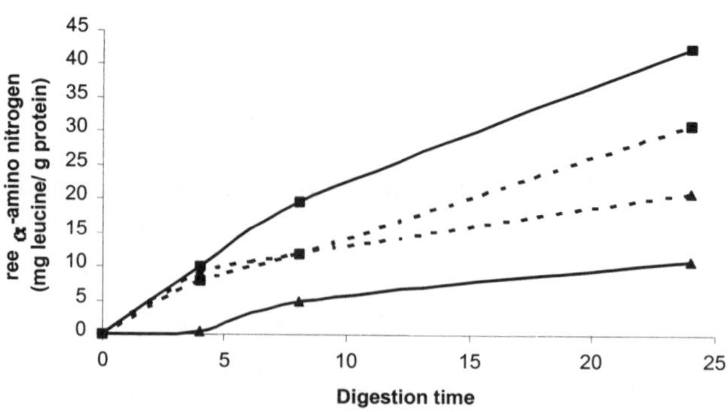

Figure 1 *Increase of free α-amino nitrogen of gluten proteins (■) and secalins (▲) as a function of time by adding an aspartic proteinase (—) and a cysteine proteinase (---) enzyme fraction at pH 4.0 and 40 °C.*

3.2 Characterisation of digested secalins and gluten by SDS-PAGE

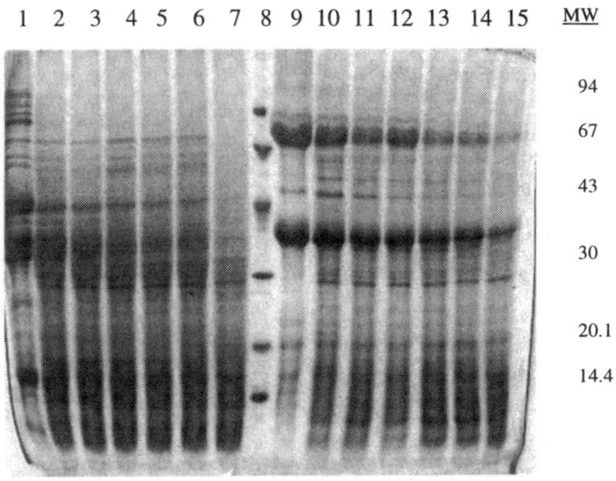

Figure 2 *SDS-PAGE patterns of digested gluten proteins (lanes 1-7) and secalins (lanes 9-15) as a function of time by a concentrated 35-60% AS fraction of rye bran. Enzyme solution and storage proteins were incubated for 0.25h (2 and 10), 2h (3 and 11), 4h (4 and 12), 6h (5 and 13), 8h (6 and 14) and 24h (7 and 15). Lanes 1 and 9 are control samples and lane 8 contains molecular mass markers.*

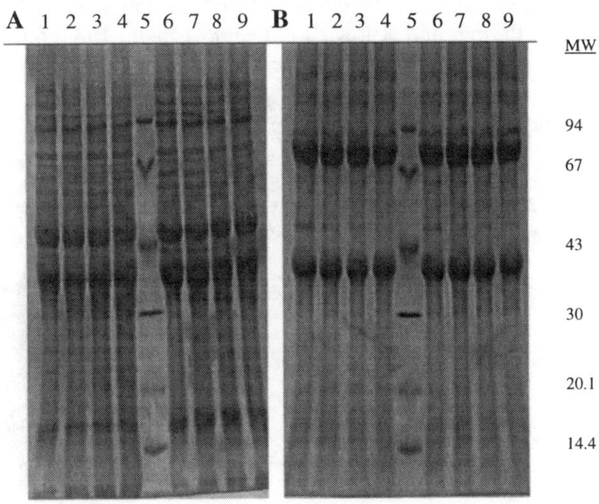

Figure 3 *SDS-PAGE patterns of digested gluten proteins (**A**) and secalins (**B**) as a function of time by an aspartic proteinase (lanes 1-4) and a cysteine proteinase (lanes 6-9) fraction isolated from the ungerminated bran fraction and from the germinated flour fraction respectively. Enzyme solution and storage proteins were incubated for 0.25h (1 and 6), 4h (2 and 7), 8h (3 and 8) and 24h (4 and 9). Lane 5 contains molecular mass markers.*

SDS-patterns after different periods of storage protein hydrolysis (Figures 2 and 3) differed for the three enzyme fractions tested. With the concentrated 35-60% AS fraction of the ungerminated rye bran, hydrolysis was very strong. For gluten, after 15 min of incubation, all HMW glutenin subunits were degraded and new proteins bands were formed with molecular masses of ca. 30 kDa. After 24h of incubation, almost all proteins with molecular masses >30kDa were hydrolysed. With secalins, hydrolysis yielded protein bands of low molecular mass.

The same results were found with aspartic and cysteine proteinase fractions from ungerminated and germinated rye. Hydrolysis was more pronounced with gluten as substrate than with secalins as substrate. Protein bands of higher molecular weight were hydrolysed and proteins of lower molecular masses were formed as a function of time. With secalins, we see a protein band with a molecular weight around 50 kDa that disappeared as a function of time.

4 CONCLUSIONS

Under the experimental conditions (including urea treatment of secalins and boiling of gluten), both rye and wheat storage proteins were degraded by enzyme fractions obtained from ungerminated and germinated rye via ammonium sulphate precipitation and partial purification by column chromatography.

References

1. K. Brijs, W. Bleukx and J.A. Delcour, *J. Agric. Food Chem.*, 1999, **47**, 3572
2. P.R Shewry, S. Parmar and B.J. Miflin, *Cereal Chem.*, 1983, **60**, 1
3. W. Bleukx, S.P. Roels and J.A. Delcour, *J.Cereal Sci.*, 1997, **26**, 183

Acknowledgments

K. Brijs wishes to acknowledge the receipt of a scolarship from the Vlaams Instituut voor de Bevordering van het Wetenschappelijk-Technologisch Onderzoek in de Industrie (Brussels).

CHARACTERISATION AND PARTIAL PURIFICATION OF A GLUTEN HYDROLYZING PROTEINASE FROM BUG (*Eurygaster* spp.) DAMAGED WHEAT

D. Sivri and H. Köksel

Hacettepe University, Faculty of Engineering, Food Engineering Department 06532 Beytepe/Ankara TURKEY

1 INTRODUCTION

Proteolytic activity associated with bug damage in wheat causes degradation of gluten proteins during the mixing and fermentation stages of breadmaking, resulting in weak dough properties and unsatisfactory bread quality.[1-2] Some species of genera *Eurygaster* and *Aelia* are responsible for bug damage in Europe, North Africa and Middle East.[3] Similar damage has been associated with another insect, *Nysius huttoni*, in New Zealand.[4-5] The effect of bug protease on gluten proteins has been well documented. Electrophoretic studies showed that bug (*Eurygaster maura*) damage caused degradation of both gliadin and glutenin[6-7]. Cressey and McStay[8] reported that bug (*Nysius huttoni*) protease had a higher specificity for the high molecular weight (HMW) glutenin subunits than the other gluten proteins. In a recent study it has been quantitatively shown that after 30 min of incubation bug (*Eurygaster* spp.) damage causes substantial (>80%) decreases in the amount of 50% propan-l-ol insoluble glutenin, which is widely considered the most important protein fraction of wheat related to breadmaking quality.[9] There is detailed information on the characterisation and purification of *Nysius* protease.[10-11] However, we are not aware of any reports of similar studies on *Eurygaster* spp. or *Aelia* spp. proteases.

2 MATERIAL AND METHODS

2.1 Materials

A bug (*Eurygaster* spp.) damaged bread wheat cultivar (cv. Bezostaya) with strong gluten strength (HRW) was obtained from the Turkish Grain Board. Undamaged and damaged kernels were separated by hand-picking. The undamaged kernel were used as control (C). Bug damaged kernels had a puncture mark, a black spot surrounded by a pale patch. The undamaged and bug damaged wheat (BDW) samples were ground using a coffee grinder to obtain wholemeal. The wholemeal (400 mg) was mixed for 10 min with 2 ml of 0.05 M phosphate buffer (pH 7.0) at 20°C and centrifuged at 12,000 xg for 10 min. The supernatant was used as the enzyme solution (ES) for the determination of optimum temperature and pH of the enzyme activity.

Crude enzyme extract (CEE) was prepared as the starting material for purification. CEE was obtained from a wheat heavily (>50 %) damaged by *Eurygaster* spp. The bug damaged wheat was milled into wholemeal on a Falling Number Mill AB (Type 120). The wholemeal (300 g) was extracted twice with distilled water (1500 ml) by magnetic stirring for 48 h and 24 h at 4 °C and centrifuging (15,000 xg, 10 min). The supernatants were pooled and freeze dried. The resulting dry material was the CEE.

2.2 Methods

2.2.1 Protease activity assays. The 50% propan-l-ol insoluble glutenin[12] method which was originally developed for the determination of gluten quality and sodium dodecyl sulphate (SDS)-protein gel[13] method were used to measure bug protease activity with some modifications as described previously.[7-14]

2.2.2 Measurement of protein. Protein was measured at 280 nm.

2.2.3 Effect of temperature on bug protease activity. The optimum temperature of protease enzyme activity was determined by assaying the protease activities[14] at various temperatures (20, 25, 30, 35, 40, 45 and 50 °C).

2.2.4 Effect of pH on bug protease activity. The optimum pH of the bug protease activity was determined by assaying the protease activities[14] in 0.05 M phosphate-citrate (pH 3.0-7.0), Tris-HCl (pH 7.0-90) and glycine-NaOH (pH 9.0-11.0) buffers.

2.2.5 Effect of protease inhibitors on bug protease activity. To measure the effect of inhibitors on bug protease activity, two different concentrations (0.01M and 0.001M) of inhibitors (Pepstatin A, p-CMB, soybean trypsin inhibitor and EDTA) were added with CEE and incubated at 35 °C for 2 h. After incubation residual protease activity was measured.[7]

2.2.6 Ammonium sulphate fractionation. CEE was disolved in 0.05 M phosphate buffer (pH 7.5) and fractionated by ammonium sulphate precipitation (20-100% ammonium sulphate concentrations, at 10% intervals). The resulting precipitates were collected by centrifugation at 10,000 xg for 10 min and redissolved in distilled water. The solutions were dialyzed against the distilled water.

2.2.7 Ion-exchange chromotography. The dialysate (0.175g) from ammonium sulphate precipitation (60-80%) was applied to a column (2.5 x 80 cm) of Sephadex G-75 (Pharmacia) previously equilibrated with 0.05 M phosphate buffer (pH 7.5). The fractions showing protease activity were pooled. For molecular weight determination, tyrosine (181), ribonuclease A (13,700), chymotrypsinogen (25,000), ovalbumin (43,000) and albumin (67,000) were used as standarts.

2.2.8 QAE-Sephadex A-50 column chromotography. The bug protease obtained by Sephadex G-75 (Pharmacia) chromotography was placed on a QAE-Sephadex A-50 (Pharmacia) column (2.5 x 10 cm) previously equilibrated with 0.05 M Tris-HCl buffer (pH 9.0). The bug protease was eluted with a linear gradient of 0 to 0.5 M KCl in 0.05M Tris-HCl buffer.

2.2.9 Isoelectric point (pI) determination. Analytical isoelectric focusing (IEF) was performed at 10 W for 1 hr in 0.2 mm thick polyacrylamide (8% w/v) gels containing 6% (v/v) Pharmalyte 3-10 and 12.5 % (v/v) glycerol as described by Every.[10]

3 RESULTS AND DISCUSSION

3.1 Properties of Bug Protease

Bug protease was stable over a broad pH range (pH 3.0-11.0) with the maximum activity being observed at pH 8.5 (Figure 1). The optimum temperature of the protease activity was 35 °C and the residual activity at 50°C was 50% of the initial activity after 2 hr incubation at pH 7.0 (Figure 2). The bug protease gave a major peak on Sephadex G-75 with a molecular weight of 15,000. Analytical IEF of bug protease showed only one band with strong protease activity at pI 8.0. Protease activity was almost completely eliminated by inhibitors of sulphydryl proteases (p-CMB) and serine proteases (soybean trypsin inhibitor), but not by the inhibitors of metallo (EDTA) or acid (pepstatin A) proteases.

3.2. Purification

The results of the purification procedures are summarized in Table 1. The protease activity was observed mainly in the fraction precipitated by 60-80% saturation of ammonium sulphate and separated as a major peak by anion-exchange chromotography on QAE-Sephadex A-50.

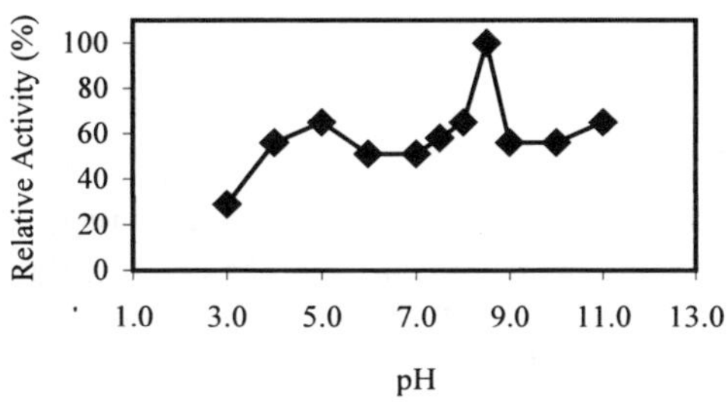

Figure 1: *Effects of pH on the bug protease activity*

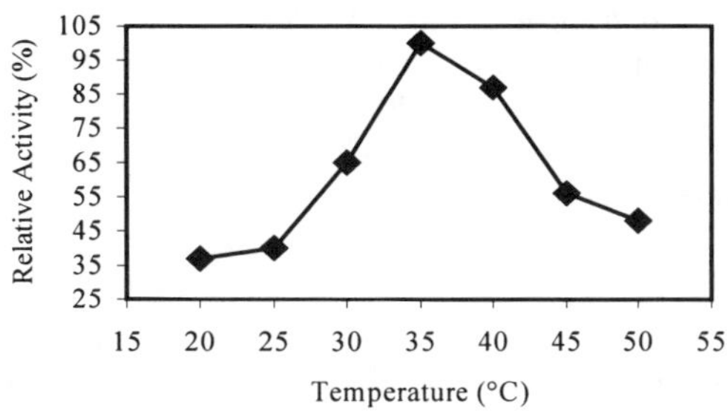

Figure 2: *Effects of temperature on the bug protease activity*

Table 1 *Purification of bug protease*

Purification Steps	Protease Activity (Ux10^3/ml)	Absorbance (280 nm)	Specific Activity	Purification Factor
Crude Enzyme Extract	5 440	6.25	886	1
Ammonium Sulphate Fractionation (60-80%)	8 360	2.28	3 667	4
Sephadex-75	18 000	0.29	62 069	70
QAE Sephadex A-50	9 710	0.03	323 667	365

4 CONCLUSIONS

In this study, the optimum pH, optimum temperature, pI and molecular weight of a bug protease (*Eurygaster* spp.) were determined on a crude enzyme extract and the protease was purified 365 fold by various purification techniques. The bug protease showed optimum activity at 35 °C and pH 8.5. Molecular weight and pI were estimated as 15,000 and 8.0 respectively. The protease activity was inhibited by both a SH-modifying reagent, p-CMB and soybean trypsin inhibitor, suggesting that the bug protease could be one of a subfamily of SH-containing serine proteases. The results indicate that the properties of *Eurygaster* spp. protease were similar to those of *Nysius huttoni* protease in various respects. However, further purification of the enzyme is required prior to further characterisation.

References

1. V. L. Kretovich, *Cereal Chem.*, 1944, **21**, 1.
2. N. P. Matsoukas and W. R. Morrison, *J. Sci. Food Agric.*, 1990, **53**, 363.
3. F. Paulian and C. Popov, in *Wheat*, ed. Häfliger, Ciba-Geigy, Basel, 1980, p. 69.
4. P. J. Cressey, J. A.Farrell and M.W. Stufkens, *N. Z. J. Agric. Res.*, 1987, **30**, 209.
5. D. Every, *J. Cereal Sci.*, 1992, **16**, 183.
6. D. Sivri and H. Köksel, in *Proceedings of the Sixth International Gluten Workshop*, ed. C. W. Wrigley, Royal Aust. Chem. Inst., Melbourne, 1996, p. 261.
7. D. Sivri, H. Köksel, and W. Bushuk, *N. Z. J. Crop. Hort. Sci.*, 1998, **26**, 117.
8. P. J. Cressey, and C.L. McStay, J. Sci. *Food Agric.*, 1987, **38**, 357.
9. D. Sivri, H. Sapirstein, H. Köksel, H. and W. Bushuk, *Cereal Chem.*, 1999, **76**, 816.
10. D. Every, *J. Cereal Sci.*, 1993, **18**, 239.
11. P. J. Cressey, *J. Sci. Food Agric.*, 1987, **41**, 159.
12. B. X. Fu and H. D. Sapirstein, *Cereal Chem.*, 1996, **73**, 143.
13. D. Every, *Anal. Biochem.*, 1991, **197**, 208.
14. D. Sivri and H. Köksel, in *Proceedings of Euro Food Chem X*, FECS-Event No:234, Budapest, 1999, Volume 3, p.740.

Acknowledgements

We thank Dr. W. Bushuk and Dr. H. Sapirstein for providing their laboratory facilities for gel filtration chromatography at the University of Manitoba, Department of Food Science, Winnipeg, Canada. This work was supported by the Turkish Scientific and Technical Research Council of Turkey (TUBITAK) under project number TOGTAG-1609.

EFFECTS OF TRANSGLUTAMINASE ENZYME ON GLUTEN PROTEINS FROM SOUND AND BUG- (*EURYGASTER SPP.)* DAMAGED WHEAT SAMPLES

H. Köksel[1], D. Sivri[1], P.K.W. Ng[2], and J.F. Steffe[2]

1. Dept. of Food Engineering, Hacettepe University, Ankara, Turkey. 2. Dept. of Food Science & Human Nutrition, Michigan State University, Michigan, USA

1 INTRODUCTION

Transglutaminase (TG) enzyme may catalyze conversion of soluble proteins to insoluble high molecular weight protein polymers through formation of non-disulphide covalent cross-links. The enzyme catalyzes acyl-transfer reactions between peptide-bound glutaminyl residues and primary amines. When ε-amino groups of protein-bound lysyl residues act as acyl acceptors, intra- or intermolecular ε-(g-glutamyl) lysyl isopeptide crosslinks are formed[1]. Berghofer *et al*[2] reported improving effects of a TG enzyme on bread properties. Gerrard *et al*[3] showed that TG had beneficial effects during breadmaking that are comparable to traditional oxidizing improvers. Preharvest bug damage to wheat caused by *Eurygaster spp.*, *Aelia spp.* and *Nysius huttoni* occurs in many countries[4, 5]. The infested grain contains a protease that breaks down gluten structure[6-9] and results in a sticky dough and poor bread quality[10-12]. This study investigated the possibility of using TG to repair the structure of gluten proteins hydrolyzed by wheat bug proteases. Effects of TG enzyme on fundamental rheological and electrophoretic properties of sound and bug-damaged wheat flours (BDF) proteins were examined.

2 MATERIALS AND METHODS

2.1 Materials

Straight-grade flours were milled from two sound wheat cultivars (Augusta, weak, and Sharpshooter, strong physical dough properties) and a Suni bug-damaged wheat cultivar (cv. Gün-91). Samples of both sound flours were treated with a bacterial TG (Ajinomoto, Teaneck, NJ) at 1.5% (w/w).

2.2 Methods

Dynamic rheological tests were performed on a Haake RS 100 rheometer (Paramus, NJ). Dough was placed between the plates, rested for 3 min and tested at strain amplitudes up to 0.2% through a frequency sweep from 0.1 to 25.1 Hz at a constant temperature of 30°C. All tests were run in the linear range of viscoelastic behavior using standard methods[13]. Doughs were tested immediately after mixing and after resting 30 and 60 min at 30°C. Overall dough properties were evaluated by comparing plots of complex modulus (G*) as a function of frequency. Bug crude enzyme (BCE) extract was prepared from BDF. It was blended with Augusta and Sharpshooter flours suspended in distilled water and incubated for 30, 60 and 120 min at 35°C. Effects of TG on electrophoretic properties of gluten proteins were determined by SDS-PAGE[13].

3 RESULTS AND DISCUSSION

3.1 Effects of TG on rheological properties

The G* values of Augusta and Sharpshooter control doughs decreased after 60 min of resting. The G* values of TG-treated Augusta and Sharpshooter doughs were comparable to those of respective control doughs at 0 min of incubation (Fig. 1; results were similar for both cultivars and only one set will be presented), however, increased significantly after 60 min of incubation. An increase in the average molecular weights of gluten proteins due to TG activity would be expected to cause an increase in the complex modulus. A recent study on TG-treated gluten confirmed the results of the present study[15]. The BDF was blended with Augusta and Sharpshooter flours at the 10% level to observe effects of bug protease on rheological properties. The G* values of BDF-blended doughs decreased significantly after 30 min of incubation (Fig. 2). Dough samples were extremely soft and sticky and impossible to handle for testing purposes after 60 min of incubation due to the proteolytic activity. Similar results have been reported in rheological studies by adding of reducing agents to dough[16]. However, G* values of TG-treated BDF-blended dough samples did not decrease, but increased significantly after 30 and 60 min of incubation (Fig. 3).

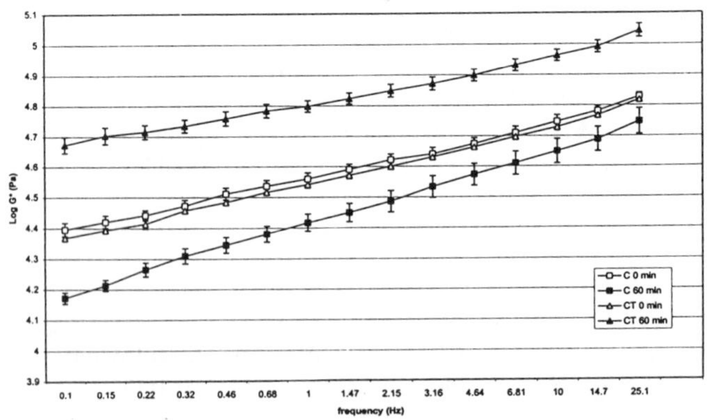

Fig. 1 Variations of G* with frequency for Augusta control and TG-treated doughs. C = Control; CT = Control-TG.

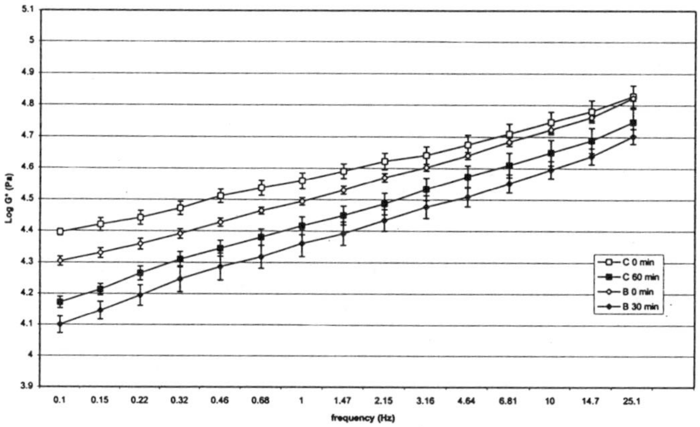

Fig. 2 Variations of G* with frequency for Augusta control and blended (see text for details) doughs. C = Control; B = Blended.

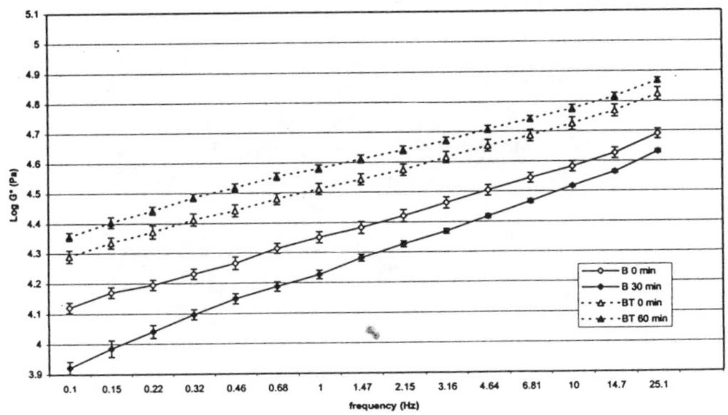

Fig. 3 Variations of G* with frequency for blended Sharpshooter with or without TG-treated doughs. B = Blended; BT = Blended-TG.

3.2 Effects of transglutaminase on electrophoretic properties

The SDS-PAGE patterns of samples treated with BCE changed as expected during incubation (Fig. 4), with the relative band intensities decreasing especially in the HMW region at all incubation periods (lanes 6-8). Similar effects of BCE on electrophoretic patterns of gluten were previously reported[7,8]. In TG-treated sound flours, relative band intensities decreased in both HMW and LMW regions with increasing incubation, with decreases more obvious in the HMW region (lanes 4 and 5). Furthermore, new protein aggregates appeared at the origins of stacking and separating gels for TG-treated samples, indicating that HMW protein polymers formed with TG treatment. Heavy streaking at the upper region of the separating gel increased with increasing incubation period. The electrophoresis sample buffer contained 7% 2-mercaptoethanol, therefore the protein polymers appearing in the gels are not disulphide cross-linked. Similar results were observed for BCE-TG-treated samples (lanes 10 and 11); the protein aggregates at the origin of the stacking gel and streaking at the upper region of the separating gel were still visible after the longest incubation period. Treatment of hydrolyzed gluten proteins with TG produced a heterogeneous population of cross-linked polymers, some of which could not enter the stacking and separating gels.

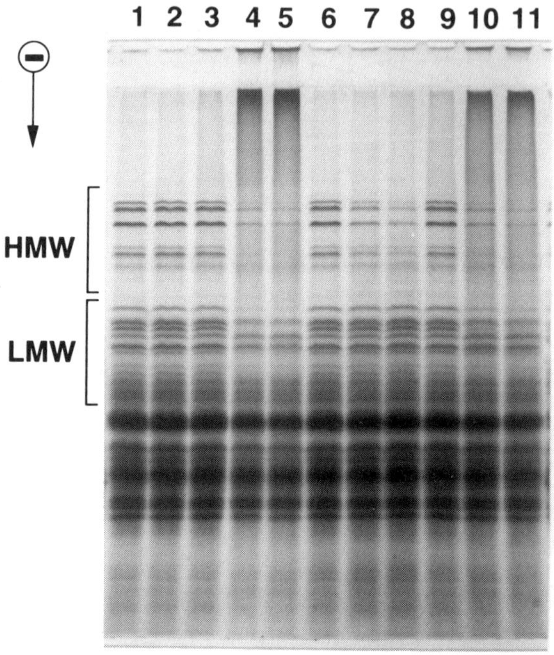

Figure 4 *Effects of TG and BCE on SDS-PAGE patterns of Sharpshooter flour proteins at various incubation periods (min). Lanes 1-11: 0, 120, TG-0, TG-60, TG-120, BCE-0, BCE-60, BCE-120, TG-BCE-0, TG-BCE-60, TG-BCE-120; HMW = high molecular weight, LMW = low MW.*

4 SUMMARY

Both rheological measurements and electrophoresis results clearly indicated that TG enzyme has substantial repairing effect on the dough hydrolyzed by wheat bug proteases.

References

1. J. E. Folk and J. S. Finlayson, in Advances in Protein Chemistry, ed. C. B. Anfinsen, J. T. Edsall and F. M. Richards, Academic Press Inc., New York, 1977, Vol. 31, p.133.
2. E. Berghofer, H. Bogner, and R. Schönlechner, in XVII. ICC Conference Abstract Book, June 6-9, 1999, Valencia, Spain, 1999, p.156.
3. J. A. Gerrard, S. E. Fayle, A. J. Wilson, M. P. Newberry, M. Ross and S. Kavale, *J. Food Sci.*, 1998, 63, 472.

4. F. Paulian. and C. Popov, in Wheat, ed. Häfliger, Ciba-Geigy, Basel, 1980, 69.
5. P. J. Cressey, J. A. K. Farrell and M. W. Stufkens. *N. Z. J. Agric. Res.*, 1987, 30, 209.
6. P. J. Cressey and C. L: McStay, *J. Sci. Food Agric.*, 1987, 38, 357.
7. D. Sivri, H. Köksel, in "Gluten 1996" ed. C. W. Wrigley, 1996, Sydney, Australia, p. 545.
8. D. Sivri, H. Köksel, and W. Bushuk, *N. Z. J. of Crop and Horticultural Sci.*, 1998, 26,117.
9. D. Sivri, H. D. Sapirstein, H. Köksel, and W. Bushuk, *Cereal Chem.*, 1999, 76, 816.
10. V. L. Kretovich, *Cereal Chem.*, 1944, 21,1.
11. N. P. Matsoukas. and W. R. Morrison, *J. Sci. and Food Agric.*,1990, 53, 363.
12. E. Karababa and A. N. Ozan, *J. Sci. Food Agric.*, 1998, 77, 399.
13. J.F. Steffe 1996. "Rheological Methods in Food Process Engineering". Second edition, Freeman Press, East Lansing, MI, USA.
14. P. K. W. Ng, and W. Bushuk, *Cereal Chem.*, 1987, 64, 324-327.
15. C. Larre, S. D-Papini, Y. Popineau, G. Deshasey, C. Desserme and J. Lefebvre, *Cereal Chem.*, 2000, 77, 32.
16. S. Berland and B. Launay, *Cereal Chem.*, 1995, 72: 48.

EXTRACELLULAR FUNGAL PROTEINASES TARGET SPECIFIC CEREAL PROTEINS

M-P. Duviau, K. Kobrehel

INRA, Unité de Biochimie et Biologie Moléculaire des Céréales, 2, Place Viala, 34060 Montpellier Cedex 02, France. Tel : (33) 4 99 61 23 88 Fax : (33) 4 99 61 23 48
e-mail : kobrehel@ensam.inra.fr

1 INTRODUCTION

Extracellular fungal proteinases (EFPs) were shown to be required for fungal attachment to host cells and subsequent invasion or infection of host cells by fungi. Many EFPs were characterized as "aspartic" proteinases. Some of these enzymes are involved in the initiation of "pourriture noble" (noble rot), while others are responsible of the common "pourriture grise" (wet rot).

In this study, the proteolytic action of a set of extracellular "aspartic" proteinases on the proteins of different cereals, principally on wheat proteins, was investigated. Prior to their use, the EFPs from the different fungi were purified.

2 MATERIALS AND METHODS

2.1 Materials

2.1.1. Fungal aspartic proteinases. Five extracellular fungal proteinases, isolated from *Aspergillus saitoi, Aspergillus sojae*, *Aspergillus oryzae*, *Rhizopus* and *Mucor*, respectively, were assayed against different cereal proteins.

2.1.2. Cereal samples. Two wheat samples, a bread wheat (*Triticum aestivum*) cv. Thésée and a durum wheat (*Triticum durum*) cv. Néodur and among the other cereals triticale, barley, sorghum, rice and corn were studied. Flours or semolina were obtained using a laboratory pilot mill.

2.2 Methods

2.2.1. Extraction of proteins. Total cereal proteins were extracted with dilute acetic acid. The specific cereal protein fractions, albumins, globulins, gliadins or prolamins (hordeins, kafirins and zeins) and glutenins or glutelins, were obtained by using a sequential extraction procedure based upon the Osborne method.

2.2.2. Enzyme assays. In most of the cases, enzyme assays were carried out in 0.01M sodium acetate-acetic acid buffer at pH 4.7, temperature : 37°C, incubation time : 30 min.

Aliquots were applied to the gel and SDS-PAGE analyses were performed at pH 8.5 by using the method of Laemmli. The proteolytic activity of the enzymes was detected on the Coomassie blue stained polyacrylamide gels by comparing the electrophoretic patterns of the assayed samples to the controls. The quantification of the proteolytic activity of the enzymes was obtained by densitometric scanning of the gels.

3 RESULTS AND DISCUSSION

3.1 Effects of main parameters

The effects of the main parameters on the action of EFPs was determined. The effects of enzyme/substrate ratio, incubation time, pH and temperature were investigated. Results obtained by the hydrolysis of gliadins by the EFP of *Asperpillus saitoi* are illustrated in Figures 1, 2, 3 and 4.

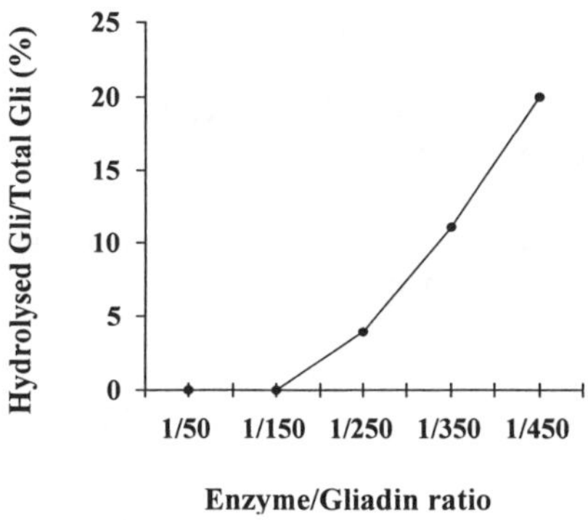

Figure 1 : *Influence of Enzyme/Gliadin ratio. Experimental conditions : incubation time 30 min, temperature 37°C, pH 4.*

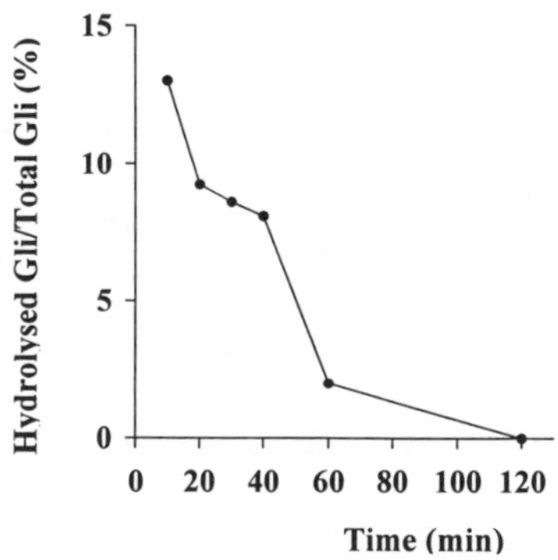

Figure 2 : *Influence of Incubation Time. Experimental conditions : enzyme/gliadin ratio 1/250, temperature 37°C, pH 4.*

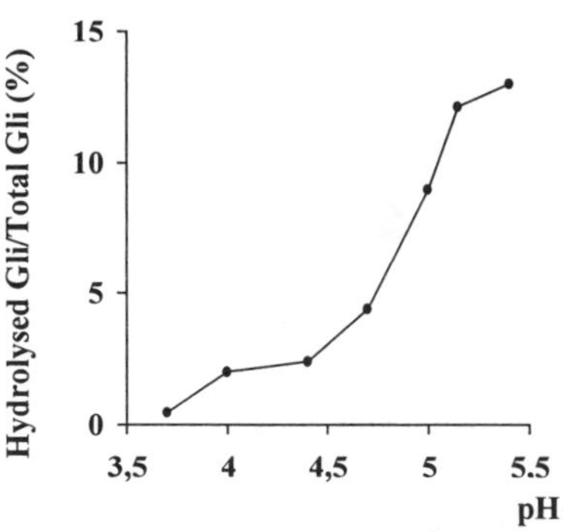

Figure 3 : *Influence of pH.Experimental conditions : enzyme/gliadin ratio 1/250, incubation time 30 min, temperature 37°C.*

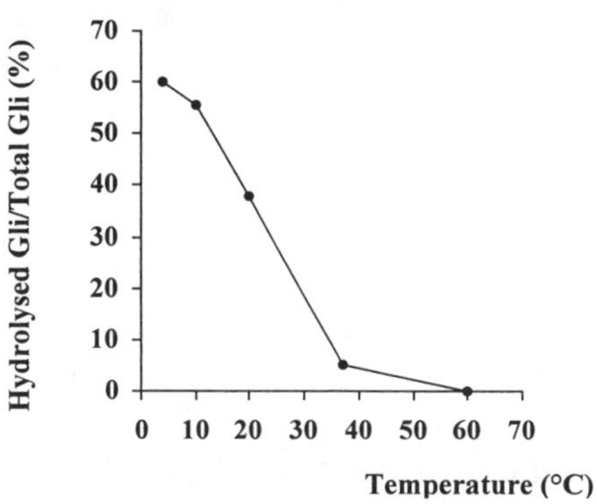

Figure 4 : *Influence of Temperature. Experimental conditions : enzyme/gliadin ratio 1/250, incubation time 30 min, pH 4.*

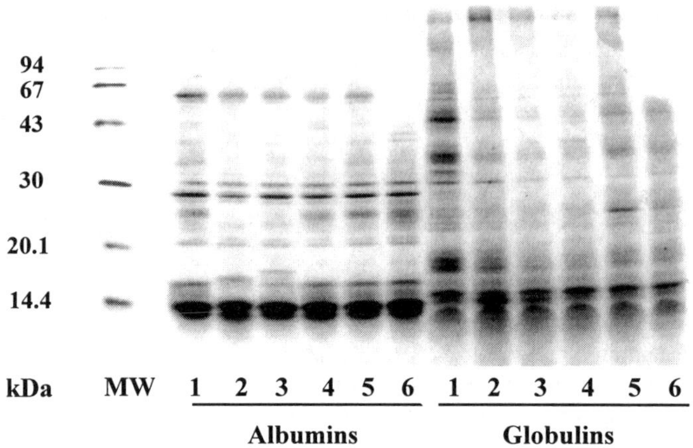

Figure 5 *SDS-PAGE analysis. Effects of EFPs on wheat albumins and globulins. 1. Control (proteins); 2-6. Proteins plus proteinases obtained from* Aspergillus saitoi *(2),* Rhizopus *(3),* Mucor *(4),* Aspergillus sojae *(5),* Aspergillus oryzae *(6).*

Most of the wheat albumin fractions, as detected on SDS-PAGE, resisted the proteolytic activity of all the five EFPs tested (Figure 5). However, differences were found between enzymes regarding their activity and specificity.

Globulins, the other group of metabolic wheat proteins, showed much higher sensitivity to all the EFPs assayed (Figure 5). Differences were found between the specificities of the enzymes, but to a lesser extent than for albumins.

3.2 Wheat storage proteins

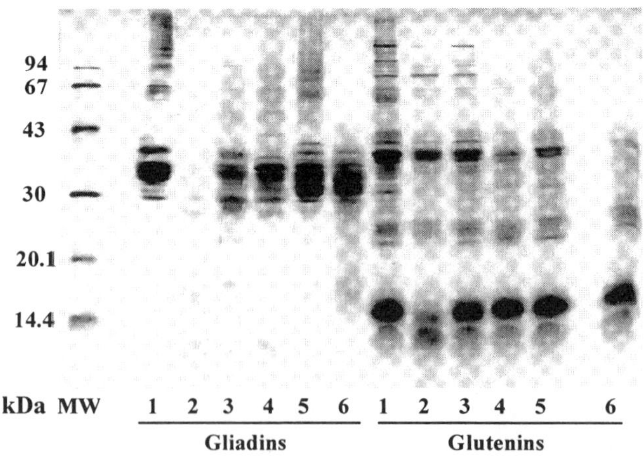

Figure 6 *SDS-PAGE analysis. Effects of EFPs on wheat gliadins and glutenins. 1. Control (proteins); 2-6. Proteins plus proteinases obtained from* Aspergillus saitoi *(2),* Rhizopus *(3),* Mucor *(4),* Aspergillus sojae *(5),* Aspergillus oryzae *(6).*

In general, gliadins were very sensitive to all of the five EFPs assayed (Figure 6). However, great differences were found between the activity and specificity of the enzymes. Each proteinase reacted with specificity, although, some similarities were observed between the electrophoretic patterns of the polypeptides obtained. The proteinase of *Aspergillus saitoi* was the most effective, as shown by electrophoresis, as the typical gliadin bands almost completely disappeared after proteolysis. As in the case of albumins and globulins, the proteinase of *Aspergillus oryzae* targeted more specifically the proteins that had higher apparent M_r than about 50,000.

Compared to gliadins, all the proteinases assayed were less efficient in digesting glutenins, with the exception of one (proteinase of *Aspergillus oryzae*), which, in the case of the other protein fractions, preferentially targeted higher M_r proteins (Figure 6). The specificity of this proteinase towards high M_r protein fractions, including both metabolic and storage proteins, suggest considerable functional differences between this proteinase and the other proteinases investigated.

3.3 Effects on different cereal proteins

The effects of EFPs on the proteins of other cereals (barley, triticale, corn, sorghum and rice) were also investigated.

As for wheat, the activity towards proteins of the albumin and globulin fractions of the different cereals was specific, however, none of the fungal proteinases studied showed great effects on these proteins.

Partially purified prolamins of triticale and hordeins were digested by all five proteinases studied. The proteinase of *Aspergillus saitoi* was the most efficient against zeins. The storage proteins of sorghum, both kafirin and glutelin, were very resistant, conversely, rice storage proteins were efficiently digested by all the proteinases assayed.

Prolamins, were, in general, very sensitive to proteolysis when they were partially purified. In contrast, they were practically undigested by the proteinases when total protein extracts were used for the assays, suggesting the presence of natural inhibitors in the total protein extracts.

4 CONCLUSIONS

The EFPs assayed targeted specific proteins, mostly gliadin and glutenin fractions. Depending on the sources, the specificities of the enzymes were different.

The hydrolysis of specific gliadin and glutenin fractions by the different EFPs studied may be of particular practical interest. In fact, the results suggest the possibility of obtaining modified gluten preparations for specific uses.

The metabolic protein fractions of most of the cereals appear to contain natural inhibitors towards the EFPs.

Note : This work was initiated in Berkeley, the purified enzymes used for this study were obtained in Berkeley (T. J. Leighton, Department of Biochemistry and B. B. Buchanan, Department of Plant Science and Microbiology, University of California, Berkeley, U.S.A.).

STUDY OF THE TEMPERATURE TREATMENT AND LYSOZYME ADDITION ON FORMATION OF WHEAT GLUTEN NETWORK : INFLUENCE ON MECHANICAL PROPERTIES AND PROTEIN SOLUBILITY

B. Cuq[1], A. Redl[1], and V. Lullien-Pellerin[2]

1. UFR "Technologie des Céréales et des Agro-polymères", 2. Unité "Biochimie et Biologie Moléculaire des Céréales" ENSA – INRA Montpellier, 2 place Viala, 34060, Montpellier, Cedex 1, France.

1 INTRODUCTION

The manufacture of many food products based on wheat (breads, biscuits, snack bars, etc.) requires the preparation of a dough which depends partly on the formation of a gluten network. Knowledge of the rheological properties of the gluten network and of its evolution during processes is generally considered as a critical key for quality of the wheat products. Modifications of the viscoelastic properties of "wheat dough" during thermal treatments depend mainly on the physico-chemical characteristics of the wheat gluten and in particular on its capacity to establish intra- and inter-molecular interactions. Thermal treatments (between 100 and 150°C) of wheat flours are favourable for the formation of protein interactions stabilised by intermolecular disulphide bonds[1]. Indeed, formation of disulphide bonds and SH/SS interchanges during dough development are supposed to play a central role in the quality of wheat products[2]. It has been hypothesised that some wheat low molecular weight cysteine-rich proteins that are reduced by a thioredoxin system could be involved in the gluten network as reticulating agents[3].

Recently, the formation and characterisation of wheat gluten networks were studied for edible or biodegradable material applications[4]. In the present work, we used a "thermoplastic" process that combines simultaneously the formation of the protein network under low plasticizer content conditions and thermal treatment in order to investigate the effects of temperature increase and the addition of a low molecular weight cysteine-rich protein.

2 MATERIALS AND METHODS

Formation of thin gluten networks were achieved according to a previously described protocol[5]. Briefly, 1g wheat gluten (with eventually 45 mg lyophilised lysozyme added) was mixed with 0.4g glycerol in a mortar with a pestle. The homogeneous blend was pressed at 20 MPa for 10 min at a defined temperature (80-135°C). Mechanical properties of the films were according to ISO 5A 527-2 procedure at 20°C and 60-65% relative humidity. Buffers used for the protein solubility were 50 mM sodium phosphate (pH 7.0) with or without 2% SDS or 2% SDS + 10 mM 2-mercaptoethanol (2-ME).

3 RESULTS AND DISCUSSION

3.1 Temperature effects on the mechanical properties of the wheat gluten network

The effects of thermal treatments on the mechanical properties of wheat gluten network were investigated in a previous study[5]. The main data from the experimental stress-elongation curves as a function of the process temperature are reported in Table 1. Increasing the processing temperature from 80 to 135°C induces an increase in the mechanical resistance of the gluten network (tensile strength σ from 0.26 to 2.04 MPa) and a decrease in the elongation ratio (λ from 5.68 to 3.36). There is also a large increase in the Young's modulus (E from 0.09 to 6.39 MPa).

Table 1: *Main data from the stress-elongation ratio curves as a function of the processing temperature. σ_b is stress at break, λ_b is elongation ratio at break, E is Young's modulus, M_C is the molecular size between crosslinks as derived from the young's modulus according to Eq (1&2), C1 and C2 are the Mooney Rivlin constants of Eq (3). Numbers in brackets are the standard deviations of five replicates.*

T (°C)	σ_b (MPa)	λ_b (L/L0)	E (MPa)	M_c (g/mol)	C1 (kPa)	C2 (kPa)
80	0.26 (0.09)	5.68 (0.89)	0.09 (0.03)	31000	41	-15
95	0.42 (0.09)	5.51 (0.33)	0.17 (0.02)	17000	52	-6
110	1.00 (0.13)	5.05 (0.39)	1.01 (0.15)	3000	84	111
120	1.45 (0.25)	4.15 (0.31)	2.94 (0.56)	1000	158	104
135	2.04 (0.45)	3.36 (0.21)	6.39 (1.23)	500	259	211

Variation in the mechanical properties of the wheat-gluten network as a function of the temperature was shown to follow a sigmoidal shape[5] that could be fitted with the model described by Peleg[6] and used to determine a characteristic inflexion point at 116°C. Thus, increasing the processing temperature from 95°C to 125°C involves a significant increase of the film cohesion that could be explained by a cross-linking effect of the thermal treatment. The cross-linking density of the network or concurrently the molecular size between crosslinks can be estimated using the theory of rubberlike elasticity, assuming that:

$$E = 3G \quad (1) \qquad G = \frac{\rho RT}{M_c} \quad (2)$$

where ρ is the density of the material ($\rho = 1.2\ 10^6$ g/m^3), M_c is the molecular weight of network strands between cross-links (g/mol), R is the universal gas constant (R = 8.314 J/mol K) and T is the temperature (K). The molecular weight of network strands obtained with Eq 1 ranges from M_c = 31 kg/mol to M_c = 0.5 kg/mol (Table 1). These values compare well with those obtained from synthetic rubbers such as slightly crosslinked butyl rubber and polybutadiene, (M_c = 8.5 kg/mol and 2.5 kg/mol respectively[7]) but are lower than those observed in a previous work[8] for extruded gluten/glycerol materials (M_c = 40-150 kg/mol). In order to describe the behaviour of cross-linked rubbers in unidirectional extension an empirical formulation, known as the Mooney Rivlin equation[9], is commonly used:

$$\sigma = 2\left(\lambda - \frac{1}{\lambda^2}\right)\left(C1 + \frac{C2}{\lambda}\right) \qquad (3)$$

with σ stress, λ elongation ratio ($\lambda = L/L_0$) and C1, C2 characteristic constants. This model fits our experimental data well (Figure 1), the corresponding model constants are given in Table 1. Of special interest is the coefficient C1, which increases from 41 kPa to 259 kPa. Although a molecular interpretation of the coefficients C1 and C2 is not possible (due to the empirical origin of the model) the constant C1 was related to the degree of vulcanisation for a series of vulcanised rubber compounds[9].

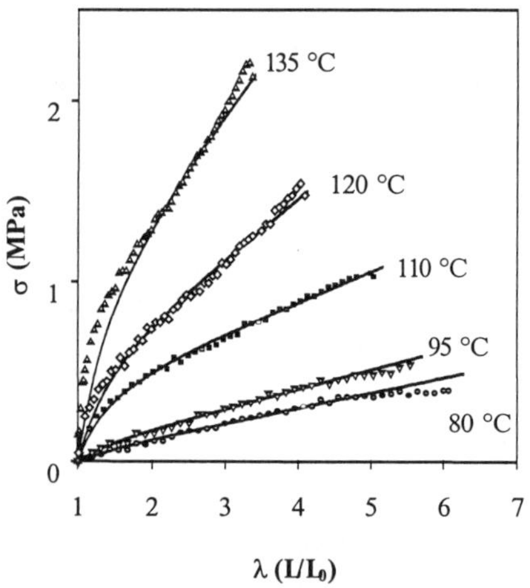

Figure 1. Stress versus elongation ratio of thermomolded wheat gluten films. Numbers indicate the processing temperature. Dots are experimental data, lines represent the Mooney Rivlin Model with constants as specified in Table 1.

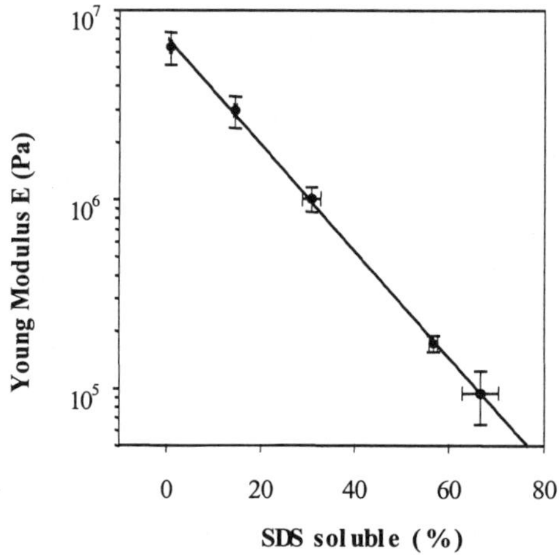

Figure 2. Young's modulus versus extractability in 2% SDS buffer. Error bars represent the standard deviation of five replicates.
Regression line is $y = 7.1\ 10^6 \exp(-0.065\ x)$, r^2=0.99.

3.2 Temperature effects on the protein solubility of the wheat gluten network

The nature of the interactions in the gluten-based network were studied through the protein solubility in different solvents. The solubility properties of wheat gluten proteins as a function of thermal treatment of films are presented in Table 2.

Table 2: *Solubility in different solvents of the proteins from the gluten network with or without lysozyme added. Numbers in brackets are standard deviations.*

T		GLUTEN		GLUTEN +6% LYSOZYME
(°C)	Buffer	SDS	SDS + 2-ME	SDS
80	5.1 (0.1)	68.2 (1.1)	64.1 (3.4)	72.4 (1.4)
110	3.0 (0.2)	28.0 (1.4)	62.7 (0.9)	30.2 (3.0)
135	2.0 (0.1)	0 (1.2)	59.2 (4.6)	2.0 (1.1)

At 80°C, only 5 % of proteins are solubilized in buffer whereas nearly 70% are solubilized in 2% SDS. The solubility in 2% SDS and 2-ME is not very different from that observed in only SDS. It seems to indicate strong participation of hydrophobic interactions in the stabilisation of the gluten network made at 80°C. However, about 35% of proteins remain insoluble in 2% SDS + 2-ME therefore a number of the interactions remain inaccessible to the disruptive agents during the solubilisation. Increasing the temperature from 80 to 135°C induces a very large reduction (to 0%) in the SDS solubility of the proteins that could be recovered if 2-ME is added to the solubilisation buffer. This indicates that disulphide bonds are mainly involved in the temperature induced cross-linking. The temperature effect on the protein solubility in SDS could be modelled with the equation of Peleg[6] and shows an inflection temperature of 108°C that is not different from the temperature found in the modelling of the mechanical properties[5]. Furthermore, the protein solubility in SDS is strongly correlated with the Young's modulus and concurrently with the corresponding molecular size between entanglements (Figure 2).

3.3 Lysozyme addition effect on the gluten network

Because the temperature-induced network of gluten films seems to result mainly from disulphide crosslinks, the role of a low molecular cysteine-rich protein in the potential cross-linking of the gluten network was investigated. In theory, a cysteine-rich protein should enhance the reaction kinetics of disulphide bonds and therefore lower the inflection point of the SDS solubility versus temperature curve. We used a commercial protein, lysosyme, that shows the same characteristics as known wheat proteins of this type, i.e. a molecular weight between 5 and 15 kDa and a content of between 8 and 14 cysteines. However, addition of up to 6% lysozyme relative to the gluten protein content did not show any significant effect on mechanical properties or protein solubility in our assay. As shown in Table 2, the protein solubility in SDS is not affected by the lysozyme addition. We suggest that lysozyme is passively trapped in the disulphide gluten network.

4 CONCLUSIONS

Thermoplastic processed wheat gluten films, plasticized with glycerol, behaved rubberlike, their tensile properties could be modelled very well with the Mooney Rivlin theory. Increasing processing temperature above a critical level (in our conditions between 108-110 °C) induced cross-linking reactions that were reflected by an increase in the elastic modulus and a decrease in solubility in 2% SDS buffer; the solubility in 2%

SDS + 2-ME remaining constant. A very close relationship was observed between the elastic modulus and the protein solubility in 2% SDS. Therefore, we conclude that the temperature-induced crosslinks are disulphide bonds. However, the addition of up to 16 moles of cysteine from lysosyme per 100 moles of cysteine in gluten did not significantly change the gluten properties.

References

1. L.P. Hansen, P. H. Johnston and R.E. Ferrel, *Cereal Chem.*, 1975, **52**, 459.
2. P. Shewry and A.S. Tatham, *J. Cereal Sci.,.* 1997, **25**, 207.
3. P. Joudrier, V. Lullien-Pellerin, R. Alary, J. Grosset, A. Guirao and M-F. Gautier, *DNA Seq.,* 1995, **5**, 153.
4. B. Cuq, N. Gontard, J.L. Cuq and S. Guilbert, *Nahrung/Food*, 1998, **42,** 260.
5. B. Cuq, F. Boutrot, A. Redl and V. Lullien-Pellerin, *J. Agric Food Chem.*, 2000, accepted.
6. M. Peleg, *Biotechnol Prog.*, 1984, **10**, 652.
7. J.D. Ferry 'Viscoelastic Properties of Polymers', John Wiley & Sons, New York, 1980.
8. A. Redl, M.H. Morel, Bonicel J., B. Vergnes and S. Guilbert. *Cereal Chem.,* 1999, **76**, 361.
9. L.R.G. Treolar, *'The Mechanics of Rubber Elasticity'*, Clarenson Press, Oxford, 1975.

Quality Testing, Non-Food Uses

A RAPID SPECTROPHOTOMETRIC METHOD FOR MEASURING INSOLUBLE GLUTENIN CONTENT OF FLOUR AND SEMOLINA FOR WHEAT QUALITY SCREENING

H.D. Sapirstein and W.J. Johnson.

Department of Food Science, University of Manitoba, Winnipeg, MB, Canada R3T 2N2.

1 INTRODUCTION

The quality of glutenin for breadmaking is mainly a function of its molecular size distribution which varies depending on the composition of constituent subunits and their different capacities to cross-link via disulphide bonds. While determining the molecular size distribution of glutenin is a non-trivial problem, the amount of "functional glutenin" of high molecular weight (HMW) in a given wheat or flour sample can be measured more or less routinely in at least three ways: 1) from the amount of SDS-protein gel after high-speed centrifugation of flour dispersions in SDS solutions, 2) by size exclusion chromatography of sonicated SDS-insoluble protein residue, and 3) from the amount of insoluble protein remaining in the residue after extraction with a number of suitable solvents. Of these three approaches, the latter is the least complex and its effectiveness to discriminate wheats of varying breadmaking quality is well supported in the literature[1-5].

At the last Gluten Workshop we reported[4] preliminary results of a rapid small-scale spectrophotometric procedure to measure the amount of insoluble or HMW glutenin in a flour sample by peptide bond absorption (214 nm) of a fraction of 50% propan-l-ol insoluble (50PI) protein. Very high correlations were obtained between this measure of HMW glutenin and dough mixing properties for the samples evaluated in the initial study. Quantification of 50PI protein by another rapid method[5] can also be performed by combustion nitrogen analysis (CNA). A key difference between the CNA method (and previous insoluble protein measurement methods) and the spectrophotometric procedure is that, in the latter, total 50PI protein is not determined. Rather, only 50PI protein that is extractable by a propan-l-ol solution containing the reductant, dithiothreitol (DTT) is measured. As was previously reported[6], the residue after propan-l-ol/DTT extraction contains a considerable amount of (non-glutenin) protein which is not related to breadmaking quality.

The spectrophotometric procedure has been simplified and refined to increase its effectiveness for wheat quality screening. To this end, sample size, DTT concentration, extraction temperature and time have been optimized, the requirement of a buffered reductant has been eliminated, and protein calibration has been simplified. The procedure was validated using several sets of wheats, and was used to evaluate genotype and environment effects on the ability of the procedure to screen for breadmaking quality of common (bread) wheats, and gluten strength of durum wheats.

2 MATERIALS AND METHODS

2.1 Wheat Samples and Technological Quality Assessment.

Three Canadian wheat cultivars (Glenlea, Katepwa and Harus) of diverse quality were used for method optimization. Glenlea and Katepwa are hard red spring bread wheats with extra strong and moderately strong dough mixing properties, respectively. Harus is a weak soft white winter wheat. The quality characteristics of these samples have been described[6] . These wheats and 11 other Canadian cultivars[4] were used to evaluate the accuracy of the spectrophotometric procedure relative to Kjeldahl protein (N x 5.7) determinations.

Method validation was performed using different sets of wheat varying more or less widely in protein quality and technological performance; typical results are shown in this paper based on a sample set of 88 Canadian registered cultivars and advanced wheat breeder lines evaluated in the 1997 Prairie Registration Recommending Committee for Grain (PRRCG) “C Test Coop”. The PRRCG is responsible for the testing and evaluation of grain crop candidate cultivars for registration in western Canada. The PRRCG wheats were a very diverse set of genotypes spanning seven commercial classes including durum wheat (18 genotypes). Each wheat tested was a composite of many (6-11) locations.

Dough mixing characteristics of the common wheat flours were evaluated using a 2 g computerized Mixograph at constant water absorption (60%). Alveograph properties of durum semolinas were determined by the ICC standard method 121[7] .

2.2 Extraction of Monomeric Protein and Soluble LMW Glutenin

Flour or semolina (100 mg) is extracted twice with 1 mL 50% (v/v) propan-l-ol (solution 'A', Certified grade) for 30 min at room temperature (23°C) in a 1.5 mL microcentrifuge tube with intermittent vortexing (every 10 min for 5 sec). After the first extraction, the mixture is centrifuged for 3 min at 2,200 g in a table top centrifuge (Biofuge A, Heraeus-Christ). The supernatant can be discarded, or used to similarly quantify 50% propan-l-ol soluble protein (Sapirstein and Lukie ref these proceedings). Liquid remaining in the centrifuge tube was carefully removed with a Pasteur pipette so as not to disturb the pellet which was resuspended in 1 mL of solution 'A'. A micro-spatula was used to facilitate disruption of the starchy pellet which is quite dense and hard. After the second extraction, the mixture was centrifuged for 3 min at 15,000 g. The supernatant is decanted and any liquid remaining in centrifuge tube is removed with a Pasteur pipette.

2.3 Extraction of Insoluble Glutenin

The 50PI residue, free of monomeric protein and propan-l-ol soluble glutenin is reduced with 1 mL of solution 'A' containing 0.1% (w/v) DTT for 30 min at 55°C in a heating block. Samples are vortexed at 2 min, and 14 min intervals thereafter. Vortexing after samples have been heated for 2 min facilitates complete suspension of the 50PI residue. Subsequently, the mixture is centrifuged for 3 min at 15,000 g. The microcentrifuge tube is inverted once to obtain a homogeneous supernatant, and placed in a rack. After centrifugation of the partially reduced glutenin, it is important not to delay the dilution step more than necessary, particularly for extra strong mixing wheats whose glutenin tended to re-aggregate and precipitate (results not shown). Accordingly, no more than 20 to 30 min should be allocated for dilution of all samples to ensure maximum quantitative reproducibility. An aliquot of the supernatant is diluted 100-fold in a fresh microcentrifuge tube with solution 'A', and the solution is thoroughly mixed by vortexing (5 sec) to obtain the sample for UV absorbance measurement.

2.4 UV Absorbance Measurement and Protein Calibration

A 1 mL aliquot of solution 'A' was used as the blank for absorbance measurements at 214 nm. Absorbance of samples, based on a 10 mm path length cuvette, were in the range 0.250 to 0.700, depending on genotype. This absorbance range corresponded to a concentration of 50PI glutenin in the range 15 to 35 mg/mL or 1.5 to 3.5% of flour (14% mb). A calibration curve can probably be prepared from wheat flour or semolina of any source, although we prepared a curve that was a composite of 50PS protein (extracted once as described above) from four different wheats. The protein contents of aliquots (0.5 mL) of the resulting supernatants were determined by the Kjeldahl procedure. The remaining 50PS extract was used to produce a dilution series (in 50% propan-l-ol) in the appropriate protein concentration range. The calibration was perfectly linear throughout the range of absorbance that was tested (results not shown). There was also near perfect agreement (R^2=0.990) between 50PI glutenin protein measurements obtained by Kjeldahl and spectrophotometric procedures. Based on an analysis of 12 different wheat cultivar samples and three separate determinations for each, the average measurement error for the spectrophotometric procedure (CV=5.9%) was higher than that for the Kjeldahl procedure (CV=3.3%), but still indicated good reproducibility. Random pipetting errors are the most probable source of protein measurement variability in the UV absorbance procedure.

3 RESULTS

3.1 Effects of Experimental Parameters

The effects of various pertinent experimental parameters and procedural steps on the yield and repeatability of measurement of 50PI glutenin were examined. The key experimental variables were extraction time, temperature and DTT concentration. 50PI glutenin increased rapidly with increasing extraction time up to about 5 min (Fig. 1A), after which time values began leveling off. There was little or no change in the yield of 50PI glutenin beyond about 10 min extraction time. We chose a 30 min extraction time as a matter of convenience to permit ease of handling of a group of samples at one time, e.g. 20-40.

The effect of extraction temperature on the amount of extracted glutenin was considerable (Fig. 1B). We adopted 55°C, as this was in the temperature range with the most stable response of glutenin yield, and maximum differentiation of 50PI glutenin content. Higher extraction temperatures would likely be equally effective in differentiating wheat quality by this procedure, as the different response curves were highly correlated. Clearly, precise control of extraction temperature is an important requirement.

As 0.1% DTT was used as reductant, the supernatant contains only partially reduced glutenin. SDS-PAGE analysis indicated that this extract, when fully reduced, contained glutenin of very high purity (see Fig. 2 in reference 6). Figure 1C shows the effectiveness of using 0.1% DTT (or even lower concentrations) for extraction of glutenin from the 50PI residue. Using 1% DTT for extraction in 50% propan-l-ol produced erroneous absorbance measurements, as Kjeldahl analysis of these extracts gave no increase in N-content compared to results obtained using 0.1% DTT (results not shown). The increase in UV absorbance for the 1% DTT extracts (Fig. 1C) is probably due to the absorbance of excess unoxidized DTT.

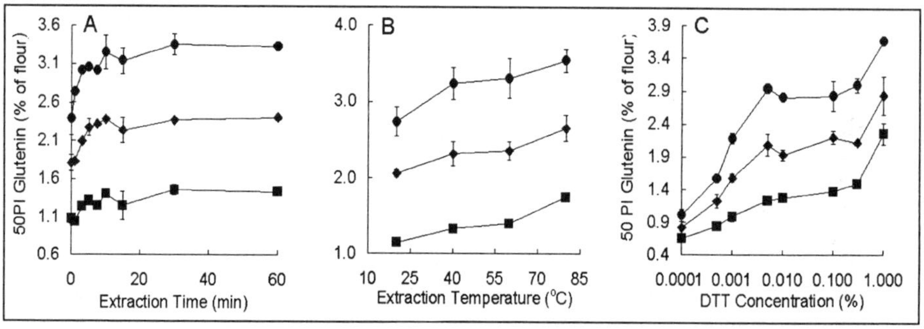

Figure 1 *Effect of extraction time (A), extraction temperature (B) and DTT concentration (C) on the amount of protein extracted from the 50% propan-l-ol insoluble residue.*

In terms of general procedural details, pipetting is the single most important source of analytical error, owing to the high sensitivity of protein absorbance determination at 214 nm. Only two pipetters should be used in the procedure, each capable of dispensing volumes of 1 mL and 10 µL (or 100 µL for two-step dilution of the 50PI glutenin). Pipetters should be regularly tested for accuracy and repeatability as even very small errors in pipetting of the extracted 50PI glutenin will be magnified in the UV absorbance results. Pre-rinsing each new pipette tip with each partially-reduced glutenin sample also significantly improved measurement repeatability. Another source of pipetting-related error is in the handling of the propan-l-ol solvent in the context of glutenin aliquot dilution; special care must be taken when pipetting the 50% propan-l-ol for sample dilution as this reagent is accessed many times, and inadvertent tip contamination from a protein sample in a microcentrifuge tube will affect all following results. To mitigate this potential problem, working solutions of 50% propan-l-ol should be prepared daily from the stock 100% propan-l-ol solvent.

3.2 Prediction of Dough Strength

For the full set of PRRCG common wheats (partially charted in Fig. 2A), the range of 50PI glutenin content was 12 to 30% of flour protein, and 1.3 to 3.4% of flour. For these wheats, the relationship between flour protein content (range: 9.6 to 14.2%) and 50PI glutenin content was relatively weak ($R^2 = 0.35$). The PRRCG durum wheats (Fig. 2B) had a narrower range of 50PI glutenin content:19 to 28% of flour protein, and 2.3 to 3.4% of semolina, which possibly reflected a much narrower range of semolina protein content (12.0 to 13.3%) compared to the corresponding range of common wheat flour protein. There was no correlation between semolina protein and semolina 50PI glutenin content.

The result shown in Fig. 2A is typical of the strong relationships we have invariably found between spectrophotometric determination of flour 50PI glutenin content and dough mixing requirements for diverse common wheat genotypes. A comparable result for Canadian hard red spring (HRS) wheat cultivars, more narrow in genotypic composition, but grown in different locations is also reported in another paper in these proceedings (refer to reference 8). However, in that study, 50PI glutenin values (flour basis) were normalized relative to flour protein contents in order to obtain an acceptably high R-square (R^2=0.83) for dough strength prediction. It is important to point out that these high correlations would have been significantly lower had the relationships been based on the total 50PI protein fraction, i.e. a combination of 50PI glutenin and the remaining residue protein. The latter fraction (reference 8), varied widely in amounts from 17 to 28% of total flour protein depending on genotype and growing location, and was negatively correlated with dough mixing time and work input (r = -0.15).

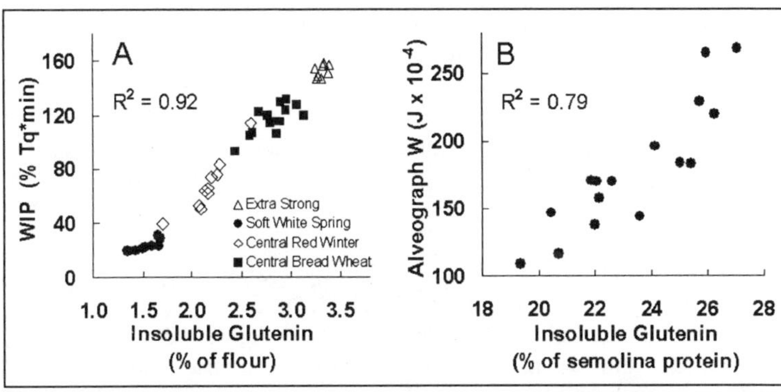

Figure 2. Relationship between spectrophotometric measures of insoluble glutenin and dough strength of Canadian common wheats of different classes (A) and durum wheats (B) determined as Mixograph work input to peak development (WIP), and Alveograph deformation energy (W), respectively.

For the durum semolina samples (Fig. 2B), we obtained good prediction of dough strength in relation to 50PI glutenin expressed as a percentage of semolina protein (R^2=0.79). Alveograph W values were less well predicted on the basis of 50PI glutenin in semolina (R^2=0.67). The prediction of durum semolina dough strength as measured by the Alveograph method, was less strong compared to analogous results for the common wheats. The use of constant dough mixing times and low water absorptions in the standard Alveograph procedure may have contributed to the lower R-square results. More work is still required to assess the quality screening performance of the spectrophotometric procedure with other durum wheat samples and different technological measures of durum wheat gluten strength. However, if future results are comparable to that depicted in Fig. 2B, then the spectrophotometric procedure would offer an attractive option as a high throughput small-scale test of durum wheat protein quality.

4 SUMMARY AND CONCLUSIONS

A robust small-scale procedure was developed to isolate and measure HMW glutenin in common wheat flour and durum semolina by spectrophotometry (A214) of partially reduced extracts of 50% propan-l-ol insoluble glutenin. While this fraction of polymeric protein is a quantitatively minor constituent of wheat endosperm, results underscore the importance of even small variation in the concentration of this key protein fraction to wheat end-use quality. Results showed that differences in dough strength among diverse common wheats, in particular, were almost completely attributable to differences in amounts of 50PI glutenin protein, thus supporting the concept that glutenin molecular size, glutenin solubility and glutenin functionality are inter-related

Compared to alternate methods of measuring insoluble or HMW glutenin protein or wheat protein quality in general, the advantages of the spectrophotometric procedure include the following: capability to directly measure protein as peptide bond absorbance; extracts only glutenin from the 50PI residue, highly effective to differentiate protein quality and predict end-use quality; minimal reagent and equipment needs, no special precautions for reagent handling; low cost; very small scale; efficient and convenient handling of many samples in a short time. This small-scale test of protein quality should be beneficial in variety or sample selection by plant breeders and processors, respectively.

References

1. Orth, R.A. and Bushuk, W. *Cereal Chem.,* 1972, **49**, 268
2. Tanaka, K. and Bushuk, W. *Cereal Chem.,* 1973, **50**, 590
3. Orth, R.A. and O'Brien, L.A. *J. Aust. Inst. Agric.,* 1976, **42**, 122
4. Sapirstein, H.D. and Johnson, W.J.'Spectrophotometric method for measuring functional glutenin and rapid screening of wheat quality', Proceedings of the Sixth International Gluten Workshop, C.W. Wrigley, ed. Royal Aust. Chem. Inst., Melbourne,1996, p.494.
5. Bean, S.R., Lyne, R.K., Tilley, K.A., Chung, O.K. and Lookhart, G.L. *Cereal Chem.*, 1998, **75**, 374
6. Sapirstein, H.D. and Fu, B.X. *Cereal Chem.,* 1998,**75**, 500
7. ICC 1995. International Association of Cereal Science and Technology. Standard 121. Method for using the Chopin Alveograph. The Association; Vienna, Austria.
8. Sapirstein, H.D. and Lukie, C. 'Effects of environment on the gluten protein composition of strong mixing wheats and the ability to predict breadmaking quality by small-scale tests'. Paper in these proceedings.

Acknowledgements

The financial assistance provided by the Natural Sciences and Engineering Research Council of Canada is gratefully appreciated. We thank the PRRCG and all the wheat breeders of the Wheat, Rye and Triticale Subcommittee for granting permission to use their wheat lines in this study. We also thank Randy Roller for his expert technical assistance.

PREDICTION OF WHEAT PROTEIN AND HMW-GLUTENIN CONTENTS BY NEAR INFRARED (NIR) SPECTROSCOPY

D.G. Bhandari, S.J. Millar and C.N.G. Scotter

Campden & Chorleywood Food Research Association, Chipping Campden, Gloucestershire GL55 6LD, United Kingdom

1 INTRODUCTION

Near infrared (NIR) spectroscopy is a widely applied to the measurement of cereal quality and cereal product composition. The rapid assessment of wheat lots is routinely performed using NIR to measure protein and moisture. In general, NIR can predict these parameters with a high degree of accuracy, as the relevant spectral regions show reasonably clear changes with changing sample composition. Recently, it has been reported that NIR may be applied to the assessment of wheat plant tissue during development to predict both the yield and the final protein content of the grain[1,2]. Such measures may then be used to decide on the nutritional status of the crop and the need (if any) for subsequent fertiliser application.

The aim of this study was to develop a means of predicting the final quality in breadmaking wheat through monitoring the high molecular weight glutenin subunit (HMW-G) levels in developing grain, and through NIR spectroscopy of immature and mature grain.

2 MATERIALS AND METHODS

2.1 Materials

Growing trials were conducted over two seasons (1997 and 1998) involving six sites in the UK, three breadmaking varieties (Hereward, Caxton and Rialto), and a range of ammonium nitrate and foliar urea fertiliser treatments.

2.2 Methods

Immature grains were freeze-dried at growth stage (GS) 75 for measurement of protein content, HMW-G levels by SDS-PAGE, and molecular weight distribution by SE-HPLC. NIR spectra were acquired using a Foss NIRSystems 6500 monochromator. Harvest material was subjected to standard breadmaking quality evaluation, SDS- PAGE and gel densitometry and NIR analysis. The proteins were resolved by SE-HPLC into three main fractions (peaks 1, 2 and 3). Peak 1 corresponds mainly to high M_r glutenin polymers, peak 2 to a mixture of medium M_r polymers and monomers, and peak 3 mainly to monomers with some M_r polymers.

3 RESULTS AND DISCUSSION

By applying multivariate statistical techniques to the NIR spectral data, it was possible to discriminate between samples on the basis of grain maturity and growing location and to predict a range of quality parameters. Canonical variates analysis allowed the samples to be grouped on the basis of site (Figure 1).

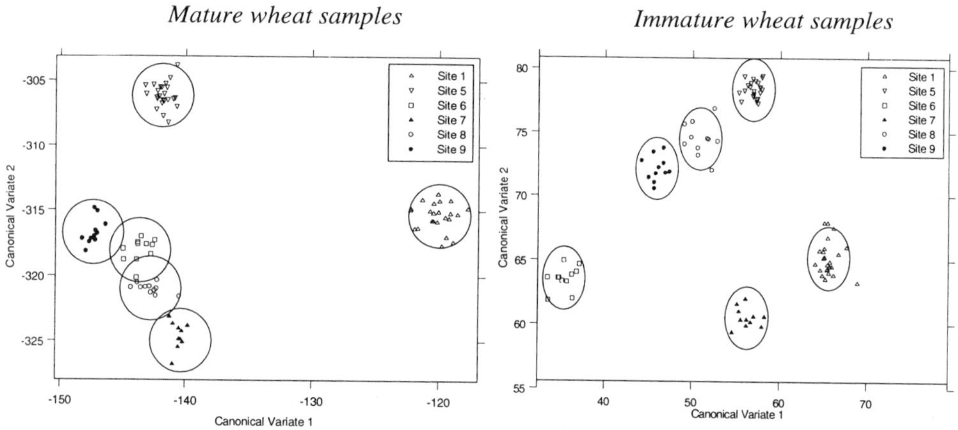

Figure 1 *Score plots of canonical variate 1 versus variate 2 (first derivative spectral data with a datapoint gap of 4nm, datapoint smooth of 4nm, no secondary smooth (1,4,4,1) and Multiplicative Scatter Correction (MSC)).*

An NIR calibration was produced for HMW-G levels (% total protein) in immature and mature samples (Figure 2) using modified partial least squares regression (MPSLR) analysis (R^2 of 0.88, SEC (standard error of calibration) of 0.60 and SECV (standard error of cross validation) of 0.77).

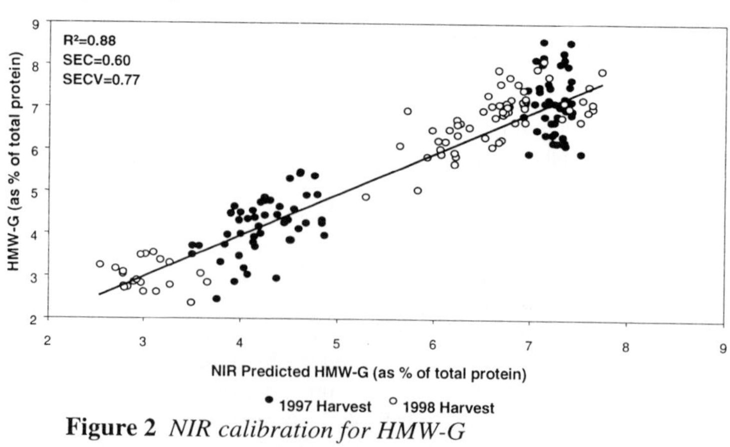

Figure 2 *NIR calibration for HMW-G*

This relationship was independent of protein content and is thought to reflect the functional properties of both immature and mature wheat. The spectral loadings used to produce the calibration for HMW-G featured a number of peaks/troughs in the region 2000 – 2500 nm (Figure 2b); a region previously associated with glutenins and gliadins[3].

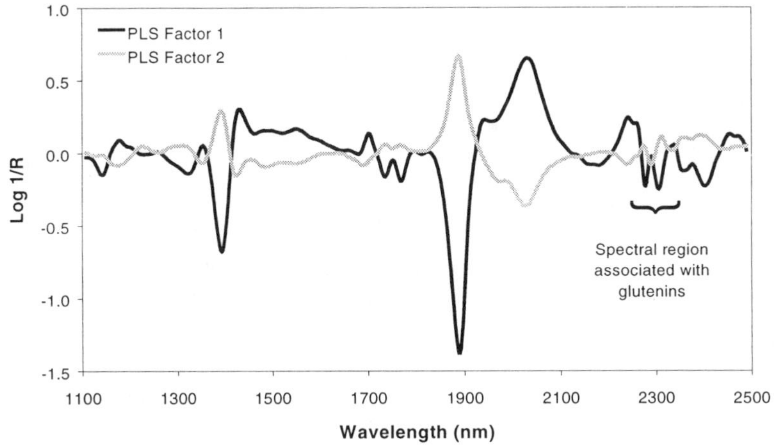

Figure 3 *Modified Partial Least Squares Regression Factors 1 and 2 for the NIR calibration*

Two other parameters related to flour protein quality (SE-HPLC peak 3) and gel protein elastic modulus (G') also could be predicted satisfactorily by NIR (Table 1).

Table 1 *NIR regression results*

Constituent	**MPLSR**	
	R^2	**SECV**
Loaf volume	0.49	36.57
HPLC Peaks 1+2	0.19	4.54
HPLC Peak 3	0.84	1.73
HMW-G	0.88	0.60
Gel Protein weight	0.74	1.07
Gel Protein G′	0.80	4.39

The sample set was also split such that NIR spectra from the immature samples were related to the protein contents of the mature samples. A calibration was developed using MPLSR analysis that predicted the mature protein content from immature samples with acceptable accuracy (R^2 of 0.88 and SECV of 0.50) for samples from two harvest years (1997 and 1998), irrespective of variety (Figure 4).

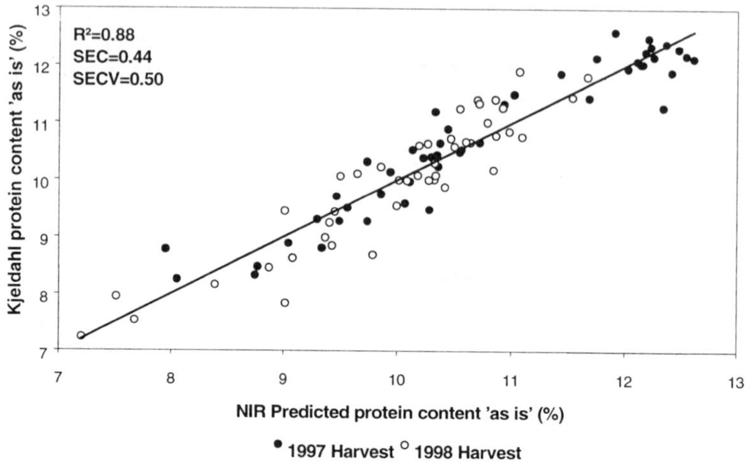

Figure 4 *NIR calibration for mature wheat protein content using immature wheat spectra*

4 CONCLUSIONS

This study has related NIR spectral data to wheat quality, and has demonstrated that NIR spectroscopy can be used to:

- discriminate between samples on the basis of their maturity and growing locations
- predict with acceptable accuracy the harvested grain protein content from developing grain samples taken at GS 75, irrespective of variety
- predict with acceptable accuracy a range of parameters (gliadin and HMW-G content and gel protein G) related to breadmaking quality.

These preliminary findings suggest that NIR technology has a potential role in the rapid assessment of the nitrogen fertiliser needs of wheat cultivated in the UK.

References

1. G.D. Batten, V.B. McGrath, S. Ciavarella and A.B. Blakeney. *Cereal Foods World*, 1993, **38**, 620.
2. V.B. McGrath, A.B. Blakeney, G.D. Batten and O.W. Boland. *Proc. 45th Australian Cereal Chemistry Conference, Adelaide 1995*, Eds. Y.A. Williams and C.W. Wrigley, 1995, p538.
3. I.J. Wesley, R.S. Uthayakumaran, G.B. Anderssen, G.B Cornish, F. Bekes, B.G. Osborne and J.H. Skerritt. *Journal of Near Infrared Spectroscopy*, 1999, **7**, 229.

Acknowledgements

This work was fully funded by the *Home-Grown Cereals Authority and by NABIM. The authors wish to acknowledge Levington Agriculture who conducted the growing trials, and Professor Peter Shewry and IACR staff for their assistance.

**The full report of this study (Project Report No. 219) may be obtained directly from HGCA, whose e-mail address is publications@hgca.com*

LABORATORY MILL FOR SMALL-SCALE TESTING

J. Varga[1], D. Fodor[2], J. Nánási[2], F. Békés[3,6], M. Southan[4,5], P.Gras[3,6], C. Rath[4], A. Salgó[1] and S. Tömösközi[1]

1. Department of Biochemistry and Food Technology, Budapest University of Technology and Economics, Budapest, Hungary. 2. INTER-LABOR – METEFEM LTD, Budapest, Hungary. 3 CSIRO Plant Industry, Grain Quality Research Laboratory, North Ryde, NSW, Australia. 4. CSIRO Plant Industry, Canberra, ACT, Australia. 5. BRI Australia, North Ryde, NSW, Australia. 6. Quality Wheat CRC LTD, North Ryde, NSW, Australia

1 INRODUCTION

Different miniaturized analytical methods are available in order to detect chemical, physico-chemical, rheological and breadmaking characteristics of wheat. For example, mixing studies on the 2g Mixograph, micro-valorigraph or micro baking are becoming essential tools both in early selection of lines for quality traits in breeding programs and as research tools [1,2]. Small and representative wheat sample size is needed for introduction of the reduced-scale methods. The milling step has been a limiting factor in these studies, requiring the production of flour from 5-10g grain. The aim of this study was to compare the quality of flours obtained by recently developed micro-scale laboratory mill (FQC-2000, Inter-Labor, Hungary) with flours obtained by conventional method. Milling yield, sample size distribution and mixing properties of different flours were compared in these experiments.

2 MATERIALS AND METHODS

2.1. Materials

Five bread wheat and two durum wheat samples cultivated in Hungary (Agricultural Research Institute, Martonvásár, Hungary) were selected for comparative study (Table 1.)

2.2. Instruments

The conditioned samples were milled on FQC-2000 micro-scale laboratory mill and on conventional-scale QC-109 experimental labmill, both produced by Inter-Labor Ltd., Hungary.

The QC-109-type mill is a compact and standardized unit equipped with four rolls ensuring three grinding passes. The optimal amount of wheat seed is more than 100 grams.

Table 1 *Characterisation of investigated wheat lines*

Wheat species	Protein content of whole grain (%, N*5.83)	Quality characteristics of wheat species
MV 17	11.6	Breadmaking quality is satisfactory, average gluten content <30%
MV 25	12.8	Good quality wheat for breadmaking, average gluten content >28%
MV Irma	12.8	Breadmaking quality is weak, soft wheat, average gluten content <28%
Mv Mezőföld	13.7	Good quality wheat for breadmaking, average gluten content is 28-32%
Mv Magvas	13.6	Breadmaking quality is excellent, hard wheat, average gluten content >30%
MVTD 1299	14.2	Durum wheat
Odmadur 2	12.7	Durum wheat

The recently developed FQC-2000 micro-scale laboratory mill produces flour, semolina and bran fraction with acceptable yields. The seed is fed to finely grooved steel rolls rotating with different speed in opposite directions (Figure 1). In order to avoid the slip effects of rolls the mill has a special drive with a cogged belt. The milling products can be fractionated by appropriate sizes of sieves. The typical sample size for wheat milling is higher than 3 grams[3].

Figure 1 *Constructional sketch of FQC 2000 micro-scale laboratory mill*

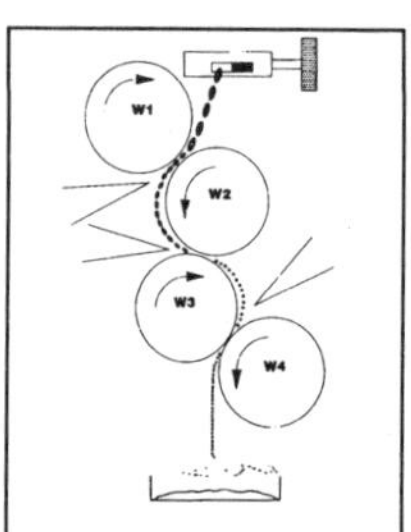

Technical parameters:
- Optical moisture content
  for durum wheat: 16%
  for aestivum heat: 15%
- Milling yields: 55–70%
- Power supply: 230 V, 50-60Hz
- Mass: 17.5 kg
- Dimension: 270x210x350mm
- Sample size: > 3 g

2.3. Methods

The wheat samples were conditioned to 15.5% (for bread wheat) or to 16.0% (for durum wheat) moisture content overnight.

500g grain for macro-scale milling and 5g grain for micro-scale milling were milled in triplicate. The size distribution of milling fractions obtained was studied with a Retsch AS-200 sieving machine equipped with 500, 315 and 200μm sieves for separation. Fractions with particle size <315μm defined as experimental flour.

Mixing tests were executed by a prototype of recently developed micro-scale Z-arm mixer (Figure 2). In these experiments 4g sample/test was used. The evaluation of micro-scale curves was similar to the original standard procedure[4]. The Dough Development Time (DDT, min), Dough Resistance (DR, min) and Dough Softening Value (DSV in micro-Valorigraph Unit, mVU) were calculated and compared.

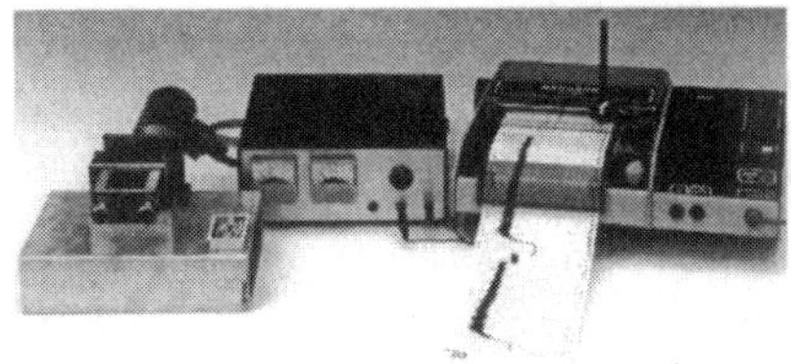

Figure 2
Prototype of micro-scale Z-arm mixer

3 RESULTS AND DISCUSSION

Comparing the effectiveness of different milling procedures (Table 2) the following tendencies were observed: the yields of flour obtained from the macro-scale mill varied in the range of 54-74% and significantly higher values were obtained for durum wheat varieties compared to bread wheats. The micro-mill produced a significantly narrower range of flour yield (51-64%) for the same sample population. The differences in yield between macro- and micro methods were 3-9% for bread wheats and 8-20% for durum wheats. The relationship between the milling yields determined on the two mills ($r=0.98$, $p<0.05$, except of MVTD 1299 durum wheat samples, see later) indicates that the milling yield might be estimated from micro-milling results. These conclusions are consistent with the results of an earlier comparative study where a Buhler test mill was used[5].

The flour fractions obtained with the micro mill were also separated with a 200µm sieve for semolina (> 200µm) and fine flour (<200µm) fractions. The bread wheat samples gave approx. 50% semolina and 50% flour while the proportions of these fractions for the durum samples were about 70% and 30%, indicating the higher particle size of milled durum product.

Table 2 *Milling yields of aestivum and durum wheat samples using macro- and micro-scale laboratory mill*

Varieties	Flour yield (%)	
	Macro-scale milling (QC 109 labmill)	Micro-scale milling (FQC-2000 labmill)
Bread Wheats		
MV 17	62.6 ± 2.3	56.5 ± 2.6
MV 25	58.0 ± 1.7	55.4 ± 2.3
MV Irma	54.3 ± 2.6	51.4 ± 1.9
Mv Mezőföld	72.5 ± 1.8	63.8 ± 2.5
Mv Magvas	67.2 ± 2.2	61.6 ± 1.8
Durums		
MVTD 1299	73.7 ± 4.1	53.9 ± 2.3 (!)
Odmadur 2	68.9 ± 28	61.4 ± 3.4

The chemical composition and the mixing properties of flours obtained by different milling procedures are very similar. The strong correlation ($r= 0.81$ for DDT, $r=0.83$ for DR and $r=0.94$ for DSV, $p<0.05$) between appropriate values also show that the quality of flours obtained with the new micro-scale labmill is similar to the flours originated from conventional laboratory milling tests.

Table 3 *Comparison of mixing properties of different wheat flours determined with micro-scale Z-arm mixer*

Samples	'Macro-mill' flours					'Micro-mill' flours				
	Ash (%)	Protein N*5.7 (%)	DDT (min)	DRV (min)	D SV (mVU)	Ash (%)	Protein N*5.7 (%)	DDT (min)	DRV (min)	D SV (mVU)
Bread Wheats										
MV 17	0.50	9.8	1.5	3.5	8.0	0.54	9.7	1.5	2.5	9.0
MV 25	0.53	12.5	2.5	4.0	4.0	0.55	12.1	1.5	2.0	11
MV Irma	0.63	11.4	1.5	3.5	6.0	0.57	11.2	1.5	3.5	7.0
MvMezőföld	0.51	12.6	1.5	2.0	11.0	0.55	12.1	2.0	2.5	11.0
Mv Magvas	0.51	12.1	1.5	4.0	5.0	0.53	12.0	2.0	4.5	5.0
Durums										
MVTD1299	1.12	13.5	2.0	2.5	7.0	0.88	13.1	2.0	3.0	6.0
Odmadur 2	0.95	11.3	1.0	1.5	10	0.91	11.4	1.0	1.5	11.0

4 CONCLUSION

The milling results and the quality of flours obtained with the micro-scale laboratory mill show that the recently developed FQC-2000 labmill could be an excellent tool for micro-scale research work and for earlier selection of wheat varieties. Additionally, the micro-mill seems to be suitable for milling durum wheat varieties.

References

1. P. W. Gras and L.O'Brien, *Cereal Chem.*, 1992, **69**, 254
2. F. Békés and P. W. Gras, Proc. Of AACC Meeting, Minneapolis, USA, Cereal Food World, 1998, **44**, 580.
3. D. Fodor, J. Varga, S. Tömösközi, A. Salgó, J. Nánási and Gy. Veres, 'Procedure and instrument for small-scale milling', Patent, 1999. (under evaluation).
4. International Standard, 'Determination of water absorption and rheological properties using a valorigraph', ISO 5530-3, 1988, Part 3
5. F. Békés, M.S. Southan, S. Tömösközi, J. Nánási, P.W. Gras, J. Varga, J. McCorquodale, B. Osborne, 'Comperative studies on a new micro scale laboratory mill', In: A.W. Tarr, A. S. Ross and C.W. Wrigley (eds.): Proc. 49th RACI Conference, 1999, Melbourne, Australia (in press).

Acknowledgement

This research work was supported by the National Committee for Technological Development (Project No.:96-97-68-1354).

SCALE DOWN POSSIBILITIES IN DEVELOPMENT OF DOUGH TESTING METHODS

S. Tömösközi[1], J. Varga[1], P.W. Gras[2,5], C. Rath[3], A. Salgó[1], J. Nánási[4], D. Fodor[4] and, F. Békés[2,5]

1. Department of Biochemistry and Food Technology, Budapest University of Technology and Economics, Budapest, Hungary. 2. Quality Wheat CRC LTD, North Ryde, NSW., Australia. 3. CSIRO Plant Industry, Canberra, ACT, Australia. 4. INTER-LABOR – METEFEM LTD, Budapest, Hungary. 5. CSIRO Plant Industry, Grain Quality Research Laboratory, North Ryde, NSW., Australia

1 INRODUCTION

We report on the development and application of a new, small scale Z-arm mixer, analogous to the Valorigraf (TM) and Farinograph(TM). Instruments based on this design can provide reproducible estimates of mixing time and useful measures of the water absorption of flour samples. Being the smallest Z-arm mixer currently available, requiring 4g of flour, the new equipment has the potential to be applied as both a breeding- and a research tool.

2 MATERIALS AND METHODS

Flour samples of 20 Australian wheat varieties (Table 1) were milled on a Buhler laboratory mill. Water absorption and mixing properties of doughs were also determined with the Farinograph (Brabender GmbH, Germany) and Valorigraph (INTER_LABOR, Hungary) using the appropriate standard methods[1,2].

Micro-scale mixing tests were performed a recently developed prototype Micro-Z-arm mixer using 4g flour per test. Mixes were performed using water absorption values obtained from the conventional macro-scale Valorigraph test.

The effects of protein content and composition on mixing properties and water absorption were investigated using flours of cultivars Eradu, Hartog and Vulcan. Gluten, starch, gliadin and glutenin were separated from the flours[3] and flour plus starch or flour plus gluten blends were produced by altering the protein contents of the original flours by factors of 0.8, 0.9, 1.0, 1.1 and 1.2. The ratio of polymeric to monomeric proteins in the Eradu flour was systematically altered by supplementing the flours with glutenin, gluten or gliadin in amounts resulting in 10 and 20 percent increases in the original protein contents of each flours. Blends were mixed at two water absorption levels differing by approx 5% and the expected water absorption values at 500VU were calculated by interpolation. Similarly, estimates of mixing requirement at maximum dough resistance were calculated by interpolation.

Table 1 *List of the investigated wheat flours*

Wheat species	Protein content of flour (%, N*5.7)	Wheat species	Protein content of flour (%, N*5.7)
Sunvale	11.7	Amery	,9.2
Leichardt	11.9	VK412	7.8
QT6515	12.2	VG12714	10.9
SUN276 a	13.3	WAWHT2168"	7.7
WW2463	12.6	WAWHT2175	7.7
WW2455	11.5	Gutha	11.0
Janz	9.2	Tammin	10.1
Banks	10.3	Tincurrin	8.4
WAWHT2229	11.9	Tincurrin	6.8
Carnamah	12.2	WAWHT2193	7.1
Eradu	Protein: 13.9%; glutenin: 40.79%; gliadin: 52.25%		

3 RESULTS AND DISCUSSION

3.1. Development of small-scale Z-arm mixer

In contrast to the usual torque balance design, the micro-scale Z-arm mixer is driven by a servo-driven DC motor and is used to measure the power required to mix the dough (Figure 1A). The measurement of power can be directly related to the torque applied to the dough by using a windlass to winch up a series of known weights instead the mixer blades. The results show that the power recorded is directly proportional to the mass on the windlass, that is, the torque generated by the motor. This relation is linear in the range of interest to the limits of measurement (approx.1 in 1000 from 0 to 500VU).

Figure 1 *The prototype of Micro Z-arm mixer (A) and the structure of mixing bowl (B)*

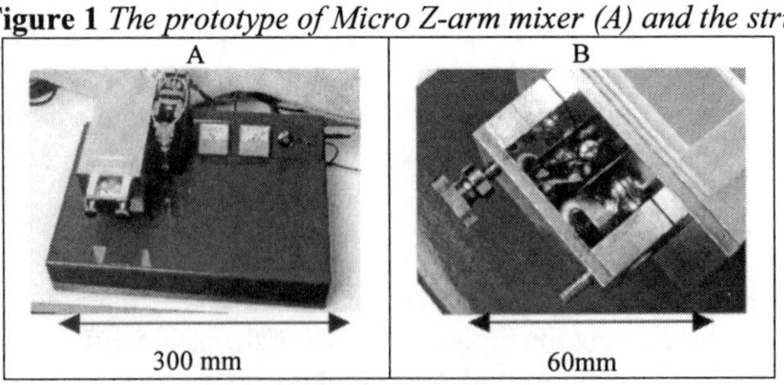

For mixing, the standard error of the mix time was 12 seconds, and the standard error in the maximum resistance was 15 Valorigraph units, or 0.3% of the maximum height of the curve. Since the reproducibility of water addition using a pipettor is at best 0.01 ml, or 0.25%, the reproducibility appears to be excellent. When mixed at 22 °C on the Micro Z-arm mixer using the standard Valorigraf water requirement, the correlation between the mixing times observed on the two machines was remarkably good (r^2=0.91). The mixing times recorded on the Micro Z-arm machine were considerably shorter than those observed on the Valorigraph, but the curves obtained exhibited a general similarity to the conventional Valorigraph (Figure 2).

Figure 2 *Mixing curves registered with Micro Z-arm mixer (Left) and conventional Valorigraph (Right)*

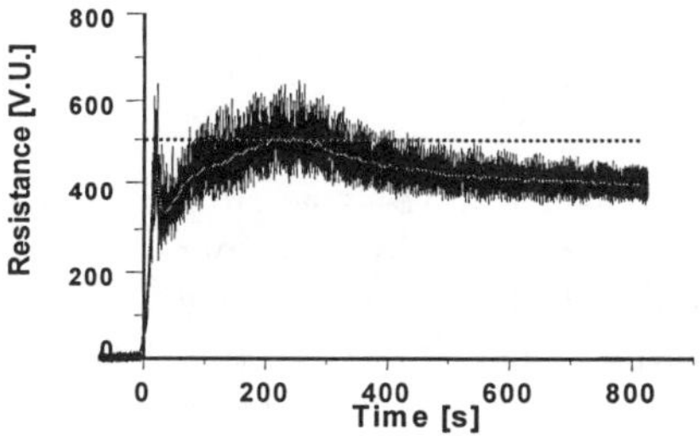

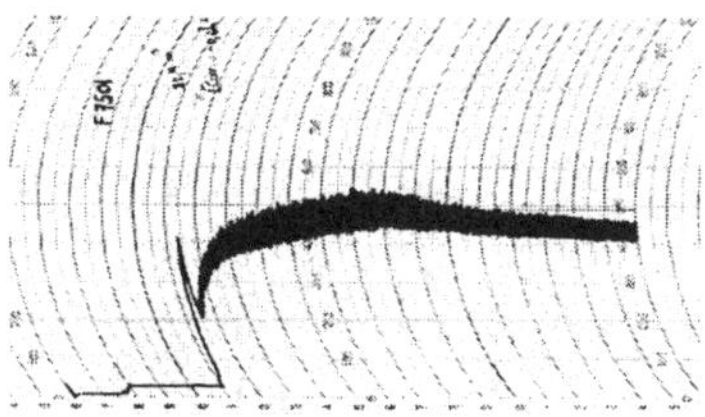

The relationship between maximum dough resistance and water content was quite linear for the three flours tested (r^2=0.91, 0.87 and 0.94, respectively) and the slopes were effectively parallel (Figure 3). Examination of the relations observed between water and maximum dough resistance on the blends of flour and flour components (i.e. starch/gluten/gliadin/glutenin) showed similar linear dependence on water content. It is thus feasible to estimate the water absorption accurately to better than +/-1% from a single measurement. More accurate assessment requires multiple measurements, but the limiting effect of accuracy of water delivery limits the potential accuracy to about +/-0.3%.

Figure 3 *Relationship between resistance and water absorption*

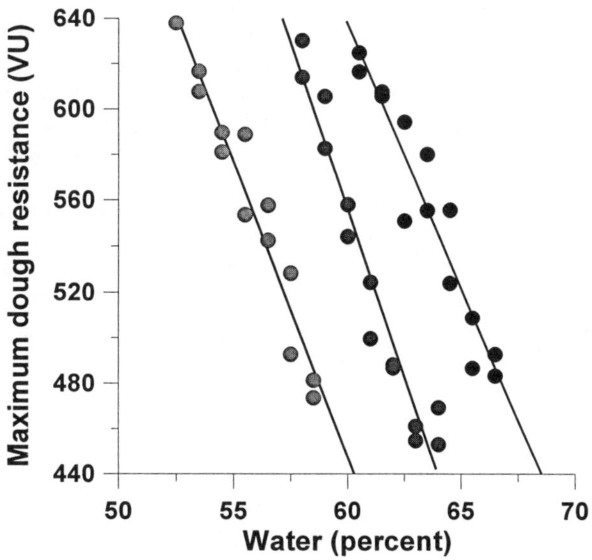

Another potential advantage of the use of a servo controlled DC-motor is the easy of alteration of the mixer speed. This simplifies analysis of the effects of mixing speed on dough water absorption and development time.

All the above observations indicate that the Z-mixer can be used in cases where measurements similar to the Farinograph or Valorigraph are needed and a limited amount of sample are available. The Z-mixer can be applied in breeding programs for early selection for water absorption and for mixing properties.

3.2. Investigation of the structure-function relationships

As a research tool, the micro-mixer can be applied to investigate structure/function relationships, providing essential information on the role of individual flour components defining water absorption and/or mixing properties. In this way, the effects of protein content or glutenin to gliadin ratio on water absorption and on mixing properties (Figure 4) can be studied independently, keeping all of other chemical parameters constant.

Figure 4 *The effects of protein content and gliadin/glutenin ratio on Water Absorption and Dough Devopment Time (type of flour: Eradu)*

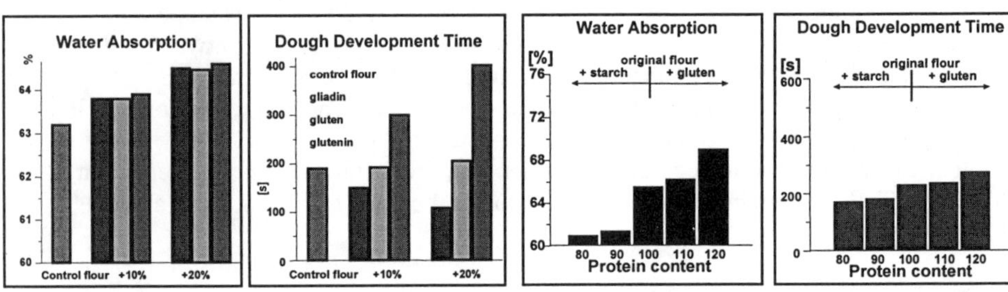

Results show that both water absorption and dough development time increase if the protein content has been increased by gluten addition and decrease if the protein content has been decreased by starch addition. There was no difference between the extent of increase in water absorption if the protein content was increased by supplementing with gliadin, gluten or glutenin. Dough development time was changed significantly depending on whether gliadin, gluten or glutenin was added. Gliadin addition decreased while glutenin addition increased the mixing requirement.

These findings are analogous to those obtained from studies carried out on pin mixers[4]. However, it is known that the rheological properties of doughs produced on either pin mixers (e.g. Mixograph) or Z-arm mixers (e.g. Valorigraph) show significant differences. The new equipment can provide essential data for investigation of the differences in the rheological properties of doughs produced in different mixers.

4. CONCLUSION

The recorded curves and the calculated parameters obtained with the small-scale Z-arm mixer are very similar to the Farinograph or Valorigraph values indicating that the micro-mixer could be a very effective tool in breading programs and in basic and applied research work. The versatile micro-scale mixer is capable of measuring water absorption, mixing requirement and breakdown with only four grams of flour.

References

1. International Standard, 'Determination of water absorption and rheological properties using a farinograph', ISO 5530-1, 1988, Part 1
2. International Standard, 'Determination of water absorption and rheological properties using a valorigraph', ISO 5530-3, 1988, Part 3
3. S. Uthayakumaran, P.W. Gras, F. Stoddard and F. Békés, *Cereal Chem*, 1999, **76**, 389.
4. P. W. Gras and L.O'Brien, *Cereal Chem.*, 1992, **69**, 254

Acknowledgement

This research work was supported by National Committee for Technological Development (Project No.:96-97-68-1354)

QUALITY TEST OF WHEAT USING A NEW SMALL-SCALE Z-ARM MIXER

J. Varga[1], S. Tömösközi[1], P.W. Gras[2,5], C. Rath[3], J. Nánási[4], D. Fodor[4], F. Békés[2,5], and A. Salgó[1]

1. Department of Biochemistry and Food Technology, Budapest University of Technology and Economics, Budapest, Hungary. 2. Quality Wheat CRC LTD, North Ryde, NSW., Australia. 3. CSIRO Plant Industry, Canberra, ACT, Australia. 4. INTER-LABOR – METEFEM LTD, Budapest, Hungary. 5. CSIRO Plant Industry, Grain Quality Research Laboratory, North Ryde, NSW., Australia

1 INRODUCTION

Small-scale dough testing instruments are required for the early selection of wheat varieties on the basis of rheological properties and/or for basic studies of structure-function relationships of wheat flour. Recently, the 2g Mixograph has been developed[1] and successfully applied both as breeding and research tool[2,3]. However, most current dough testing methods employ mixers with Z-arm action, such as Farinograph™ and Valorigraph™. There is a considerable evidence that the rhelogical properties of dough produced on either Mixograph-type pin or on Z-arm mixers show significant differences. We report here the development and application of a computer controlled Micro Z-arm mixer with electronic recording which requires only four grams of flour.

2 MATERIALS AND METHODS

2.1. Materials

Eight Hungarian and Australian type bread wheat and two durum wheat varieties were used in experiments. The grains were milled on a recently developed small-scale laboratory mill (FQC 2000) produced by Inter-Labor Ltd., Hungary[4] (Table 1).

2.2. Instruments

The prototype of small-scale Z-arm mixer was designed with servo-driven DC motor and the power required to mix to dough was measured electronically. The mixing curves were drawn with a mechanical recorder (Figure 1a). The equipment is suitable to measure 4g samples (Figure 1b). Recently, the system has been completed with an automatic syringe pump for water addition, with a thermostat and with the whole system controlled and the electronic signal evaluated with a PC (Figure 1c). In the comparative studies, a conventional 50g Valorigraph and a 35g Farinograph instrument were used.

Table 1 *Characterisation of investigated wheat flour*

Wheat species	Protein (%, N*5.7)	Wheat line	Protein (%, N*5.7)
MV 17	9.7	Janz	9.2
MV 25	12.1	Hartog	11.9
MV Irma	11.2	Eradu	13.9
Mvmezőföld	12.1	MVTD 1299 (durum)	13.1
Mv Magvas	12.0	Odmadur 2	11.4

Figure 1 *The prototype (a), the mixing bowl (b) and the first generation (c) of micro-scale Z-arm mixer*

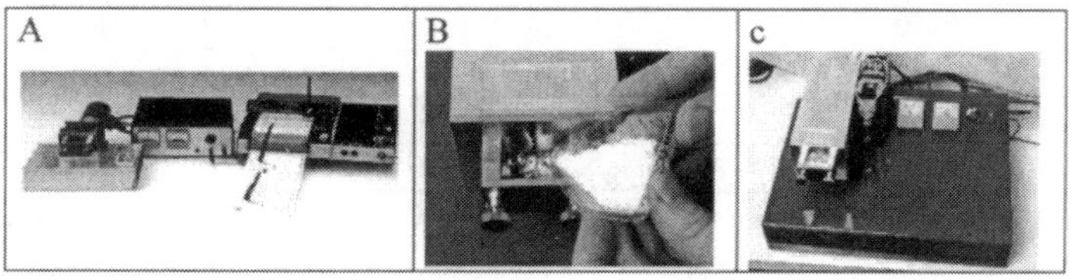

2.3. Methods

Standard Farinograph and Valorigraph tests were used for determination of water absorption values of flours and mixing properties of dough[5,6]. For micro-scale testing, both Micro Z-arm mixers were used. Mixing trials were performed using water absorption values obtained with the conventional Valorigraph test. The evaluation procedure of micro-scale curves was similar to the standard procedure.

3 RESULTS AND DISCUSSION

The first comparative study was carried out with a 50g Valorigraph and with the prototype small-scale Z-arm mixer equipped with a mechanical registration unit. The water absorption value, Dough Development Time (DDT), Dough Resistance (DR) and Dough Softening Value (DSV) were calculated and compared (Table 2). The similarity in the mixing curves (Figure 2) and the significant correlations between appropriate values obtained with micro- and macro- procedures (Table 2) indicate that the small-scale method could be used as an alternative for testing flours.

The relationship between the amount of added water and the Maximum Dough Resistance (MDR) was studied using the computerized Micro Z-arm Mixer. The linear dependence of MDR values on water content (Figure 3a) provides a good basis to estimate the "correct" water absorption requiring only two simple mixing trials with about 5% difference in added water. (Figure 3b).

It was found that the Valorigraph water absorption values were somewhat higher than those obtained with the Farinograph using the same flours. However, the very strong correlation between Valorigraph and Farinograph values (r^2= 0.96) allows the water absorption values to be converted.

Table 2. *Quality parameters of wheat flour samples measured with the Valorigraph (50g sample) and with Micro Z-arm mixer (4g sample)*

Samples	'Macro-Valorigraph'				'Micro Z-arm mixer'		
	Water Abs. (%)	DDT (min)	DRV (min)	D SV (mVU)	DDT (min)	DRV (min)	D SV (mVU)
Aestivums							
MV 17	56.0	2.5	5.0	110	1.5	3.5	8.0
MV 25	60.4	3.5	4.5	100	1.5	4.0	4.0
MV Irma	60.2	2.0	5.0	90	1.5	3.5	6.0
MvMezőföld	64.2	3.0	3.0	135	2.0	2.0	11.0
Mv Magvas	66.6	2.5	5.5	60	2.0	4.0	5.0
Durums							
MVTD1299	70.0	2.0	3.0	175	2.0	2.5	7.0
Odmadur 2	62.9	1.5	1.5	240	1.0	1.5	10
Correlation with micro test (r^2=)		0.65	0.34	0.98	-	-	-

Figure 2. *Comparison of mixing curves measured with Valorigraph (50g sample) and with Micro Z-arm mixer (4g sample)*

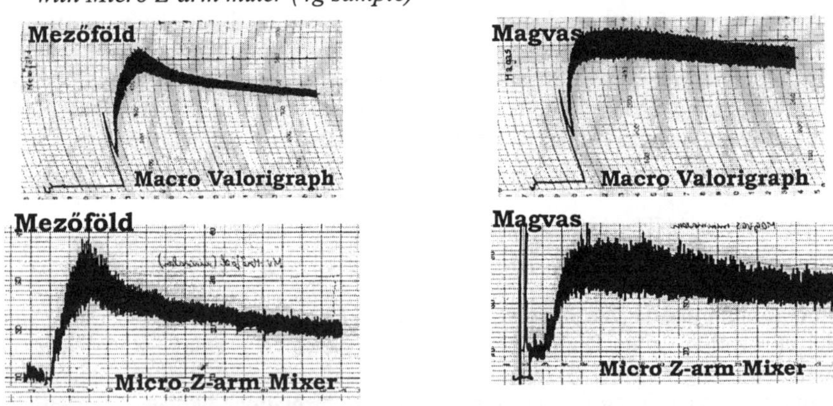

Figure 3. *Estimation of water absorption with Micro Z-arm mixer*

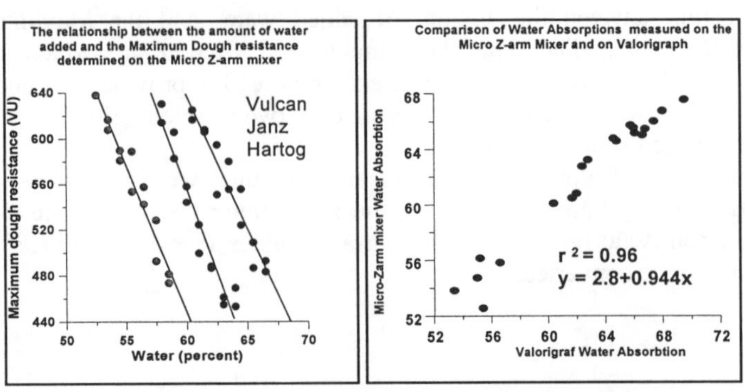

The good sensitivity and the small sample size make it possible to use the small-scale mixer as a research tool. As an example, the effects of protein content and protein composition on the mixing properties and water absorption can be studied (Figure 4) by systematic alteration of flour composition using isolated flour components. Results show that the micro-scale system is sensitive enough to detect the small changes in water absorption and in dough development time.

Figure 4 *The effects of protein content on Water Absorption and Dough Development Time measured on the Micro Z-arm Mixer*

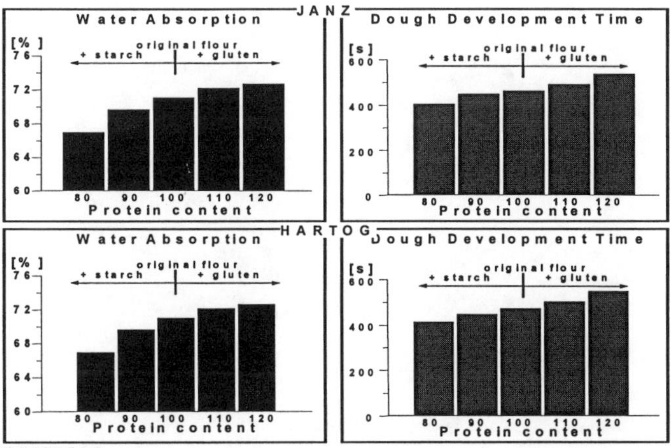

4 CONCLUSION

The recently developed Micro Z-arm mixer is able to provide rheological information on as little as 4g flour. The mixing curves from the micro test closely resemble those obtained with the conventional Valorigraph or Farinograph procedures. The utilisation of the new instrument allows the selection of new wheat varieties for mixing properties and water absorption at least one generation earlier in wheat breeding programmes. The Micro Z-arm Mixer can also be used as a research tool for investigating structure/function relationships in flour and/or the effects of different ingredients on rheological parameters.

References

1. C. Rath, P.W.Gras, C.W. Wrigley and C.E. Walker, *Cereal Foods World*, 1990, **35,** 572

2. M. J. Sissons, J.M. Skerit and F. Békés, 'Purification of low molecular weight glutenin subunits and role in dough functionality', In.:C.W. Wrigley (ed.), Proc Sixth Gluten Workshop, Royal Australian Chem. Ins., Melbourne, 1996, 485.

3. P. W. Gras and L.O'Brien, *Cereal Chem.*, 1992, **69**, 254

4. D. Fodor, J. Varga, S. Tömösközi, A. Salgó, J. Nánási and Gy. Veres, 'Procedure and instrument for small-scale milling', Patent, 1999. (under evaluation).

5. International Standard, 'Determination of water absorption and rheological properties using a farinograph', ISO 5530-1, 1988, Part 1

6. International Standard, 'Determination of water absorption and rheological properties using a valorigraph', ISO 5530-3, 1988, Part 3

Acknowledgement

This research work was supported by National Committee for Technological Development , Hungary (Project No.:96-97-68-1354)

EFFECTS OF PROTEIN QUALITY AND PROTEIN CONTENT ON THE CHARACTERISTICS OF HEARTH BREAD

E.M. Færgestad[1], P.Baardseth[1], F.Bjerke[1], E.L.Molteberg[2], A.K.Uhlen[3], K.Tronsmo[1], A.Aamodt[1] and E.M.Magnus[1]

1 MATFORSK, Norwegian Food Research Institute, Oslov. 1, 1430 Ås, Norway. 2 PLANTEFORSK, The Norwegian Crop Research Institute, Apelsvoll Research Centre, 2849 Kapp, Norway. 3 Dept of Horticulture and Crop Science, Norwegian Agricultural University, Box 5022, 1432 Ås, Norway

1 INTRODUCTION

Effects of flour quality on hearth bread are likely to differ from its well known effects on pan bread, as there is no side wall support for the dough during proving and baking in the oven. Therefore, the ability of the dough to retain a proper form during proving and baking is critical. Furthermore, the volume of hearth loaves is a more complex property than the volume of pan loaves, as it is a function of expansion of the dough in three dimensions; the height, the width and the length.

Hearth bread is commonly made in many countries. However, in the scientific literature on wheat functionality, little attention has been paid to hearth bread. As hearth bread is more complicated than pan bread, exploration of the influence of flour quality on hearth loaves may give additional insight to the fundamental properties of the gluten proteins.

We have carried out a number of baking experiments, using different materials and baking processes, to explore how wheat flour quality affect the hearth bread characteristics, and to understand the underlying mechanisms. The aim of the present report is to present effects of protein quality and protein content on the characteristics of hearth bread as investigated using a small scale laboratory baking experiment and a commercial scale baking experiment. For the commercial scale experiment we aimed to investigate the effects of changing only the flour quality, with the recipe and the process parameters being the same as normally used at that bakery.

2 MATERIALS AND METHODS

The materials tested consisted of two sets of wheat samples:

1. 17 wheat samples; 10 Norwegian grown varieties, where 7 of the varieties were selected at 2 protein levels. The samples were milled on a laboratory mill.
2. 6 wheat samples of the same varieties as above selected at large quantities (30 ton) to be milled at commercial scale.

Three ppm of ascorbic acid was added to the flour.

The samples of both materials were selected to show variability in both protein quality and protein content. The first material represents orthogonal variability in protein quality vs. protein content as described by Færgestad et al.[1]. For the second material, the characteristics of the flours are presented in Table 1.

Table 1 *Flour quality parameters for the flours of material no 2*

Wheat variety	Protein Content (%)	HMW glutenin subunits	Water content (%)	Water absorption (ISO 5530-1)
Folke	10.6	2*, 2+12, 6+8	14.6	62.0
Folke	11.5	2*, 2+12, 6+8	15.5	60.8
Polkka	12.5	2*, 2+12, 7+8	13.0	57.8
Portal	11.2	1, 5+10, 7+9	14.5	62.8
Portal	12.1	1, 5+10, 7+9	14.2	66.7
Bastian	13.0	2*, 5+10, 7+9	13.1	60.5

The baking experiment of material no 1 was performed at laboratory scale as described by Færgestad et al.[1], whereas the baking of material no 2 was performed at a commercial bakery named Nøtterøy Bakeri and Konditori AS. The recipe on a flour basis was 2.6% margarine (Paltex, AS Pals, Oslo, Norway), 2% S-5000 (Puratos, Belgium; containing wheat flour, emulsifier E472e, soy flour, sugar and ascorbic acid), 1.88% salt, 2.6% bakers yeast (Idun Industri A.S, Oslo, Norway) and 52% water. Mixing time was fixed for all flours; 5 min at low speed and 5 min at high speed in a spiral mixer (Glimex, Sweden) and dough temperature after mixing was 27-29 °C. The doughs were rounded and moulded on a rounding and moulding table (Glimex, Sweden). Proving was performed at 37 °C and 76 % RH in a proving cabinet (Lillnor A/S, Odder, Denmark). The proving time was set individually for each dough according to the bakers' judgement during proving as this is the normal procedure of the bakery. The proving times were 55 min for the two samples of Portal, 43 min for Bastian, 35 min for the two samples of Folke and 30 min for Polkka. The loaves were baked 30 min in a rotating hearth oven equipped with a fan (Danbak, Denmark). Live steam was injected during the first 35 s of baking, and the temperature was reduced from 280 to 230 °C immediately after the loaves were put in the oven. The baking experiments were performed in random order.

Loaf volume was measured by rapeseed displacement, and height and width of loaves was measured using a PAV-caliper (ABC Maskin AS, Skien, Norway). Form ratio was calculated as the ratio loaf height/loaf width.

3 RESULTS AND DISCUSSION

For hearth loaves the form ratio is a critical product characteristic. When using fixed proving time, dough resistance as measured by extensogram maximum height correlates positively with the form ratio (the height/width relation) of the loaves, whereas dough extensibility correlates negatively (Table 2). The positive relation between protein quality and form ratio for hearth loaves baked at fixed proving time agrees with the results from baking experiments performed in our laboratory on blends of wheat[2].

Table 2 *Simple correlations between protein content, dough extensibility and dough resistance on one side and hearth bread characteristics on the other for 17 wheat samples baked at laboratory scale. Mixing time was optimized whereas proving time was fixed.*

Wheat material		Height	Width	Height/ Width	Volume
17 wheat samples baked at small scale at MATFORSK, fixed proving time	**Protein %**	-0.34	**0.55***	-0.48	**0.51***
	Extensibility	-0.42	**0.75*****	**-0.62***	**0.72****
	Resistance	**0.77 *****	-0.39	**0.68****	0.19

*** = $p < 0.001$, ** = $p < 0.01$, * = $p < 0.05$

When proving is adjusted according to the individual flour, an important aspect would be to obtain acceptable form ratio. In general, there is a negative relationship between proving time and form ratio. To obtain similar form ratio with flours of variable protein quality, dough made from flours of strong protein quality should be proved longer than those made from flours of weak protein quality[1]. The proving time should also be adjusted according to the extensibility of the dough to avoid excessive flow of the dough during proving and oven spring.

In the baking experiment performed at commercial scale, dough of flours of strong protein quality behaved similarly to the commercial flour normally used, and the loaves had high quality. In contrast, flours of weak protein quality gave poor dough handling properties and poor loaf characteristics. In particular, form ratio was low for flour of weak protein quality (Figure 1).

Figure 1 *Protein content vs. form ratio for 6 wheat flours of variable protein quality as investigated using baking experiment at commercial scale. The samples are grouped according to the presenceHMW glutenin subunits 5+10 and 2+12, respectively.*

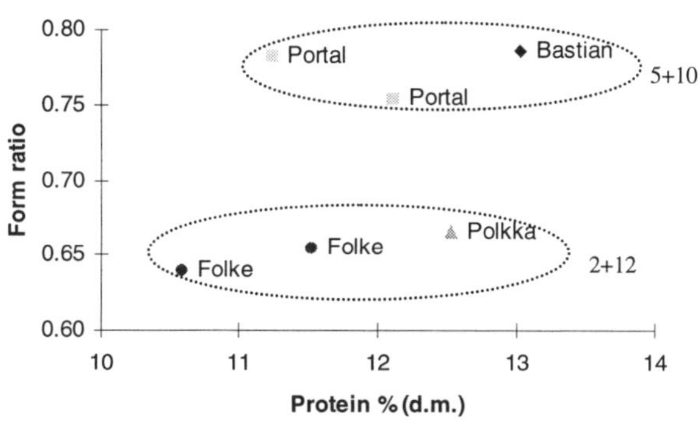

As shown in Figure 1, the form ratio is affected by protein quality but not protein content. Thus, although proving time was adjusted for each flour in this experiment, the form ratio was higher for loaves of the strongest protein quality. This shows that doughs made from flours of the weakest protein quality exhibited more flow during oven spring than expected by the baker.

4 CONCLUSIONS

The baking experiments both at small laboratory scale and at commercial scale have shown that the form ratio, an important characteristic of hearth loaves, is positively affected by protein quality, whereas protein content has no positive effect.

To optimise the baking process one must adjust the proving time according to the individual flour to achieve the desired form ratio. Doughs of strong protein quality should then be proved longer than doughs of weak protein quality to achieve similar form ratio. However, the optimisation is difficult as doughs of weak protein quality flours flow more during oven spring than doughs of strong protein quality flours.

To explore the fundamental properties of wheat dough, hearth loaves appears to be a better experimental system than pan loaves as the effects of protein quality and protein quantity can be better distinguished. The protein quality contributes to the resistance of the dough and thereby prevents the dough from excessive flow during proving and oven spring. A major challenge is to understand how dough resistance and extensibility, governed by the protein quality and quantity, interact when using different baking processes.

References

1. E.M.Færgestad, E.L.Molteberg and E.M.Magnus, *J. Cereal Sci.*, 2000, **31**, in press.
2. T. Næs, F. Bjærke and E.M.Færgestad, *Food Qual. Pref.,* 1999, **10**, 209.

RELATIONSHIPS OF SOME FUNCTIONAL PROPERTIES OF GLUTEN AND BAKING QUALITY

E. M. Magnus, K. Tronsmo, Å. Longva and E. M. Færgestad

MATFORSK - Norwegian Food Research Institute, Osloveien 1, N-1432 Ås, Norway

1 INTRODUCTION

The objective of the study was to determine if analytical methods commonly used to characterise the functionality of proteins in food systems, such as water and oil absorption, foaming properties, and emulsifying properties, could provide information about the wheat storage proteins that could supplement sedimentation properties and physical dough properties in predicting baking properties. We also wished to determine if oil-or water-absorption, emulsifying or foaming properties of gluten could be related to the breadmaking quality of wheat flours and, thus, provide an additional criteria to distinguish between wheats of varying breadmaking quality.

2 MATERIALS AND METHODS

2.1. Materials

The wheat material consisted of nine wheat varieties varying in hardness and protein composition. For seven of the varieties, samples at two protein levels were included and for the remaining two varieties samples at one protein level only included. The wheat samples were milled on an experimental mill.

2.2. Methods

2.2.1 Sedimentation tests. Zeleny (ISO Method 5529) and SDS[1] sedimentation tests were carried out.

2.2.2 Glutenin macropolymer. The content of content of glutenin macropolymer[2] (GMP) was determined.

2.2.3 Physical dough testing. Physical dough testing by farinograph (ISO Method 5520-1), mixograph[1] and extensigraph (ISO Method 5520-2) was performed.

2.2.4 Experimental baking. Small-scale experimental baking were carried out. The experimental baking included pan loaves and hearth loaves, fixed and optimised mixing, and varying proofing time[8]. The bread loaves were objectively and subjectively evaluated.

2.2.5 Preparation of glutens. Glutens were isolated from the same flours by a handwashing procedure[3], and the isolated glutens were freeze dried, ground and the protein content analysed.

2.2.6 Functional properties. Simple testing of some of the functional properties of the isolated glutens was performed. The tests included oil absorption capacity[4] (OAC), water retention capacity[1](WRC), foaming properties[5], and emulsification capacity[6,7].

2.2.7 Statistical analyses. The data were analysed by Principal Component Analysis (PCA) using Unscrambler (Camo AS, Trondheim, Norway). Simple regression was carried out using the software package MINITAB (MINITAB Inc.)

3 RESULTS AND DISCUSSION

3.1. Relationships of physical dough properties and functional properties of glutens

The relationships between the physical properties of the doughs and the characteristics of glutens are shown in Figure 1. Mixograph peak time was correlated with OAC (r=0.73; p=0.001)) and WRC (r=0.72; p=0.002). No other close relationships were found between the physical properties of the doughs and the characteristics of glutens. The relationships between the physical properties of the doughs and the characteristics of the bread loaves obtained by different baking methods have been discussed elsewhere[8].

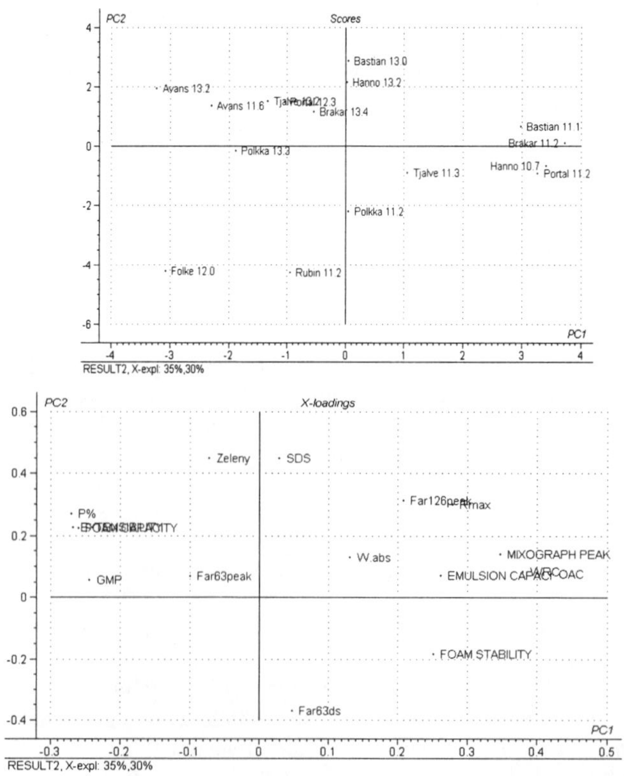

***Figure1** Score and loading plots for physical dough properties and gluten characteristics.*

3.2. Relationships of bread loaf properties and functional properties of glutens

The plots in Figures 2-4 show the relationships between bread loaf characteristics and sedimentation volumes, water and oil absorption, emusifying and foaming properties of glutens.

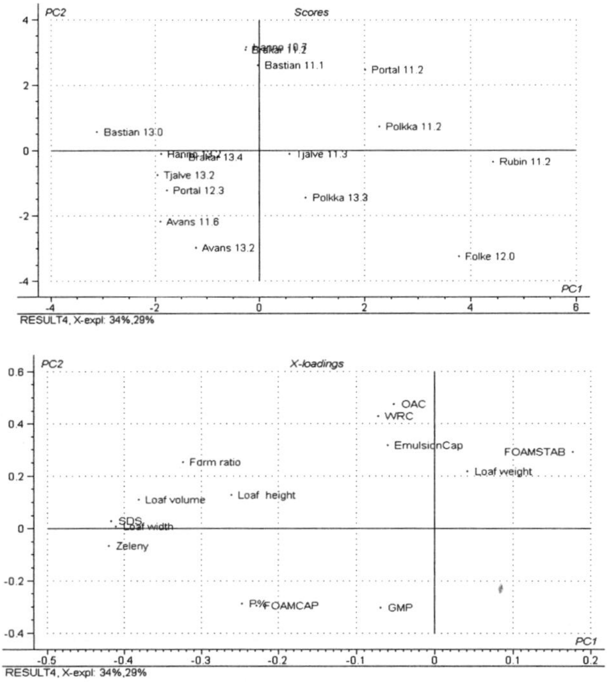

Figure 2 *Score and loading plots for gluten characteristics and properties of pan loaves.*

Simple correlations between protein characteristics and bread loaf properties are shown in Table 1. The volumes of pan loaves were most closely related to the SDS and Zeleny sedimentation volumes. For hearth loaves loaf volume and SDS and Zeleny sedimentation volumes were more strongly correlated when the doughs were mixed at low speed (r= 0.56 and 0.76) than when the doughs were mixed at high speed (r=0.45 and 0.60). The OAC of the glutens isolated from the flours were correlated with the form ratio of hearth loaves. A stronger postive relationship was found for hearth loaves obtained from doughs mixed at high speed than for loaves from doughs mixed at low speed. The content of GMP was poorly correlated to the characteristics obtained by all three baking baking procedures.

The stronger correlation between loaf volume and Zeleny sedimentation volume compared to SDS sedimentation volume could be due to protein content influencing both volume and Zeleny sedimentation volume. SDS and Zeleny sedimentation volumes were poorly correlated with the form ratio of hearth loaves from doughs mixed at high and low speed (results not shown). The form ratio of hearth loaves has been shown to be strongly dependent on the protein quality. The relationship between form ratio and OAC indicates a positive relationship between baking quality, expressed by form ratio, and protein quality. The findings showed that whereas the sedimentation volumes were relatively well

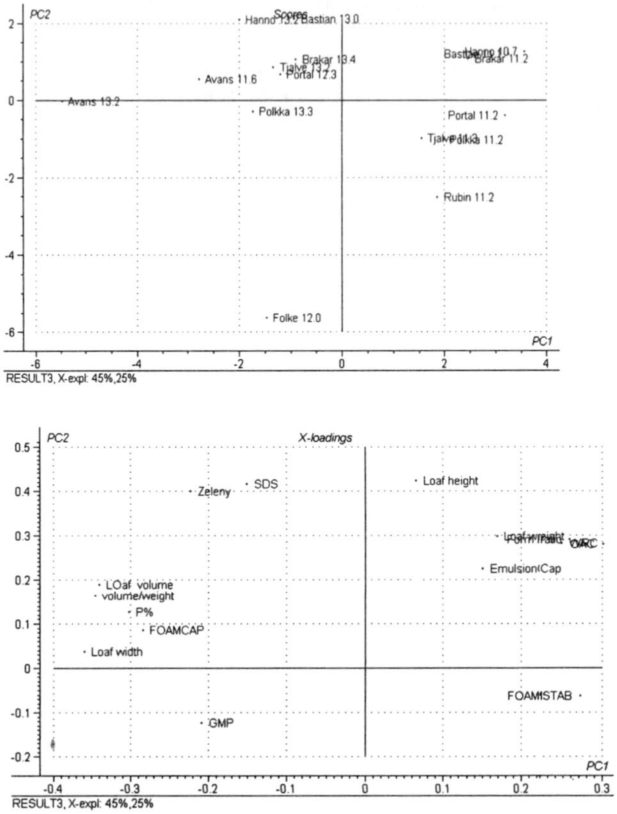

Figure 3 *Score and loading plots for gluten characteristics and properties of hearth loaves from doughs mixed at low speed.*

correlated with loaf volume, other tests might be more useful that the sedimentation tests to predict the form ratio of hearth loaves.

Table 1 *Simple correlations (r) between protein characteristics and bread loaf properties.*

	GMP and loaf volume	Zeleny sedimentation volume and loaf volume	SDS sedimentation volume and loaf volume	OAC and form ratio
Pan loaves	0.05	0.64***	0.64***	0.58*
Hearth loaves, slow mixing	0.32	0.76***	0.56**	0.74***
Hearth loaves, fast mixing	0.20	0.60**	0.45	0.84***

* p<0.05; ** p>0.01; *** p<0.001

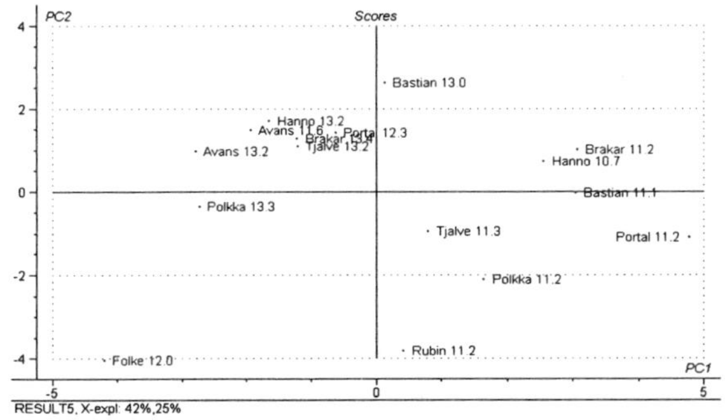

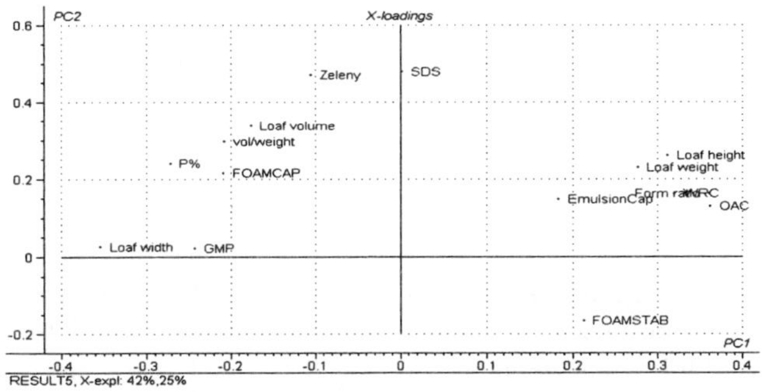

Figure 4 *Score and loading plots for gluten characteristics and properties of hearth loaves from doughs mixed at high speed.*

4 CONCLUSIONS

The results from the present study indicated that among the analytical methods commonly used to characterise the functionality of proteins in food systems, such as water and oil absorption, foaming properties, and emulsifying properties, oil absorption capacity appeared to provide information about the wheat storage proteins that could supplement sedimentation properties and physical dough properties in predicting baking properties.

References

1. American Association of Cereal Chemists. Method 54-40A, Method 56-20, Method 56-70.Approved Methods of the AACC, 9th ed.1995. The Association: St.Paul, MN.
2. A. Graveland, P. Bosveld, W. J. Lichtendonk, H. H. E. Moonen and A. Scheepstra, *J. Sci. Food Agric.,* 1982, **33,** 1117.
3. P. C. Dreese and R. C. Hoseney, *Cereal Chem.*, 1990, **67**, 400.
4. M.J.Y Lin, E. S. Humbert and F. W. Sosulski, *J. Food Sci.*, 1974, **39**, 368.
5. B. Mohanty, D. M. Mulvihill and P. F. Fox, *Food Chem.*, 1988, **28**, 17.
6. O. J. Cotterill, J. Glauert and H. J. Bassett, *Poultry Sci.*, 1976, **55**, 544.
7. S. E. Hill, Emulsions. In *Methods of Testing Protein Functionality*, G.M. Hall (ed). Blackie Academic Professional, London, 1996. Chapter 6, p 153.
8. E. M. Færgestad, E.L. Molteberg and E. M. Magnus, *J. Cereal Sci.* 2000, (in press).

THERMAL PROPERTIES OF GLUTEN AND GLUTEN FRACTIONS OF TWO SOFT WHEAT VARIETIES

Falcão-Rodrigues, M.M., Beirão-da-Costa, M.L.

Instituto Superior de Agronomia, Lisboa, Portugal

1 INTRODUCTION

The wheat endosperm is composed by four principal components: gluten (10-12%), starch (75-80%), lipids (1-2%) and a water-soluble fraction (4-5%). All of these components are important for the functional properties of the batters used in the breadmaking process, with the gluten and the starch forming the matrix bases for analysis of the functional properties of multicomponent systems. The choice of gluten as a matrix base is justified by its role in giving viscoelastic batters appropriate expansibility, gas retention and texture.

In this study differential scanning calorimetry (DSC) was used to evaluate gluten and its protein fractions. This study was conducted at different water contents and on two varieties of wheat showing different baking properties.

2. MATERIALS AND METHODS

2.1 Materials

Gluten, gliadins and glutenins (soluble and insoluble in acetic acid solution) were extracted from two varieties of wheat (Amazonas and Sorraia), selected by Estação Nacional de melhoramento de Plantas de Elvas (Portugal).

2.2 Methods

Trials were performed on a Shimadzu DSC using a TA 50 SI thermal analyser. For each trial about 7 mg of sample was used. Gluten thermograms were obtained at a programme rate of 10ºC/min. Gluten was analysed at an aw=0.05. Gliadin and glutenin thermograms were obtained at a programme rate of 2ºC/min.

Gluten fractions were analysed at an aw=0.79. An empty cell was used as reference. All samples were milled in a Falling Number 3100 mill to pass a 0.5 mm sieve. All the results are the average of at least four runs.

Thermal transition was defined in terms of onset temperature (expressed as Tm) and peak of denaturation (Td). Heat of transition or enthalpy, ΔH ($mJ.g^{-1}$) was evaluated from peak areas.

The denaturation kinetics were studied by DSC using the heat evolution method of Borchardt and Daniels (1). The method assumes that the reaction obeys the relationship:

$$d\alpha/dt = k(1-\alpha)^n$$

Where $d\alpha/dt$ = reaction rate; K = rate constant (sec^{-1}); n = reaction order.

The reaction rate $d\alpha/dt$ is obtained by dividing the peak height at a temperature T by the total area, and the fraction unreacted ($1-\alpha$) obtained by measuring the ratio of the partial area at temperature T to the total peak area, and then subtracting this value from unity. The method also assumes that the dependence of the reaction rate follows the Arrhenius expression:

$$V = v_0\, e^{Ea/RT}$$

Where $v = d\alpha/dt$; v_0 = constant rate; Ea = activation energy (J/mol); R = gas constant; and T = absolute temperature. Width at half-peak height ($\Delta T_{1/2}$) was also recorded, as a measure of cooperativity of the denaturation phenomenon.

2.3 Statistical Analysis

Principal Component Analysis (PCA) and Cluster Analysis were applied to study the relationship between gluten, gliadins and glutenins from the two varieties of soft wheat, and also to study their denaturation characteristics using the software "Statistica™", version 5, from Statsoft, USA.

3. RESULTS AND DISCUSSION

3.1 DSC analysis of gluten

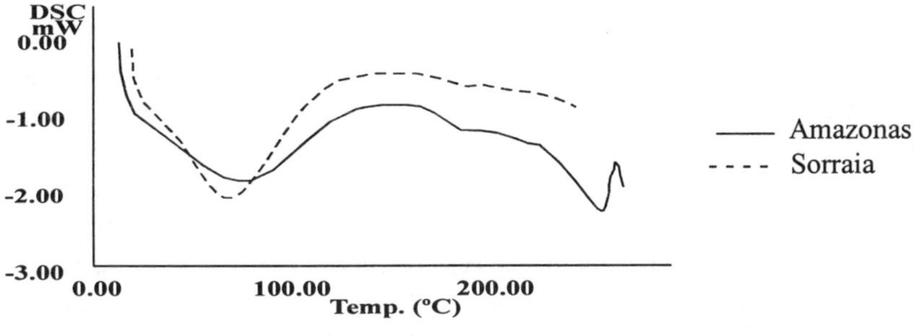

Figure 1 *Gluten thermograms from Amazona and Sorraia wheat at a heating rate of 10°C/min.*

Three distinguishable transitions were observed in gluten from the two soft wheat varieties. The average temperatures and the average enthalpy of the three transitions were estimated as shown in Figure 2.

The gluten from Sorraia wheat behaves stonger rheologically than that from Amazonas and this characteristic appears to be associated with differences in the second endotherm. With respect to denaturation enthalpy, the two first endotherms of Sorraia wheat show higher enthalpic values and consequently a stronger relationship with viscoelasticity.

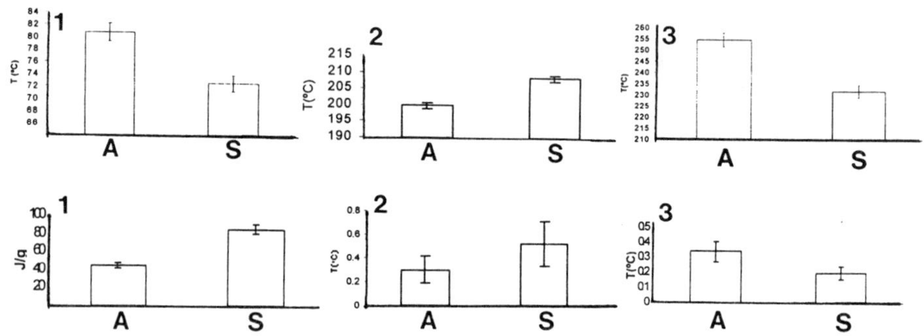

Figure 2 *Average temperatures of transitions (top) and average enthalpy of transitions (bottom) of glutens from* cvs *Amazonas (A) and Sorraia (S). 1, 2 and 3 are the* 1st, 2nd *and* 3rd *endotherms, respectively.*

3.1.1. Thermal denaturation kinetics of Amazonas and Sorraia wheat gluten. The $\Delta T_{1/2}$ values of glutens from Amazonas and Sorraia wheat are shown in Table 1.

Table 1 - Values of $\Delta T_{1/2}$ (°C)

Gluten	Average $\Delta T_{1/2}$ of 1st Endotherm	Average $\Delta T_{1/2}$ of 2nd Endotherm	Average $\Delta T_{1/2}$ of 3rd Endotherm
Amazonas	56	8.8	10
Sorraia	49	6.7	14

The denaturation of a protein is a cooperative phenomenon which is accompanied by a significant increase of heat, seen as an endothermic peak in the DSC thermogram (2). The cooperativity of protein unfolding has been studied (3). It is suggested the use of $\Delta T_{1/2}$ (width at half peak height) can be used to measure this cooperativity. A low $\Delta T_{1/2}$ is considered to be indicative of a highly cooperative transition. The progressive sharpening of the endotherm peak, demonstrated by the increase in Tm and decrease in $\Delta T_{1/2}$, suggests that the protein would denature in a highly cooperative way. Analysing the $\Delta T_{1/2}$ values of gluten, it can be concluded that the two varieties are similar in respect to their denaturation cooperativity, with the third endotherm requiring more energy for denaturation. These results are in agreement with the activation energy values presented in Table 2 (with the highest values being for the third endothermic transition). The denaturation kinetic constants of gluten are presented in Table 2.

Table 2 - Denaturation kinetic constants of gluten

Gluten	Activation Energy 1st end (J/molK^{-1})	Activation Energy 2nd end (J/molK^{-1})	Activation Energy 3rd end (J/molK^{-1})	Reaction Order (n) 1st end	Reaction Order (n) 2nd end	Reaction Order (n) 3rd end
Amazonas	1373	42528	68836	1.2	0.8	1.2
Sorraia	888	64655	72180	0.7	0.8	1.4

The Sorraia gluten shows high activation energies for the 2nd and 3rd endotherms and higher cooperativity ($\Delta T_{1/2}$) of denaturation. As the 2nd endotherm show high activation energy and low cooperativity, it seems that this endotherm is most closely related to baking quality.

3.2. DSC analysis of gliadin and glutenin fractions

Thermograms for the different gluten fractions are shown in Figure 3.

Comparison of the average temperature and average enthalpy of the first, second and third endotherms (not shown) shows differences between the protein fractions from the two cultivars. The principal difference is that gliadin from Amazonas wheat has only two endothermic transitions. The average temperatures of the three endotherms are similar for all the fractions studied with the exception of the acid-soluble glutenins from Amazonas wheat (which showed higher temperature values). With respect to the average enthalpy of the first endotherm, three groups can be recognised: the gliadins from Sorraia wheat, the gliadins from Amazonas wheat and the glutenins from Amazonas and Sorraia wheats. Analysing the second endotherm shows differences between the acid-insoluble glutenins and the gliadins from Amazonas wheat and the other fractions. With respect to the third endotherm, the acid-soluble glutenins from Amazonas wheat, the acid-insoluble glutenins and the gliadins from Sorraia wheat needed more energy for denaturation.

Principal Component Analysis (PCA) (Figure 4) and Cluster Analysis (Figure 5) were applied to study the relationships between the gluten fractions from the two different varieties of soft wheat and the denaturation characteristics.

Analysing the eigen values and respective variances of the principal components it can be concluded that the first two components account for 88.5% of the total variability. On the basis of the loadings of variables on the first and second principal components it can be concluded that the Td values from both the first and second endotherms are related to the second principal component and the enthalpy to the first principal component.

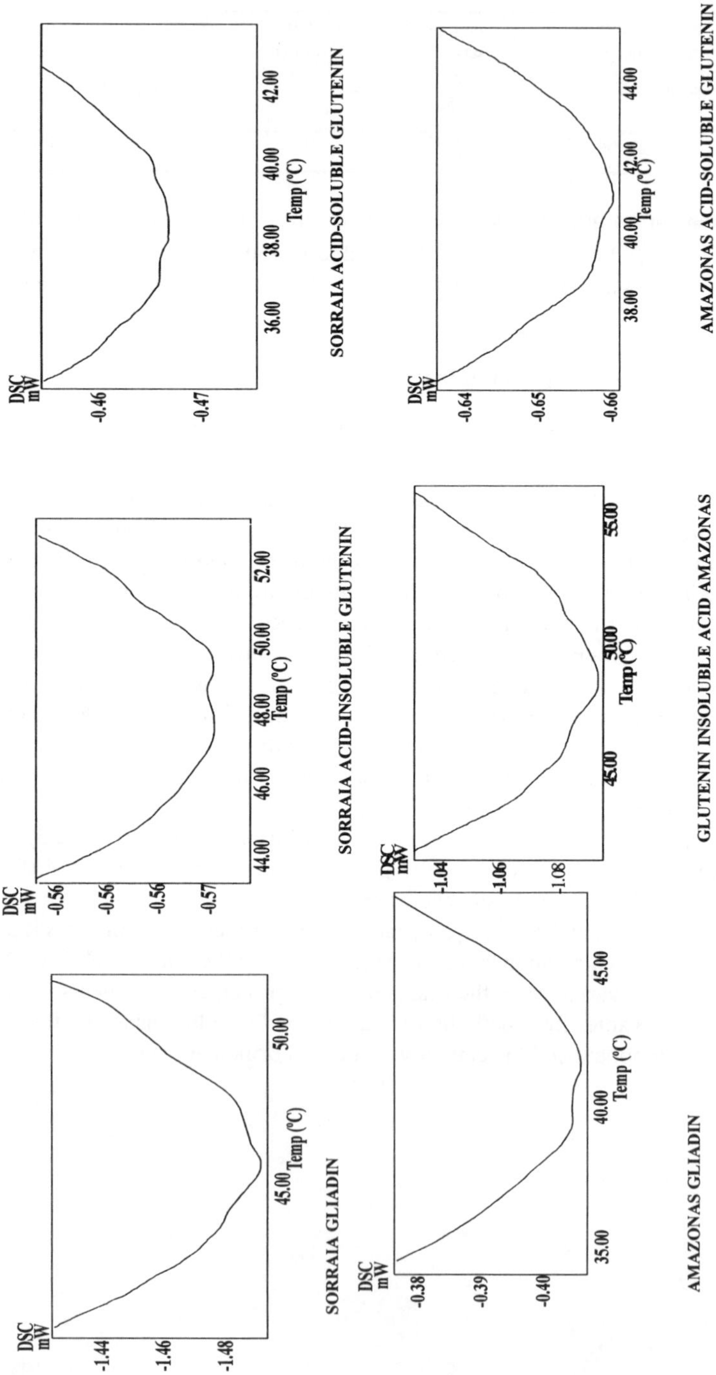

Figure 3 *Gluten fractions thermograms at a heating rate of 2°C/min.*

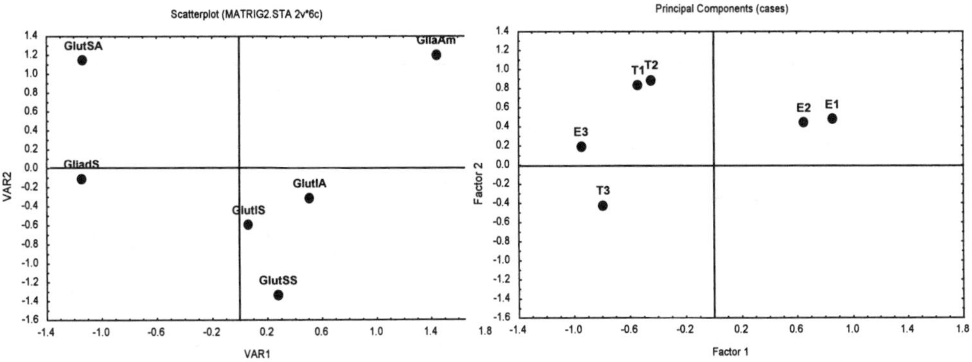

Figure 4 *Loadings of variables on the first and second principal components.*

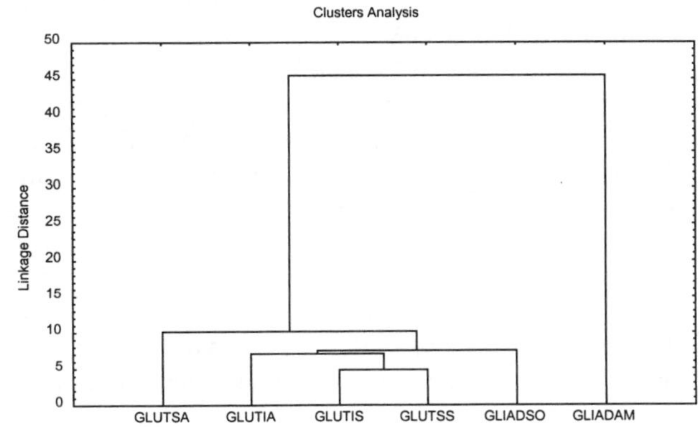

Figure 5 *Clusters analysis of gluten fractions*

A sharp separation exists between the gliadin from Amazonas wheat and all the other fractions. In addition, the gliadin from Sorraia wheat forms a separate group from the other fractions.

3.2.1. Thermal denaturation kinetics of gluten fractions. The $\Delta T_{1/2}$ values of the gluten fractions are shown in Table 3 and their denaturation kinetic constants in Table 4.

Table 3 - Values of $\Delta T_{1/2}$ (°C)

Gluten Fractions	Average $\Delta T_{1/2}$ of 1st Endotherm	Average $\Delta T_{1/2}$ of 2nd Endotherm	Average $\Delta T_{1/2}$ of 3rd Endotherm
Amazonas Gliadins	1.4	1.25	
Sorraia Gliadins	0.88	0.94	1.05
Amazonas acid-soluble Glutenins	1.14	1.285	0.636
Sorraia acid-soluble Glutenins	0.74	0.61	0.5
Amazonas acid-insoluble Glutenins	1	1.6	0.05
Sorraia acid-insoluble Glutenins	0.7	0.14	0.9

Table 4 - Denaturation kinetic constants of gluten fractions

Gluten Fractions	Act. Energy 1st end (J/molK^{-1})	Act. Energy 2nd end (J/molK^{-1})	Act. Energy 3rd end (J/molK^{-1})	Reaction Order 1st end	Reaction Order 2nd end	Reaction Order 3rd end
Amazonas Gliadins	772.9	1083.8		1.16	0.98	
Sorraia Gliadins	1234.4	2732.7	2349.1	0.56	0.98	0.74
Acid-soluble Amazonas Glutenins	1937.3	913.8	1656.8	1.91	1.29	0.77
Acid-soluble Sorraia Glutenins	1182.9	1383.2	1326.2	0.88	0.89	0.6
Acid-insoluble Amazonas Glutenins	1261.3	1157.6	2374.1	1.04	0.76	0.58
Acid-insoluble Sorraia Glutenins	1163.4	950.9	697.4	2.17	0.41	0.74

The gluten endotherms are completely different from those of the gluten fractions. Gluten shows three endotherms with high denaturation temperatures, which means that the gluten fractions interact giving high thermal resistence to gluten.

The gliadins from Sorraia wheat show three endotherms while those from Amazonas show two. The two endotherms from Amazonas have higher enthalpic values and lower activation energy values than those of Sorraia. This means that the major difference is in the third endotherm. The Sorraia wheat shows high cooperativity of denaturation.

4. CONCLUSIONS

The 2nd endotherm of gluten seems to be most closely related to the thermal properties. The absence of the third endotherm in Amazonas gliadins marks the difference between the gluten fractions.

References

1. A.N. Danielenko *et al*, Studies on the stability of 11S globulin from soybeans by differential scanning microcalorimetry. *Int. J. Biol. Macromol.* **7**: 109-112 (1985).

2. D.J. Wright, Thermoanalytical methods in food research. In: *Biophysical Methods in Food Research.* Chan. H W S, ed Blackwell Scientific, Oxford (1984).

3. P.L. Privalov, N.N. Khechinashvili and B.P. Atanasov Thermodynamic analysis of thermal transitions in globular proteins. I. Calorimetric study of chymotrypsinogen, ribonuclease and myoglobulin. *Biopolymers* **10**: 1865 (1971).

USE OF RECONSTITUTION TECHNIQUES TO STUDY THE FUNCTIONALITY OF GLUTEN PROTEINS ON DURUM WHEAT PASTA QUALITY

M. Sissons[1,2] and C. Gianibelli[3]

1. NSW Agriculture, RMB 944 Calala Lane, Tamworth NSW 2340 Australia 2. Quality Wheat CRC, North Ryde NSW 1670 3. CSIRO Plant Industry, Grain Quality Research Laboratory, North Ryde NSW 1670

1 INTRODUCTION

Differences in pasta cooking quality are believed to depend on both protein quantity and quality[1,2]. However, this relationship is complex and these factors alone do not explain all the variation in cooking quality. Although there have been several studies describing the relationship between protein content and gluten strength on pasta textural properties, all have relied on a statistical approach. Usually different varieties are grown at one or more locations/years, quite often bulk samples are used which decreases the variation in the cooking quality measurement and when evaluating the effect of gluten strength, protein content typically varies between samples. Fractionation and reconstitution allows the functionality of components to be assessed under controlled conditions. These procedures have been used in leavened and Arabic bread work[3,4]. By comparison, there have been few attempts to use this approach in durum wheat[5]. This work explores the relationship between gluten strength, protein content and pasta quality using reconstituted flours.

2 MATERIALS AND METHODS

2.1 Isolation of components and reconstitution of flours.

Bulk quantities of four fractions from commercial semolina (gluten, starch, soluble proteins and a residue remaining after filtration of the starch and soluble proteins) were isolated[3]. Gluten was also isolated from selected hexaploid and tetraploid wheat. Fractions were freeze-dried, ground and a bulk of the tailings, starch and soluble proteins prepared to their original proportions. Using different glutens, all reconstituted flours were recombined to have approximately the same protein content (11.99±0.22%). To investigate the effect of varying the total protein, reconstituted flours were prepared from the commercial semolina (RGF) with protein in the range 9.2-20.3%.

2.2 Preparation and evaluation of pasta.

Spaghetti was prepared from the reconstituted samples using a micro-scale extruder (variation of AACC method 66-42[6]) and then dried under controlled temperature and humidity conditions. For each sample, three replicate batches of pasta were made. The optimum cooking time was determined for each sample, then tested for firmness (AACC method 16-50[6]) and stickiness[7] using a TA.XT2 texture analyser. Parameters for firmness obtained are: maximum cutting force = peak height, total work to cut = area under the curve. For stickiness: peak force to separate probe from the sample surface = peak height; work of adhesion = area. Cooking loss (CL%) was measured by the method of Matsuo[8]

2.3 Analytical methods.

Mixing tests were carried out with a 2-g Mixograph. Parameters recorded were time to peak dough resistance (or mixing time, MT) peak dough resistance (PR) and resistance breakdown (RBD). Extension test was performed with a prototype 2-g extension tester. Maximum resistance (R_{max}) and extension before rupture were calculated using specially written software[9].

3 RESULTS AND DISCUSSION

3.1 Evaluation of mico-scale pasta processing and reconstitution methods

The small-scale pasta processing method was evaluated against our macro-scale reference method. Optimum flour:water ratios, dough mixing times, extrusion pressure and holding period were established to ensure pasta made by either process had the same texture. The fractionation and reconstitution process should retain functionality. We assumed that if the mixing properties of the reconstituted flour (RGF) and the firmness, stickiness and CL% of the pasta prepared from flour were the same as the unreconstituted flour (UGF), then functionality was maintained. There were no significant differences in properties between RGF and UGF (Tables 1-3).

3.2 Effect of gluten source (strength) on pasta quality

Several different sources of gluten were used for this study to provide a range in dough strengths: Triller (Australian soft wheat); Sunbrook and Hartog (Australian prime hard wheats); Cadoux (Udon noodle wheat); Glenlea (extra strong Canadian wheat); Waxy wheat; Yallaroi and Kamilaroi (Australian durum wheats); 1349 (*Triticum persicum* Aus # 3549); 1348 (*Triticum polonicum* Aus # 22342A); 1351 (*Triticum fungicidum* Aus # A3917). There was a large variation in gluten strength as indicated by MDDT, PR and Rmax (Table 1). The soft wheat Triller, Waxy wheat and the tetraploids all had short mix times and low Rmax. The durum cultivars and Glenlea had the lowest RB whilst the other hexaploid wheats had the highest RB. Glenlea and the durum cultivars had the strongest gluten type. MDDT was correlated to Rmax (r^2 0.52) and strongly correlated to pasta firmness (r^2 0.90) and less so for PR (r^2 0.41). Rmax showed a weaker correlation with pasta firmness (r^2 0.55).

Pasta texture is affected by flour protein content, therefore it is necessary to keep protein constant to evaluate the effect of gluten type on pasta cooking quality. All the reconstituted samples had protein contents within 1% of each other (Table 2). For pasta firmness there were a number of differences between the samples with those from the weaker ?? gluten sources (Triller, Waxy wheat, 1349 and 1351) with lower pasta firmness. The RGF sample had comparable firmness to Yallaroi and Kamilaroi. Glenlea

Table 1 *Effect of interchanging gluten on the mixing properties*

Source of Gluten	MDDT (sec)	PR (au)	RBD (%)	Rmax (au)
UGF	476[a]	346[a]	5[a]	554[a]
RGF	467[a]	328[a]	4[a]	690[b]
Sunbrook	337[b]	414[b]	10[c]	611[a]
Triller	235[c]	356[a]	9[c]	447[d]
Hartog	515[d]	406[b]	8[c]	446[d]
Cadoux	459[a]	390[c]	8[c]	374[e]
Glenlea	838[e]	434[d]	3[a,b]	785[g]
Waxy	222[c]	334[a]	6[b,c]	254[h]
99/1349	300[f]	214[e]	4[a,b]	169[i]
99/1348	410[g]	310[f]	2[a]	386[d,e]
99/1351	215[c]	241[g]	5[b]	209[h]
Yallaroi	597[h]	326[a]	3[a]	871[i]
Kamilaroi	412[i]	328[a]	3[a]	695[b]

gluten produced the firmest pasta whilst the Waxy wheat gluten pasta had the lowest firmness. Only Cadoux, Glenlea and Waxy wheat produced pasta that was significantly more sticky than the RGF pasta (Table 2). The other bread wheat glutens produced pasta of similar stickiness to the durum gluten. There were no significant differences in the CL% between samples except for the Waxy wheat starch sample. The CL% of this sample was very low because of the almost complete absence of amylose in the starch. This is expected since the assay measures iodine binding, and amylose is the main substance released from the pasta during cooking.

Table 2 *Effect of interchanging gluten on pasta quality*

Gluten Source	Flour Protein (%)[1]	Firmness (peak height)	Stickiness (peak height)	CL%
UGF	11.8	389[a]	26.1[a]	6.7[a]
RGF	12.1	388[a]	27.8[a,c]	6.6[a]
Sunbrook	12.5	348[b,c]	27.1[a]	6.4[a]
Triller	12.1	332[b,c]	27.1[a]	6.4[a]
Hartog	12.1	379[a]	27.1[a]	6.1[a]
Cadoux	11.9	365[a]	32.2[b,d]	6.6[a]
Glenlea	12.0	431[d]	31.8[c,d]	6.7[a]
Waxy	11.9	277[e]	33.6[d]	6.1[a]
1349	11.9	324[c]	28.7[a,b,c]	6.5[a]
1348	11.9	363[a]	27.7[a,c]	6.4[a]
1351	11.5	305[c,e]	29.8[a,c,d]	6.3[a]
Yallaroi	12.1	390[a]	28.4[a,d]	6.3[a]
Kamilaroi	12.1	379[a]	27.8[a,c]	6.5[a]

[1] Protein corrected to moisture content of UGF

3.3 Effect of altering protein content on the pasta making quality

The protein content of the RGF flour was set as 100%. Flour protein was adjusted to cover the range of 68 to 152%, equivalent to 9-20% protein (Table 3). There was a linear increase in pasta firmness with increasing protein (r^2 =0.87). Protein explained 87% of the variation in firmness. This relationship is much stronger than found in studies where protein composition differences confound the result. Resilience increased with protein but the data were more variable, and reached a plateau at 120%. Pasta stickiness decreased with increasing protein content (r^2 =0.48) but clearly other factors are involved. There was no significant difference in stickiness above a protein content of 130% (17.4% protein). It is interesting to note that the very high protein flour (152%) had the greatest firmness, least stickiness and CL% compared to pastas made from different gluten types. The 17% protein flour was also superior, suggesting that to obtain the highest quality, grain protein levels of 18% are desirable with good gluten strength (mainly controlled by glutenin composition).

Table 3 *Effect of changing total protein content on pasta making quality.*

Relative to RGF	Protein (%)	Firmness (Peak height)	Firmness (peak area)	Stickiness (Area 3:4)	CL%
69%	9.20	302 ^{a}	139 ^{a}	8.6 ^{a}	6.1 ^{a}
81%	10.88	328 ^{a}	150 ^{a}	7.8 ^{a,b}	5.9 ^{a,b}
91%	12.24	380 ^{b,c}	188 ^{b}	7.3 ^{b,c}	5.8 ^{a,b}
100% (RGF)	13.39	357 ^{b}	180 ^{b}	7.1 ^{b,c,d}	5.9 ^{a,b}
110%	14.74	395 ^{b,c,d}	198 ^{b,c}	7.4 ^{b,c}	6.0 ^{a,b}
120%	16.10	416 ^{c,d}	224 ^{c,d}	6.2 ^{c,d,e}	5.8 ^{a,b}
130%	17.39	440 ^{d,e}	250 ^{d,e}	6.0 ^{c,d,e}	5.7 ^{b}
143%	19.18	467 ^{e}	270 ^{e}	6.2 ^{d,e}	5.8 ^{b}
152%	20.35	534 ^{f}	316 ^{f}	5.9 ^{e}	5.8 ^{a,b}
UGF	12.15	373 ^{b,c}	185 ^{b}	7.4 ^{b}	6.0 ^{a}

4 CONCLUSIONS

Using reconstitution methods the relationship between gluten strength and protein content has been related to pasta texture under carefully controlled conditions. Stronger gluten flours were associated with firmer pasta but the range found was not large. Glenlea produced the firmest pasta. Most samples had similar stickiness and while there were significant differences, the range was narrow. Variation in protein content had a much bigger effect on pasta firmness and stickiness. Low protein produced pasta as poor in quality as the waxy wheat gluten. Protein above 17% produced very firm and low stickiness pasta. The data suggest breeding programs should attempt to increase grain protein once the gluten composition has been optimised.

References

1. J.C. Autran, J. Abecassis and P. Feillet, *Cereal Chem.*, 1996, **63,** 390.
2. M.G. D'Egidio, B.M. Mariani, S. Nardi, P. Novaro and R. Cubadda, *Cereal Chem.*, 1990, **67**, 275.
3. F. MacRitchie *J. Cereal Sci* 1985, **3**, 221.
4. I. Toufeili, B. Ismail, S. Shadarevian, R. Baalbaki, B.S. Khatkar, A.E. Bell and J.D. Schofield *J. Cereal Sci.* 1999, **30**, 255.
5. J.E Dexter and R.R. Matsuo, *Cereal Chem.* 1979, **56**, 190.
6. AACC, St. Paul, MN, USA,1995, Methods 16-50, 66-42.
7. J. Smewing *Cereal Foods World* 1997, **42,** 8.
8. R.B. Matsuo, L.J.Malcolmsom, N.M. Edwards and J.E. Dexter. *Cereal Chem.*, 1992, **69**, 27.
9. C. R. Rath, P. W.Gras, Z. Zhen, R. Appels, F. Bekes and C.W. Wrigley *Proceedings of the 44th Australian Cereal Chemistry Conference* 1994 p 122.

Acknowledgements

Funding for the work, as part of a Quality Wheat CRC project and GRDC (postdoctoral fellowship to C.G.) is gratefully acknowledged. We also thank Goodman Fielder Mills Ltd. Tamworth for the nitrogen analyses and semolina.

THERMAL PROPERTIES AND PROTEIN AGGREGATION OF NATIVE AND PROCESSED WHEAT GLUTEN AND ITS GLIADIN AND GLUTENIN ENRICHED FRACTIONS

V. Micard*, M.-H. Morel, J. Bonicel and S. Guilbert

U.F.R. Technologie des Céréales et des Agropolymères, AGRO.M/I.N.R.A.
2 Place P. Viala, 34060 Montpellier cedex 1, France

1 INTRODUCTION

Wheat gluten is a renewable and biodegradable material which, like other proteins, posesses thermoplastic properties. A wet process, leading to film formation ("casting"), and dry processes, using these thermoplastic properties (extrusion, thermomolding), can be used to make materials based on proteins[1]. The glass transition temperature (T_g) is generally governed primarily by the nature and structure of the proteins and can be classically depressed when the amount of plasticizer increases[1,2]. Wheat gluten undergoes glass transition[3] and some studies report glass transition of gliadin and glutenin[4,5].

At T_g, a change in heat capacity (ΔCp) of proteins occurs, which is related to the weight fraction affected by the transition[6]. It is generally admitted for synthetic polymers that T_g increases and ΔCp decreases when strong cross-linkages or others intermolecular links, restraining chain mobility of the polymers chains, occur[2,7,8]. An increase in T_g and a decrease in ΔCp values have been registered after heat polymerisation of wheat gluten[9]. While some studies demonstrated that the network creation, operating when temperature is applied to the polymer, modifies its T_g and ΔCp, no study has been attempted to compare the changes in glass transition parameters as a function of treatments applied to protein.

In the present study, thermal properties of native and processed gluten and its fractions were investigated using modulated differential scanning calorimetry (MDSC). An attempt to relate changes in thermal properties with composition of the fraction, or biochemical changes occurring in gluten network during process, is presented.

2 MATERIALS AND METHODS

2.1 Materials

Vital wheat gluten (76.5% of proteins) was provided by Amylum (Aalst, Belgium). Glutenin and gliadin enriched fractions (Table 1) were prepared at INRA Nantes (UBTP) from the same gluten following a fractionation procedure developped by Bérot *et al.*[10].

Table 1 *Composition of native gluten and its fractions determined by size exclusion chromatography (% of total protein)*

	Native gluten	Gliadin-rich	Glutenin-rich
Extracted by sonication	17.6	2.0	39.1
MM*>500,000	16.7	17.0	8.9
70,000<MM<500,000	16.1	23.0	12.5
15,000<MM<70,000	49.6	58.0	39.5

*MM : Molecular mass

2.2 Preparation of processed glutens

Casted gluten films were prepared according to Gontard *et al.*[11]. Gluten was mixed with water (30%) using a Plasticorder W 50 (Brabender, Germany) thermostated at 20°C, with a filling ratio of 80%. A specific mechanical energy of 15.2 kJ/g was provided to mixed gluten. Gluten (10g, preliminary equilibrated at 85% relative humidity) was molded with a thermomolder (Techmo PL 10T; 0-250 bars), during 2.5 min at 100 bars and 130°C.

2.3 DSC measurements

T_g and ΔCp measurements were performed on a T.A. instrument 2920 CE modulated DSC (New Castle, USA), with use of indium and aluminium oxide for calibration. Before DSC analysis, all the samples were equilibrated over P_2O_5. T_g was recorded on the first scan, from the inflexion point of the change in heat capacity on the reversing heat flow signal. A 5°C/min heating rate, 60 s period and 0.796°C amplitude were used for gluten and materials made from gluten, and a 2°C/min heating rate, a 60s period and 0.5°C amplitude for gluten, gliadin and glutenin rich fractions.

2.4 Total extracted proteins

Proteins were extracted 80 min with a sodium phosphate buffer that included sodium dodecyl sulphate (1%), sonicated and fractionated by SE-HPLC as described by Redl *et al.*[12].

3 RESULTS AND DISCUSSION

T_g and ΔCp of native gluten and its gliadin and glutenin-rich fractions are presented in Table 2. The gliadin-rich fraction has a T_g 10°C lower than the T_g of the native gluten and the glutenin-rich fraction. As gliadin-rich fraction still contains large-size extractable polymers (Table1), it gave a T_g slightly higher than the T_g of pure commercial gliadin[4].

An intermediary ΔCp value was obtained for native gluten (Table 2). As ΔCp decreased when aggregation or re-organization occurred[9,13,14], the low ΔCP obtained with glutenin-rich fraction is probably due to the high content of large-size extractable polymers, compared to the whole gluten and gliadin-rich fraction. The ΔCp of gliadin, dependent on the degree of compactness of the molecule, gave a value close to that found for the less compact gliadin (ω)[5]. It indicates a high mobility of the protein chains in this fraction across the glass transition.

Table 2 *Glass transition temperature and ΔCp of native gluten and its fractions*

Samples	T_g (°C)	ΔCp ($jg^{-1}°C^{-1}$)
Native gluten	173 ± 1	0.41 ± 0.03
Glutenin-rich fraction	175 ± 2	0.32 ± 0.02
Gliadin-rich fraction	162 ± 1	0.50 ± 0.08

T_g and ΔCp of native gluten and processed gluten are presented in Table 3. No clear relation between the process applied to gluten and evolution of T_g has been found, except that casting led to a lower T_g than dry processes. This led us to think that the film is a less reticulated network. Multiple range analysis (Duncan's test) of ΔCp distinguished two homogeneous groups constituted by, 1) the native and casted gluten, and 2) all the dry processed gluten samples. It clearly shows that the molecular mobility of the polymers in dry processed gluten is lower.

Biochemical changes in processed gluten samples (TEP, Table 3), showed that proteins in films with cross-linking agent (formaldehyde) or thermally treated proteins (thermomolding) were no longer extractable in sodium dodecyl sulphate. Thermomolding, characterized by a thermal treatment, would lead to glutenin polymerisation as classically reported with temperature treatment[15.16]. In contrast, mixing of gluten at low temperature but with intense energy input induced minor changes in protein extractability.

Table 3 *T_g, ΔCp and total extracted protein (TEP) of native and processed gluten*

Samples	Tg (°C)	ΔCp ($jg^{-1}°C^{-1}$)*	TEP (%)
Native gluten	172.4 ± 0.9	0.37 ± 0.02 a	97.8 ± 2.0
Film with formaldehyde	154.2 ± 2.5	0.47 ± 0.03 a	0.2 ± 0.3
Mixed gluten 15.2 kJ/g	167.8± 0.8	0.24 ± 0.02 b	92.5 ± 2.5
Thermomolded gluten	164.3 ± 0.2	0.21 ± 0.08 b	1.2 ± 0.3

*(Duncan's multiple range test, $p<0.05$)

4 CONCLUSION

T_g and ΔCp of gluten and its gliadin and glutenin-rich fractions gave results in agreement with biochemical analysis of proteins. In contrast, analysis of the gluten network resulting from different processes has led to different conclusions depending on whether calorimetric parameters or protein solubility was taken into account. Indeed, film that gave a high ΔCp value, indicating an open network, had proteins which were highly cross-linked. Mixing of gluten, which gave a lower ΔCp value, could indicate that it created network reticulation equivalent to that of thermomolded gluten, which is not consistent with the results of biochemical analysis. Different linkages strengthening the gluten could be measured by the different approaches.

References

1. B. Cuq, N. Gontard and S. Guilbert, *Cereal Chem*, 1998, **75**, 1
2. M. Song , D.J. Hourston, H.M. Pollock and A. Hammiche, *Polymer*, 1999, **40**, 4763
3. R.C. Hoseney, K. Zeleznak and C.S. Lai, *Cereal Chem.*, 1986, **63**, 285
4. E.M . De Graaf, H. Madeka, A.M. Cocero and J.L. Kokini, *Biotechnol. Prog.*, 1993, **9**, 210
5. T.R. Noël, R. Parker, S.G. Ring and A.S. Tatham, *Int. J. Biol. Macromol.*, 1995, **17**, 81
6. R.E. Wetton, 'Developments in polymer characterization', Elsevier Applied Science, 1986, p.179
7. E.J. Donth, 'Relaxation and thermodynamics in polymers : glass transition', Akademie Verlag Gmbh, Berlin, 1992, p.207
8. D.W. Van krevelen, 'Properties of polymers', Elsevier Science, Amsterdam, 1997, p.71
9. G. Sartor and G.P. Johari, *J. Phys. Chem.*, 1996, **100**, 19692
10. S. Bérot, S. Gautier, M. Nicolas, B. Godon and Y. Popineau, *Int. J. Food Sci. and Technol.*, 1994, **29**, 489
11. N. Gontard, S. Guilbert and J.-L. Cuq, *J. Food Sci.*, 1992, **57**, 190
12. A. Redl, M.-H. Morel, J. Bonicel, S. Guilbert and B. Vergnes, *Rheol. Acta*, 1999, **38**, 311
13. M.T. Kalichevsky, J.M.V. Banshard, R.D.L. Marsh, 'The glassy state in foods', Nottingham University Press, 1993, p. 133
14. J.L. Kokini, A.M. Cocero, H. Madeka and E. De Graaf, *Trens Food Sci.Technol.*, 1994, **5**, 281
15. T.D. Strecker, R.P. Cavalieri, R.L. Zollars and Y. Pomeranz, *J. Food Sci.*, 1995, **60**, 532
16. P.L. Weegels and R.J. Hammer, 'Interactions : the keys to cereal quality. Tempearture-induced changes of wheat products', AACC, St. Paul, 1998, p. 95

Acknowledgements

This work is a part of the ERBFAIRCT 96 1979 project supported by E.C.

WHEAT GLUTEN FILM : IMPROVMENT OF MECHANICAL PROPERTIES BY CHEMICAL AND PHYSICAL TREATMENTS

V. Micard*, M.-H. Morel and S. Guilbert

U.F.R. Technologie des Céréales et des Agropolymères, AGRO.M/I.N.R.A.
2 Place P. Viala, 34060 Montpellier cedex 1, France

*Contact author. Email : micard@ensam.inra.fr-Phone : 33-4-99-61-28-89. Fax : 33-4-67-52-20-94

1 INTRODUCTION

A variety of proteins have been studied due to their film-forming ability. Wheat gluten films possess some properties of interest as their relative resistance to water and their selective O_2 and CO_2 barriers properties. However, their mechanical properties, strongly affected by the presence of plasticizer, must be improved to make them generally useful. The cross-linking of proteins from several sources has already been attempted as a method to obtain stronger films. Chemical, thermal and radiation treatments have been applied on film forming solution (pre-treatment) and, for some of them, as post-treatment (on the pre-formed film)[1-6]. Few studies present a comparison of several treatments on the same protein source, making a comparative evaluation of the treatments very difficult.

The objective of our study was to compare the effects of chemical and physical pre- and post-treatments on the mechanical properties (stress, elongation at break and Young's Modulus) of films made from wheat gluten.

2 MATERIALS AND METHODS

2.1 Films casting

A film forming solution of wheat gluten (N x 5.6= 76.5%)(Amylum Group, Aalst, Belgique) was prepared according to Gontard *et al.*[7], with sodium sulphite (30 mg) as reducing agent. The film-forming solution was poured and spread onto a crystal PVC plate and dried for 20 h at 25°C in a ventilated oven before peeling of the support.

2.2 Film treatments

Post-treatments were applied on films conditioned for 48 h at 20°C and 60% RH. Formaldehyde (37% v/v) was added in the film forming solution (0.8 ml/100 ml) or applied during 30 h on films as vapours (1 l of an ethanolic solution of formaldehyde (10% v/v) in a hermetic box). UV treatment were applied 1) on the film solution during drying (2h) by using a 125 W lamp, 2) on film with an UV oven (254 nm) with two radiation doses (0.25 and 1 J/cm^2). Heat treatment (95, 110, 125, 140°C) was performed during 15 min as a post-treatment with a molder (Techmo, France) without any pressure application. Eventual loss of film plasticizer (glycerol) due to heat treatment was determined by high performance liquid chromatography. Treatment by gamma radiation (10 and 20 kGy) was performed by the Commissariat à L'Energie Atomique (Cadarache, France) using ^{60}Co source.

2.3 Mechanical properties

Mechanical properties of films were investigated on dumbbell shape specimens (5A type, standard ISO 527-2: 1993 (F)) with a TAXT2 (Rheometer, Champlan, France) at 20°C and 60% RH. Tensile strength (σ), elongation at break (ε) and Young's modulus (E) values are the mean of 10 replicates. Prior to mechanical properties evaluation, films were always stored 72 h at 20°C and 60% RH. Properties of treated films were always compared with non treated films ("control") with the same aging. Properties of "control" films from individually prepared and cast film-forming solutions were reproducible (ratio standard deviation to mean value<20%).

2.4 Total extracted proteins

Protein films were extracted 80 min with a sodium phosphate buffer that included sodium dodecyl sulphate (1%), sonicated and fractionated by SE-HPLC.

3 RESULTS AND DISCUSSION

Typical tensile stress-strain curves obtained by some of the pre- and post-treatments (temperature, formaldehyde, UV 0.25 J/cm^2) are presented in Figure 1. All curves displayed a characteristic S-shape classically observed with rubber-like material. However, important differences of mechanical properties at break were recorded between treatments. Values at break are reported in Table 1.

The action of formaldehyde as pre- or post-treatment clearly decreases elongation (giving values from 38 to 61% of the initial control value) and increases the tensile strength (from 402 to 474% of the initial control value). An increase of tensile strength has already been reported on pea[5], cottonseed flour[6] and zein[8] films pre- and/or post-treated by formaldehyde. Concerning elongation, it was commonly decreased for pea and zein films treated by formaldehyde, but increased for cottonseed flour. In our experiment, vapours are more efficient than pre-treatment by formaldehyde, especially concerning the stiffness of the film (E). Gueguen *et al.*[5], comparing the effect of formaldehyde as pre- and post-treatment (soaking) on pea films also concluded that post-treatment was more efficient. Changes in mechanical properties of formaldehyde-treated films were due to reinforcement of the network covalent linkages, which led to a drastic insolubilisation of proteins (< 1% and 94% of soluble proteins in formaldehyde treated and 'control' films, respectively).

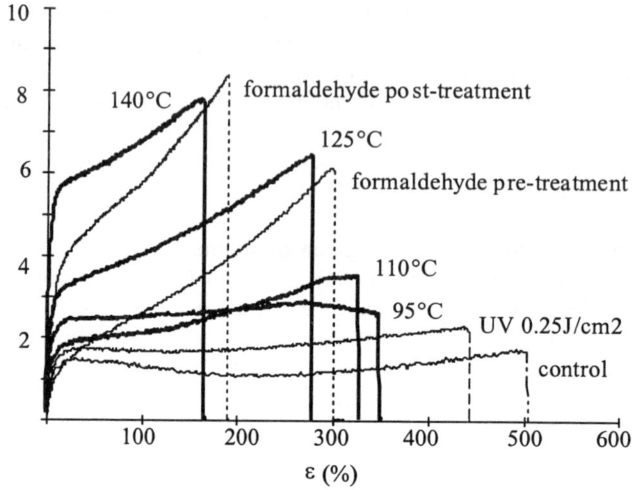

Figure 1 *Stress-strain curves of treated films*

Due to heating treatment, an increase in tensile strength (150 to 432% of initial control value) and a decrease in elongation (from 77 to 34% of initial control value) was observed when the temperature increased from 95 to 140°C. Such changes may result from protein aggregation[1,3], as heating at 140°C led to an almost total insolubilisation of the proteins (2.3 % of soluble proteins). Tensile strength mainly changed when temperature increased from 110°C to 125°C, while elongation decreased progressively. The inverse relationship observed between stress and strain (Figure 1) could be attributed

Table 1 *Mechanical properties of treated gluten films[a] (% of initial value of control film)*

Treatments		Mechanical properties		
		σ	ε	E
Formaldehyde	Vapours	474 ± 16 e	38 ± 2 e	541 ± 66 d
	on casting solution	402 ± 18 cd	61 ± 1 d	97± 5 a
Temperature	95°C	150 ± 6 ab	77 ± 5 bc	231 ± 29 bc
	110°C	184 ± 8 b	66 ± 4 cd	285 ± 24 c
	125°C	369 ± 15 c	55 ± 2 d	303 ± 60 c
	140°C	432 ± 22 d	34 ± 2 e	751 ± 39 e
UV radiations	on Films : 0.25 J/cm^2	117 ± 5 a	85 ± 6 ab	140 ± 15 ab
	1 J/cm^2	120 ± 6 a	95 ± 4 a	113 ± 9 a
	on casting solution	145 ± 7 a	77 ± 2 bc	227 ± 22 bc
γ radiations	10 kGy	140 ± 10 a	68 ± 7 cd	180 ± 17 ab
	20 kGy	126 ± 6 a	90 ± 4 ab	128 ± 6 a

[a]Reported values are means of 10 replicates ± standard deviation. Any two means in the same column followed by same letter are not significantly (p>0.05) different by Duncan's multiple range test.

to differences in plasticizer content[9]. However, the glycerol loss of heat cured films never exceeded 11% of its initial content in our study. Referring to the results of Gontard *et al.*[10], such a difference in glycerol concentration of wheat gluten films would have little effect on film mechanical properties, which confirms the effect of thermal treatment.

Heating films at 140°C during 15 min led approximately to the same improvement of their mechanical properties as treatment by formaldehyde vapours. However, thermally-treated films presented a higher Young's Modulus.

γ and UV radiation slightly modified the mechanical properties of wheat gluten films whatever the dose used, in comparison with formaldehyde and high temperature treatment. UV was more efficient when applied as a pre-treatment, due probably to higher molecular mobility of the gluten components. The efficiency of UV radiation depends on the protein source. Soy proteins rich in tyrosine and phenylalanine are sensitive to UV radiation[4], in contrast to pea[5] and gluten (our study) films. For γ radiation, increasing the dose from 10 to 20 kGy reduced the effects demonstrated on the tensile strength, elongation and Young's Modulus. This could be attributed to a decrease in insoluble glutenin polymers, by depolymerisation and /or breakdown of covalent linkages, as shown by Köksel *et al.*[11] on wheat flour. The positive effect of 10 kGy radiation could be explained by the formation of dityrosine[2].

4 CONCLUSION

Mechanical properties of wheat gluten films can be substantially modified by treatment with formaldehyde, especially used as vapours. Short time (15 min) heat curing, with temperatures above 110°C, also improves the mechanical properties of films. Such changes in mechanical properties are due to high aggregation of the network proteins. In comparison, radiation (UV and γ) led to little or no change in the mechanical properties of the films.

References

1. Y. Ali, V.M. Ghorpade and M.A. Hanna, *Ind. Crops and Prod.* , 1997, **6**, 177
2. D. Brault, G. D'Aprano and M. Lacroix, *J. Agric. Food Chem.*, 1997, **45**, 2964
3. A. Gennadios, V.M. Ghorpade, C.L. Weller and M.A. Hanna, *Trans. ASAE*, 1996, **39**, 575
4. A. Gennadios, J.W. Rhim, A. Handa, C.L. Weller and M.A. Hanna, *J. Food Sci.*, 1998, **63**, 225
5. J. Gueguen, G. Viroben, P. Noireaux and M. Subirade, *Ind. Crops and Prod.*, 1998, **7**, 149
6. C. Marquié, C. Aymard, J.-L. Cuq and S. Guilbert, *J. Agric. Food Chem.*, 1995, **43**, 2762
7. N. Gontard, S. Guilbert and J.-L. Cuq, *J. Food Sci.*, 1992, **57**, 190
8. N. Parris and D.R. Coffin, *J. Agric. Food Chem.*, 1997, **45**, 1596
9. A.C. Sánchez, Y. Popineau, C. Mangavel, C. Larré and J. Gueguen, *J. Agric. Food Chem.*, 1998, **46**, 4539
10. N. Gontard, S. Guilbert and J.-L. Cuq, *J. Food Sci.*, 1993, **58**, 206
11. H. Köksel, H.D., Sapirstein, S. Celik and W. Bushuk, *J. Cereal Sci.*, 1998, **28**, 243.

Acknowledgements

This work is a part of the ERBFAIRCT 96 1979 project supported by E.C.

Viscoelasticity, Rheology and Mixing

DO HIGH MOLECULAR WEIGHT SUBUNITS OF GLUTENIN FORM 'POLAR ZIPPERS'?

Peter S. Belton, Klaus Wellner, E.N. Clare Mills, Alex Grant and John Jenkins

Institute of Food Research, Norwich Research Park, Colney Lane, Norwich, NR4 7UA.

1 INTRODUCTION

Dough made from wheat flour has unusual viscoelastic properties, which depends upon the proteins of wheat flour and in particular on the high molecular weight (HMW) subunits of wheat glutenin[1]. HMW subunits of wheat glutenin comprise around 8-10% of the total extractable flour protein, and have M_rs by SDS-PAGE of about 80,000-146,000, although the true M_rs are much lower (~ 67,000-88,000). A central repetitive domain makes up 75-85% of the protein sequence, comprising repeating tri -, hexa - and nonapeptide repeat motifs. Typical repeating sequences are PGQGQQ and GYYPTSGQQ[2]. Disulphides formed by the cysteine containing N- and C-terminal domains are also known to be critical for breadmaking quality[3]. The low molecular weight (LMW) subunits of glutenin also contain proline and glutamine rich repetitive sequences, which are different from those of the HMW subunits, and may contain longer consecutive stretches of glutamines[4].

Although the formation of disulphides is necessary, elasticity has been proposed to depend upon the existence of two forms of gluten with different degrees of extension for which the names "loops" and "trains" have been proposed[5]. Secondary structure prediction suggests that the repetitive sequences of the HMW subunits have a high probability of forming β-turns and circular dichroism spectroscopy of related peptides suggests either β-turns or random coil in solution[6]. Thus it is likely that the "loops" are formed of structures which are rich in β-turns such as the β-spiral, which has been proposed as a structure of the HMW subunits in their "native" state[7,8]. Here we propose a molecular model of the "trains" based upon the "glutamine zipper" proposal of Perutz et al.[9] and show that this model predicts the observed effects of hydration on the structure of gluten and its elasticity.

2 RESULTS AND DISCUSSION

2.1 The Glutamine Zipper

Although poly-L-glutamine has an extremely stable β-sheet structure, glutamine is not normally regarded as a strongly β-forming residue in algorithms predicting the secondary structure of globular proteins and the insertion of a single glutamine into a peptide will not strongly induce the formation of β-sheet. However, Perutz and co-workers[9] noted that sequences with many glutamines can form β-sheets in which the glutamines on neighbouring strands hydrogen bond and used circular dichroism to show that the peptide Asp_2-Gln_{15}-Lys_2 formed β-sheet. The hydrogen bonding can be achieved with all the glutamines in the same (most favoured) conformation, if necessary optimised by rotations about χ_3 and small movements of main chain, which lead to large displacements of the side chain amide. The sequence -Gln-Gln- can naturally interact with itself forming two side chain to side chain hydrogen bonds in addition to those of the standard β-sheet. As well as hydrogen bonding, there is also some burial of hydrophobic surface as the Cβ and Cγ atoms of the glutamine side-chains become less accessible. However, if the sequences interacting are - Gln - X_1 - Gln - and - Gln - X_2 - Gln - more hydrophobic surface can be buried if χ_1 angles are rotated by 120° in the diagonally opposed glutamines as shown in figure 1, reducing the distance between side chains from 7 Å towards 4.5 Å. β-sheet formation by these sequences has some of the co-operativity normally seen in tertiary structure formation.

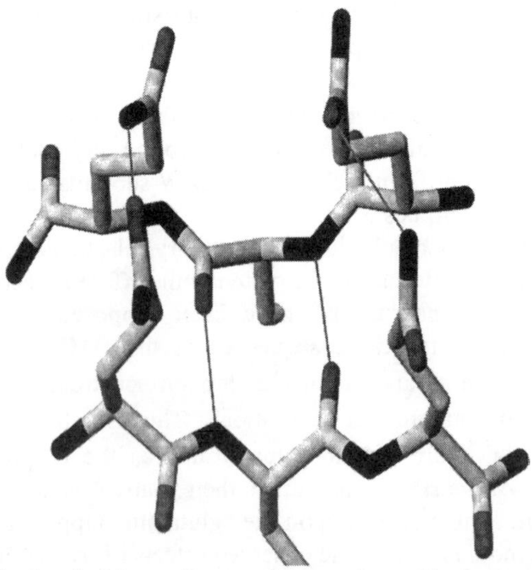

Figure 1 *Burial of hydrophobic surface and formation of hydrogen bonds by interaction of two Gln-X-Gln sequences. Carbon atoms are light gray, and the more polar nitrogens and oxygen atoms are dark grey. Hydrogen bonds are shown as thin lines.*

2.2 The Secondary Structures of HMW and LMW Subunits

The obvious feature of the repetitive sequences of these proteins is the very high proportion of glutamine (whose name derives from gluten), proline and, in the HMW subunits, glycine. Secondary structure prediction algorithms necessarily ignore "tertiary" interactions between regions separated in the primary sequence such as that shown in Figure 1 and do not initially predict glutamine rich sequences as β-sheet. In fact, the repetitive sequences of HMW subunits in isolation or at very low concentration in circular dichroism experiments would not be expected to form β-sheet, while the polyglutamine sequences which may be present in some of the LMW subunits probably would form β-sheet as does polyglutamine itself.

Although proline is treated in secondary structure prediction as a strong β-breaker, proline can be accommodated in an outside β-strand in alternate positions. The fixed ϕ angle of proline even implies that constraint to the β-conformation cost less entropy for proline than for other residues. However, the sequence -Pro-Pro- would clearly prevent the formation of the characteristic hydrogen bonding. Despite this, proline is sometimes found in the β-sheets of globular proteins.

The sequence Gln-X_1-Gln can form a reasonably stable anti-parallel β-sheet with another Gln-X_2-Gln sequence with most residue types including glycine. However, glycine is normally considered a β-breaker as it has many possible conformations and is normally selected by evolution for sites in globular proteins requiring unusual Ramachandran angles. The optimal X for the stabilisation of the β-conformation would be a large residue such as Leu, Phe or Tyr or naturally Gln itself.

2.3 Elasticity

As the HMW subunits of wheat glutenin are believed to play a critical role in the elasticity of dough, it is likely that these molecules (or some other component of dough) have at least two "structures", with the more stable structure being more compact than the extended structure. It is likely that the secondary structure of the repetitive regions of HMW subunits of wheat glutenin is an equilibrium between extended β-sheet structure and a more compact structure dominated by β-turns. Formation of the β-sheet conformation is inherently co-operative and is thus automatically favoured as the protein concentration increases. If the repetitive sequences of the HMW subunits are able to interact with other proteins, any β-sheets with exposed edges able to hydrogen bond or exposed single strands with a high propensity to form β-sheet will favour sheet formation. The LMW subunits of glutenin may also form β-structure and as these molecules are present in the dough at relatively high concentrations, intermolecular β-sheets will probably be formed with the side-chain hydrogen bonds of the "zipper". Experimental evidence from FTIR for β-sheet in the hydrated solid state of many gluten proteins shows that other repeat sequences without long polyglutamine sequences can also form β-structures[10].

Clearly, mechanical extension will favour the formation of extended structures and thus the formation of anti-parallel hydrogen bonding between strands. Mechanical force effectively takes the role of large residues in restricting the conformational flexibility of the peptide chains. The role of the disulphides is critical in allowing an externally applied force to generate extension of the repetitive regions rather than simply viscous flow of the molecules. However, there is a further co-operative element in that intermolecular β-sheet may also prevent molecules from moving and thus force them to extend.

The formation of β-sheet involves the protein forming hydrogen bonds with itself rather than with water. A structure dominated by β-turns is less likely to form protein - protein hydrogen bonds without extensive annealing and β-turns in crystal structures of globular proteins typically show several bound waters. Thus hydration will favour β-turn formation. Hydration will also favour structures resembling β-sheet but in which some of the regular hydrogen bonding has been replaced by bridging waters. Thus increased hydration will strengthen the influence of the glycine and proline residues in favouring less regular structures whereas dehydration will increase the importance of the glutamine - glutamine hydrogen bonds. Water also acts as a "lubricant" allowing hydrogen bonds to interchange. Thus a small quantity of water might allow two hydrogen bonded β-strands to slide past each other but significantly more might be needed to allow the strands to separate.

Dough can be contrasted with the apparently similar case of elastin in the effect of hydration. Elastin is formed from a repetitive sequence of pentapeptides such as Val-Pro-Gly-Val-Gly. On heating above 30° C, elastin contracts and releases water, probably with the formation of hydrophobic clusters of valines[11]. However, wheat gluten absorbs water on heating[12], suggesting that elasticity arises in a different way. The model for -Gln-Gly-Gln- above suggests that in gluten the extended structure may bury hydrophobic surface more effectively than the β-turn structure as well as leaving fewer hydrogen bonding groups available to interact with water.

3 CONCLUSIONS

The elasticity of bread dough requires that the energy input by mechanical work is stored in the conformations of proteins or aggregates of proteins. The "loop and train" model[5] assumes that extensions the equilibrium towards the more extended form, "trains", and that elasticity arises from the restoration of the equilibrium. The "glutamine zipper" is the best available model for an extended form of the HMW subunits of gluten and may also be the most stable form of the LMW subunits. Formation of "glutamine zippers" with some burial of hydrophobic surface and release of hydrogen bonded water explains the observed differences in the relationship of elasticity to hydration between elastin and gluten and may explain the "non-entropic" nature of gluten elasticity. It also makes the new prediction that protein - protein interactions should strengthen on extension.

Acknowledgements

This research was supported by the BBSRC (UK) and EU project FAIR-CT96-1170.

References

1. P.I. Payne, M.A. Nightingale, A.F. Krattiger, and L.M. Holt *J. Sci. Food. Agric.* 1987, 38, 51

2. J. Forde, J.M. Malpica, N.G. Halford, P.R. Shewry, O.D. Anderson, F.C, Greene and B.J. Miflin, *Nucleic Acids Res.* 1985, 13, 6817
3. M.P. Lindsay and J.H. Skeritt, *J. Agric. Food Chem.* 1998, 64, 3447
4. E.G. Pitts, J.A. Rafalski and C. Hedgcoth, *Nucleic Acids Res.* 1988, 16, 11376
5. P.S. Belton, *J. Cereal Sci.* 1999, 29, 103
6. Tatham, A.S., Shewry, P.R., and Belton, P.S. Advances in Cereal Science and Technology, 1990, 10, 1
7. M.J. Miles, H.J. Carr, T.C. McMaster, K.J. Lanson, P.S. Belton, V.J. Morris, J.M. Field, P.R. Shewry and A.S. Tatham, *Proc. Nat. Acad. Sci. USA*, 1991, 88, 68
8. P.R. Shewry, N.G. Halford and A.S. Tatham, *J. Cereal Sci.* 1992, 15, 105
9. M. F. Perutz, T. Johnson, M. Suzuki and J.T. Finch, *Proc. Natl. Acad. Sci. USA,* 1994, 91, 5355
10. M. Pézolet, S. Bonenfant, F. Dousseau and Y. Popineau, *FEBS Lett.* 1992, 299, 247
11. H. Reiersen, A.R. Clarke and A.R. Rees, *J. Mol. Biol.* 1998, 283, 255-264
12. A. Grant, P.S. Belton, I.J. Colquhoun, M.L. Parker, J.J. Plijter, P.R. Shewry, A.S. Tatham and N. Wellner, *Cereal Chem.* 1999, 76, 219

WHAT CAN NMR TELL YOU ABOUT THE MOLECULAR ORIGINS OF GLUTEN VISCOELASTICITY?

E. Alberti[1], A.S. Tatham[2], S.M. Gilbert[2] and A.M. Gil[1]

1. Department of Chemistry, University of Aveiro, 3810-193 Aveiro, Portugal 2. Department of Agricultural Sciences, University of Bristol, Institute of Arable Crop Research, Long Ashton Research Station, Long Ashton, Bristol BS 18 9AF, UK

1 INTRODUCTION

Gluten proteins play an important role in conferring viscoelasticity to dough[1]. In order to understand the molecular origins of gluten viscoelasticity it is important to gain some insight about the molecular structure and behaviour of the system particularly upon its hydration. Much interest has been given to the fraction of gluten identified as high molecular weight (HMW) glutenins and their role in viscoelasticity, since clear correlations are found between the type of HMW subunit present in wheat and breadmaking ability: the pairs 1Dx5/1Dy10 and 1Dx2/1Dy12 have been associated with good and poor baking performances, respectively[2]. Due to the insolubility of gluten proteins, the use of solid state NMR spectroscopy has become a powerful tool to study this type of samples[3,4]. Here we show that both ^{13}C NMR and ^{1}H NMR may provide structural information about wheat proteins and their response to hydration. The application of ^{13}C cross-polarisation and magic-angle-spinning (CPMAS) NMR to study the behaviour of 1Dx5, 1Dx2, 1Dy10 and 1Dy12 proteins upon hydration is presented. The more complex system of hydrated gluten was investigated by ^{1}H MAS NMR spectroscopy. The application of Rheo-NMR to the study of gluten response to mechanical stress is also briefly described, providing a direct correlation between molecular properties and rheology for the gluten system.

2 MATERIALS AND METHODS

The unalkylated HMW subunits were purified from wheat as described elsewhere[5]. The samples were dried under vacuum at 35°C, until constant weight, and were hydrated by standing over D_2O. The water content was determined gravimetrically and expressed in g water/100 g dry protein. Gluten was obtained from a Portuguese soft winter wheat (*Triticum aestivum* L.) defatted flour of the Amazonas variety using the fractionation method described previously[6]. Gluten doughs were prepared by mixing the freeze-dried powder with D_2O in a 1:1 ratio and leaving the samples to stand for 8 hours.

All ^{13}C and ^{1}H MAS experiments were carried out on a Bruker MSL 400P NMR spectrometer operating at 400 MHz for protons, using a 4 mm double bearing MAS probe and a rotor with an inner insert. The ^{13}C CPMAS spectra were recorded using 90° pulses of 4-6 μs, contact time of 1 ms and spinning rates (SR) of 6-8 kHz. The ^{1}H NMR spectra were recorded using 90° pulses of 4-5 μs, recycle time of 4 s and SR of 15 kHz. Rheo-NMR experiments were performed on a Bruker AMX300 spectrometer equipped with a micro imaging attachment, at 25 °C. A miniature concentric Couette cell (inner and outer diameters respectively 2.0 mm and 3.5 mm) was used to study shearing effects and a bi-axial extension device was used to investigate compression effects. These experiments have been described in more detail elsewhere[7].

3 RESULTS AND DISCUSSION

3.1 ^{13}C CPMAS study of individual HMW subunits and pairs of subunits

Figure 1 shows the ^{13}C CPMAS spectra of 1Dx5 subunit as a function of hydration. Similar spectra were registered for the remaining three subunits (1Dx2, 1Dy10 and 1Dy12). Generally, a decrease in signal to noise ratio is observed as hydration increase. This reflects a decrease in the amount of hindered protein as hydration increases, accompanied by a change in CP dynamics (determined by the parameters T_{CH} and $T_{1\rho H}$). At 65% hydration, a few peaks arising from proline (25, 48 ppm) and glycine (42ppm) decrease relatively to those of glutamines which remain significantly intense (31, 52 and 177ppm). It has been suggested that these glutamines remain hindered at high hydration due to their involvement in H-bonds, particularly affecting the terminal $C_{\delta}ONH_2$ groups. By normalising the CPMAS spectra, a semi-quantitative comparison of the spectra may be made. In particular, the 1Dx5 subunit shows a gradual decrease of the "estimated % rigid protein" (Figure 2A) which is expected in view of the increase in protein mobility. For 1Dx2, 1Dy10 and 1Dy12, the estimate of rigid protein decreases at 35% D_2O and tends to increases at 65% D_2O. This suggests the occurrence of some structural change e.g. molecular aggregation/association which may result in a more hindered network at 65% D_2O. Estimation of the % rigid glutamines (result not shown) gives an indication of the amount of glutamines involved in H-bonds. In general, the "estimated % rigid glutamines" tends to follow the trend observed for "estimated % rigid protein" which indicates that hydrogen bonds between glutamines are broken as molecular mobility increases.

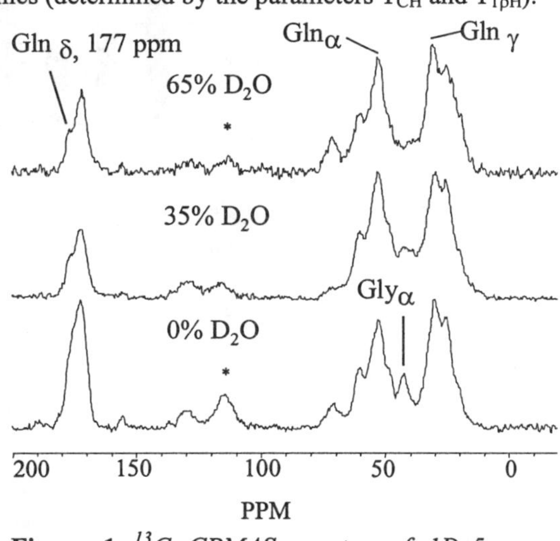

Figure 1 *^{13}C CPMAS spectra of 1Dx5 as a function of hydration.*

Similar calculations of "estimated % rigid protein" applied to the 1Dx5/1Dy10 and 1Dx2/1Dy12 pairs (Figure 2B) show that the latter pair has a different behaviour than that predicted by the behaviour of the individual subunits (expressed by the sum *S* of the trends obtained for individual subunits). This indicates that the pair 1Dx2/1Dy12 exhibits significant synergism towards hydration whereas the pair 1Dx5+1Dy10 does not, observation which may potentially be at the basis of the different technological behaviour of the two subunit pairs.

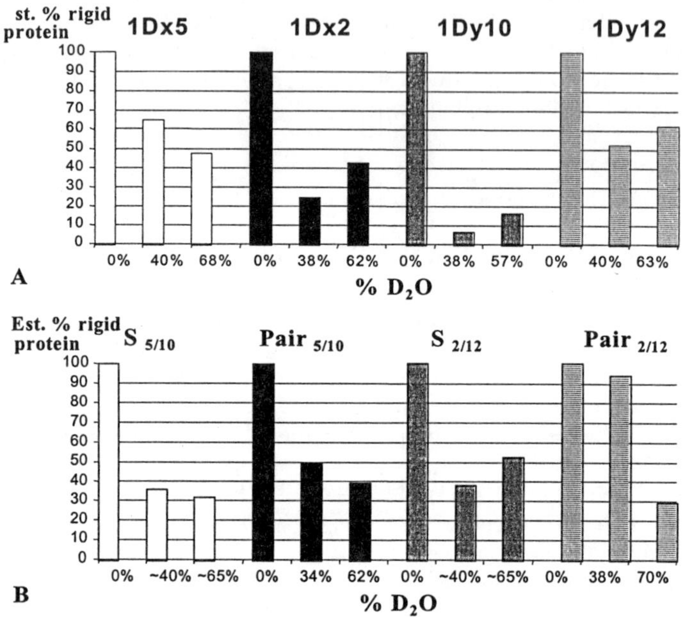

Figure 2 *Estimated % of rigid protein at different hydrations for the four subunits and the two subunit mixtures.*

3.2 ^{1}H NMR study of hydrated gluten

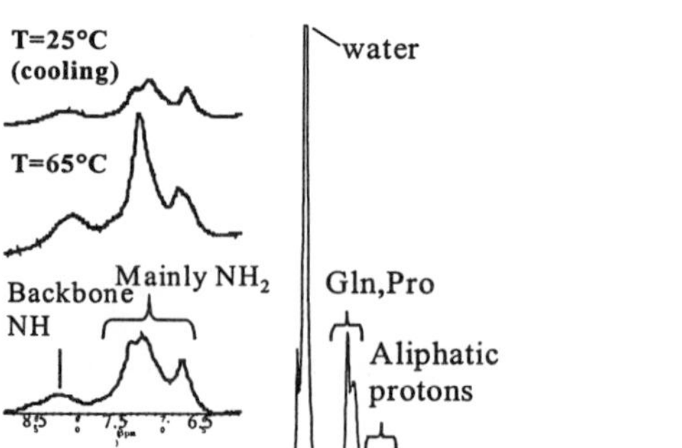

Figure 3 *^{1}H MAS spectra of hydrated gluten. Insets show NH region at different temperatures.*

The ^{1}H MAS spectrum of hydrated gluten (Figure 3) shows a number of narrow peaks and no broad spectral component, indicating that a mobile protein network is formed by hydration. Heating is a very important factor in dough performance, resulting in significant changes in the molecular and rheological properties of gluten. The insets in Figure 3 show the changes occurring in the NH$_2$ region upon heating at 65 °C and cooling back to 25 °C. The intensity increase of the peak at 7.3 ppm, at 65°C, is consistent with breakage of hydrogen bonds since it should arise from glutamine NH$_2$ groups with increased mobility. Subsequent

lowering of the temperature recovers the initial spectrum, indicating reformation of the initial hydrogen bonded network.
The coupling of standard rheometric techniques and NMR spectroscopy (rheo-NMR) has enabled, for the first time, a direct link to be established between the application of stress (shear and extensional) and the molecular structure of gluten[7]. The principle of this technique is that ^{1}H NMR spectra are recorded *while* mechanical stress is applied to the gluten sample, thus giving information about the structural changes resulting from the application of stress. It is interesting to note that the spectral changes occurring when shear and bi-extensional stress are applied to gluten are similar to those observed upon heating, that is, indicative of breakage of H-bonds[7].

4 CONCLUSIONS

It has been shown that solid state NMR is a powerful technique to study gluten protein systems. Two possible approaches have been described in order to understand the behaviour of wheat proteins upon hydration. Firstly, ^{13}C CPMAS was applied to study the structural changes of individual HMW subunits with hydration. The hydration of individual subunits was compared with that of subunit pairs and synergism was detected for 1Dx2/1Dy12 but not for 1Dx5/1Dy10. Secondly, the use of ^{1}H MAS and ^{1}H Rheo-NMR was shown to allow the behaviour of H-bonds in hydrated gluten to be followed. The application of thermal, shear and extensional stress is shown to disrupt H-bonds between Gln NH_2, suggesting that these bonds may be an important factor in determining gluten rheology.

References

1. B.J. Miflin, J.M. Field, P.R. Shewry, in *Seed Proteins,* ed. J. Daussant, J. Mossé, & J. Vaughan, Academic Press, London, 1983 p. 255
2. P.I. Payne, K.G. Corfield, J.A. Blackman, *J. Sci. Food Agric.*, 1981, **32**, 51
3. P.S. Belton, A.M. Gil, A. Grant, E. Alberti, A.S. Tatham, *Spectrochimica Acta Part A*, 1998, **54**, 955
4. A.M. Gil, K. Masui, A. Naitô, A.S. Tatham, P.S. Belton, H. Saitô, *Biopolymers*, 1997, **41**, 289
5. P.R. Shewry, J.M. Field, A.J. Faulks, S. Parmar, B.J. Miflin, M.D. Dietler, E.J. Lew, D.D. Kasarda, *Biochimica et Biophysica Acta*, 1984, **788**, 23
6. Z. Czuchajowska, Y. Pomeranz, *Cereal Chem.*, 1993, **70**, 701
7. P.T. Callaghan, A.M. Gil, *Rheol. Acta*, , 1999, **38**, 528.

Acknowledgements

The authors acknowledge the European Commission for funding under the project EU FAIR CT96-1170 on "Improved EU Wheats for Food" and the Fundação para a Ciência e Tecnologia for funding under the PRAXIS/PCNA/BIO/0703/96 project. The authors also thank Dr. C. Brites and Dr. F. Bagulho, from the National Plant Breeding Station (ENMP), Elvas, Portugal for providing the flour from which gluten was prepared.

BACK TO BASICS: THE BASIC RHEOLOGY OF GLUTEN

S. Uthayakumaran[1], M. Newberry[1,2] and R. Tanner[1,2]

1. Department of Mechanical and Mechatronic Engineering, The University of Sydney, NSW 2006, Australia. 2. Quality Wheat Cooperative Research Centre Ltd., Locked Bag 1345, North Ryde, NSW 1670, Australia.

1 INTRODUCTION

Rheological properties are important in determining the behaviour of wheat flour doughs during mechanical handling in addition to their influence on the quality of the finished product. The automation of baking industry requires the knowledge of rheological behaviour and dough properties. It has been established that gluten proteins play an important role in determining the differences in bread making performance[1]. A number of research studies have been performed on fundamental gluten rheology[2,3]. Much research work on gluten has been focused on the dynamic shear properties, however this represents only one, idealised, flow condition experienced by bread doughs during processing and baking. The rheological behaviour of bread doughs is very complex; both shear and elongation studies are required for extensive characterisation of its rheological behaviour. In this paper we investigated the shear and elongation properties of flour and gluten doughs in order to compare them.

2 MATERIALS AND METHODS

2.1 Samples

Commercial wheat flours, strong (14% protein) and Baker's flour (12.5% protein) were obtained from Weston Milling, Sydney, Australia. A standardised procedure[4] was adopted for the extraction of gluten from the two flours. Gluten was then freeze dried and ground to a powder. Doughs (16.0 g) were prepared using each of the flours and the glutens extracted from the flours. The amounts of water to be added were calculated from the protein content and the moisture of the flour using the standard method[5]. In the case of gluten, 60% of water (w/w) was added to the gluten. Flour/ gluten and water were mixed in a 10-g Mixograph to peak dough development and rheological measurements were then carried out.

2.2 Rheological tests

The elongational properties of the doughs were studied using a constant-strain-rate extension test performed on a United Testing Machine, Model SSTM 5000 (United Calibration Corp, Huntington Beach, CA). The dough (6 g) was compressed between a fixed and a moving upper grip both with a diameter of 30 mm coated with super glue. The dough sample was then rested for 20 min before testing. Moisture loss was prevented by applying a layer of food-grade petroleum jelly (free of ethanol residue) around the edge of the sample. After the 20 min rest, the dough was pulled apart at an exponentially increasing speed to maintain a constant strain-rate (0.01, 0.1 and 1 s^{-1}) in the dough sample. Control of the top plate speed and collection of data were performed by a desktop computer running a programme written in QuickBasic version 1.1 (Microsoft Corporation, Seattle, WA, 1992). Force and distance data collected by the computer were used to calculate the rheological parameters of strain and elongational viscosity (Pa·s).

Strain, defined here as Hencky strain, was calculated using $\varepsilon = \ln(l/l_0)$, where l_0 is the original length of the sample, i.e. the initial plate separation (5 mm). The stress, σ is given by $\sigma = F / A$, where F is the force exerted by the sample on the load cell, and A is the minimum cross-sectional area of the sample (usually at the midpoint of the sample). Preliminary investigations showed that at strains above 1, the elongated dough sample took the shape of a cylinder. The minimum cross-sectional area of the dough sample was calculated, assuming that during elongation the volume of dough sample did not change and was cylindrical in shape. The extensional viscosity η_E, is given by $\eta_E = \sigma / \dot{\epsilon}$, where $\dot{\epsilon}$ is the strain rate in the sample. The tests were performed in an air-conditioned laboratory with a variation of $\pm$0.5°C in the 24°C ambient temperature.

The linear visco-elastic limit of the flour/ gluten doughs was determined using stress sweep experiment. The dough (3 g) was mounted on a controlled stress rheometer (Reologica Stresstech, Reologica Instruments AB, Lund, Sweden) in the parallel plate configuration (25 mm diameter and with 2 mm gap). The plates were glued with sand paper (100 cv). The edge of the sample was coated with food-grade petroleum jelly. Before starting the measurement, the dough was allowed to rest for 20 min. A stress sweep (1 Pa – 100 Pa) was carried out at frequency of 1Hz (pre-shearing was carried out at 1Pa for 10 sec prior to the test). Stress sweep experiments were also carried out on two-component gluten and starch (commercial starch) mixtures. These mixtures were prepared by diluting gluten with starch such that the concentration of starch varied from 0% to 100% in increments of 20%.

A frequency sweep was also carried out (0.01 Hz – 20 Hz) at a constant strain of 0.0005 (strain within the linear visco-elastic limit). Pre-shearing was carried out at 1 Pa for 10 sec prior to the test. The tests were performed at 25°C.

Four grams of dough was mounted on a Bohlin VOR rheometer in the parallel plate configuration with a 2mm gap. The edge of the sample was coated with food-grade petroleum jelly to prevent moisture loss. The dough was rested for 20 min. viscosity measurements were carried out at 3 shear rates (0.0105, 0.105 and 1.05 s^{-1}) at 25°C. All basic rheological measurements were conducted in duplicate.

3 RESULTS AND DISCUSSION

The elongational viscosity of dough remained constant at low strains or extension levels.

At strains above unity, the elongational viscosity started to increase rapidly. This sharp increase in elongational viscosity with increasing strain levels is known as strain hardening. A maximum viscosity value was reached during this strain-hardening stage, at which point the dough sample ruptured between the plates. The viscosity and strain (elongation) at which the dough sample broke or ruptured (elongational rupture viscosity and rupture strain) were useful simple measures of dough strain-hardening properties. Gluten dough increased the strain-hardening properties of the dough, as measured by higher elongational rupture viscosity compared to flour dough. This indicates that gluten contributes to the strength and stability of the flour. The rupture strain of gluten and flour dough was similar and was not significantly different. This was observed at all strain rates.

Stress sweep experiments showed that the linear visco-elastic limit for flour dough was below 0.001 and for gluten 0.03. In the two-component gluten and starch mixtures, increase in the amount of starch led to decrease in the linear visco-elastic limit (Figure 1), which confirms previous observations[6].

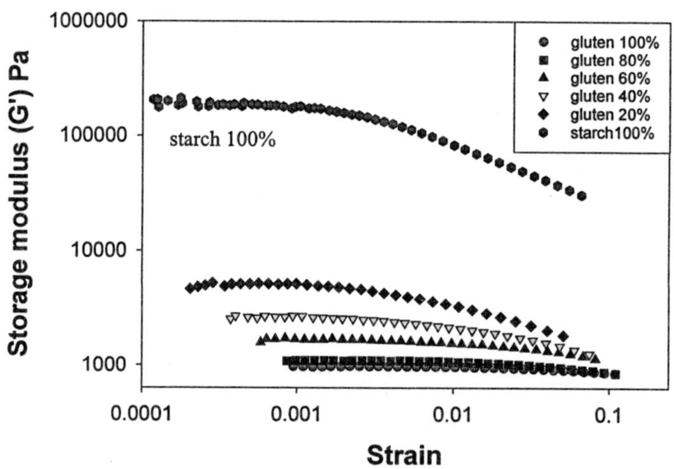

Figure 1 *Stress sweep experiments on two-component mixtures gluten and starch*

The linearity of flour and gluten dough has been investigated by several authors[6,7,8] and various results have been obtained. Though there is no agreement on an exact strain value at which the linear visco-elastic behaviour ceases to exist, the results from these studies show that for flour doughs it is less than 0.2%. The present results on flour dough linearity, confirm these observations. Gluten dough has been found to have a longer linear visco elastic limit of up to 2.1%[6], which is similar to the results we obtained. On the contrary Wang and Kokini[3] conducted strain sweeps from 0.01 to 100% at a frequency of 10 rad s^{-1} and found that the gluten doughs studied were linear up to 10% strain. Khatkar *et al.*[2] found the linear region of gluten to be at the strain value of up to 15%.

Stress sweep experiments also indicated that flour dough had a higher values for G*, G'', G' and dynamic viscosity η* compared to gluten dough. Frequency sweep experiments at 1Hz also showed that glutens had lower G*, G'', G' and dynamic viscosity

η^* than flour doughs. Similar effect has been observed in dynamic viscosity measurements of strong bakers and weak biscuit flours, where dynamic properties decreased with increasing protein content. The strong bakers flour had a lower viscosity than the weak biscuit flour at high strains[9].

During viscometry, dough never reached a steady shear state. Instead, the viscosity increased with shearing, reaching a maximum at which point the sample fractured. Hence the maximum viscosity was used to compare the different treatments. At all three shear rates, the gluten doughs had the highest viscosity.

4 CONCLUSION

The linear visco-elastic region for both wheat flour and gluten is below the strain of 0.001 (0.1%) and 0.03 (3%) respectively. Addition of starch to gluten reduced the linear visco-elastic limit. Flour dough has a higher G', G", G* and dynamic viscosity values compared to gluten dough in dynamic tests. Gluten dough has a higher elongational viscosity and shear viscosity compared to flour dough.

References

1. S. Uthayakumaran, P. W. Gras, F. L. Stoddard, and F. Bekes, *Cereal Chem.*, 1999, **76**, 389

2. B. S. Khatkar, A. E. Bell, and J. D. Schofield, *J. Cereal Sci.*, 1995, **22**, 29

3. C. F. Wang, and J. L. Kokini, *J. Rheol.*, 1995, **39**, 1465

4. F. MacRitchie, *J. Cereal Sci.,* 1985, **3**, 221

5. American Association of Cereal Chemists, Approved Methods of the AACC. Mixograph Method 54-40A., 1988, The Association: St. Paul, MN

6. J. R. Smith, T. L. Smith, and N. W. Tschoegl, *Rheol. Acta*, 1970, **9**, 239

7. G. E. Hibberd, and W. J. Wallace, W. J. *Rheol. Acta*, 1966, **5**, 193

8. L. L. Navickis, R. A. Anderson, E. G. Bagley, and B. K. Jasberg, *J. Texture Stud.,* 1982, **13**, 249

9. M. Safari-Ardi, and N. Phan-Thien, *Cereal Chem.,* 1998, **75**, 80

RHEOLOGY OF GLUTENIN POLYMERS FROM NEAR-ISOGENIC WHEAT LINES

A W J Savage[1], P Rayment[2], S B Ross-Murphy[2], P R Shewry[1] and A S Tatham[1]

1. IACR-Long Ashton Research Station, Department of Agricultural Sciences, University of Bristol, Long Ashton, Bristol BS41 9AF, UK. 2. Biopolymers Group, King's College London, Division of Life Sciences, Franklin-Wilkins Building, 150 Stamford Street, London SE1 8WA.

1 INTRODUCTION

Gluten proteins are important components of flour, providing the structural framework created when flour is mixed with water to form a dough. The two major classes of gluten proteins, gliadins and glutenins, are thought to confer viscosity and elasticity, respectively. The high molecular weight (HMW) subunits of glutenin are implicated in the formation of an extensive polymeric network during dough development with variation in the composition of the HMW subunits being associated with differences in dough elasticity.

In order to elucidate the roles of different proteins in the rheological properties of doughs, many studies have been performed on doughs or glutens with or without added proteins. Fewer studies have been performed on the rheological properties of the individual proteins or polymers. In this work we have investigated the rheology of glutenin polymers isolated from near-isogenic wheat lines, which differed only in the number and types of HMW subunits.

2 MATERIALS AND METHODS

2.1 Polymer isolation

Polymers were isolated from seven near-isogenic wheat lines[1] with different HMW subunit compositions (Table 1). Flour was de-fatted with chloroform; gliadins, albumins and globulins were then removed by extraction with 70% (v/v) aqueous ethanol. Polymers were isolated by extraction with 50% (v/v) propan-1-ol containing 1% (v/v) acetic acid and 10 mM N-ethylmaleimide at 60°C, dialysed against 1% (v/v) acetic acid and freeze-dried. The protein composition of the polymers was analysed using SDS-PAGE on 10% acrylamide gels using a Tris-borate buffer system[2], under reducing conditions.

2.2 Rheological measurements

Polymers were hydrated to 50% moisture content with deionised water in an airtight container overnight at 25°C. Measurements were taken with a Rheometrics Fluids Spectrometer (RFSII, Rheometrics Inc., USA), with a 25 mm diameter parallel plate. Experiments used an arbitrary radius method[3], and were performed at 25°C in the presence of a solvent trap.

Table 1 HMW subunit composition of near-isogenic wheat lines

Wheat line	HMW subunits present		
L88-6	1	17+18	5+10
L88-10		17+18	5+10
L88-14	1		5+10
L88-18	1	17+18	
L88-31		17+18	
L88-25	1		
RG37	no HMW subunits		

3 RESULTS AND DISCUSSION

A typical mechanical spectrum (Figure 1) shows that the samples have a solid-like spectrum with complex viscosity showing power-law behaviour. Complex viscosities for all the polymer samples are similar (Figure 2), showing gel-like behaviour, and are quite different from a gliadin sample, included as a control, which is typically liquid-like. There is a trend for the samples containing a greater number of HMW subunits, and for those containing subunits associated with highly elastic doughs (5+10), having more liquid-like characteristics. SDS-PAGE patterns of isolated polymers and of proteins remaining in the residue after extraction (Figure 3) showed that both polymer and residue fractions contained HMW subunits. It is possible, therefore, that unextracted polymers remain in the residue and that these could also contribute to the rheological behaviour of doughs.

4 CONCLUSIONS

Polymers isolated from wheat lines containing different numbers and types of HMW subunits show very similar rheological behaviour which is in contrast to the different rheological behaviour of doughs and glutens containing different numbers and types of HMW subunits. The reasons for this are unclear, but the results imply that interactions of glutenin polymers with other proteins and/or dough components are important in determining their role in grain processing properties.

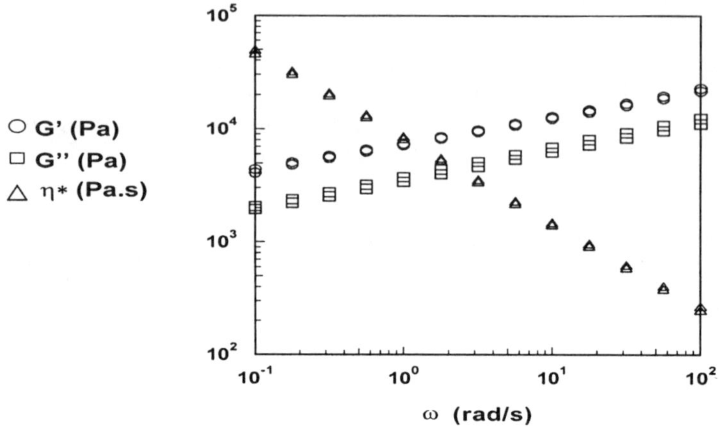

Figure 1 *Mechanical spectrum of polymers from wheat line L88-25 hydrated to 50% moisture content. Replicate measurements are shown.*

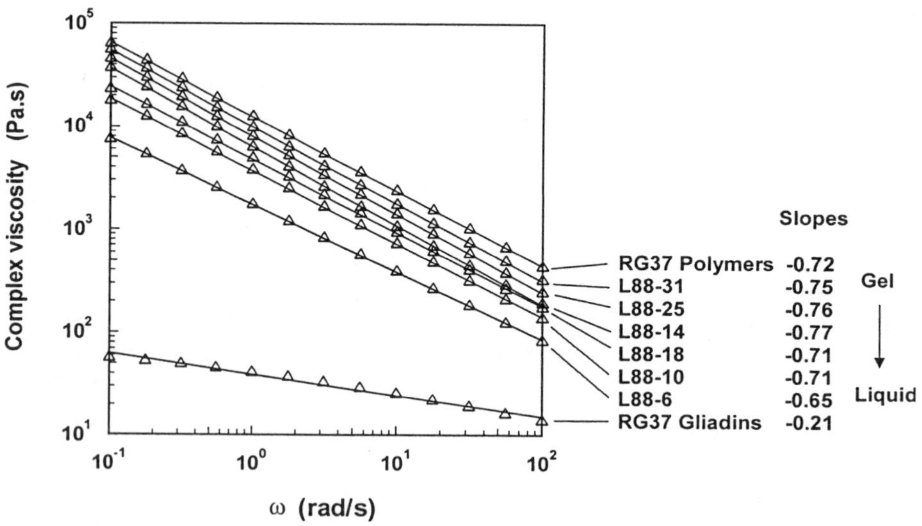

Figure 2 *Complex viscosity versus radial frequency profile of polymers from near-isogenic wheat lines hydrated to 50% moisture content. The calculated slopes show a trend from gel-like to liquid-like character.*

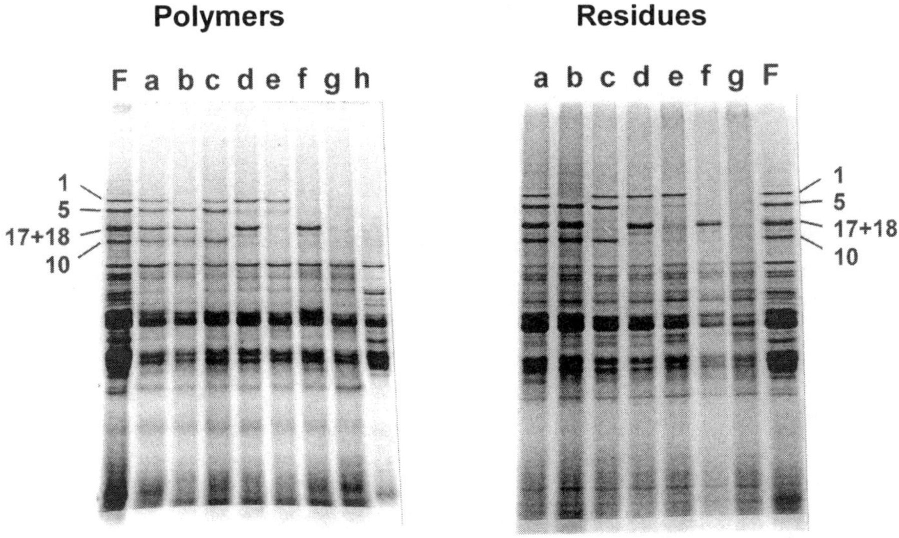

Figure 3 *SDS-PAGE page patterns of proteins from isolated polymers and residual flour proteins from near-isogenic wheat lines: (a) L88-6; (b) L88-10; (c) L88-14 (d) L88-18; (e) L88-25; (f) L88-31 and (g) RG 37. (F) L88-6 flour and (h) RG 37 gliadins.*

Acknowledgements

We thank Rudi Appels (CSIRO, Canberra) for supplying the near-isogenic wheat lines. IACR receives grant-aided support from the Biotechnology and Biological Sciences Research Council.

FERMENTATION FUNDAMENTALS: FUNDAMENTAL RHEOLOGY OF YEASTED DOUGHS

M. Newberry [1,2], N. Phan-Thien[1,2], R. Tanner[1,2], O. Larroque[1,3], S. Uthayakumaran[2]

1. Quality Wheat Cooperative Research Centre Ltd., Locked Bag 1345, North Ryde, NSW 1670, Australia. 2. Department of Mechanical and Mechatronic Engineering, The University of Sydney, NSW 2006, Australia. 3. CSIRO Plant Industry, Grain Quality Research Laboratory, P.O. Box 7, North Ryde, NSW 1670, Australia.

1 INTRODUCTION

Measuring the rheological changes occurring in yeasted bread doughs presents several challenges. Firstly the material is constantly changing during fermentation whilst the decreasing density and fragility of fermenting doughs make them difficult to handle. Furthermore any such studies will be as time consuming as bake testing, since both require fermentation of the doughs. It is for these reasons that dough rheology is almost exclusively conducted on non-yeasted bread doughs. Consequently most of our knowledge of dough rheology relates to studies of wheat flour doughs that do not contain any yeast. Whilst the non-yeasted dough rheology has taught us much about bread dough rheology, and continues to do so, direct investigation of yeasted dough rheology will provide more knowledge of the affects of fermentation. Such knowledge will allow for more accurate conversion of the non-yeasted rheology into the bakery.

2 MATERIALS AND METHODS

2.1 Sample Preparation

A standard commercial bakers flour (12 % protein, Weston Foods Ltd, Australia) was mixed on a 10 g Mixograph to peak dough development. The formulation contained sugar 0.8 %, salt 1 %, 1.2 % dried yeast, and 63% water by flour weight. The doughs were incubated for 0, 30, 60 and 90 minutes at 37°C, before being frozen in liquid nitrogen and then gradually thawed at –20°C overnight and then thawing to room temperature. The samples were then stored at –20°C until the rheological properties were measured. The freezing and thawing process was designed to prevent the fermentation of the samples during rheological measurements, which if this occurred would confound the interpretation of the rheological information.

2.2 Fundamental Rheological Tests

Previous studies have shown that high rather than low strain rheology is needed to distinguish between functionally different flours[1,2]. Thus the yeasted doughs were studied with shear viscometry and uniaxial elongation tests, which deform the dough to high strain levels.

The elongational properties of the dough samples were studied under uniaxial extension at a constant strain rate on a United Universal Testing Machine (United Calibration Crop, Huntington Beach, CA, USA). The samples were glued to 20 mm diameter plates with a cyano-acetate glue to ensure the samples ruptured between the plates. The samples were then pulled apart at an exponentially increasing speed equivalent to a constant strain rate of 0.1 s^{-1}. Shear viscometry measurements were conducted on a Bohlin VOR controlled strain rheometer at a shear rate of 0.105 s^{-1} using a parallel plate configuration of 40 mm diameter and a gap of 2 mm. Sandpaper (100 cv grade) was mounted to both plates to prevent slipping of the samples. The loaded samples were trimmed, coated with paraffin oil to prevent drying out of the sample surfaces and rested for 20 min before testing. All the elongation and shear measurements were conducted at a temperature of 25°C and were performed in triplicate.

2.3 Size Exclusion Analysis

The composition of the proteins within the fermenting doughs was determined using size-exclusion high performance liquid chromatography. The protein samples were prepared for analysis using the methods of Batey et al[3] and Gupta et al[4]. Analysis was conducted using the method of Larroque et al[5].

3 RESULTS AND DISCUSSION

When a dough sample is subjected to uniaxial elongation the elongation viscosity remains relatively constant at low strains. Then as the dough is extended beyond a strain of one the viscosity then starts to increase rapidly. This rapid increase in viscosity at higher strains is known as strain hardening behaviour. Flours with superior baking performance exhibit a greater degree of strain hardening than flours that bake poorly[6,7]. Elongation viscosity reaches a maximum at which point the dough starts to rupture. The maximum viscosity and the strain at which this occurs provide a useful estimate of the strain hardening properties of dough samples under uniaxial elongation. Doughs fermented for longer periods exhibited a decrease in the maximum viscosity and strain at which this maximum occurred.

Shear viscometry of bread dough does not reveal any steady state[2], where a constant viscosity is observed. Instead the viscosity of the sample increases with shearing to a maximum at which point the sample undergoes edge fracture.

Fermented doughs possessed maximum shear and elongational viscosities lower than doughs not subjected to fermentation. Continuing fermentation results in lowered elongational and shear viscosities. The lowered elongational viscosities correspond to decreased strain hardening properties of the fermenting doughs. Likewise the lowered shear viscosities also point to a decrease in the rheological shear properties with fermentation. Thus fermentation decreases the rheological strength of bread doughs.

Analysis of protein composition of the yeasted doughs showed that the soluble protein decreased whilst the insoluble protein increased as fermentation proceeded. In studies of flours of different baking quality greater levels of insoluble proteins correlates to stronger flours with better baking performance[8]. Thus the increase in insoluble proteins during fermentation would seem to be evidence for fermentation induced development of dough structure. However, the decline in shear and elongation viscosities indicates the opposite effect. One explanation would be that as the gas bubbles inflate during fermentation this leads to an aligning of the protein structure aiding in the development of larger protein aggregates. However, as these gas bubbles expand, the ability of the protein structure to maintain a three dimensional network decreases, as the protein network is squeezed into essentially two-dimensional areas of dough surrounding each gas bubble. This reduced interconnection between the proteins could then account for the reduced rheological 'strength' of the fermented dough.

4 CONCLUSIONS

Large strain measurements reveal that fermentation lowers the maximum elongational and shear viscosities of yeasted dough. These rheological changes are indicative of a weakening of dough structure with fermentation. Size exclusion HPLC analysis of the protein composition reveals an increase in the insoluble protein suggesting fermentation contributes to development of protein structure. Whilst protein structure developed with fermentation this development may be limited by the growth of gas bubbles that reduce the interconnection of the protein network leading to rheologically weakened doughs.

References

1. N. Phan-Thien and M. Safari-Ardi, *J. Non-Newt. Fluid Mech.,* 1998, **74,**137.
2. M. Safari-Ardi and N. Phan-Thien, *Cereal Chem.,* 1998. **75**, 80.
3. I. L. Batey, R. B. Gupta and F. MacRitchie, *Cereal Chem.,* 1991, **68**, 207.
4. R. B. Gupta, K. Khan and F. MacRitchie, *J. Cereal Sci.*, 1993, **18**, 23-41.
5. O. R. Larroque, S. Uthaykumaran and F. Bekes, in: *Proceedings of the 47th Australian Cereal Chemistry Conference,* eds. A. W. Tarr, A. S. Ross and C. W. Wrigley, Royal Australian Chemical Institute, North Melbourne, Victoria, Australia, 1997, p. 439.
6. T. van Vliet, A. M. Janssen, A. H. Bloksma and P. Walstra, *J. Texture Stud.*,1992, **23**, 439.
7. T. van Vliet, A. J. J. Kokelaar and A.M. Janssen, in: *Food Colloids and Polymers: Stability and Mechanical Properties.* eds. E. Dickinson and P. Walstra, The Royal Society of Chemistry, Cambridge, UK, 1993, p. 272.
8. F. MacRitchie, in: *Advances in Food and Nutrition Research.,* 1992, **36**, 1.

A FRESH LOOK AT WATER: ITS EFFECT ON DOUGH RHEOLOGY AND FUNCTION

H.L. Beasley[1,2], S. Uthayakumaran[3], M. Newberry[2,3], P.W. Gras[1,2] and F. Bekes[1,2]

1. CSIRO Plant Industry, Grain Quality Research Laboratory, PO Box 7, North Ryde, NSW 1670, Australia. 2. Quality Wheat Cooperative Research Centre Ltd., Locked Bag 1345, North Ryde, NSW 1670, Australia. 3. Department of Mechanical and Mechatronic Engineering, Building F07, The University of Sydney, NSW 2006, Australia.

1 INTRODUCTION

Water is an essential component in bread making. Too much water and flour turns into a thin viscous paste, too little and the flour components will not form a cohesive dough mass. The relationship between dough's visco-elastic properties and end product quality can be substantially altered by small variations in water addition above and below a determined optimum. Understanding the complexities of flour/water interactions provides insight into the basis for the intuitive corrections to water addition (already practised by bakers) and an avenue for predictable manipulation of water to achieve optimum industry parameters. In this study fundamental rheological techniques were examined which could provide well defined, simple physics measurements (with dimensions in standard units) with which to study and define water-induced changes in the visco-elastic properties of dough. Small-scale model doughs (defined at the level of starch granule size) were also used to investigate relationships between water, simplified flour components and functional properties.

2 METHODS

2.1 Basic Rheological Methods

Basic rheological measurements were applied to the study of flour doughs prepared with standard Baker's flour, mixed in an experimental direct drive 10-g pin-mixer to peak dough development. Three levels of water addition were used, that calculated by the AACC standard method[1] and 5% above and below this value. Two types of rheological tests were performed after a 20 min resting period.

A. Uni-axial elongation tests[2] were carried out at three different strain rates (0.01, 0.1 and 1.0 s^{-1}) in a United Testing Machine (Model SSTM 5000).

B. Frequency sweeps between 0.1 Hz- 20 Hz were conducted on a Controlled Stress Rheometer (Reologica Stresstech) at a strain of 0.0005.

2.2 Model Dough Studies

2.2.1 Model Dough Concept. The small-scale model dough system is a sophisticated reconstitution method[3] using gluten proteins fractionated to the level of glutenin and gliadin, and starch to "A-type" (>10 micron diameter) and "B"-type granules (<10 micron diameter). This level of fractionation allows the flexibility to alter glutenin to gliadin ratio, total protein content and starch granule size distributions in the model dough.

2.2.2 Components of the Model Dough. Model doughs were built up from functional glutenin, gliadin, water solubles and starch fractions. Starch, water solubles and gluten were prepared by gluten washing. Gluten was further separated into glutenin and gliadin-rich fractions by mild pH treatment[4]. The protein contents of all fractions were determined by Dumas Combustion (Leco Inc, St Joseph, MI, USA) enabling the total protein content to be controlled. In these experiments all model dough fractions were kept constant except the one of interest, which could be interchanged.

2.2.3 Fractionation of Starch. A commercial starch (provided by Goodman Fielder), containing 67% of the larger A-type granules and 33% of the B-type granules (with minimal starch damage, 2%) was separated by countercurrent sedimentation[5] to give fractions rich or poor in the larger A-type granules. Starch particle size distributions were analysed using a Malvern Particle Sizer. Fraction II showed an enrichment of A-granules (84%) and a small proportion of residual B-granules (16%). Fraction III contained 26% A-granules and a larger proportion (74%) of the B-granules. Water was added at levels between 43-75%, to duplicate models prepared at constant protein content and glutenin to gliadin ratio. Model doughs were mixed in a 2-g Mixograph™ (TMCO, Lincoln, NE, USA).

2.2.4 Altering glutenin to gliadin ratio. Glutenin and gliadin fractions were analysed by size-exclusion HPLC to determine the monomeric to polymeric protein ratio, which was used as a close approximation of the glutenin to gliadin ratio. By keeping the total protein constant (10%) and varying the proportion of glutenin and gliadin added, three ratios (mg glutenin / mg gliadin) were tested (1.25, 0.83 and 0.54) against four levels of water addition (45%, 47.5%, 50% and 52.5%).

3 RESULTS AND DISCUSSION

3.1 Effects of water on dough rheological properties

Uni-axial elongation tests showed that increases in water content led to a decrease in dough elongational viscosity (Figure 1), at all three strain rates. Results of frequency sweep tests also showed a decrease in dynamic viscosity (Figure 2a) with increase in water content. Increased water content also led to decreases in dynamic rheological properties such as dynamic modulus, storage modulus (Figure 2b), loss modulus and phase angle.

3.2 Effects of water on dough functional properties

Model dough rich in A-granule starch had low water requirements and gave long mixing times (time to peak dough development, MT) but formed workable doughs. Dough made from starch consisting of 73% B-granule starch had very different properties, with much higher water requirements and shorter MT, forming dough with a putty-like consistency. These results reflect those with classical gluten/starch reconstitutions using commercially separated A and B starch granules (Gras and Asp, unpublished). Models with a higher percentage of A-granule starch were more tolerant of higher/lower than optimum water additions. Decreasing A-granule starch in the model led to lower water addition requirements, shorter MT, increased dough breakdown and greater bandwidth at peak dough development. Over all mixes, MT was the Mixograph parameter most influenced by the combined effects of water addition and starch size distribution. Varying water addition had proportionately less effect on Mixograph peak resistance.

Altering the glutenin to gliadin ratio of dough, while keeping the total protein content constant, can greatly effect functional properties, particularly dough strength and stability[6]. Model doughs clearly demonstrated this relationship across three levels of glutenin to gliadin ratio and all four levels of water addition (Table 1). Doughs with high glutenin to gliadin ratios (1.25) were characterised by long MT and low rates of breakdown. They had an exaggerated response to water addition, particularly at the lower water levels, while doughs with the lowest glutenin to gliadin ratio (characterised by short

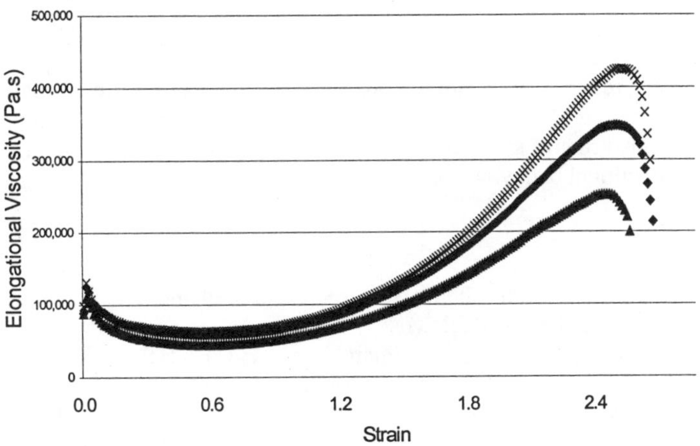

Figure 1 *Dough Elongational Viscosity changes with water addition at 61%* (×), *66%* (◆) *and 71%* (▲) *at a strain rate of 1.0* s^{-1}.

Table 1 *The combined effects of water addition and altered glutenin to gliadin ratio on key Mixograph parameters Mixing Time (MT) and Peak Resistance (PR).*

	MT (sec)				PR (AU)			
	% water addition				*% water addition*			
	45.0	*47.4*	*50.0*	*52.5*	*45.0*	*47.4*	*50.0*	*52.5*
Glutenin to Gliadin Ratio:								
1.25	543	460	402	381	3	5	5	8
0.83	281	228	234	237	13	13	17	16
0.33	199	143	156	122	21	24	27	34

MT and poor stability) were more tolerant of different water additions. Models with a glutenin to gliadin ratio of 0.83 were the most stable (ie mixing parameters fluctuated least) across the levels of water used in this experiment.

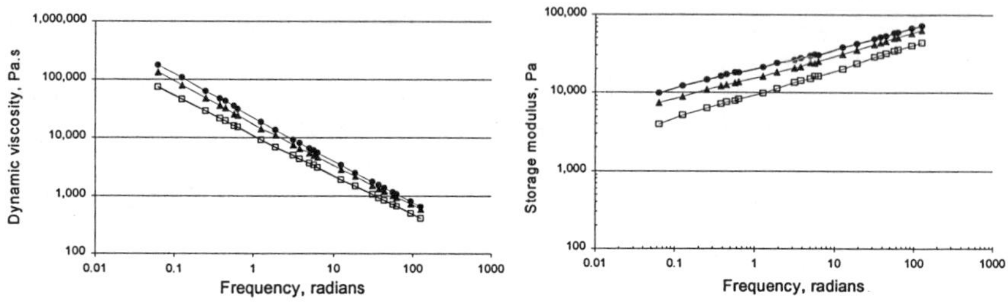

Figure 2 *Shifts in (**a**) dynamic viscosity and (**b**) storage modulus during frequency sweep experiments at 59% (•), 64% (▲) and 69% (□) water addition.*

4 CONCLUSIONS

Increased water content reduced both elongational and dynamic viscosities. Small-scale model doughs demonstrated the interaction of both protein composition (glutenin to gliadin ratio) and starch granule size on water absorption properties of doughs which translated to altered functional behaviour.

References

1. American Association of Cereal Chemists, 1988. Approved method of the AACC. Method 54-40A, 1988. AACC, St Paul, MN
2. S. Uthayakumaran, M. Safari-Ardi, N. Phan-Thien and F. Bekes. in: *Proc.48th Aust.Cereal Chem. Conf.,* Cairns, QLD, RACI Melbourne, 1998, 83
3. C.L. Blanchard, D. Rylatt, H.P. Manusu, C.W. Wrigley, P.W. Gras, J.H. Skerritt, and F. Bekes, in *Proc.48th Aust.Cereal Chem.Conf.,* Cairns, QLD, RACI Melbourne, 1998, 17
4. F. MacRitchie, *J. Cereal Sci.,* 1985, **3**, 221
5. P. Meredith, *Starch*, 1981, **33**, 40
6. S. Uthayakumaran, P. W. Gras, F. L. Stoddard and F. Bekes, *Cereal Chem.*, 1999, **76**, 389

GLUTEN QUALITY VS QUANTITY: RHEOLOGY AS THE ARBITER

K.M. Tronsmo[1,2], E.M. Færgestad[1], E.M. Magnus[1] and J.D. Schofield[2]

1. MATFORSK - Norwegian Food Research Institute, Osloveien 1, N-1432 Ås, Norway
2. The University of Reading, Department of Food Science and Technology, Reading RG6 6AP, UK

1 INTRODUCTION

The effects of protein content and gluten quality are often hard to separate, and many of the current flour quality tests show results that are influenced by both factors. Since protein content and protein quality have different effects on baking performance of wheat flours, there is a need for methods that can distinguish between these factors.[1-3]

In this study, results from large and small deformation rheological analyses were compared with baking data to find which parameters were related to gluten quality rather than to protein content or a mixture of these two factors. The baking experiment was production of hearth bread, which is made without a tin. For such products, protein quality is particularly important due to the need for the dough to retain a proper shape during proving and baking.

Large deformation rheological tests were performed on dough, as the stress applied by these methods is comparable to stresses experienced by the dough during mixing and proving[4]. On the other hand, small deformation rheology can be related to molecular weight distribution[5,6]. Earlier studies[7] have shown that small deformation tests on dough give poor correlations with baking data as the non-linear behaviour of starch masks the effect of the gluten proteins, the latter having the most important influence on breadmaking quality. Small deformation tests on gluten have been shown to give much better correlations with baking data, and in this study the small deformation tests have thus been performed on freshly washed gluten from the respective flours.

2 MATERIALS AND METHODS

2.1 Wheat material

Four Norwegian wheat cultivars were chosen to span a relevant range of gluten quality and quantity: cultivars Folke and Polkka as weak wheat varieties, Portal and Bastian as strong ones. Protein contents ranged from 10.6% to 13.0% (d.m.). For Portal and Folke, samples of two different protein levels were included in the study. The samples were milled in a commercial mill at a scale of 30 tons and 30ppm of ascorbic acid were added. Table 1 shows the high molecular weight (HMW) subunit composition of each variety and the protein contents in percent of dry matter (d.m.) of the flour samples.

Table 1 *Variety, high molecular weight glutenin subunit composition and protein content of the flours studied*

Variety	HMW subunits	Protein (% of d.m.)
Bastian	2*, 7+9, 5+10	13.0
Portal	1, 7+9, 5+10	11.2
"	"	12.1
Polkka	2*, 7+8, 2+12	12.5
Folke	2*, 6+8, 2+12	10.6
"	"	11.5

2.2 Baking

Experimental baking of hearth bread was performed in a commercial bakery (Nøtterøy Bakeri og Konditori AS) as described by Færgestad et al.[2] The standard recipe and process parameters of the bakery were used, varying only the flour quality. Loaf volume was measured by rapeseed displacement, and height and width of loaves were measured using a PAV-caliper (ABC Maskin AS, Skien, Norway). Form ratio was calculated as the ratio loaf height/loaf width.

2.3 Large deformation rheology

Uniaxial extension measurements were carried out using the Stable Microsystems/Kieffer dough and gluten extensibility rig for the SMS Texture Analyser (Stable Micro Systems, Godalming, UK)[8]. Flour-water doughs were prepared with water absorptions to give Farinograph consistencies of 500BU, and were mixed to peak in a 10-g Mixograph. Doughs with 2% salt (based on flour weight) were prepared using 2% less water than the flour-water doughs, but mixed for the same time.

Biaxial extension was performed with the Dobraszczyk/Roberts Dough Inflation System[4,9] on flour-water doughs with water absorptions to give Farinograph consistencies of 500BU and were mixed to peak time in a Farinograph with a 300-g bowl.

2.4 Small deformation rheology

For small deformation tests gluten was freshly prepared by the Glutomatic washer (Falling Number, Huddinge, Sweden) using 5.2mL of distilled water for 10g of flour, mixing for 1min and washing with distilled water for 10min. Fifteen minutes after the end of washing, gluten was loaded onto a Stress Tech dynamic oscillatory rheometer (ReoLogica Instruments AB, Lund, Sweden) with a plate-plate geometry and a 2 mm gap. A bath of silicon oil was created around the sample to prevent it from drying out. After an equilibration time of 60 min after loading, the sample was subjected to a frequency sweep from 0.006Hz to 10Hz at a strain of 0.02 and 25°C, followed by a creep recovery test at a stress of 50Pa for 100 s and a recovery period of 20 min.

3 RESULTS AND DISCUSSION

3.1 Baking

In accordance with earlier studies[1-3], the baking of hearth bread from these flours showed a clear relationship between gluten quality and form ratio (loaf height/loaf width). Cultivars Bastian and Portal gave high form ratios in the range 0.75 to 0.79, while loaves from cvs Folke and Polkka gave lower form ratios, ranging from 0.64 to 0.67. Protein content had no positive effect on form ratio. On the other hand, loaf volume was influenced both by protein content and gluten quality.

3.2 Large deformation rheology

Kieffer resistance to extension for flour-water doughs showed significantly higher values for the strong varieties Bastian and Portal than for the weak varieties Polkka and Folke, as shown in Figure 1a. There was no significant correlation with protein content. Extension at peak and extensibility (distance at rupture) were influenced both by protein content and gluten quality.

Kieffer resistance to extension for doughs without salt gave a better correlation with loaf form ratio than doughs containing 2% salt. Salt increased the peak force for all samples except Portal-12.1, which was not significantly affected by the addition of salt.

Dough Inflation measurements on flour-water doughs also gave good correlations with loaf form ratio, especially for the parameters maximum pressure (P), area under the pressure-time curve (W), and strain hardening index (Fig. 1b). These methods enabled a distinction to be made between the group of strong varieties (Bastian and Portal) from the weak ones (Polkka and Folke).

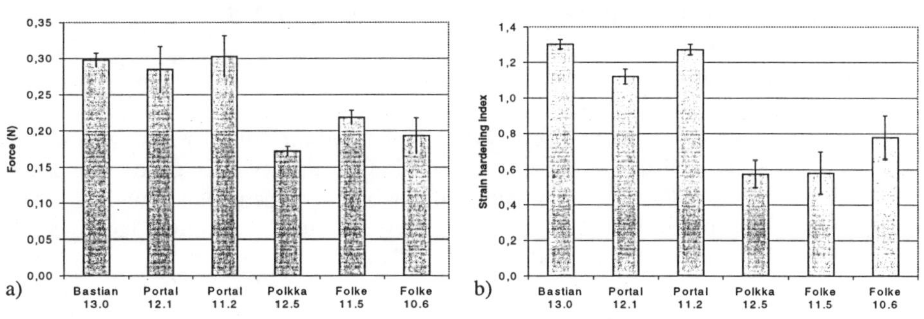

Figure 1 a) Resistance to Extension as measured by the SMS/Kieffer rig and b) Strain Hardening index as calculated from Dough Inflation measurements.

3.3 Small deformation rheology

Frequency sweeps showed that the elastic modulus (G′) of Portal-11.2 was significantly higher and G′ of Polkka significantly lower than those of the remaining samples, which all showed similar values for G'.

Creep recovery tests showed higher rapid recovery, lower delayed recovery and lower viscous loss for glutens from strong varieties than for glutens from weak varieties. The samples were grouped according to quality as strong (containing the HMW subunits 5+10) and weak (HMW 2+12). Analysis of variance of the creep recovery data showed a significant difference between these groups for all three parameters: rapid recovery (p=0.01), delayed recovery (p=0.015) and viscous loss (p=0.000).

4 CONCLUSIONS

Large deformation rheological tests (Kieffer rig and Dough Inflation) gave good correlations with baking performance for the sample set. For small deformation tests, the differences between strong and weak samples were less explicit, but further studies will determine the significance of these tests and their abilities to distinguish between glutens of different baking qualities. Studies involving larger samples are also in progress.

References

1. E.M. Færgestad, E.L. Molteberg and E.M. Magnus, *J. Cereal Sci.*, 2000, **31**, in press .
2. E.M. Færgestad, P. Baardseth, F. Bjerke, E.L. Molteberg, A.K. Uhlen, K. Tronsmo, A. Aamodt and E.M. Magnus, in *Gluten 2000*, Royal Society of Chemistry, Cambridge, 2000 (these proceedings).
3. T. Næs, F. Bjerke and E.M. Færgestad, *Food Quality and Preference*, 1999, **10**, 209.
4. B.J. Dobraszczyk, *Cereal Foods World*, 1997, **42**, 516.
5. A.A. Tsiami, A. Bot, W.G.M. Agterof and R.D. Groot, *J. Cereal Sci.*, 1997, **26**, 15.
6. A.A. Tsiami, A. Bot and W.G.M. Agterof, *J. Cereal Sci.*, 1997, **26**, 279.
7. B.S. Khatkar, *Functional and Dynamic Rheological Properites of Wheat Gluten*, PhD thesis, The University of Reading, UK, 1996.
8. R. Kieffer, H. Wieser, M.H. Henderson and A. Graveland, *J. Cereal Sci.*, 1998, **27**, 53.
9. B.J. Dobraszczyk and C.A. Roberts, *J. Cereal Sci.*, 1994, **20**, 265.

Acknowledgements

We would like to thank Dr. Pernille Baardseth who headed the project within which these wheat samples were collected, milled and baked. Our thanks also go to the bakery *Nøtterøy Bakeri og Konditori AS* for performing the baking experiments. Funding for the rheology work from the Norwegian Research Council is gratefully acknowledged.

THE HYSTERETIC BEHAVIOUR OF WHEAT-FLOUR DOUGH DURING MIXING

R. S. Anderssen[1] and P. W. Gras[2]

1. CSIRO Mathematical and Information Sciences, GPO Box 664, Canberra ACT 2601.
2. CSIRO Plant Industry, PO Box 7, North Ryde, NSW 1670

1 INTRODUCTION

In a traditional examination of the rheology of a wheat-flour dough, the currently accepted strategy is to first mix the dough to peak dough development, to then perform a suitable rheological experiment on the dough, and finally, to formulate a qualitative or a simple mathematical model to analyse and interpret the observed behaviour of the dough. However, in such situations, the dough is being treated separately from the process by which it has been manipulated and produced, as if it were not a "*living system*"[1]. An alternative approach is to directly utilize the information in the measurements obtained from a recording mixer, such as a Mixograph™. This allows one to monitor the changing behaviour of a dough during mixing. In this way, the rheology of the dough is simulated as a part of the overall modelling of the mixing, and the constitutive relationship, which defines the changing rheology of the dough, is encapsulated implicitly within the modelling. The first step in a project to examine the feasibility and merit of performing such a study of the '*elongate-rupture-relax*' mixing action of a Mixograph™ has been discussed previously[2]. Normally, this is done as two separate steps where the dough is first mixed and then a sample of it is tested in an extension tester. In essence, a Mixogram is a record of a series of extension tests which result from the '*elongate-rupture-relax*' action which a Mixograph™ performs on the dough. The significance of this fact has been discussed in the recent literature[3-5]. In particular, it has been established that the bandwidth of a Mixogram measures the changing extensional viscosity of a dough as it is being mixed[5].

Consequently, the flow and deformation occurring within a dough can be viewed as a repetitive loading process consisting of a sequence of '*elongate-rupture-relax*' events. Chemically and physically, the elongation relates to the nature of the protein unfolding and cross-linking which the extension of the dough induces, while the relaxation relates to the partial refolding of the proteins, which occurs after the rupture, until the start of the next extension. Because the unfolding and refolding will occur at different rates, partially in response to the intensity of the cross-linking being stimulated by the mixing, one is led naturally to the hypothesis that "*the behaviour of wheat-flour dough during mixing is hysteretic in nature*". The purpose of this paper is to present evidence that supports, confirms and exploits this hypothesis. As explained in an earlier paper[6], high resolution Mixograms can be used to determine partial Bauschinger plots of the changing stress-strain behaviour of the dough during mixing as well as to perform rainflow counting on the 'elongate-rupture-relax' events. In this paper, the focus is the analysis and interpretation of the partial Bauschinger plots. The advantage of such methodologies is that they give one the opportunity *to see to* the molecular level of the changing rheology of a dough during mixing, which is important for the optimization of industrial mixing and plant breeding investigations.

2 MATERIALS AND METHODS

Details of the materials and methods have already been published[7].

3 RESULTS AND DISCUSSION

3.1 Some Background

The importance of high resolution Mixograms in understanding and analysing the mixing of wheat-flour dough has been previously identified and examined[6-8]. Such Mixograms (Figure 1) contains greater detail about the mixing than those previously published. The evolving rheology of dough will be determined not only by its current configuration, but also by the nature, size and duration of the forces to which it has been subjected and the way in which the dough has responded to such forces[4]. There is a counterpart with metals, which over short periods, behave like elastic solids, but which over longer periods, exhibit elasto-plastic behaviour. The importance of this observation dates from the seminal work of Bauschinger[12]. Though the mixing of wheat-flour dough is more complex, it is a repetitive (extensional) loading process. Consequently, the modelling and decision-making already developed for the fatigue analysis of metals represents a starting point for how one might model and analyse the mixing of wheat-flour dough.

3.2 Rainflow Counting

A high-resolution Mixogram can be reinterpreted as a loading curve by simply replacing it by the piecewise linear curve which connects its successive minima and maxima. When applied to a loading curve, rainflow counting maps that loading curve into an equivalent rainflow matrix and a unique irreducible string[13]. Rainflow counting has been applied recently to some high resolution Mixograms[6].

3.3 Bauschinger Plots

Bauschinger plots are essentially stress-strain plots. The steps involved in their construction will be outlined elsewhere. The Bauschinger plots obtained from typical Mixograms change throughout mixing, passing through the following four stages of mixing: (a) Hydration, (b) Maximum Bandwidth, (c) Peak Dough Development, and (d)

Overmixed (Figure 2). These plots show the changing stress-strain patterns in the dough during mixing much more explicitly and compactly than the high-resolution Mixogram from which they were derived. During hydration, the Bauschinger plot shows quite erratic behaviour that develops into a well defined pattern as the dough passes through maximum bandwidth and peak dough development. As the dough becomes more and more overmixed, dough progressively collapses to a state approximating viscous drag. The dough resists elongation resulting from the relative motion between the fixed and moving pins. It is quite weak during hydration and becomes stronger and more viscoelastic during maximum bandwidth and peak dough development and looses strength and viscoelasticity as overmixing progresses. The extent of viscoelasticity developed can be assessed qualitatively by the extent of non-linearity of the Bauschinger plots. Independently, it is known how extensional information about a polymer, like the Bauschinger plots of Figure 2, is determined by the level of cross-linking within that polymer[14]. In terms of such techniques, the Bauschinger plots of Figure 2 show how the cross-linking in the dough is increasing from hydration to peak dough development and then decreasing.

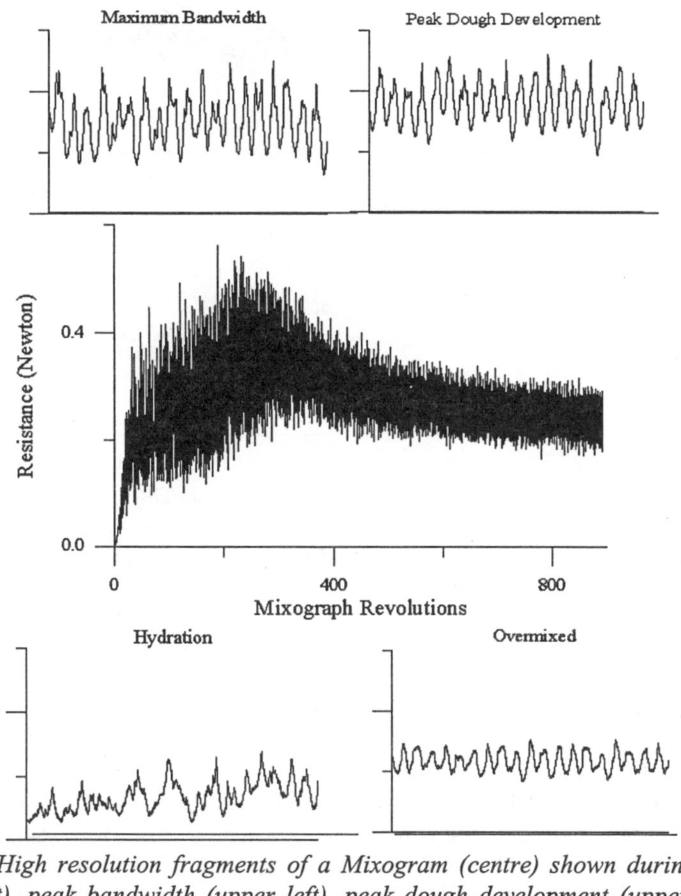

Figure 1 *High resolution fragments of a Mixogram (centre) shown during hydration (bottom left), peak bandwidth (upper left), peak dough development (upper right) and overmixed dough (lower right).*

4 CONCLUSIONS

Historically, cereal chemistry has lived on the success of the invention of recording mixers for more than 60 years. This is now changing under the challenges of modern molecular biology and genetic engineering as well as improved plant breeding strategies and the rationalisation of food technology. The impacts of cDNA technology, electrophoresis and HPLC on cereal science, and their successes with the identification of glutenin subunits, the starch biosynthetic pathways and the identification of grain hardness genes are already familiar. The real challenge for cereal science is to understand how all these different molecular moieties combine to produce a wheat-flour dough. Rainflow counting and the cross-linking information hidden in Bauschinger plots have a key role to play in meeting this challenge.

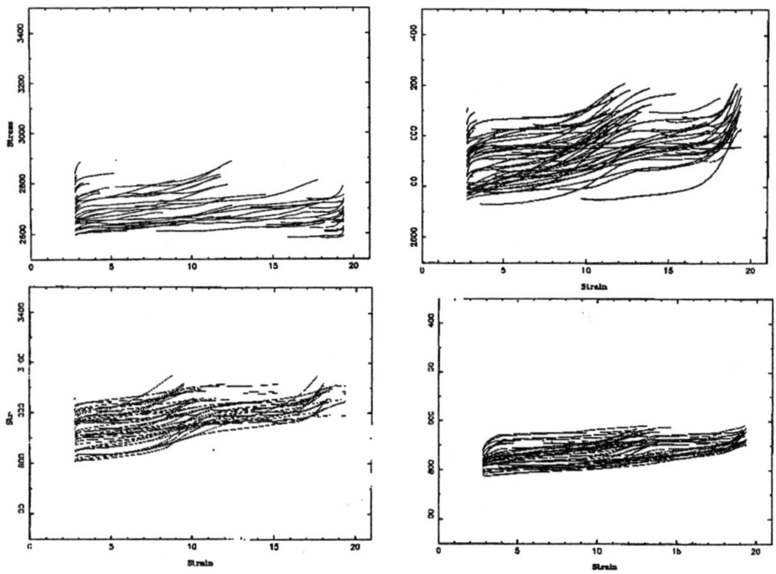

Figure 2 *Bauschinger plots of short intervals of dough mixing from a Mixogram. The plots show Stress (ordinate) against Strain (abscissa) and data from hydration (top left), peak bandwidth (top right), peak dough development (lower left) and overmixed dough (lower right).*

References

1. J. Meisner and J. Hostettler, *Rheol. Acta*, 1994, **33**, 1.
2. P. W. Gras, F. MacRitchie and R. S. Anderssen, 'MODSIM 97, Proceedings of International Congress on Modelling and Simulation', Modelling and Simulation Society of Australia, ANU, Canberra, ACT, 1998, p. 512.
3. R. S. Anderssen, P.W. Gras and F. MacRitchie, *Chem. in Aust.*, 1967, **64**, 3.
4. R. S. Anderssen, I. G. Gotz, and K.-H. Hoffmann, *SIAM J. Appl. Math.*, 1998, **58**, 703.
5. P. W. Gras, H. C. Carpenter and R. S. Anderssen, *J. Cereal Sci.*, 2000, **31**, 1.

6. P. W. Gras and R. S. Anderssen, 'Proceedings of the 49th Annual RACI Cereal Chemistry Conference', 2000.
7. R. S. Anderssen, P. W. Gras and F. MacRitchie, *J.Cereal Sci.,* 1998, **27**, 167.
8. M. Shogren, J. L. Steele and D. L. Brabec, 'Proceedings of the Internat. Wheat Quality Conf.', Grain Industry Alliance, Manhattan, Kansas, 1997, p. 105.
9. R. H. Buchholz, *The Mathematical Scientist,* 1990, **15**, 7.
10. R. S. Anderssen, P. W. Gras and F. MacRitchie, 'Proceedings of the 6th International Gluten Workshop: Gluten '96', RACI, North Melbourne, VIC, 1990, p. 249.
11. P. W. Gras, G. E. Hibberd and C. E. Walker, *Cereal Foods World,* 1990, **35**, 572.
12. H. J. Mughrabi, *Metallkde.,* 1996, 77, 703
13. M. Brokate, K. Dressler and P. Kreji, *Eur. J. Mech., A/Solids*, 1996, 15, 705
14. L. E. Nielsen, 'Mechanical Properties of Polymers and Composites', Marcel Dekker, New York, 1974.

QUANTITY OR QUALITY? ADDRESSING THE PROTEIN PARADOX OF FLOUR FUNCTIONALITY

S. Uthayakumaran[1,2], M. Newberry[1,2], F.L. Stoddard[1,3], F. Bekes[1,4]

1. Quality Wheat Cooperative Research Centre Ltd., Locked Bag 1345, North Ryde, NSW 1670, Australia. 2. Department of Mechanical and Mechatronic Engineering, Building F07, The University of Sydney, NSW 2006, Australia. 3. Plant Breeding Inst., Woolley Bldg. A20, The University of Sydney, NSW 2006, Australia. 4. CSIRO Plant Industry, Grain Quality Research Laboratory, PO Box 7, North Ryde, NSW 1670, Australia.

1 INTRODUCTION

There are several approaches to answer the question "what constitutes the basis of baking quality?" One approach is to measure the compositional parameters such as the amount of protein, glutenin or gliadin in a flour and to search for correlation with bread making performance. Another is to study the relationship between composition and functionality by fractionation and reconstitution[1]. In this method, the functionality of each of the separated flour components is evaluated by varying its amount in a given flour or by interchanging separated fractions between flours of different baking quality, while holding all other components constant. If the protein composition of the wheat is to be used to predict dough and bread quality, the details of the relationship between the quality and composition should be better understood. Therefore the effect of protein content (quantity) and glutenin-to-gliadin ratio (quality) on dough properties has been studied in detail using both empirical and basic rheological methods. The study shows that both protein quantity and composition play an important role in determining bread dough functionality.

2 MATERIALS AND METHODS

2.1 Sample

Flours from cultivars Banks, Hartog, Rosella and Sunbri (protein contents of 13.0%, 12.4%, 8.2%, 14.7% respectively) were used. Flours were defatted, then gluten and starch were prepared from them. Glutenin- and gliadin-rich fractions were prepared from each of the glutens using hydrochloric acid precipitation[2]. Blends of each of the base flours with fractions isolated from that flour were prepared. To vary the protein content at constant glutenin-to-gliadin ratio, flour and its gluten were blended to give 110%, 120% and 130% of the base flour protein content and flour and its starch were blended to give 70%, 80% and 90% of the base protein content. The glutenin-to-gliadin ratios of the flours were varied by adding isolated gluten, glutenin or gliadin to the parent flour and in this experiment the protein content was maintained at 120% of the parent flour. The amount of water to be added to the blend was calculated using a standard method[3].

2.2 Functional properties

2.2.1 Mixing, extension and baking studies. All formulations were mixed in a 2-g Mixograph (TMCO, Lincoln, NE, USA). The parameters determined were mixing time (MT), peak resistance (PR) and resistance breakdown (RB). Doughs for extension testing were mixed to peak dough development. Extension testing was carried out on a micro-extension tester using 1.7 g dough[4]. The extension parameters determined were extensibility (Ext) and maximum resistance to extension (Rmax). Test baking was carried out on 2.4 g dough[4] and the loaf height (LH) was measured. All tests were carried out in triplicate.

2.2.2 Basic rheological tests. The elongational properties of dough were studied using a constant-strain rate extension technique. All formulations were mixed in a 10-g Mixograph. The dough was mixed to peak dough development and suspended between a fixed and a moving grip both having a diameter of 30 mm. The dough sample was rested for 45 min before testing and moisture loss was prevented by applying a layer of petroleum jelly around the edge of the sample. The dough was pulled apart exponentially in a Universal Testing Machine (United Calibration Corp, Huntington Beach, CA), at a constant strain-rate (0.01 s^{-1}). Force and distance data collected by the computer were used to calculate the rheological parameters of strain and elongational viscosity (Pa·s). The tests were performed in an air-conditioned laboratory with a variation of ±0.5°C in the 24°C ambient temperature.

The viscometric properties of the doughs were studied using shear viscometry. The mixed dough was mounted on a controlled stress rheometer (Reologica Stresstech, Reologica Instruments AB, Lund, Sweden) in the parallel plate configuration (25 mm diameter). Sandpaper was glued to the plates to prevent slippage. The edge of the sample was coated with food-grade petroleum jelly. Before starting the measurement, the dough was allowed to rest for 45 min. A constant shear rate of 0.9644 s^{-1} was applied to the sample and the viscosity was plotted against time. The maximum viscosity (Pa·s) during shear was determined. The temperature was maintained at 24°C ± 0.5°C.

3 RESULTS AND DISCUSSION

MT, PR, Rmax, LH and Rupture viscosity all increased with increases in both protein concentration and glutenin-to-gliadin ratio (Figure 1). In cv. Sunbri increasing the protein content caused an almost linear increase in MT. In the others, the MT increased only at higher levels (120% and 130% of parent flour) of protein. The PR at any specific protein content appeared to be determined by variety.

The effect of protein content on RB was cultivar-specific (Figure 1). Banks and Rosella showed a general increase in RB with increase in protein content, in Hartog RB decreased as protein increased and in Sunbri there was a initial decrease in RB followed by an increase. Increasing the ratio of glutenin to gliadin led to decreased RB. Ext increased with protein content but decreased with increase in glutenin-to-gliadin ratio.

Increasing protein content increased the strain-hardening properties (elongational rupture viscosity and rupture strain). For each cultivar, the elongational viscosity curves for the various protein levels differed only at the point of rupture. Elongational rupture viscosity and rupture strain increased linearly with increasing protein levels in all cases. Increases in glutenin-to-gliadin ratio increased the strain-hardening properties as indicated by increasing elongational rupture viscosity. Nevertheless, the greatest effect was observed with the addition of gliadin, which resulted in a considerably decreased elongational rupture viscosity. Increasing the protein content of doughs lowered the measured shear viscosity and maximum viscosity for all the cultivars. Increasing the glutenin-to-gliadin ratio had the opposite effect, increasing the shear viscosity and the maximum viscosity. Rosella had the lowest viscosity and Hartog the highest.

The use of basic rheological techniques provides a greater level of information on the elongation and shear properties of bread doughs than conventional, empirical techniques have allowed. The results obtained from small-scale empirical extension testing (Ext and Rmax) and basic rheological extension testing (elongational rupture viscosity and strain) were strongly correlated (Table 1), showing that they measured very similar parameters. There was, however, a certain amount of scatter around the regression line, attributable to variation within the samples as well as to differences in accuracy between the two types of instruments. Nevertheless, this comparison confirms the validity of these basic rheological measurements. The results obtained by using a small-scale empirical extensigraph-like device showed many significant correlations with the basic rheological results. The extensibility measured by the small-scale extension tester (Ext) and the uniaxial elongational rupture strain were highly correlated, with r values of 0.924 where protein content was varied and 0.903 where glutenin-to-gliadin ratio was varied. The maximum resistance to extension and the elongational rupture viscosity also had high correlations (0.757 and 0.719) in the two experiments. Both maximum resistance to extension and elongational rupture viscosity had very strong correlations with maximum shear viscosity and negative correlations with Ext and elongational rupture strain. Both correlations were stronger when the glutenin-to-gliadin ratio was modified than when protein content was altered. The Mixograph resistance breakdown showed a strong positive correlation to Ext and elongational rupture strain in the glutenin-to-gliadin ratio experiment and almost as strong a negative correlation in the protein content experiment. Loaf height was positively correlated with elongational rupture strain and Ext and negatively correlated with maximum shear viscosity when the protein content was varied, but no correlations were significant when glutenin-to-gliadin ratio was varied.

Mixing time, Peak resistance, Rmax, Loaf height, Rupture viscosity		Resistance breakdown		Extensibility, Rupture strain		Maximum shear viscosity	
↑	↑	↑↓	↓	↑	↓	↓	↑
Pro	glu:gli	Pro*	glu:gli	Pro	glu:gli	Pro	glu:gli

* Direction of effect depends on cultivar

Figure 1 *Effects of increases in protein concentration (Pro) or glutenin-to-gliadin ratio (glu:gli) on mixing, extension and baking characteristics of wheat flours*

4 CONCLUSION

Protein content and protein composition play different roles in determining the various dough properties. The greatest contrast was in extensibility, which was positively correlated with protein content and negatively with glutenin-to-gliadin ratio. Fundamental rheology has confirmed the value of traditional, empirical dough rheology.

Table 1 *Correlation matrix for Mixograph, Extensograph and baking parameters with fundamental rheological parameters, for samples varying in protein content (below the diagonal, df = 22) or in glutenin-to-gliadin ratio (above the diagonal, df = 10)*

	Mixograph			Extensograph		Baking	Elongation		Shear
	Mixing time	Peak resistance	Resis-tance break-down	Maximum resistance to extension	Extensi-bility	Loaf height	Rupture strain	Rupture viscosity	Maximum viscosity
Mixing time		-0.103	-0.886**	0.701*	-0.913**	-0.108	-0.881**	0.107	0.611*
Peak resistance	-0.573**		-0.045	0.495	0.009	0.108	-0.100	0.786**	0.554
Resistance breakdown	0.541**	-0.247		-0.763**	0.946**	-0.185	0.794**	-0.320	-0.624*
Maximum resistance to extension	0.503*	0.632	0.358		-0.778**	0.173	-0.715**	0.719**	0.963**
Extensi-bility	-0.832**	0.551**	-0.693**	-0.477*		0.065	0.903**	-0.221	-0.673*
Loaf height	-0.131	0.278	-0.547**	-0.061	0.595**		0.333	0.534	0.121
Rupture strain	-0.717**	0.640**	-0.599**	-0.276	0.924**	0.657**		-0.124	-0.684**
Rupture viscosity	-0.116	0.843**	-0.088	0.757**	0.177	0.241	0.349		0.713**
Maximum shear viscosity	0.339	0.178	0.430	0.766**	-0.652**	-0.590**	-0.549**	0.457*	

*, **: $P < 0.05$, 0.01 respectively

References

1. S. Uthayakumaran, P. W. Gras, F. L. Stoddard and F. Bekes, *Cereal Chem.*, 1999, **76**, 389.
2. F. MacRitchie, *J. Cereal Sci.*, 1985, **3**, 221.
3.American Association of Cereal Chemists. Approved Methods of the AACC. Mixograph Method 54-40A, 1988, The Association: St. Paul, MN.
4. P. W. Gras and F. Bekes, *Proceedings of the 6th International Gluten Workshop*, ed. C. W. Wrigley, Royal Australian Chemical Institute, North Melbourne, Vic., 1996, p. 506.

EFFECT OF PROTEIN FRACTIONS ON GLUTEN RHEOLOGY

C. E. Stathopoulos, A. A. Tsiami, J. D. Schofield

The University of Reading, Department of Food Science and Technology, Whiteknights, PO Box 226, Reading, RG6 6AP, UK

1 INTRODUCTION

The rheological behaviour of wheat gluten can be pictured as the sum of the behaviour of its components. In this study a number of gluten subfractions of narrow molecular weight range was used. Flour proteins of the cultivar Hereward (good breadmaking quality) were fractionated, using a salt precipitation technique[1]. The fractions obtained have distinct molecular weight distribution ranging from high molecular weight to low molecular weight[2,3]. The fractions have been characterised rheologically and as expected the HMW fractions showed a more gel-like behaviour while the LMW fractions had a more viscous-like behaviour.

The aim of this study is to investigate how the rheological profile of total gluten is affected when the proportion of its subfractions is changed with the addition of a fraction of known rheological properties.

2 MATERIALS AND METHODS

2.1 Materials

The flour used was of the good breadmaking variety Hereward.

2.2 Methods

2.2.1 Fractionation and characterisation. The flour was fractionated using a sequential salt precipitation technique[1,3]. The fractions obtained from this procedure have been characterised in terms of molecular weight distribution with the use of Flow Field Flow Fractionation[3]. The polypeptide composition of the fraction was examined using SDS-PAGE.

2.2.2 Sample preparation. The samples subjected to the rheological characterisation were obtained after adding fractions to flour at a level of 1% (w/w), and subsequently extracting gluten using the Glutomatic.

2.2.3 Rheological tests. Small scale oscillation rheological tests were carried out, over a range of temperatures between 20 °C and 95 °C. The frequency used was 1 Hz, and the peak strain was 0.02.

3 RESULTS AND DISCUSSION

The fractions obtained from the salt precipitation were characterised in terms of molecular weight and rheological behaviour[2,3], and the results are presented in Table 1, while the polypeptide characterisation is presented in the gels in Figures 1and 2.

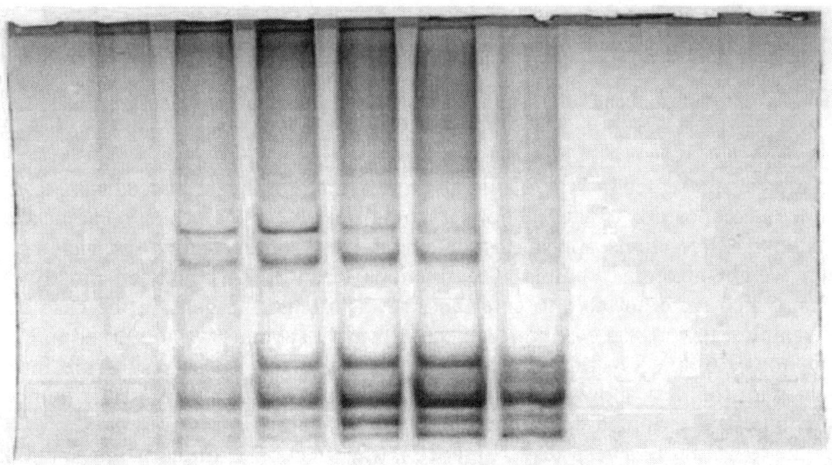

Figure 1, *SDS-PAGE of unreduced Hereward fractions*

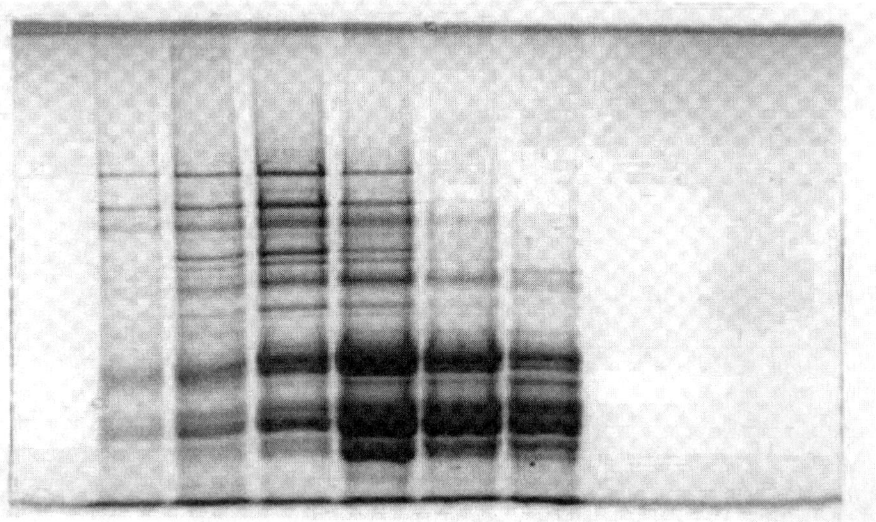

Figure 2, *SDS-PAGE of reduced Hereward fractions*

Table 1 *Tan δ and molecular weight values for Hereward fractions measured at 1 Hz, 0.02 peak strain and at 20 °C.*

Sample	Molecular weight		tan δ	Sample	tan δ
R2	$9*10^7$	$1*10^9$	0.212	Control	0.594
	(0.5)[1]	(0.5)		C+ Gluten	0.592
R3	$7*10^8$		0.232	C + R2	0.567
R4	$6*10^4$	$5*10^7$	0.551	C + R3	0.531
	(0.5)	(0.5)			
R5	$1*10^5$	$1*10^7$	1.316	C + R4	0.569
	(0.9)	(0.1)			
R6	$8*10^4$		3.081	C + R5	0.626
R7	$6*10^4$		6.836	C + R6	0.639
				C + R7	0.667

[1]The number in brackets is the cumulative weight fraction of the injected polymer with a particular M_w

From the gels presented in Figure 1 it can be seen that, for the unreduced sample, there is a change in composition of the fractions in going from R2 to R7, which contains almost only gliadins. All fractions appear to contain at least some gliadins, as well as low molecular weight glutenin subunits. The fractions R2, R3, R4 contained mostly large molecules, which were not able to enter and travel through the gels. In the case of the reduced samples, the presence of high molecular weight subunits was observed inR2-R4, while fractions R6 and R7 contain proteins of smaller sizes. These results are in good agreement with the data shown in Table 1, where the molecular weight distribution of those fractions, as determined with Flow Field Flow Fractionation, is presented.

From the temperature sweep tests (Figure 2), the effect of addition of the fractions is clear. The fractions biggest in size increase the elastic character of the gluten, while the lower molecular weight fractions increase the viscous character of the gluten.

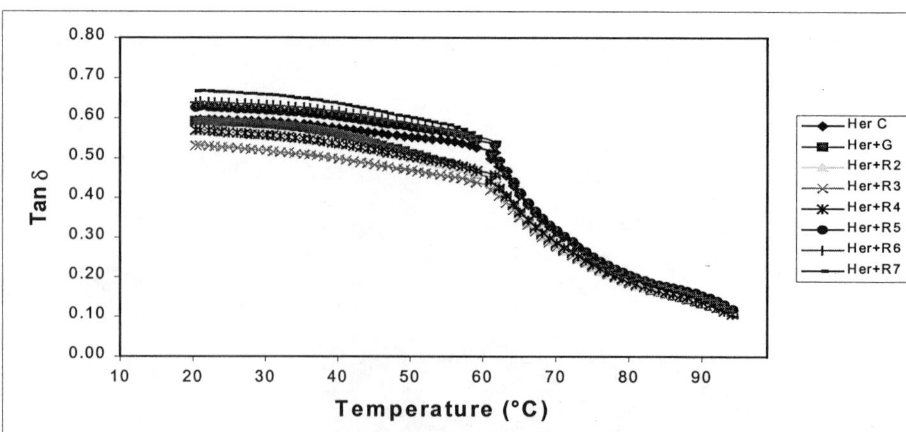

Figure 3, *Tan δ of Hereward glutens with various additions (conditions as in Table 1).*

4 CONCLUSIONS

It is possible to alter the rheological profile of gluten by addition of protein fractions.

Addition of fractions containing mainly HMW glutenin resulted in an increase in strength of the total gluten, while addition of LMW fractions mainly containing gliadins resulted in a weakening of the gluten network.

References

1. A. Graveland, M.H. Henderson, M. Paques, and P.A. Zandbelt. In *Wheat Structure Biochemistry and Functionality,* ed. J.D.Schofield, Royal Society of Chemistry, London, 2000, p.90.
2. A.A. Tsiami, A. Bot, W.G.M. Agterof and R.D. Groot, *J. Cereal Sci.* 1997, **26**, 15.
3. A.A. Tsiami, C.E. Stathopoulos and J.D.Schofield. In *4th FFF Symposium*, Paris, September 1999.

Acknowledgements

Funding for this work from MAFF (part of LINK project CSA 4580) is gratefully acknowledged.

EFFECTS OF HMW AND LMW GLUTENIN SUBUNIT GENOTYPES ON RHEOLOGICAL PROPERTIES IN JAPANESE SOFT WHEAT

T. Nagamine[1], T. M. Ikeda[1], T. Yanagisawa[2] and N. Ishikawa[1]

1. Chugoku National Agricultural Experiment Station, 6-12-1 Nishifukatsu, Fukuyama, Hiroshima 721-8514, Japan. 2. National Agriculture Research Center, 3-1-1 Kannondai, Tsukuba, Ibaraki 305-8666, Japan.

1 INTRODUCTION

Japanese soft wheat cultivars have been mainly used for Japanese white salty noodles which require medium strength gluten with the optimal viscoelastic properties.

The genotypes of the HMW and LMW glutenin subunits have reported to have critical effects on the rheological properties of gluten[1-7]. In this study, we clarified the allelic variation of glutenin loci and their effects on rheological properties in Japanese soft wheat.

2 MATERIALS AND METHODS

2.1 SDS-PAGE analysis for *Glu-3* and *Glu-1* alleles and N-terminal Amino Acid Sequencing of Major LMW Subunits

The SDS-PAGE band patterns of the LMW glutenin alleles of Japanese soft wheat were identified by allelism analysis of three sets of DH lines derived from (Norin 61 x Chinese Spring) F_1, (Norin 61 x Asakazekomugi) F_1 and (Norin 61 x Gogatsukomugi) F_1. In addition to the analysis of DH lines, the screening of genetic resources of modern cultivars and landraces in southern Japan also detected alleles that lacked major LMW subunits. SDS-PAGE analysis was carried out using the method described by Singh *et al.*[8] except the use of 15% separation gel. The *Glu-1* genotypes were identified by the band pattern of HMW subunits according to Payne and Lawrence[9].

N-terminal amino acid sequences of major LMW subunits were determined using a protein sequencer (Model 476 A, Applied Biosystems).

2.2 Rheological Measurements and Statistical Analysis

To study the effects of glutenin genotypes on rheological properties, 61 F_6 breeding lines which were tested in an advanced yield trial in 1996 were measured with farinograph test and a gluten compression test.

In the gluten compression test, hydrated gluten was compressed and relaxed in a rheometer to calculate the gluten relaxation coefficient (GRC) which was the ratio of the depth of non-recovery to the compression depth. The GRC is smaller for stronger gluten that recovers to a larger degree.

The effects of each glutenin allele on the valorimeter value (VV) and GRC were analyzed with a model integrating the effects of six glutenin loci and protein content.

3 RESULTS AND DISCUSSION

3.1 Allelic Variation of *Glu-1* and *Glu-3* in Japanese Soft Wheat

SDS-PAGE analysis of DH lines and Japanese soft wheat cultivars showed that genetic variation in glutenin loci of Japanese soft wheat was small. Two alleles each were found for *Glu-B1*, *Glu-D1* and *Glu-D3* loci, and three each for *Glu-A1, Glu-A3* and *Glu-B3*. The major subunits useful for identification of *Glu-3* alleles are indicated in Figure 1. The *Glu-3* alleles are described in this paper as the names of standard cultivars. The frequency of each *Glu-1* and *Glu-3* allele in modern cultivars, landraces and F_6 breeding lines is shown in Table 1.

Based on *N*-terminal amino acid sequences, the largest LMW subunits of Norin 61 and Asakazekomugi, which were controlled by the *Glu-A3* allele, were shown to belong to the LMW-s type[11]. The major subunit controlled by *Glu-D3* alleles of Norin 61 belonged to the gamma-gliadin type[11].

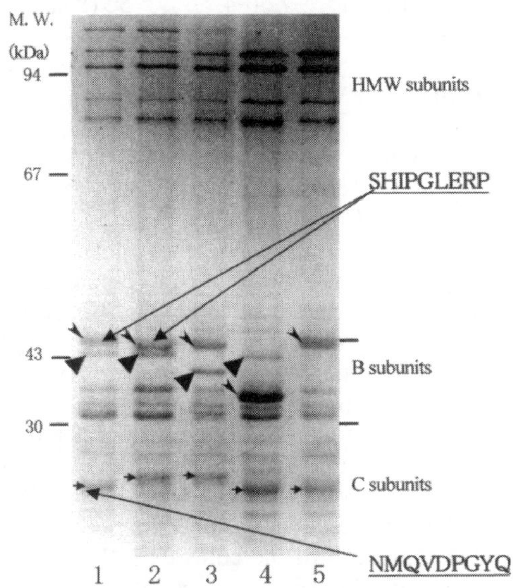

Figure 1 *SDS-PAGE of glutenin fractions of Japanese standard cultivars Arrows (➢, ▼, ➔) indicate major LMW subunits of Glu-A3, Glu-B3, and Glu-D3 alleles. Alphabets show the N-terminal sequences of LMW subunits (Lane 1). Lane; Cultivar, Genotypes of Glu-A3, Glu-B3, and Glu-D3. 1; Norin 61, N61-N61-N61; 2; Asakazekomugi, Asakaze-N61-Asakaze; 3; Gogatsukomugi, Asakaze-Gogatsu-Asakaze; 4; Aobakomugi, Aoba-N61-N61; 5; Toyohokomugi, N61-Toyoho-N61.*

Table 1 *Allelic variation of glutenin subunits in Japanese soft wheat*

Locus	Allele (Subunits)	Number of Landraces (%)	Modern cultivars (%)	F_6 breeding lines (%)
Glu-A1	*b* (2*)	10 (60.0)	25 (29.4)	12 (19.7)
	c (null)	15 (40.0)	60 (70.6)	49 (80.3)
Glu-B1	*b* (7+8)	25 (100.0)	76 (89.4)	20 (32.8)
	c (7+9)	0 (0.0)	7 (8.2)	41 (67.2)
	e (20)	0 (0.0)	2 (2.4)	0 (0.0)
Glu-D1	*a* (2+12)	18 (72.0)	35 (41.2)	7 (11.5)
	f (2.2+12)	7 (28.0)	50 (58.8)	54 (88.5)
Glu-A3	N61 allele	16 (64.0)	53 (62.4)	17 (27.9)
	Asakaze allele	6 (24.0)	24 (28.2)	37 (60.6)
	Aoba allele	3 (12.0)	8 (9.4)	7 (11.5)
Glu-B3	N61 allele	3 (12.0)	38 (44.7)	47 (77.0)
	Gogatsu allele	20 (80.1)	40 (47.1)	14 (23.0)
	Toyoho allele	2 (8.0)	7 (8.2)	0 (0.0)
Glu-D3	N61 allele	3 (12.0)	13 (15.3)	9 (14.8)
	Asakaze allele	22 (88.0)	72 (84.7)	52 (85.2)

3.2 Effects of Glutenin Alleles on VV and GRC Variations

The VVs of 61 F_6 breeding lines ranged from 29 to 87 and GRCs from 62.55 to 76.19. Most lines were classified as soft or medium type, but a few had higher VV or GRC values of hard types such as No. 1 Canadian white.

In the mean VV and GRC estimated for each glutenin allele, *Glu-A1b* and *Glu-B1c* had a higher VV than allelomorphic *Glu-A1c* and *Glu-B1b* (Figure 2). *Glu-A1b* and the Aoba allele of *Glu-A3* had a lower GRC than the allelomorphic *Glu-A1b* and the Asakaze allele of *Glu-A3*. The *Glu-D1f* allele reported to be distributed at a higher frequency in East Asian cultivars[10] did not differ significantly from *Glu-D1a* in both GRC and VV.

Despite these significant differences between the alleles, the total determination coefficients (R^2) of the model including the effects of six glutenin loci and protein content were much smaller than those reported for cultivars from Western countries. The R^2 in our experiment were 41.2% for VV and 38.3% for GRC. However, the *Glu*-score determined by the HMW glutenin genotype has been reported to account for the largest variation in quality-related characters of the cultivars in U. K. (59.8% for breadmaking quality[3]), in Spain (68% for the Zeleny test[4]), and in Canada (75% for breadmaking quality[5]).

Because of the small effects of glutenin genotype on rheological variation in Japanese soft wheat cultivars, we assume the quality requirement is the major reason. Japanese wheat has been mostly used for Japanese white salty noodles which are preferably made from soft, tender dough rather than the stronger dough desirable for breads[12]. The lack of strong glutenin alleles such as *Glu-D1d* may reduce the relative effects of the glutenin genotype in the variation of rheological properties.

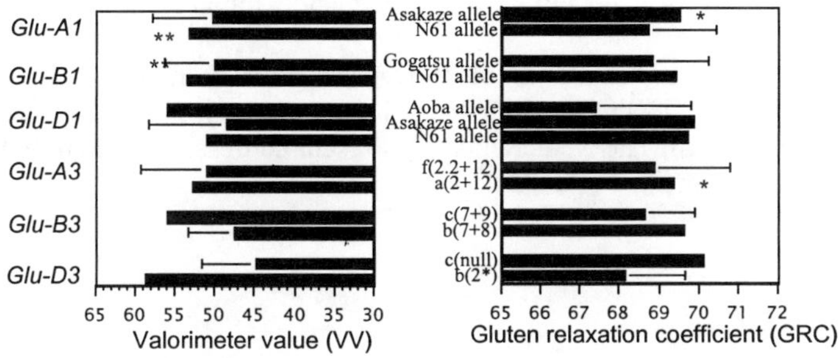

Figure 2 *Estimated mean valorimeter value and gluten relaxation coeffiicient for each glutenin genotype. Bars indicate least significant differences at 5 % level. Asterisks indicate significant differences between alleles at (*) 1 % and (**) 5 % lelels.*

4 CONCLUSIONS

The allelic variation in *Glu-1* and *Glu-3* loci among Japanese soft wheat cultivars was small. Only two alleles were found for *Glu-B1*, *Glu-D1,* and *Glu-D3* loci, and three alleles for *Glu-A1, Glu-A3,* and *Glu-B3* loci. Allelic differences were significant for *Glu-A1* and *Glu-B1* loci in farinographic VVs and for *Glu-A1* and *Glu-A3* loci in gluten compression tests. The relative effect of glutenins on total variation of these rheological features was lower than that reported for foreign cultivars. The lack of strong glutenin alleles such as *Glu-D1d* in Japanese soft wheat was considered the major reason.

References

1. P.I. Payne, A.N. Mark, A.F. Krattiger and L.M. Holt, *J. Sci Food Agric*, 1987, **40**, 51.
2. P.I. Payne, L.M. Holt, A.F. Krattiger and J.M. Carrillo, *J. Cereal Sci,* 1988, **7**, 229.
3. O.M. Lukow, P.I. Payne and R. Tkachuk, *J. Sci Food Agric*, 1989, **46**, 451.
4. P. Feillet, O. Ait-Mouh, K. Kobrehel and J.-C. Autran, *Cereal Chem.*, 1989, **66**, 26.
5. N.E. Pogna, J.-C. Autran, F. Mellini, D. Lafiandra and P. Feillet, *J. Cereal Sci.*, 1990, **11**, 15.
6. E.V. Metakovsky, C.W. Wrigley, F. Bekes and R.B. Gupta, *Aust. J. Agric. Res.*, 1990, **41**, 289.
7. R.B. Gupta, J.G. Paul, G.B. Cornish, G.A. Palmer, F. Bekes and A.J. Rathjen, *J. Cereal Sci.,* 1994, **19**, 9.
8. N.K Singh, K.W. Shepherd and G.B. Cornish, *J. Cereal Sci.,* 1990, **14**, 203.
9. P.I. Payne and G.J. Lawrence, *Cereal Res. Commun.* 1983, **11**, 29.
10. H. Nakamura, H. Sasaki, H. Hirano and A. Yamashita, *Japan. J. Breed.* 1990, **40**, 485.
11. E.J.-L. Lew, D.D. Kuzmicky and D.D. Kasarda, *Cereal Chem.* 1992, **69**, 508.
12. S. Nagao, S. Imai, T. Sato, Y. Kaneko and H. Otsubo, *Cereal Chem.,* 197, **53**, 988.

MIXING OF WHEAT FLOUR DOUGH AS A FUNCTION OF THE PHYSICO-CHEMICAL PROPERTIES OF THE SDS-GEL PROTEINS.

August C.A.P.A. Bekkers[1,3], Wim J. Lichtendonk[1], Aris Graveland[2], Johan J. Plijter[1]

1. TNO Nutrition and Food Research, Food Technology Department, P.O. Box 360, 3700 AJ Zeist, The Netherlands. 2. Unilever Research Vlaardingen, Bakery Products Unit, P.O. Box 114, 3130 AC Vlaardingen, The Netherlands. 3. Present address: Heineken, PO Box 530, 2380 BD Zoeterwoude, theNetherlands

1 INTRODUCTION

The functional properties of flour, such as bread baking potential, are to a large extent determined by the proteins. It has been recognised for a long time that protein quantity alone is unable to explain the encountered variation in bread baking potential[1].

The gluten proteins comprise a mixture of gliadin and glutenin proteins. The glutenin proteins can be further classified into high molecular weight (HMW) and low molecular weight (LMW) glutenin subunits[2].

The glutenin fraction of flour can be isolated rather easily. Upon extraction of flour using the detergent SDS, a gel-like layer is found on top of the starch after centrifugation which consists mainly of glutenin polymer proteins[3]. These proteins are referred to as the gel proteins. It was demonstrated by various studies that the amount of these gel proteins has a better potential to predict the bread baking properties of a flour than the protein content alone[4].

In this respect, it is important to consider the bread baking potential of flour in relation to the method used for its testing. It was shown by a number of studies that bread volumes depend heavily on the amount of water and the combination of kneading times/mixer type used during dough preparation[5,6].

In this study we set out to investigate the underlying molecular basis of the kneading characteristics of wheat flour using the characteristics of its gel proteins.

2 MATERIALS AND METHODS

2.1 Mixing

Mixing was performed in a 50 g. Farinograph, analogous to[6]. In the preparation of the doughs, 2% NaCl and water were added and the mixing speed was 63 rpm. The amount of water added was determined according to ICC procedure 115/1[7].

2.2 Physico-chemical characterisation of gel proteins

The preparation of the gel proteins was performed according to Graveland. Quantification of gel proteins and derived fractions was done using Kjeldahl analysis (N=5.7). For rheological characterisation of the gel proteins, 1 g of gel was carefully scraped off from the top of the gel and transferred into a Bohlin VOR strain controlled rheometer equipped with a 30 mm plate-plate geometry and a gap of 1 mm. Measurements were performed at 20 C. For comparison of the rheological properties of the gel proteins derived from different flours, we use G' measured at 2% strain at 0,15 Hz (linear range). Fractionation of gel proteins was performed by twice weighing 1 of two separately prepared gel-protein fractions into a test tube and adding ethanol up to a concentration of 70% (v/v). After homogenisation, the mixture was centrifuged for 30 minutes at 40.000 g. The resulting supernatant containing the gliadins and glutenin III polymers was decanted, leaving the glutenin I polymers of a relative high molecular weight in the pellet.

3 RESULTS

The amount of gel protein that can be isolated from different flour types is not related to the mixing characteristics of these flours (data not shown).

In order to study the relationship between the dough properties and the polymeric nature of the glutenin polymers, dynamic rheological measurements were performed using the top layers of the gel proteins. As shown in Figure 1, a very high and positive correlation was present between G' of the gel and the dough development time. From these results we conclude that the kneading behaviour of flour under the conditions applied is indeed related to the polymeric nature of the glutenin proteins as present in the gel proteins, rather than to the total amount of gel proteins in flour. The rheological properties of the gel proteins depend on the concentration of glutenin polymers in the gel.

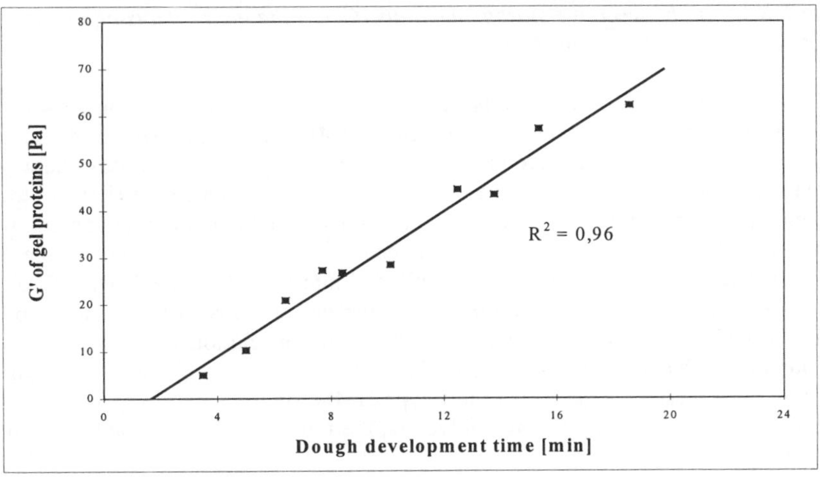

Figure 1 *Relation between the dough development time of a wheat flour and the storage modulus (G') of its gel proteins.*

The gel proteins still contain small amounts of gliadins and glutenin III polymers in addition to the glutenin I polymers. For the flours tested, analysis of the gel proteins using SDS-PAGE under reduced and non-reduced conditions showed that the different gel-protein samples indeed contain variable amounts of gliadin and glutenin III polymers [results not shown]. From polymer chemistry it is known that in materials containing polymers of different molecular weights, the polymers of higher molecular weight contribute most to the elastic properties[8]. Relating this concept to gluten polymers, it is envisaged that the glutenin I polymers contribute mainly to the elasticity of the gel proteins. Therefore, the gel proteins were fractionated using 70% ethanol. The top layer of the gel proteins was divided into two fractions: the supernatant containing the gliadins and glutenin III polymers with relative low molecular weight and the pellet containing the glutenin I polymers of high molecular weight[3]. The results presented in Figure 2 show that G' of the gel-protein fractions correlates strongly and positively with their contents of glutenin I polymers.

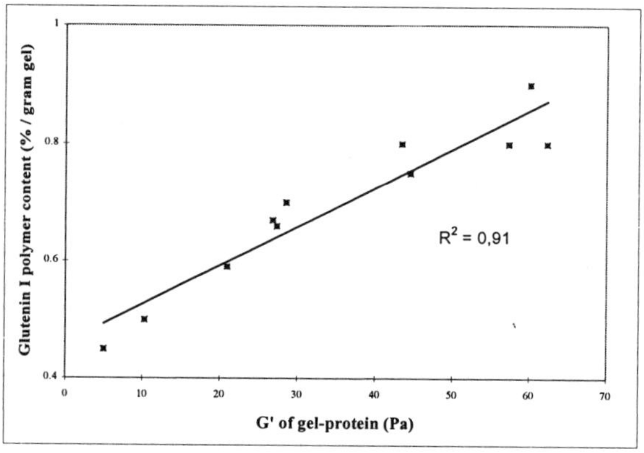

Figure 2 *Relation between the storage modulus (G') of the gel proteins and their content of high polymeric glutenin I proteins.*

The amount of HMW subunits was determined using capillary electrophoresis. Figure 3A shows the correlation between the amount of HMW subunits present in the glutenin I isolated and the G' of the gel proteins. Again, a clear positive relationship between the two characteristics is shown. Since the HMW glutenin subunits are considered to form the backbone of the linear glutenin polymers it can be imagined that a higher content of these subunits results in polymers of higher molecular weight and size with more elastic properties. The HMW subunits are further divided into x- and y-types[9]. Therefore, it was decided to relate the rheological properties of the gel proteins to the presence of these specific HMW subunit types. Figure 3B shows a strong relationship between G' of the gel-protein and the content of x-type HMW subunits within the high polymeric glutenin I polymers. These results suggest that the x-type HMW subunits play a special role (chain "extenders") in the nature of glutenin polymers; the function of the y-type HMW subunits is unclear.

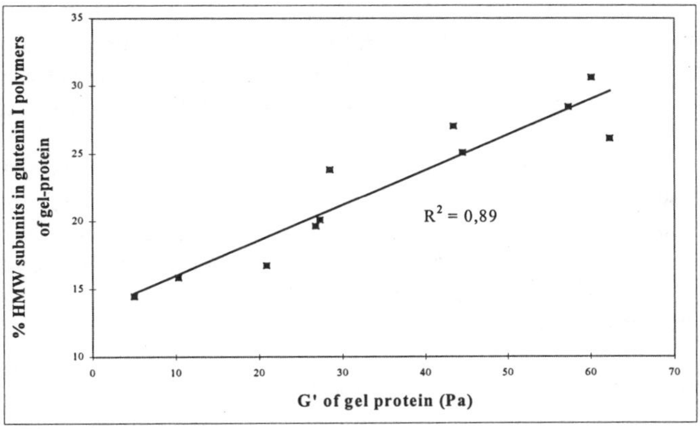

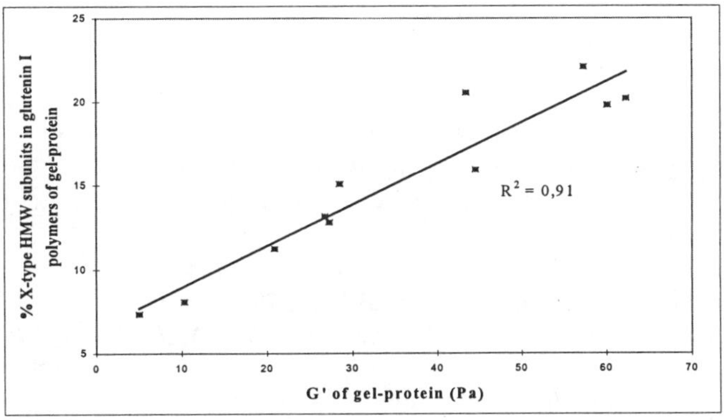

Figure 3 *Relationship between the rheological properties of the gel proteins and the content of HMW glutenin subunits (A) and the specific x-type HMW glutenin subunits (B) within the glutenin I polymers.*

To summarise, we can conclude that the dough properties of wheat flour are a function of the highly polymeric glutenin polymers, which in turn relates to the presence of (x-type) HMW subunits.

4 DISCUSSION

In order to explain the optimal kneading times of dough on a molecular level, we studied the glutenin polymers in the form of gel proteins from a range of different flours. It was shown that the total amount of protein in the gel layers was not related to the kneading behaviour in a Farinograph mixer. In contrast, the dough development time of a flour could be accurately predicted by studying the elastic modulus (G') of the gel proteins. Biochemical analysis of the gel proteins demonstrated that G' is determined by the concentration of the highly polymeric glutenin I polymers, thereby emphasising the functional importance of these polymers. Our study further showed a positive relationship between G' of the gel proteins and the concentration of HMW subunits, especially the x-type, in the glutenin I polymers. These observations may in due time contribute to the development of new tests for the prediction of flour quality based on quantification of x-type subunits only. These tests can be used in breeding and flour testing for baking and/or gluten/starch separation.

Glutenin polymers originating from different flours can differ qualitatively in their molecular size distribution. It has been proposed that only the fraction of glutenin polymers with a relative high molecular mass is capable of forming effective molecular interactions (such as entanglements) and thereby contributes to dough strength[10]. Our work adds to this suggestion and shows that the properties of the functional glutenin polymers can be studied by analysis of the gel proteins.

Acknowledgements

The authors would like to acknowledge Senter (Dutch Ministry of Economic Affairs) for financial support of the "Wheat Chain Project".

Literature Cited

1. Weegels, P.L., Hamer, R.J. and Schofield, J.D. *J. Cereal Sci.* **1996**, 23, 1.
2. Khatkar, B. S. and Schofield J.D. *J. Food Sc. Techn.* **1997**, 34(2), 85.
3. Graveland, A., Bosveld, P., Lichtendonk, W.J., Marseille, J.P., Moonen, J.H.E. and Scheepstra, A. *J. Cereal Sci.* **1985**, 3, 1.
4. Pritchard, P.E. and Brock, C.J. *J. Sci. Food Agric.* **1994**, 64, 401.
5. Oliver, J.R. and Allen, H.M. *J. Cereal Sci.* **1992**, 15, 79.
6. Kieffer, R., Wieser, H., Henderson, M.H. and Graveland, A. *J. Cereal Sci.* **1998**, 27, 53.
7. *Standard Methods of the ICC* (International Association for Cereal Science and Technology) **1992**. Schäfer, Detmold, Germany.
8. Doi, M. and Edwards, S.F. In: *The theory of polymer dynamics.* Clarendon Press, Oxford, U.K., **1995**.
9. Payne, P.I. and Lawrence, G.J. *Cereal Res. Commun.* **1983**, 11, 29.
10. MacRitchie, F. and Lafiandra, D. *In: Food proteins and their applications* (edited by Damodaran, S. and Paraf, A., published by: Marcel Dekker Inc. New York, USA) **1997**, 293-324.

EFFECTS OF ADDING GLUTEN FRACTIONS ON FLOUR FUNCTIONALITY

U.G. Purcell, B.J. Dobraszczyk, A.A. Tsiami and J.D. Schofield

The University of Reading, Department of Food Science and Technology, Whiteknights, Reading, RG6 6AP, UK

1 INTRODUCTION

It has been suggested that an optimum relationship between high and low molecular weight (MW) non-soluble protein (gluten) fractions is required in a good bread making flour, where the high MW fractions are mainly responsible for providing the elasticity (i.e. resistance to deformation) and thereby gas retention, and the low MW fractions, including gliadins, for the necessary extensibility of the dough during proving and oven rise[1,2]. Rheological analysis of doughs from good and poor breadmaking wheat flours, to which gluten protein fractions and total gluten were added, was carried out to test this hypothesis.

2 MATERIALS AND METHODS

Six protein fractions (R2 to R7), ranging from high to low MW, were obtained from Hereward (medium-strong) and Riband (weak) flours by separating the defatted flours by salt precipitation[3,4] as well as their respective glutens. For the large deformation tests, flour plus 1% (w/w) total gluten or protein fraction, where relevant, was mixed with 60% (w/w) of a 2% NaCl solution (on a flour basis) in a 10-g Mixograph. The dough was relaxed for 60 min before dough strips were tested on a Stable Micro Systems Kieffer extensibility rig until breaking point. For the small deformation tests, flour and fractions were mixed with 60% distilled water (on a flour basis) in a 2-g Mixograph. Following relaxation, frequency sweep (0.1Hz to 20Hz at 1% peak strain) tests were performed on a StressTech rheometer.

3 RESULTS AND DISCUSSION

3.1 Large Deformation Tests

Figure 1 shows that there is no significant difference in maximum force when adding large MW fractions to Hereward. The differences between the low MW fractions, R6 and R7, and almost all the other additions are significant. Riband follows closely the pattern of Hereward except for Riband plus total gluten, where the value is lower than that for Riband only. From R4 to R7 there appears to be a trend in both flours that dough elasticity decreases with decreasing MW of the fractions added.

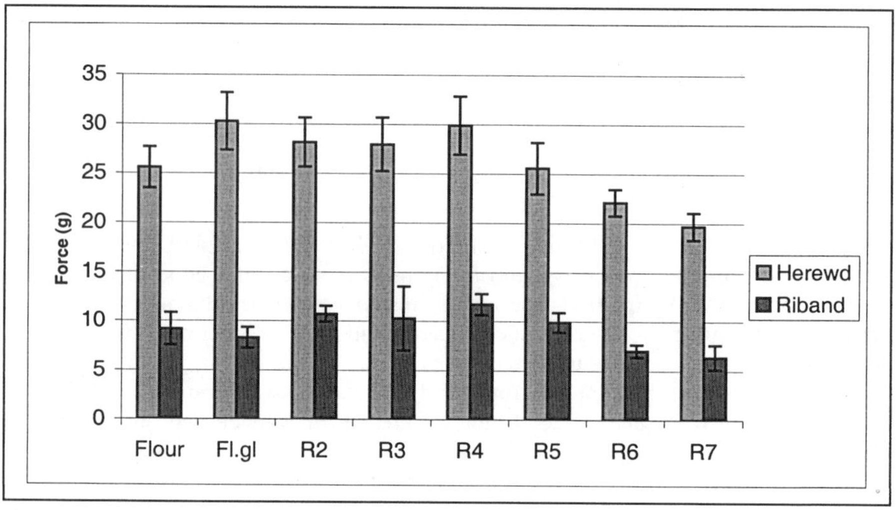

Figure 1 *Kieffer test: maximum force for Hereward and Riband flours with added protein fractions (means of 2 to 4 repeats).*

In Figure 2 it can be seen that there is no difference in extensibility when adding R2 to R4 to either flour, whereas from R4 to R7, an inverse trend to that observed in maximum force is apparent. The difference between adding R7 and any other fraction to Hereward is significant; however, this is not so in the case of Riband.

3.2 Small Deformation Tests

The moduli (G' = storage modulus, G" = loss modulus) of the frequency sweeps in Figure 3 approximately follow the pattern of maximum force in the Kieffer tests. As these moduli represent resistance to deformation the results are in agreement with expectation. The slopes of the frequency sweeps for G' and G" and tan δ (G"/G') (not shown) are inversely related to the moduli for both flours, which is also consistent with expectation.

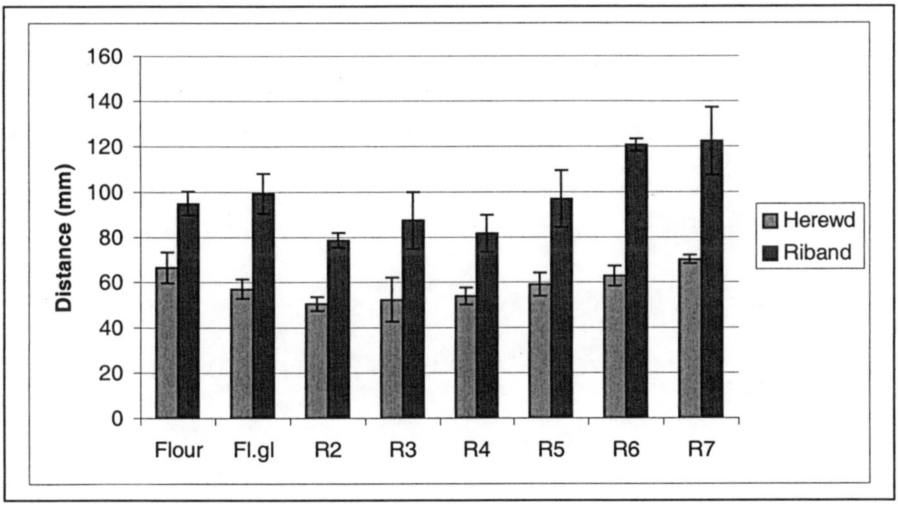

Figure 2 *Kieffer test: extensibility for Hereward and Riband flours with added protein fractions (means of 2 to 4 repeats)*

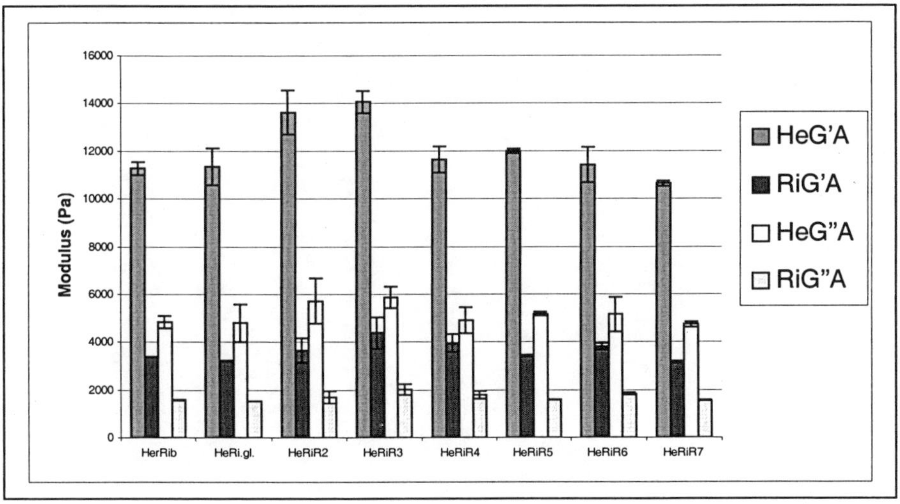

Figure 3 *Modulus of frequency (mean of 2 or 3 repeats) at 1Hz for G' and G" for Hereward and Riband with added protein fractions.*

4 CONCLUSIONS

The results are consistent with the concept that dough rheological properties are related to the MW distribution of glutenin polymers within the gluten complex, higher MW polymers conferring greater elasticity and lower MW components greater viscous character.

References

1. A. Graveland, M.H. Henderson, M. Pâques and P. Zandbelt, in *Wheat Biochemistry*, ed. J.D. Schofield, Royal Society of Chemistry, Cambridge, 1997.
2. A.A. Tsiami, A. Bot and W.G.M. Agterof, in 1st International Symposium on Food Rheology and Structure, Zürich, 1999.
3. A.A. Tsiami, A. Bot, W.G.M. Agterof and R.D. Grot, *J. Cereal Sci.*, 1997, **26**, 15
4. P.L. Weegels, R.J. Hamer and J.D. Schofield, *J. Cereal Sci.*, 1996, **23**, 1.

Acknowledgements

Funding for the work by MAFF (Contract CSA4580) as part of an EU LINK project is gratefully acknowledged.

METHODS FOR INCORPORATING ADDED GLUTENIN SUBUNITS INTO THE GLUTEN MATRIX FOR EXTENSION AND BAKING TESTS

S. Uthayakumaran[1], F. L. Stoddard[1,2], P. W. Gras[1,3], and F. Bekes[1,3]

1. Quality Wheat Cooperative Research Centre Ltd., Locked Bag No. 1345, North Ryde, NSW 1670, Australia. 2. Plant Breeding Inst., Woolley Bldg A20, The University of Sydney, NSW 2006, Australia. 3. CSIRO Plant Industry, Grain Quality Research Laboratory, PO Box 7, North Ryde, NSW 1670, Australia.

1 INTRODUCTION

Glutenin is the major protein class contributing to the strength and stability of dough. Glutenin polymers are composed of high molecular weight glutenin subunits (HMW-GS) and low molecular weight glutenin subunits (LMW-GS) linked by disulphide bonds. Both quantitative and qualitative differences within these groups of proteins contribute to intercultivar variation in bread-making quality[1]. In order to study the functional properties of glutenin subunits added to dough, they need to be incorporated into the glutenin polymer. This requires partial reduction, to open the polymer, followed by oxidation, to incorporate the added monomer into the polymer. The existing method for incorporating glutenin subunits[2] is suitable only for studies on mixing properties. Whilst direct assessments of mixing properties address the need to understand the mechanisms underlying dough mixing, they do not address the effects of specific proteins on the extensibility or baking properties of the dough. Therefore, new techniques for estimation of the effects of incorporated glutenin subunits on extension and baking properties had to be developed. The aim of this study was, therefore, to optimise the reduction/ oxidation conditions for extension and baking.

2 MATERIALS AND METHODS

Flour from cultivar Banks (13% protein content) was obtained from BRI Australia Ltd., North Ryde, NSW. Doughs for the tests were mixed on a 2-g Mixograph™ (TMCO, Lincoln, NE, USA).

2.1 Extension testing

The optimum reduction/ oxidation conditions for mixing studies[2] did not produce extension curves similar to controls (Figure 1). The following conditions play an important role in determining the optimum condition:
1) relaxation time of dough between mixing and extension testing; 2) concentration of reducing agent; 3) time allowed for reduction reaction to take place; 4) concentration of

oxidising agent; 5) time allowed for oxidation to take place; and 6) the optimum mixing time.

Each reduction-oxidation step was tested with certain variations and four replications. A fully factorial experiment would have required 7200 treatments, so instead, each treatment was varied individually and all other treatment conditions were maintained at standard levels (45 min relaxation time, 2 mg/mL reductant (Dithiothreitol, DTT), 5 min reduction time, 5 mg/mL oxidant (KIO_3) and 5 min oxidation time). The relaxation time was tested at 0 (dough was pulled immediately after mixing), 15, 30, 45 and 60 min. Three concentrations of DTT (0.2, 1 and 2 mg/mL) were tested at one reduction time (4 min) and four reduction times (1, 2, 3 and 4 min) at two concentrations (0.2 and 2 mg/mL). Six concentrations (2, 2.5, 3, 5, 7.5 and 10 mg/mL) of KIO_3 were tested with 5 min oxidation time and five oxidation times (1, 2, 3, 4, and 5 min) were tested at 5 mg/mL oxidant concentration. The mixing time was tested at 70, 80, 90, and 100% of the maximum dough development time using 0.2 mg/mL DTT, 1min reduction time and the other standard conditions.

The total quantity of water to be used was calculated using the protein and moisture contents of ingredients[3]. Dough was prepared by mixing the flour, 450 μL of DTT solution and water for 30 seconds and it was then allowed to rest. In the last few seconds of the resting period, 250 μL of oxidant solution was added and mixing resumed for 30 seconds before the dough was allowed to rest further. The dough was then mixed to the pre-determined proportion of the peak dough development time (including the initial 2 x 30 sec mixes). Dough samples (1.7 g per test) were moulded into cylinders approximately 6 mm diameter, mounted on a sample carrier and rested at 30°C and >90% rh for 45 minutes before extension testing[4]. Extension was carried out in quadruplicate on a micro-extension tester with a 19 mm gap and 6 mm hook operating at 1 cm / sec. Recordings of the dough resistance and the sample carrier position were taken every 0.01 sec and recorded on a personal computer, using LabTech Notebook software. Maximum resistance to extension (Rmax, N) and extension before rupture (Ext, cm) were calculated[5]. After extension, all doughs were frozen in liquid nitrogen and freeze dried for protein size distribution analysis by SE-HPLC and FFF.

2.2 Microbaking

The concentration of the reductant and time of its application were varied, with two concentrations (0.2 and 2 mg/mL) being tested at one time (4 min) and four times (1, 2, 3 and 4 min) at 2 mg/mL. Four concentrations of oxidant (2.5, 5, 7.5 and 10 mg/mL) were tested at one time (5 min) and five times (1, 2, 3, 4 and 5 min) at 2.5 mg/mL.

The total quantity of water to be used to prepare doughs for microbaking was calculated using the protein and moisture contents of ingredients[3]. Flour, 450 μL of DTT solution and the water were placed in the mixing bowl. The dough was mixed for 30 seconds and allowed to rest. In the last few seconds of the resting period, 250 μL of oxidant solution (KIO_3) was added along with the required amount of yeast solution (10 g compressed yeast, 8 g salt, 2 g improver in 100 mL water). Mixing was resumed for 30 seconds and the dough allowed to rest further. The dough was then mixed to the peak dough development time (including the initial 2 x 30 sec mixes). Loaves were prepared from 2.4 g of the resulting dough which was moulded, rested for 20 minutes at 40°C in a small airtight container, then remoulded, proofed for 45 min (40°C and 90% rh) and baked at 200°C for 17 min[4]. Loaf height was measured with vernier calipers. Baking tests were carried out in triplicate.

3 RESULTS AND DISCUSSION

3.1 Extension testing

The reduction-oxidation treatments had profound effects on the shape of the extension curve. Relaxation times of 0 min, 15 min and 30 min gave shorter extensibility than the controls. A 45 min relaxation time gave the greatest extensibility and is also the conventionally used value so this time was used for further experiments. For 0.2 mg/mL, 1.0 mg/mL and 2.0 mg/mL of reductant, shorter reduction times provided greater Rmax than longer times. At the 1 min reduction time, 0.2 mg/mL reductant provided greater Rmax than higher concentrations although extensibility was reduced. Higher reductant concentrations were also associated with dough stickiness. The optimum reduction condition for further experiments was therefore 0.2 mg/mL for 1 min. For the oxidation step, though 10 mg/mL of KIO_3 gave the extensibility closer to that of the control, the 5 mg/mL solution was selected because with all other combinations (reductant concentration, reductant time, relaxation time and proportion of mixing time) it gave the best results. Oxidation times of 4 or 5 min were required for complete re-polymerisation and oxidation time of 1 min produced sticky dough which could not be used. Reducing the mixing time to 70% of optimum gave an extension curve closer to the control than longer mixing times. This set of optimised conditions provided extensibility and maximum resistance to extension values comparable to controls, with somewhat different overall curve shape (Figure 2) and was used for further investigations.

The control and the dough obtained by the new incorporation technique had similar contents of unextractable polymeric protein (%UPP) and average molecular sizes of insoluble protein, which indicated that the incorporation technique was valid. Incomplete incorporation would have left significant amounts of the subunits, while incomplete re-oxidation would have resulted in shorter polymers. Both of these effects would have reduced %UPP and average molecular size.

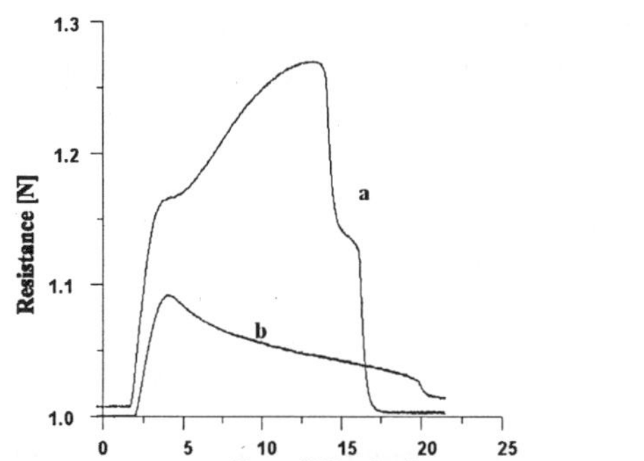

Figure 1 *Extension curves obtained using (a) water control and (b) reduction-oxidation conditions optimised for mixing studies*

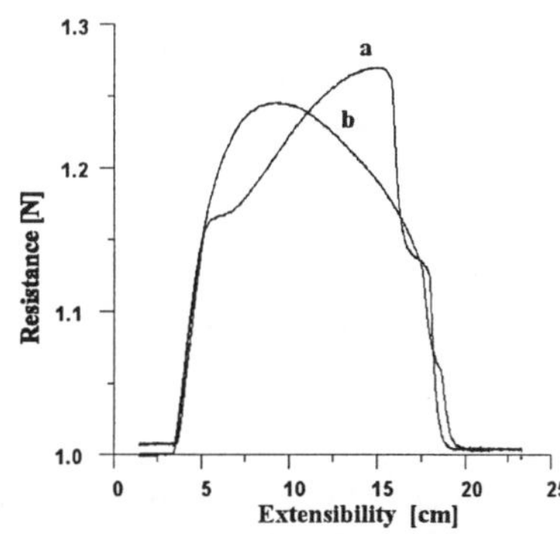

Figure 2 *Extension curves obtained using (a) water control and (b) reduction-oxidation conditions optimised for extension studies*

Table 1 *Optimum conditions for extension and microbaking*

Treatment	Extension	Microbaking
Concentration of reductant (mg/mL)	0.2	2
Reduction time (min)	1	1
Concentration of oxidant (mg/mL)	5	2.5
Oxidation time (min)	5	5
Mixing time (% of peak dough development time)	70	100

3.2 Microbaking

Using a reductant concentration of 2 mg/mL gave a loaf height closest to the control. Longer reduction times decreased loaf height and hence 1 min was selected as a suitable reduction time. For the oxidation, 2.5, 5 and 7.5 mg/mL of KIO_3 solution gave equally good results, and hence the lowest concentration (2.5 mg/mL) was chosen. The longest oxidation time, 5 min, gave a loaf height closest to the control. This set of optimised conditions was used for further investigations.

The optimised conditions for the extension testing and microbaking are given in Table 1. It was not possible to have a single incorporation technique for mixing, extension and baking with the reductant (DTT) used in these experiments. A single oxidation procedure was, however, suitable. Other reductants may allow the development of a single technique for all three tests. These observations were confirmed with two other varieties of flour.

4 CONCLUSION

A carefully selected reduction-oxidation procedure provides a method for studying the contribution of glutenin composition to functional properties of dough. The subunits were adequately incorporated into the polymer.

References

1. P. I. Payne, *Ann. Rev. Plant Physiol.*, 1987, **38**, 141.
2. F. Bekes, P. W. Gras and R. B. Gupta, *Cereal Chem.*, 1994, **71**, 44.
3. American Association of Cereal Chemists, Approved method of the AACC. Method 54-40A, 1988. The Association: St. Paul, MN, USA.
4. P. W. Gras and F. Bekes, in *Proceedings of the 6th International Gluten Workshop*, ed. C. W. Wrigley, Royal Australian Chemical Institute, North Melbourne, Vic., 1996, p. 506.
5. C. R. Rath, P. W. Gras, Z. Zhen, R. Appels, F. Bekes and C. W. Wrigley, in *Proceedings of the 44th Australian Cereal Chemistry Conference*, ed. J. F. Panozzo and P. G. Downie, Royal Australian Chemical Institute, North Melbourne, Vic., 1994, p. 122.

EFFECT OF INTERCULTIVAR VARIATION IN PROPORTIONS OF PROTEIN FRACTIONS FROM WHEAT ON THEIR MIXING BEHAVIOUR

J.M. Vereijken[1], V.L.C. Klostermann[1], F.H.R. Beckers[1], W.T.J. Spekking[1] and A. Graveland[2]

1. Agrotechnological Research Institute (ATO), P.O. Box 17, 6700 AA Wageningen, The Netherlands. 2. Unilever Research Vlaardingen, Olivier van Noortlaan 20, 3133 AT Vlaardingen, The Netherlands

1 INTRODUCTION

Wheat proteins comprise four groups of proteins. Next to the two types of non-storage proteins (albumins and globulins), the majority of the proteins consists of the storage proteins gliadins and glutenins. The latter proteins are particularly difficult to fractionate, because of their polydisperse molecular weight distribution. Therefore, only a few fractionation methods result in biochemically well-defined fractions. To obtain such fractions, one of these procedures[1] was taken as a start for the development of a fractionation method. This method resulted in three well-defined fractions:

- A+G fraction (extracted by 0.5 M NaCl), containing albumins and globulins.
- Glia fraction (extracted by SDS/ethanol), containing gliadins and glutenins III.
- Glu fraction (unextractable by SDS/ethanol), containing glutenins I and II.

Glutenins I and II consist of high and low molecular weight glutenin subunits, whereas glutenins III comprise low molecular weight subunits only. Glutenins II are, in contrast to glutenins I, soluble in SDS-solutions and are presumed to have a lower molecular weight[1,2].

Because little is known about intercultivar variation in amounts of the three fractions defined above, one of the aims of this study was to assess this variation. Furthermore, it is well known that the glutenins in particular determine to a large extent differences in baking quality between wheat varieties. Therefore, the relationship was studied between the amounts of protein fractions, especially the Glu fraction (which comprises the relevant glutenins) and two commonly used tests to assess baking performance, i.e. mixograph analysis and SDS sedimentation test.

2 MATERIALS AND METHODS

2.1 Materials

A set of 8 wheat varieties, grown at two N-fertilisation levels (245 and 359 kg N/ha; level 1 and 2 resp.) was obtained from Advanta-Van der Have, The Netherlands. The set of 10 wheat varieties was obtained from Plant Breeding International Cambridge, U.K.

2.2 Fractionation Method

The fractionation method will be described in full detail elsewhere[3]. Essentially, it comprises the following steps. Firstly a wheat sample was extracted by 0.5 M NaCl to obtain the A+G fraction. Then the remaining pellet was extracted with SDS-ethanol to obtain gliadins, glutenins III and part of the glutenins II. By lowering the temperature of this extract to 10°C, virtually all glutenins II could be precipitated. The remaining soluble fraction is termed Glia fraction. The SDS-ethanol unextractable residue together with the precipitated glutenins II is termed Glu fraction.

2.3 Analytical Procedures

Mixograph analysis was performed on flour in presence of 2% NaCl (w/w), with a 10-gram mixograph. Optimal water absorption was determined using a 50-gram Brabender farinograph. SDS-sedimentation test using whole meal samples was performed as described by Axford et al[4].

3 RESULTS AND DISCUSSION

3.1 Proportions of Wheat Protein Fractions

Despite the fact that the samples were selected for having a large range in dough development time and baking quality, the proportions of the Glia and Glu fractions show rather limited variation (Table1); particularly the stronger varieties between which this variation is nearly zero (see also 3.2). The range of the A+G fraction (data not shown) was also very limited: 18-24 %. A relatively narrow range in monomeric proteins (albumins, globulins and gliadins) was also found by Sapirstein et al[5].

By comparing the results from the 8 varieties grown at two substantially different N-fertilisation levels, it may be concluded that the variation in the relative proportions of the protein fractions is more affected by variety than by N-fertilisation level.

Table 1. *The proportions of Glia and Glu fractions (% of total protein) in wheat samples.*

Variety	Glia (%)	Glu (%)	Variety	Glia (%)	Glu (%)
Bussard[a]	40	42	Bussard[b]	42	39
Caprimus[a]	39	38	Caprimus[b]	40	37
Ritmo[a]	39	38	Ritmo[b]	42	34
Scipion[a]	38	42	Scipion[b]	38	42
Soissons[a]	37	42	Soissons[b]	37	42
Tambor[a]	40	39	Tambor[b]	40	38
Yacht[a]	35	43	Yacht[b]	36	43
Zentos[a]	40	40	Zentos[b]	38	41
Aardvark	43	36	Isengrain	36	43
Sideral	38	43	Charger	39	39
Flair	41	37	Shanto	37	40
Moulin	42	36	Bartis	38	41
Malacca	39	40	Classic	34	43

a: N-fertilisation level 1; b: N-fertilisation level 2.

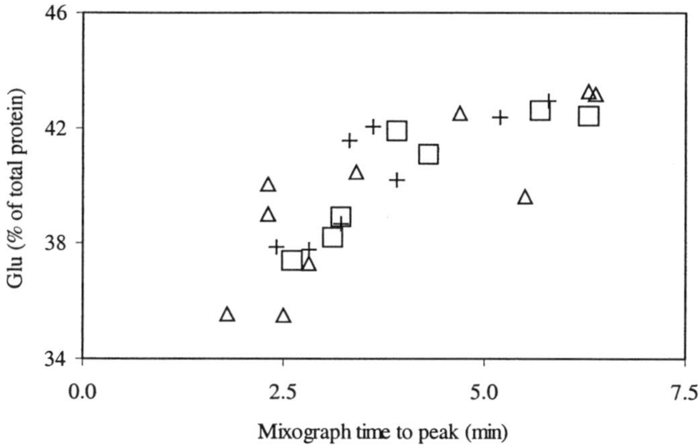

Figure 1. *The relative proportions of the Glu fraction versus mixograph time to peak. Samples: 8 wheat varieties at N-fertilisation level 1 (crosses) and level 2 (squares); additional samples (triangles).*

3.2 Mixograph Analysis

The dough strength of the wheat flours was assessed by mixograph analysis. The relation between mixograph time to peak, indicating dough strength, and the proportions of the Glu fraction are shown in Figure 1. Up to a time of about 4 min, the proportions of the Glu fraction increase; while the reverse holds for the Glia fraction (data not shown). Above a time to peak of about 4 min, the proportions of the Glu and of the Glia fraction are nearly constant; they reach levels of 43 and 36 %, respectively. Therefore, for strong and moderately strong wheat varieties, the differences in dough strength must be due to other parameters than differences in the proportions of the three proteins. One parameter could be the protein content of the wheat varieties. However, no correlation was found between protein content and mixograph time to peak. A likely parameter is the polymer size of the glutenins, especially when taking into account that the amount of glutenins II (smaller, soluble glutenins) decreases with increasing mixograph time to peak and becomes nearly zero above a time of 4 min. Our findings support the hypothesis of MacRitchie[6] that glutenins need to have a certain minimal polymer size to contribute to dough strength.

Furthermore, it can be seen from Figure 1, as expected based on our fraction results, that the N-fertilisation level does not affect the relation between the protein fractions and the mixograph time to peak. Variation in N-fertilisation only had a limited effect on the proportions of the three protein fractions.

3.3 SDS-Sedimentation Test

In the SDS-sedimentation test, the larger insoluble glutenins form the sediment. Our Glu fraction comprises these polymers. Because the proportions of this fraction are virtually constant above a mixograph time to peak of about 4 min, it is to be expected that the

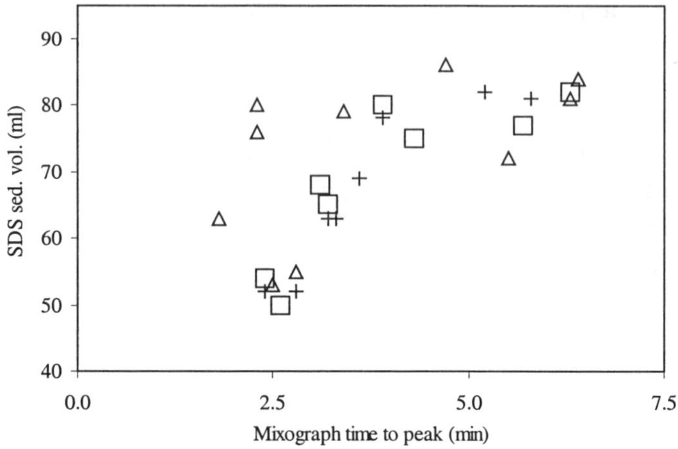

Figure 2. *SDS-sedimentation volumes of the wholemeal samples versus mixograph time to peak of their corresponding flours. Samples: 8 wheat varieties at N-fertilisation level 1 (crosses) and level 2 (squares); additional samples (triangles).*

SDS- sedimentation volumes are constant too. As can be seen in Figure 2 this was indeed found, even when using wholemeal samples and not correcting for amount of protein. The SDS-sedimentation test does not, therefore, discriminate between strong and moderately strong wheat varieties.

4 CONCLUSIONS

The main conclusions from this study are:

- There is only a limited variation in the proportions of the three wheat protein fractions.
- Above a mixograph time to peak of about 4 min, the proportions of the three wheat protein fractions are virtually constant.
- The SDS-sedimentation test does not discriminate between strong and moderately strong wheat varieties.

The findings support the hypothesis that increasing dough strength is related to increasing polymer size of the glutenins.

References

1. A. Graveland, P. Bosveld, W.J. Lichtendonk, J.H.E. Moonen and A. Scheepstra. *J. Sci. Food Agric.*, 1982, **33**, 1117.
2. A. Graveland, P. Bosveld, W.J. Lichtendonk, J.P. Marseille, J.H.E. Moonen and A. Scheepstra. *J. Cereal Sci.*, 1985, **3**, 1.
3. J.M. Vereijken, V.L.C. Klostermann, F.H.R. Beckers, W.T.J. Spekking and A. Graveland. *J. Cereal Sci.* (submitted for publication).
4. D.W.E. Axford, E.E. McDermott and D.G. Redman. *Cereal Chem.*, 1979, **56**, 582.
5. H.D. Sapirstein and B.X. Fu. *Cereal Chem.*, 1998, **75**, 500.
6. F. MacRitchie. *Adv. Food Nutr. Res.*, 1992, **36**, 1.

EVIDENCE FOR VARYING INTERACTION OF GLIADIN AND GLUTENIN PROTEINS AS AN EXPLANATION FOR DIFFERENCES IN DOUGH STRENGTH OF DIFFERENT WHEATS

H.D. Sapirstein[1] and B.X. Fu[2]

1. Department of Food Science, University of Manitoba, Winnipeg, MB, Canada R3T 2N2. 2. Canadian International Grains Institute, Winnipeg, MB, Canada R3C 3G8

1 INTRODUCTION

The transformation of hydrated flour particles into a developed dough with the “right” combination of rheological properties for optimum expansion and gas retention during proofing and baking is mainly due to the combined functionality of gliadin and glutenin proteins. Those rheological properties undoubtedly derive from a balance of viscosity and elasticity, contributed by monomeric gliadins and polymeric glutenin, respectively. Interestingly however, most correlation studies on the differences in breadmaking properties of different wheats, have strongly indicated that it is the unextractable or high molecular weight fraction of glutenin that is the principal factor related to breadmaking quality, especially dough mixing requirements[1,2,3]. Experimental evidence of the importance of gliadins to breadmaking mainly comes from fractionation-reconstitution studies where the ratio of gliadin to glutenin can be altered experimentally[4,5,6,7,8].

That most wheats, even those varying widely in breadmaking quality, appear to have very similar gliadin contents measured as monomeric proteins[9] would appear to explain why no clear relationship has been found between the gliadin fraction and inter-cultivar differences in breadmaking quality. On the other hand, recent work in our laboratory has provided evidence for genotype specific interaction of gliadin and glutenin proteins. We found that interaction to be 1) inversely related to dough strength[10,11], 2) accentuated by dough mixing during the initial water hydration phase[12] and 3) promoted by salt[13]. The objective of this study was to examine whether we could quantify the degree of interaction for wheats of widely different strength based on the relative solubilities in water, of gluten and constituent gliadin and glutenin proteins when parent glutens were prepared in salt solution.

2 MATERIALS AND METHODS

2.1 Wheat Samples and Technological Quality.

Straight grade flour of three Canadian wheat cultivars of diverse breadmaking quality was used. Glenlea and Katepwa are hard red spring wheats with extra strong and moderately strong dough mixing properties, respectively. Harus is a weak mixing soft white winter wheat that is optimally suited for cake baking. Dough mixing properties were measured using a direct drive 2 g computerized Mixograph. Representative mixograms of these three cultivar samples are shown in Fig. 1.

2.2 Preparation of Glutens and Water-Soluble and Residue Fractions.

The preparation of wet glutens, and subsequent sequential extraction with distilled and deionized water were carried out as previously described[13]. Gluten was prepared from each flour using 0.2% salt (NaCl). The 0.2% NaCl gluten was subjected to four sequential extractions with distilled and deionized water producing four water soluble fractions and an insoluble residue. The glutens and their fractions were freeze-dried. Protein contents (N x 5.7) of the dry preparations were determined by the micro-Kjeldahl method.

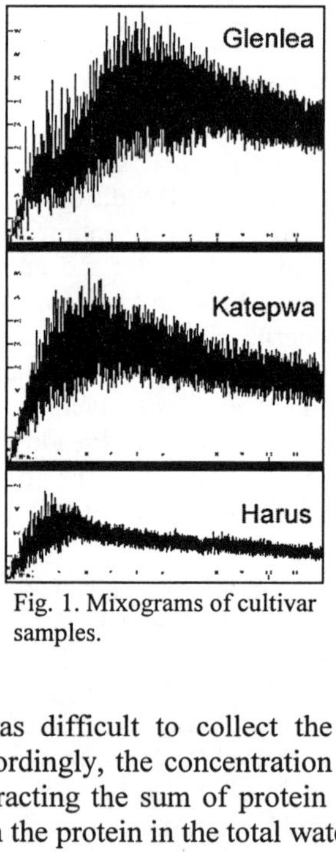

Fig. 1. Mixograms of cultivar samples.

2.3 Quantitation of Gliadin and Glutenin in Water-soluble and Residue Fractions.

Freeze-dried water-soluble fractions (10 mg) or residue (20 mg) were dispersed in 0.5 mL of 50% 1-propanol. After adding 0.34 mL of 1-propanol to bring the final 1-propanol concentration to 70% (v/v), both 50% 1-propanol soluble and insoluble protein (glutenin) were pelleted by centrifugation (15,000 g, 10 min). The precipitated glutenin was extracted twice (1 h and 30 min, respectively) with 0.75 M NaI (1 mL) to remove co-precipitated ω-gliadins[13]. The gliadins, which are soluble in 70% 1-propanol (after precipitation of glutenin) and in the 0.75 M NaI extract, were thus separated from the glutenin. The gliadins in the 70% 1-propanol supernatant and in the 0.75 M NaI extract were combined. Protein contents of the fractions were quantified by micro-Kjeldahl analysis. Because the water soluble fractions of 0.2% NaCl gluten contained only small amounts of glutenin, it was difficult to collect the 70% 1-propanol precipitate for protein quantification. Accordingly, the concentration of glutenin in each of these fractions was calculated by subtracting the sum of protein in 70% 1-propanol supernatant and 0.75 M NaI extract from the protein in the total water-soluble fraction.

2.4 Electrophoresis.

Subsamples of the original glutens and their fractions were analyzed by acid-polyacrylamide gel electrophoresis (A-PAGE)[14].

3 RESULTS AND DISCUSSION

3.1 Solubility of 0.2% NaCl Glutens in Water.

The solubilities of gluten protein and constituent gliadin and glutenin fractions of the 0.2% NaCl glutens are shown in Fig. 2. The total percentage of gluten protein extracted by the four water extractions was approximately 60%, 52% and 33% for Glenlea, Katepwa, and Harus, respectively. Clearly, the stronger the parent flour from which the gluten was isolated, the greater the proportion of protein that was solubilized. Results also showed that intercultivar differences were mainly due to the gliadin fraction. Essentially all of the gliadin protein (92%) in the gluten from the extra strong Glenlea wheat was extracted. The corresponding values for gliadin solubility in the Katepwa and Harus glutens were 76% and 51%, respectively. It was also noteworthy that gliadin from the stronger gluten was more readily water extractable than that of weaker gluten (Fig. 2). The first extraction alone solubilized 58% of gliadin in the Glenlea gluten. The equivalent values for Katepwa and Harus were 28 and 5%, respectively.

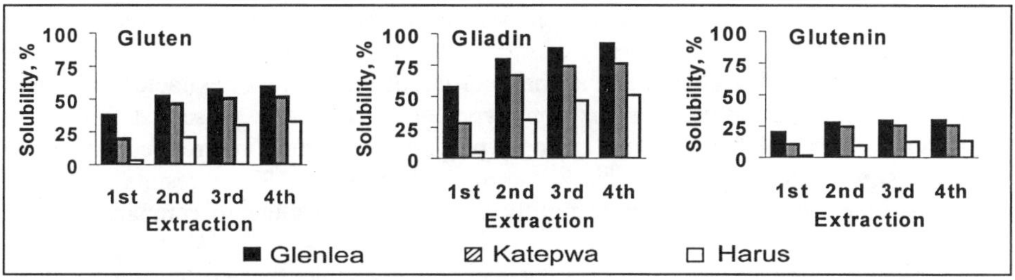

Fig. 2. Cumulative water solubilities of gluten, and gliadin and glutenin protein.

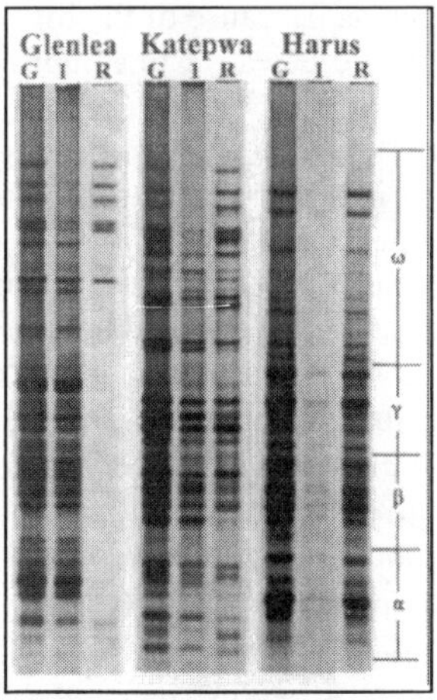

Fig. 3. A-PAGE of gliadins of gluten (G), the first water soluble fraction (1) and insoluble residue (R) from glutens of three wheats.

The nature of the proteins in the fractions was confirmed by A-PAGE of the parent 0.2% NaCl glutens, the four DDW soluble fractions, and the residues (Fig. 3). It was interesting that mainly ω-gliadins remained in the water insoluble residue from Glenlea gluten. In contrast, the gliadin pattern of the residue of the weak Harus gluten was essentially the same as that of unfractionated parent gluten. A-PAGE patterns also clearly showed the relationship (noted above) between gliadin extractability in the early water extractions and dough strength of the samples; the intensity of band staining was in the order: Glenlea > Katepwa >> Harus (compare lanes 1 in Fig. 3).

In contrast to gliadin solubility in water, a much lower amount of glutenin from 0.2% NaCl glutens was water soluble (Fig. 2). The trend observed above in relation to dough strength for gliadin solubility in the first extract was also found for glutenin in 0.2% NaCl gluten; the solubility of glutenin in the first extracts of Glenlea, Katepwa and Harus glutens were 20, 11 and 2%, respectively. The solubility in water of this fraction of glutenin could be attributed to its strong aggregation with gliadin in the gluten complex. In this case, the amount of glutenin that was water soluble depended on the amount of gliadin solubilized in water; the stronger cultivars had higher glutenin solubilities. Presumably, the solubilized glutenin was of relatively low molecular size.

4 CONCLUSIONS

We have previously shown that gluten protein solubility in water increases when the gluten was prepared in the presence of salt[13]. For gluten prepared with 0.2% NaCl, as in this study, most of the gliadin and a small portion of glutenin was extracted. The pattern of gluten protein solubility was genotype dependent; the observed intercultivar differences were related to dough strength properties; gliadins in stronger glutens were more easily extracted by water than those of weaker glutens. These results indicate that gliadins in weak glutens are more tightly associated with glutenin compared to strong glutens.

A precise explanation for these results is not yet available. It seems reasonable that the average molecular size of glutenin is directly related to gluten strength. Assuming that strong non-covalent interactions exist in gluten between gliadin and glutenin proteins, glutenin molecular size variation would affect gliadin-glutenin interactions on a purely physical basis. As previously expressed[7], the larger the glutenin, the smaller would be the ratio of surface area to mass which can affect the degree of inter-molecular protein-protein and protein-solvent interactions. On a constant molecular mass basis, proteins of larger size have smaller surface areas. Accordingly, the strength of gliadin-glutenin interactions should be lower in strong glutens than in weak glutens. Because of the higher hydrophilicity of gliadins compared with that of glutenin, it should be easier to solubilize glutenin as a gliadin/glutenin complex than as glutenin alone.

This interaction model can be extended to explain events that occur during dough formation and development. In flour, gluten proteins exist in a dispersed form. On addition of water, the originally closely-packed protein bodies hydrate and begin to interact as the dough is mixed. Dough development will result from disaggregation/depolymerization of the glutenin component and its further interaction with gliadins. In the process of dough mixing, the gliadins which interact with glutenins act as plasticizer by weakening the interaction between the glutenin aggregates. At the same time, glutenin becomes more soluble in water because the gliadin-glutenin complex is more soluble than glutenin alone. Wheats with strong gliadin-glutenin interactions would therefore require less mixing for optimum dough development, and vice versa.

The results of this study showed that significant intercultivar differences exist in the disaggregation/solubilization rate of gliadin and glutenin proteins in water. These differences are very likely due to the direct effect of intrinsic differences in the size distribution of glutenin. The indirect effect of the proposed inverse relationship between glutenin molecular size and extent of gliadin-glutenin interaction in dough may also be an equally important factor in explaining intercultivar differences in gluten strength.

References

1. Orth, R.A. and Bushuk, W. *Cereal Chem.,* 1972, **49**, 268.
2. Gupta, R.B., Khan, K. and MacRitchie, F. *J. Cereal Sci.*, 1993, **18**, 23.
3. Sapirstein, H.D. and Johnson, W.J. A paper in this volume

4. Finney, K.F. *Cereal Chem.,* 1943, **20**, 381.
5. Shogren, M.D., Finney, K.F. and Hoseney, R.C. *Cereal Chem.,* 1969, **47**, 93.
6. MacRitchie, F. *J. Cereal Sci.*, 1987, **6**, 259.
7. Sapirstein, H.D. and Fu, B.X.'Characterization of an extra-strong wheat: functionality of 1) gliadin- and glutenin-rich fractions, 2) total HMW and LMW subunits of glutenin assessed by reduction-reoxidation', Proceedings of the Sixth International Gluten Workshop, C.W. Wrigley, ed. Royal Aust. Chem. Inst., Melbourne, 1996, p.302.
8. Uthayakumaran, S., Gras, P.W., Stoddard, F.L. and Bekes, F. *Cereal Chem.* 1999, **76,** 389.
9. Fu, B.X. Sapirstein, H.D. and *Cereal Chem.,* 1998, **75**, 500.
10. Fu, B.X. and Sapirstein, H.D. *Cereal Chem.,* 1996, **73,** 143.
11. Dupuis, B., Bushuk, W. and Sapirstein, H.D. *Cereal Chem.,* 1996, **73,** 131.
12. Almonte, M.T. 'Effects of Dough Mixing and Relaxation on the Protein Solubility and Composition of Two Diverse Bread Wheats', M.Sc. Thesis, University of Manitoba, Winnipeg, 1998
13. Fu, B.X, Sapirstein, H.D. and Bushuk, W. *J. Cereal Sci.* 1996, **24**, 241.
14. Sapirstein, H.D. and Bushuk, W. *Cereal Chem.* 1985, **62,** 372.

Acknowledgement

The financial assistance provided by the Natural Sciences and Engineering Research Council of Canada is gratefully acknowledged.

RHEOLOGICAL AND BIOCHEMICAL APPROACHES DESCRIBING CHANGES IN MOLECULAR STRUCTURE OF GLUTEN PROTEIN DURING EXTRUSION

A Redl[1], M.H. Morel[1], B. Vergnes[2], S. Guilbert[1].

1. UFR "Technologie des Céréales et des Agro-polymères", ENSA.M – INRA Montpellier, 2 place Viala, 34060, Montpellier, France; 2. Ecole de Mines de Paris, CEMEF, UMR CNRS 7635, BP 207, 06904 Sophia Antipolis, France

1 INTRODUCTION

Extrusion technology is a highly efficient method for the continuous shaping of thermoplastic materials. Using this well-known technology with a renewable agricultural raw product to produce materials with low environmental impact represents a real challenge for the production of "bioplastics". Proteins, as heteropolymers, offer a large scatter of possible interactions and chemical reactions. The multitude of possible interactions might also be the reason why, in spite of the rapid development of extrusion technology in the past years, information about the molecular mechanisms of protein interactions remains limited. Extrusion of proteins was studied mainly for food uses as textured vegetable proteins, mainly based on soy proteins[1,2,3] or soy protein - carbohydrate blends[4]. The formation of the final molecular network involves the dissociation and unraveling of the macromolecules, which allows them to recombine and to cross-link through specific linkages[5]. However, the way that proteins interact with proteins in an extrusion process is still unclear.

The object of the present study is to get a better understanding of the feasibility of shaping gluten/glycerol materials by extrusion and to investigate the influence of the operating conditions on the obtained structure.

2. MATERIALS AND METHODS

Extrusion trials, rheological and biochemical analysis were performed as described in previously[6]. Commercial vital wheat gluten, plasticized with glycerol was extruded using a corotating self-wiping twin screw extruder (DS 25 Brabender O.H.G, Duisburg, Germany). The influence of feed rate, screw speed and barrel temperature on processing parameters (die pressure, product temperature, residence time, specific energy) have been examined.

Rheological measurements were carried out using a Rheometrics Dynamic Spectrometer (Mk III torsion head, Rheometrics Inc., Piscataway, USA) equipped with two parallel plates (diameter 20 mm). A time/frequency sweep was run for 60 min at □ = 0.03-30rad, γ_0 = 0.005, T = 80°C. To attempt a quantitative analysis of the obtained mechanical spectra, loss and storage moduli G' and G" were fitted by Cole-Cole distributions, as proposed by[7]. By analogy to the theory of rubberlike elasticity the plateau modulus G_N^0, derived from the Cole Cole modelling, may be related to the density of crosslinked network strands[8]:

$$G_N^0 = \frac{\rho}{M_c} RT \qquad (1)$$

where ρ is the density of the material ($\rho = 1.2\ 10^6$ g/m^3), M_c is the molecular weight of network strands between crosslinks (g/mol), R is the universal gas constant (R = 8.314 J/mol K) and T is the temperature (K).

The molecular size distribution of gluten proteins was studied by size-exclusion high-performance liquid chromatography (SE-HPLC). Extruded samples or gluten were solubilized in a 0.1 M sodium phosphate buffer (pH6.9) containing 1% sodium dodecyl sulphate (SDS).

3. RESULTS AND DISCUSSION

The main data from the extrusion trials are presented in table 1. Generally the extrudates obtained were solid, rubberlike. Depending on operating conditions, the extrudates showed very different appearances, ranging from very smooth-surfaced extrudates with high swelling to completely broken extrudates. The appearance of surface irregularities and extrudate breakup seemed to be induced by the severity of the thermomechanical treatment (high temperature, high SME). We theorize that when the mobility of the polymeric chains is reduced due to cross-linking, it inhibits the elastic recovery without rupture. High screw speed leads to important viscous heat dissipation that resulted in excessive maximum temperature of the product. Such a high temperature, together with an significant SME input, might have resulted in a more cross-linked structure, which would explain the observed extrudate rupture.

Table 1: *Main data from the extrusion trials. Tr regulation temperature, SME specific mechanical energy, T3 measured temperature in the die, ES extrudate swell (1)disrupted extrudate (2)irregular extrudate, Fi insoluble protein fraction, G_N^0 plateau modulus obtained by fitting Cole Cole Functions to experimental data, Mc molecular weight between crosslinks derived from eq (1).*

Speed (RPM)	Feed Rate (kg/h)	Tr (°C)	SME (kJ/g)	T3 (°C)	ES	Fi (%)	G_N^0 (kPa)	M_c (kg/mol)
50	1.9	80	0.64	97	1.58	5.4	23.1	151.5
100	1.9	80	1.36	111	1.38	10.1	46.2	75.8
200	1.9	80	3.70	134	n.d(1)	34.8	86.5	40.5
100	4.9	80	0.73	108	1.58	3.5	26.1	133.7
200	4.9	80	1.48	139	n.d(1)	26.5	84.5	41.8
200	8.1	80	0.88	-	1.42	7.0	37.9	92.4
100	4.9	60	0.86	101	1.85	3.7	23.4	149.0
200	4.9	60	1.55	124	n.d(2)	13.1	33.9	103.3

In order to determine the network structure and crosslink density of the extrudet materials, dynamical rheological properties and molecular size distribution were determined. The mechanical spectra and the corresponding SE HPLC elution profiles of a series of extrusion trials at fixed flow rate and increasing screw speed are shown in Figures 1 and 2.

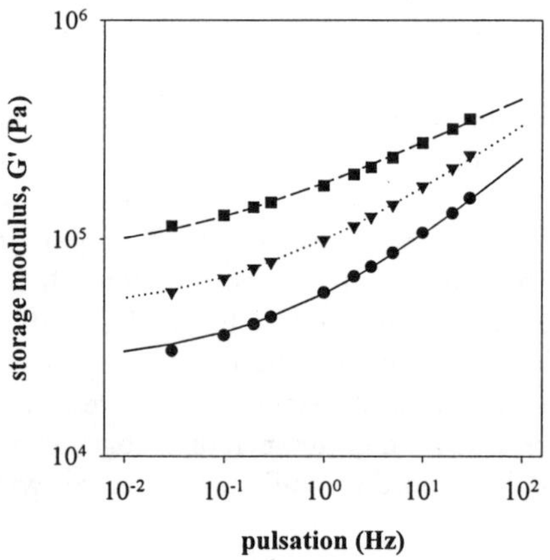

Figure 1. Evolution of storage modulus G' with the pulsation. Lines represent the fit of Cole Cole functions. Symbols : (●) N=50 rpm, (▼) N=100 rpm, (n) N = 100 rpm, Q = 1.9 kg/h; Tr = 80°C.

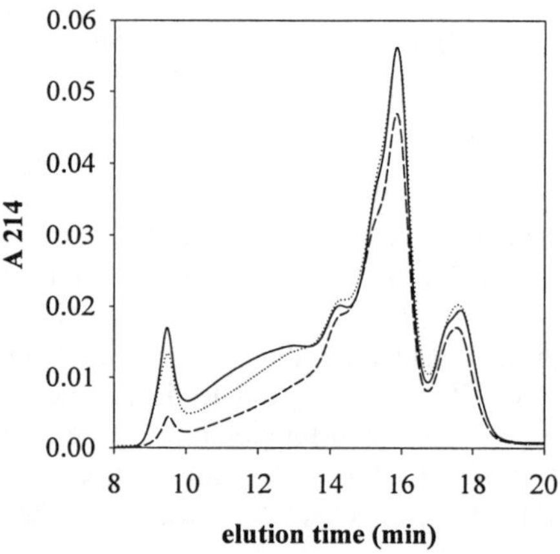

Figure 2. Elution profiles obtained by SE-HPLC chromatography of plasticized gluten extruded at different rotation speeds. Lines: continous N=50, dotted N=100 RPM, slashed N=200 RPM , Q= 1.9 kg/h, T_r = 80. °C.

The rheological properties of the samples are strongly dependent on the operating conditions. When rotation speed N was increased from 50 to 200 rpm at a flow rate Q = 1.9 kg/h and a regulation temperature T_r = 80 °C, the plateau modulus G_N^0 increased from

23 to 86 kPa and the corresponding molecular weight of network strands obtained with Eq 1 ranged from M_e= 40 10^3 g/mol to M_e= 150 10^3 g/mol (Table1). The elution profiles of the SDS soluble fraction are shown in Figure 2. The overall area diminishes when increasing the rotation speed from 50 to 200 RPM, especially the fast eluting fractions corresponding to high molecular weight fractions. Concurrently the insoluble fraction (Fi), corresponding to proteins with $M_r > 7\ 10^6$, increases from 5.4 to 34.8 %.

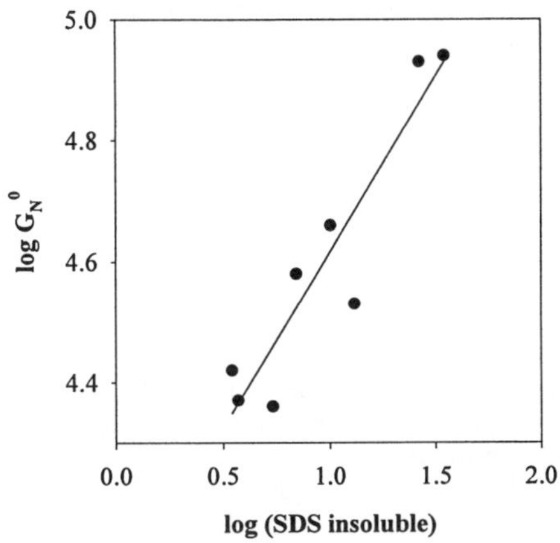

Figure 3. *Plateau modulus G_N^0, versus SDS insoluble fraction (Fi, $M_r > 7\ 10^6$), regression line is $y=4.03+0.581x$, $r^2=0.87$.*

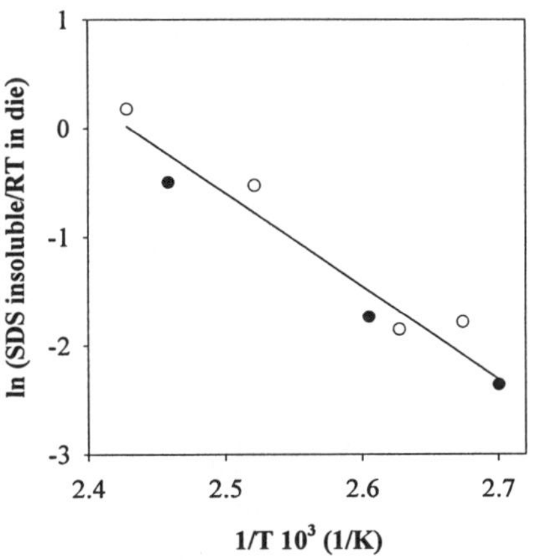

Figure 4. *Arhennius plot of the SDS insoluble fraction versus maximum temperature in the die. The residence time in the die is 57 s (●) and 22 s (○). Activation energy is 71 kJ/mol, r^2 = 0.93.*

The plateau modulus G_N^0 of the materials obtained and the corresponding molecular size between network strands appears to be correlated with the percentage of Fi, as shown in Figure 3. A linear relationship on a double logaritmic scale with a slope of 0.58 and a correlation coefficient of 0.87 can be established.

The evolution of molecular size distribution along the screw has also been determined and it appeared that the most important changes occured in the converging section of the die. The rate of protein insolubilisation was estimated by dividing the amount of the SDS insoluble fraction by the residence time in the die. In Figure 4 the corresponding Arhennius plot is shown. Consequently, an activation energy (71 kJ/mol) for the insolubilisation process can be determined. This activation energy is lower than that observed for heat denaturation of proteins in static conditions (100-125 kJ/mol[9]). Mechanical shear probably enhances the crosslinking effect of temperature. More work is needed to confirm this hypothesis

4. CONCLUSION

Extrusion of wheat proteins plasticized with glycerol appears to be feasible in steady state conditions. At a given plasticizer content, extrusion is possible in a quite narrow window of operating conditions. Depending on operating conditions, extrudates present very different aspects, ranging from very smooth surfaced extrudates with high swell to completely disrupted extrudates. We hypothesize that extrudate breakup is caused by an increasing network density, resulting in a decreasing mobility of the polymeric chains. In consequence the "protein melt" might no longer be able to support the strain experienced during its extrusion through the die. The increasing network density was reflected in the increased plateau modulus G_N^0 and the increased molecular size of protein aggregates. The molecular size of network strands as derived from the rheological properties appeared to be very well correlated with the SDS insoluble protein fraction. The crosslinking process seems to be induced by the severity of the thermomechanical treatment, with an estimated activation energy of 71 kJ/mol.

References

1. J. Camire, *JAOCS* 1991, **68**, 200.
2. Dahl, R. Villota, *Can. Inst. Food Sci. Technol. J.*, 1991, **24**, 143.
3. Horvath, B. Czukor. *Acta Aliment.*, 1993 **22**, 151.
4. Bhattacharya, M.A. Hanna, R.E. Kaufmann. *J. Food Eng.* 1988, **7**, 5.
5. Areas. *Crit. Rev. Food Sci. Nutr.* 1992, **32**, 365.
6. Redl, M.H. Morel, J. Bonicel, B. Vergnes, S. Guilbert. *Cer. Chem.,* 1999, **76**, 361.
7. Lefebvre, Y. Popineau, M. Cornec. 1993. '*Gluten Proteins*', Seibel W, Bushuk W, (eds.) Association of Cereal Research, Detmold, Germany. 180.
8. Ferry J.D. 1980. '*Viscoelastic Properties of Polymers*', 3rd Ed., John Wiley&Sons, New-York.
9. Favier, M.F. Samson, C. Aubled, M.H. Morel, J. Abecassis. *Sciences des Aliments,* 1996, **16**, 573.

EVALUATION OF WHEAT PROTEIN EXTRACTABILITY BY RHEOLOGICAL MEASUREMENTS

H. Larsson

Dept of Food Technology, University of Lund, PO BOX 124, S-221 00 Lund, Sweden

1 INTRODUCTION

The network forming properties of wheat flour dough and gluten can be investigated in oscillation small deformation rheological measurements. In dough the network formed by starch granules strongly dominates the elastic (solid) properties. However, this network is rather weak, i.e. a small linear region is observed for dough. Gluten on the other hand shows a greater linear region, and the elastic properties of gluten are about ten times smaller compared with those of dough. Both the mechanical properties rendered by the starch granules and the gluten are important during the baking process. The protein network is formed by both covalent and weaker non-covalent interactions. It is a well-known fact that some of the gluten proteins are extremely difficult to dissolve in any known solvent. The amount of extracted protein may be used for ranking solvent capacity. However, it is not only the amount, but also the quality of those proteins, which is important. The aim of the present study was to illustrate how rheology can be used to evaluate the solvent efficiency regarding the extractability of the proteins that are essential for the formation of the gluten network. Thus, the rheological properties of gluten were compared with those of the fractions obtained with different solvents.

2 MATERIALS AND METHODS

The wheat proteins were obtained through extraction with different solvents of the winter wheat Tarso (12.6 % protein dry basis (N*6.25)) milled at NordMills, Malmö, Sweden. Before the protein extraction the flour was defatted with chloroform. The defatted flour was suspended in 400 ml of the solvent (75 % v/v ethanol and 25 % v/v water or 10 mM HCl). The extraction was performed at 15°C during three hours by shaking the mixture every 15 min. The slurry was centrifuged at 48000g (15 min) to remove starch particles. The clear supernatant was collected and pH was adjusted to neutral when acid was included. The extract was added drop wise to 800 ml of acetone, and stirred allowing the protein to precipitate. Gluten or wheat protein samples were mixed with distilled water in a Bohlin Rheomix mixer with the two-gram sample geometry (Bohlin Reologi AB, Öved, Sweden). Mixing was conducted at 30°C. For samples where development of structure was observable in the mixing curve, mixing was stopped directly after the peak in torque response, otherwise mixing was continued for 10 min. Small deformation rheological measurements were performed using a StressTech 2000 (Rheologica Instruments, Lund, Sweden) controlled stress rheometer. The parallel plate geometry (15 mm, gap of 1-3 mm) was used, and the measurements were performed in the linear region at 30°C. 1000 cSt silicon oil was applied to all samples after loading to prevent water evaporation.

3 RESULTS AND DISCUSSION

The networks formed by the extracted wheat proteins have been examined in oscillating small deformation rheology and compared with the network formed by the full gluten. One unique property of the gluten protein is the swelling of the protein in water to a limited water content[1]. However, the water content of gluten may be increased if the mixing of dough is proceeded beyond the peak in mixing resistance[2]. Considering these observations suggests that the extracted wheat proteins should be investigated at a related range of water contents and under defined mixing conditions. Controlling the mixing

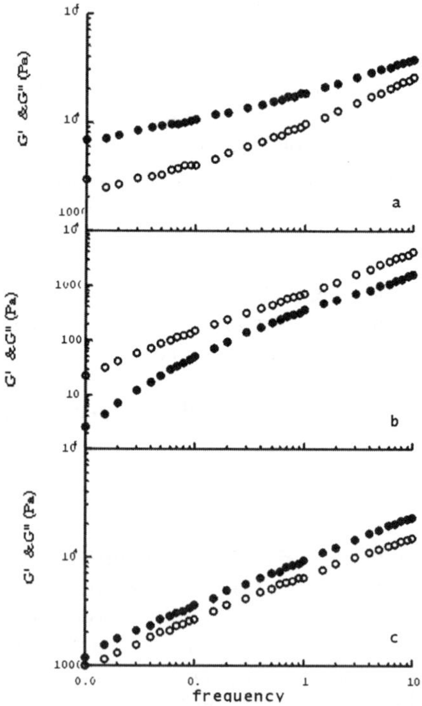

Figure 1. The frequency sweeps developed by gluten, wheat flour proteins extracted by ethanol and HCl (a, b and c, respectively) when mixed at a water content of 40% (total weight basis). G' (●) and G'' (○).

history of wheat flour proteins also seems important, as disulphide bonds are broken when dough is mixed beyond the optimum[3]. In Figure 1a, b and c the frequency sweeps for gluten and proteins extracted by ethanol and HCl, respectively, are shown at a water content of 40 % (total weight).

The full gluten show some dependence on frequency and G' > G" (Figure 1a). As expected the proteins extracted by ethanol show G' < G", i.e. liquid properties are dominating over all frequencies (Figure 1b). On the other hand, the samples formed by the proteins extracted by HCl, and mixed at a water content of 40%, show G' > G" like gluten. The frequency dependence is, however, considerable for this sample, indicating weaker gel properties (Figure 1c). Samples were also prepared at higher water contents up to 62% (Figure 2 a-c).

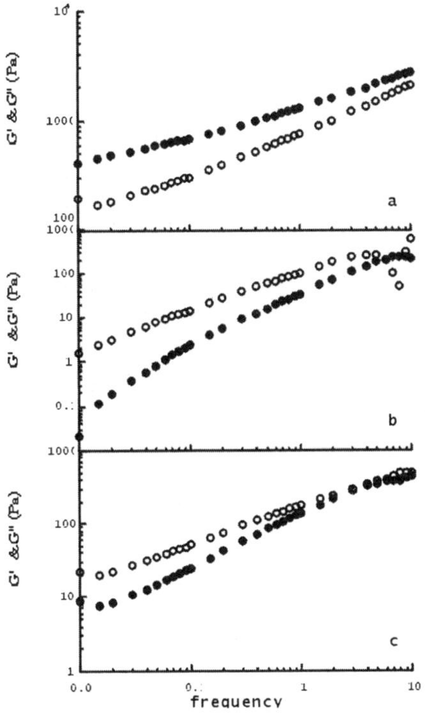

Figure 2. The frequency sweeps developed by gluten, wheat flour proteins extracted by ethanol and HCl (a, b and c, respectively) when mixed at a water content of 62% (total weight basis). G' (●) and G" (○).

In Figure 2a it can be seen that gluten shows G' > G" in the whole frequency region also at this higher water content. Only the values of G' and G" are lower. The proteins extracted by ethanol are diluted to show lower values of the moduli (Figure 2b). Due to the strong frequency dependence in Figure 1c, it may not be surprising that the network (G' > G") shown for the protein sample extracted by HCl disappears when the water content was increased. When it was mixed with 62% water, it was possible to dilute the network to result in dominating liquid behaviour (G' < G") over the whole frequency region investigated. This means that some proteins, which are important for the gel properties of gluten, are still missing in the extract when HCl is used as a solvent.

References

1. Larsson, H. and A.-C. Eliasson, *Cereal Chem.*, 1996. **73**, 25.
2. Larsson, H. and A.-C. Eliasson, *Cereal Chem.*, 1996. **73**, 18.
3. Danno, G. and R.C. Hoseney, *Cereal Chem.*, 1982. **59**, 196.

Acknowledgements

Andrew Rodd (Melbourne University, Australia) and Göte Johansson (Lund University, Sweden) are kindly acknowledged for work on the protein fractions. Financial support was obtained from the Cerealia Foundation R&D.

THE ASSESSMENT OF DOUGH DEVELOPMENT DURING MIXING USING NEAR INFRARED SPECTROSCOPY

J. M. Alava, S. J. Millar and S. E. Salmon

Campden & Chorleywood Food Research Association, Chipping Campden, Gloucestershire, GL55 6LD, UK.

1 INTRODUCTION

A near infrared (NIR) spectroscopy technique for the determination of breadmaking quality attributes and measurement of functional properties of bread dough during mixing has been developed.

2 METHODS

Near infrared measurements of the dough during mixing were taken using a Perten DA7000 spectrophotometer, with a fibre optic probe attachment. Spectra were taken approximately every 2 seconds during the mixing process.

Single variety flours covering a range of breadmaking quality were selected and a range of quality parameters was investigated. In particular, the elastic modulus (G') of the gel-protein fraction of each flour was measured using a Bohlin VOR rheometer.

The flour samples were mixed using a small-scale Do-Corder mixer under typical conditions for the Chorleywood Breadmaking Process (CBP) to determine varietal differences. The area of the peak at 1125-1180nm was plotted against time and the turning point of this curve was defined as the NIR mixing time. A CBP pilot-scale Morton Z-blade mixer was used to study the relationship between NIR mixing times and bread quality. The doughs were mixed to a range of work inputs and mixing speeds using several single variety flours. Loaf volume and crumb texture were measured by seed displacement and image analysis respectively.

3 RESULTS AND DISCUSSION

NIR spectra were processed to remove baseline variation due to movement of the dough as it mixed. Principal Component Analysis (PCA) was used to determine the spectral areas that changed most during mixing. One of the areas in the pattern associated with PC2 (1160nm) has previously been associated with water[1]; therefore, it was considered that PC2 related to the binding of water during hydration of the flour components. A linear relationship (r^2=0.85) between the NIR mixing time and gel protein G' results indicated that this region of the NIR spectrum contains relevant information on the evaluation of flour functionality (Figure 1).

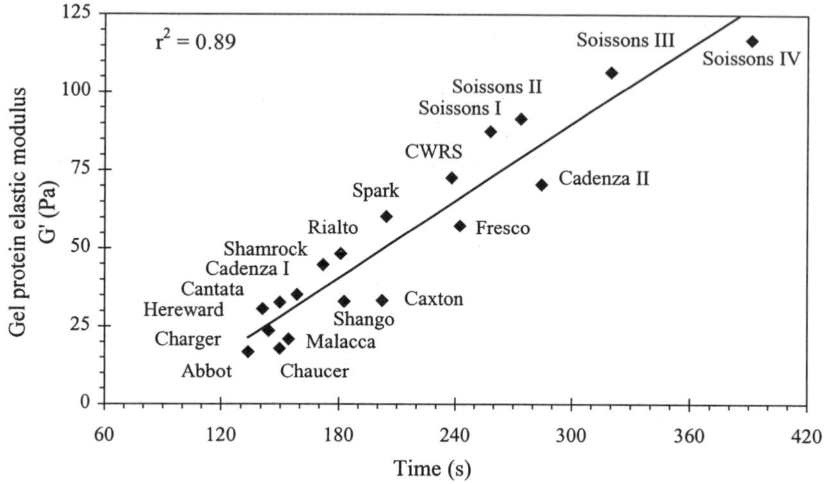

Figure 1 *Comparison of NIR mixing time against flour strength*

For bread made with a Morton mixer, the bread crumb had the finest structure for a mixing time shorter than the NIR value and the largest volume was reached after the NIR mixing time (Figure 2). This may be because the NIR mixing time resolved the moment at which the gluten network started to break down after a period of stability at an optimum mixing time.

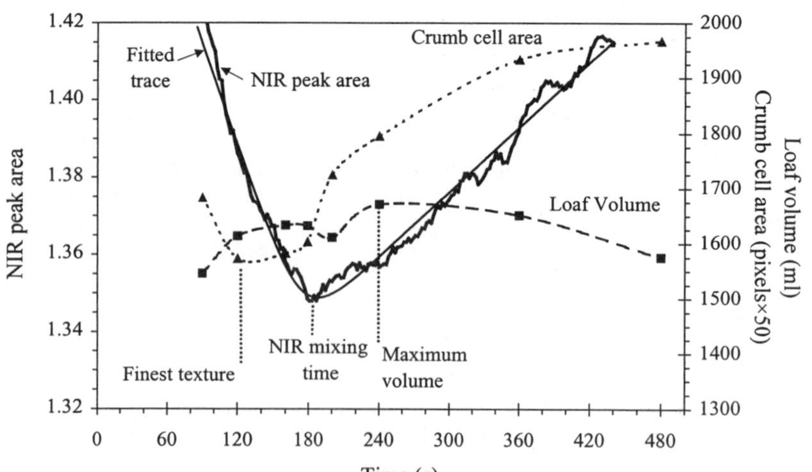

Figure 2 *Effect of dough mixing time on bread quality*

Two flours with different strengths were mixed using a Morton mixer to evaluate the effect of different rates of work-input, using blade speeds of 200, 300 and 400rpm. Figure 3 shows that the loaf volume increased for both varieties as the work input rate increased. However, the crumb texture was at its finest when the standard mixer blade speed of 300rpm was used.

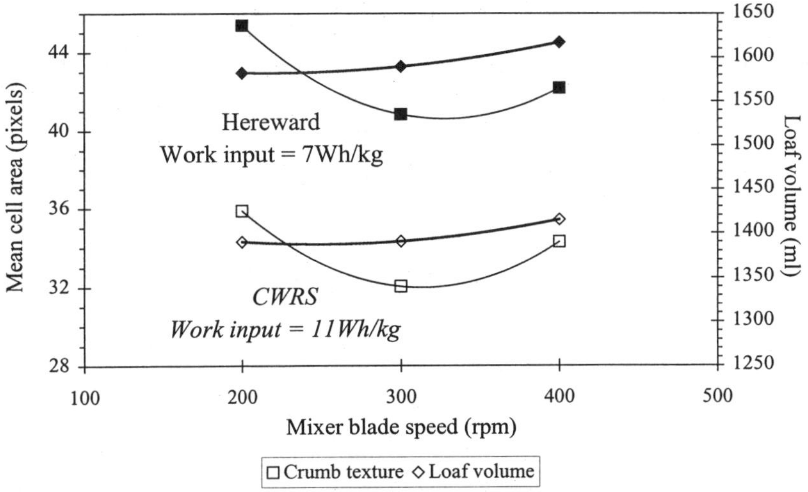

Figure 3 *Effect of variation of work-input rate on bread quality parameters*

4 CONCLUSION

Strong relationships were found between measurements taken from NIR mixing curves, bread-making flour quality parameters such as gel-protein elastic modulus and bread quality. For a range of varieties, specific mixing times are required in order to obtain a maximum loaf volume with fine crumb structure. The work input needed to achieve these conditions differs for different varieties, and the value determined by NIR showed potential for achieving this more consistently. NIR instruments can be used to help determine the mixing time or mixing performance of flour to optimise bread quality attributes and offers the potential for on-line use.

5 FURTHER WORK

Further work is being carried out at CCFRA to determine the bread parameters at a range of NIR mixing times and work inputs.

Reference

1. I.J. Wesley, N. Larsen, B.G. Osborne and J.H. Skerritt, *J. Cereal Sci.*, 1998, **27**, 61.

MEASUREMENT OF BIAXIAL EXTENSIONAL RHEOLOGICAL PROPERTIES USING BUBBLE INFLATION AND THE STABILITY OF BUBBLE EXPANSION IN BREAD DOUGHS AND GLUTENS

B.J.Dobraszczyk and J.D.Schofield

Department of Food Science & Technology, The University of Reading, P.O.Box 226, Whiteknights, Reading RG6 6AP, U.K.

1 INTRODUCTION

Baking is about the growth and stability of bubbles: their size, distribution, growth and failure during the baking process will have a major impact on the final quality of the bread both in terms of its appearance (texture) and final volume. It is considered that the limit of expansion of these bubbles is related directly to their stability, due to coalescence and the eventual loss in gas retention on bubble rupture. The rheological properties of the bubble walls will therefore be important in maintaining stability against premature failure during baking, and also in relation to gas cell stabilisation and gas retention during proving and baking, and thus to the final structure and volume of the baked product. Hence baking may in reality be governed by failure processes – the failure of the bubble walls at large deformations. Failure in materials is predictable from a knowledge of the material properties, such as stress, strain and the relevant modulus of the material.

There has long been a conviction amongst bakers that baking performance of dough is in some way related to the rheological properties of the dough, originating in the bakery practice of kneading and stretching the dough by hand to assess its quality. Although this is a very subjective method of measuring rheology, it tells us something about the sort of rheological measurements we should be making in order to predict baking performance.

Although there is a long pedigree of various empirical and fundamental methods used to measure dough rheology, correlations between these measurements on dough and baking performance have often produced inconsistent and conflicting results. One reason may be that that many of these methods measure the dough rheology at rates and conditions very different from those experienced by the dough during baking. The relevant conditions for proof and baking are those in the dough cell wall material around a slowly expanding gas cell. For example, rates of expansion of the bubble walls during proof and baking have been calculated in the range $10^{-4}s^{-1}$ to $10^{-2}s^{-1}$ (Table 1), compared with measuring rates in conventional rheological tests several orders of magnitude greater. Conventional oscillation shear rheological tests usually operate at small strains in the order of up to 1%, whilst strain in gas cell expansion during proof is in the region of several hundred percent. Furthermore, most rheological tests are carried out in shear, whilst most large-strain deformations in dough (i.e. sheeting, proof and baking) are extensional in nature. The deformation around an expanding gas cell during proof is biaxial extension. It seems intuitively reasonable to expect to make measurements of a process at conditions of strain and strain rate similar to those of the process being studied, especially

Table 1 *Typical rates and deformation conditions during bread making*

PROCESS	STRAIN RATE	DEFORMATION
MIXING	70 to $100s^{-1}$	MAINLY SHEAR
MOULDING	30 s^{-1}	SHEAR/EXTENSION
SHEETING	10 s^{-1}	EXTENSION
PROOF	10^{-3} s^{-1} to 10^{-4} s^{-1}	BIAXIAL EXTENSION
BAKING	10^{-2} s^{-1} to 10^{-3} s^{-1}	BIAXIAL EXTENSION

since we know that the rheological behaviour of dough is non-linear with strain and strain rate. Therefore, any rheological tests on dough which seek to predict baking performance should be carried out under conditions close to those of baking expansion, e.g. large deformation biaxial extension and low strain rates.

2 MATERIALS AND METHODS

There are many methods used in extensional flow measurements including: simple uniaxial tension, fibre wind-up or spinning, converging flow, capillary extrusion, opposed jets, lubricated compression and biaxial extension using inflation[1]. One of the simplest and most widely used methods for measuring biaxial extension properties of polymer melts has been the inflation technique[2-5]. Bubble inflation was first used as an empirical technique to measure wheat gluten and bread dough extensibility in the 1920's by Hankoczy[6] and Chopin[7]. This method was later developed for use as a rheological tool by Launay *et al.*[8] based on equations derived by Bloksma[9] and further developed by Dobraszczyk & Roberts[10]. This method has been developed into a new rheometer by Stable Micro Systems Ltd. (SMS) in conjunction with RHM Technology Ltd. to measure the extensional rheological properties of wheat flour doughs and gluten during bubble inflation, called the D/R dough inflation system[11]. The system inflates a sheet of dough by displacement of air using a piston driven by the standard TAXT2 crosshead. Pressure during inflation is measured by a pressure transducer (range 0-20 inches water), and the volume of the inflating dough sheet is calculated from the displacement of the piston. Inflation rate can be varied between 10 and 2000 cm^3/min, corresponding to maximum strain rates of $1x10^{-3}s^{-1}$ to $2x10^{-1}s^{-1}$, the lower limit corresponding to rates of baking expansion (Table 1). Stress (σ), Hencky strain (ε) and apparent viscosity are derived directly from a knowledge of the pressure (P) and volume (V) of the bubble during inflation, and are calculated by the instrument software, described earlier[11,12].

3 RESULTS AND DISCUSSION

Data obtained from the D/R system are normally presented as pressure against time as the bubble inflates. Figure 1 shows pressure and bubble height vs. time as the bubble inflates and the transformation into a true stress-strain curve. The stress-strain curve shows two interesting points. Firstly, the stress increases throughout inflation and shows no inflection at the peak in pressure, confirming Bloksma's[9] assertion that this peak has no rheological significance. Secondly, the stress-strain data exhibit considerable curvature up to failure, indicating that the modulus (stiffness) increases with inflation. This phenomenon is known as strain hardening, and has previously been observed in large extensional deformation of polymers[2-5], and its occurrence is known to be essential to maintain stability against premature failure in materials which stretch to large deformations, such as polymers drawn into fibres or inflated into thin films. Strain hardening allows the material to resist thinning by locally increasing resistance to further deformation. The stress-strain data can be fitted to a power-law relationship in the form of:

$$\sigma = K\varepsilon^n \qquad \text{(i)}$$

where n = strain hardening index (= slope log.stress vs. log.Hencky strain), normally obtained by fitting a power law curve to the stress-strain data (Figure 1b). Values of n greater than 1 indicate strain hardening, and the greater the value of n the greater the curvature of the stress- strain plot.

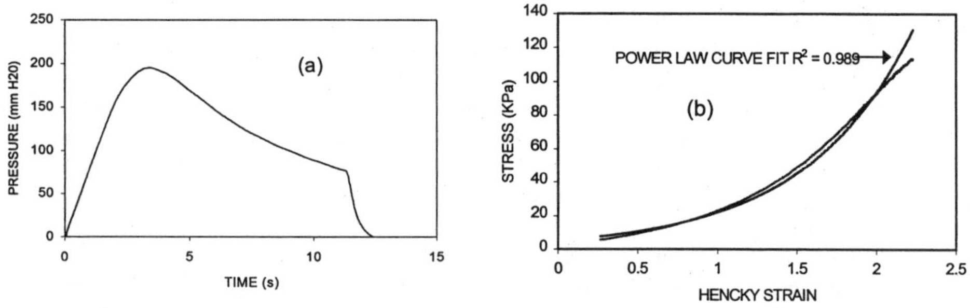

Figure 1 *(a) Pressure-time trace for an inflating dough bubble, (b) transformation into stress-strain curve (dotted line represents power law curve fit).*

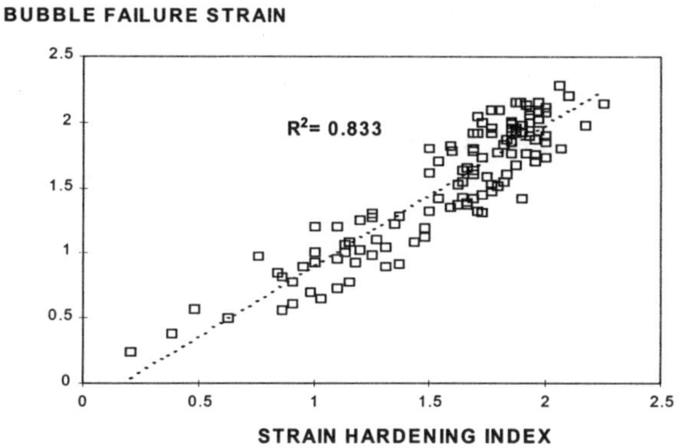

Figure 2 *Extensional strain hardening versus bubble failure strain for single wheat flour dough bubbles*

3.1 Strain hardening and baking

Recent work has shown that bread doughs exhibit strain hardening in large deformations such as bubble expansion, and that these extensional rheological properties are important in baking performance[10-14]. It has been shown that good bread-making doughs have good strain hardening properties and inflate to larger single bubble volume before rupture, whilst poor bread-making doughs inflate to lower volumes and have much lower strain hardening. It is expected that doughs with good strain hardening characteristics will expand to greater volumes before failure, give thinner cell walls and have a more even distribution of bubble sizes than doughs with poor strain hardening properties. This has been verified by Dobraszczyk & Roberts[10], who showed that the failure strain and hence the final inflation volume of single bubbles of bread dough were highly correlated with their strain hardening properties (Figure 2). Therefore, it is believed that the strain hardening characteristics of bread doughs will be critical in determining the limit of expansion of gas bubbles within dough and hence the final baking performance of bread doughs

References

1. B.J. Dobraszczyk and J.F.V. Vincent, 'Measurement of mechanical properties of food in relation to texture: the materials approach', in: *Food Texture – Measurement and Perception*, ed. A.J.Rosenthal, Aspen Publishers, 1999.
2. J.M Dealy and J. Rhi-Sausi, *Polymer Eng. Sci.,* 1981, **21**, 227.
3. C.D.Denson and R.J. Gallo, *Polymer Eng. Sci.,* 1971, **11**, 174.
4. C.D.Denson and D.L.Crady, *J. Appl. Polymer Sci.,* 1974, **18**, 1611.
5. D.D.Joye, G.W.Poehlein, and C.D.Denson, *Trans. Soc. Rheology,* 1972, **16**, 421.
6. J. Hankoczy, *Z. Gesamte Getreidewes,* 1920, **12,** 57.

7. M. Chopin, *Bull. Soc. Encour. Ind. Nal.* 1921, **133**, 261.
8. B. Launay, J. Bure, and J. Praden, *Cereal Chem.,* 1977, **54,** 1042.
9. A.H. Bloksma, *Cereal Chem.,* 1957, **34**, 126.
10. B.J.Dobraszczyk and C.A.Roberts, *J.Cereal Science,* 1994, **20**, 265.
11. B.J.Dobraszczyk, *Cereal Foods World,* 1997, **42,** 516.
12. B.J.Dobraszczyk, in: *Bubbles in Food,* eds. G.M.Campbell, C.Webb, S.S.Pandiella and K. Niranjan, American Association of Cereal Chemists, St. Paul, USA, 1999.
13. J.J.Kokelaar, T.van Vliet and A. Prins, *J. Cereal Science,* 1996, **24**, 199.
14. K.Wikström, Ph.D.thesis, Lund University, Sweden,1997.

THE EFFECT OF DOUGH DEVELOPMENT METHOD ON THE MOLECULAR SIZE DISTRIBUTION OF AGGREGATED GLUTENIN PROTEINS

K.H. Sutton, M.P. Morgenstern, M. Ross, L.D. Simmons and A.J. Wilson

New Zealand Institute for Crop & Food Research Ltd, Private Bag 4704, Christchurch 8001, NEW ZEALAND

1 INTRODUCTION

Mechanical dough development (MDD) mixing imparts a high rate of work input, dough development being accompanied by a reduction in polymeric protein size and by changes in the rheology of the dough. Dough development can also be achieved using other processes, such as sheeting, which imparts a lower rate of work input on doughs than does MDD mixing and requires much less energy.[1] Comparing the effects of these two very different methods of dough development on dough protein chemistry and molecular weight should reveal whether there are similar mechanisms operating during development and whether there are required mechanical processes for mixing.

2 MATERIALS AND METHODS

A standard 'strong' bakers flour was used. Dough preparation, sheeting procedures and rheological testing procedures have been detailed elsewhere.[2,3] Small scale MDD doughs were mixed to varying work input levels using a custom-built mixer using 10g of flour. Dough sub-samples were frozen (-200EC), freeze dried, then ground to a fine powder. Dough proteins were derivatised with monobromobimane [4] prior to extraction and analysis.[5]

3 RESULTS AND DISCUSSION

3.1 Quantification of glutenin proteins and exposed thiols

Test baking and rheological data have been presented elsewhere.[2,3] In summary, loaf volume was optimum at 40 sheeting passes, rupture stress peaked at 10 sheeting passes, whilst dough viscosity reached a maximum at 20-40 sheeting passes, similar to the loaf volume optimum.

Figure 1 illustrates the effect that the number of sheeting passes had on the SDS-soluble and SDS-insoluble polymeric glutenin proteins in the dough. These proteins have been shown to be closely linked with dough strength and quality characteristics.[5] It can be seen from Figure 1 that as sheeting progressed the quantity of the larger SDS-insoluble glutenins decreased whilst the quantity of the smaller SDS-soluble glutenins increased. It can also be seen from Figure 1 that the optimum baking performance (at ~40 sheeting passes) coincided with the levelling off of these changes in the protein composition.

Figure 2 illustrates the effect of MDD mixing on the SDS-soluble and SDS-insoluble polymeric glutenin proteins. It can be seen from Figure 2 that the changes in the quantities of both glutenin protein groups for the MDD mixing situation were similar to those observed for the sheeting experiment (Figure 1), in that the changes were beginning to level off by the time optimum mixing (7.0 Wh/kg) was reached.

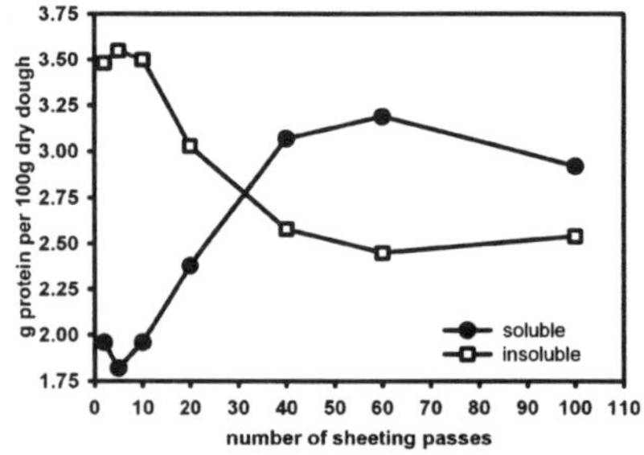

***Figure 1:** Effect of number of sheeting passes on the quantities of SDS-soluble and SDS-insoluble glutenin protein*

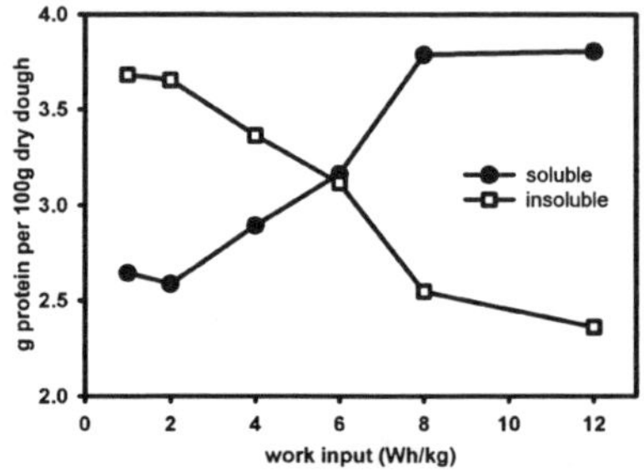

***Figure 2:** Effect of MDD mixing on the quantities of SDS-soluble and SDS-insoluble glutenin protein*

Figure 3 illustrates the effect that the number of sheeting passes had on the level of exposed thiol groups on the SDS-soluble and SDS-insoluble polymeric glutenin proteins in the dough. As sheeting progressed the relative quantity of exposed thiol groups in both of the glutenin classes decreased rapidly at first (up to 10 passes) before settling to a fairly stable value. That is, increased sheeting did not appear to result in the rupture of additional S-S bonds to form reactive thiol groups. Also, there was no coincidence

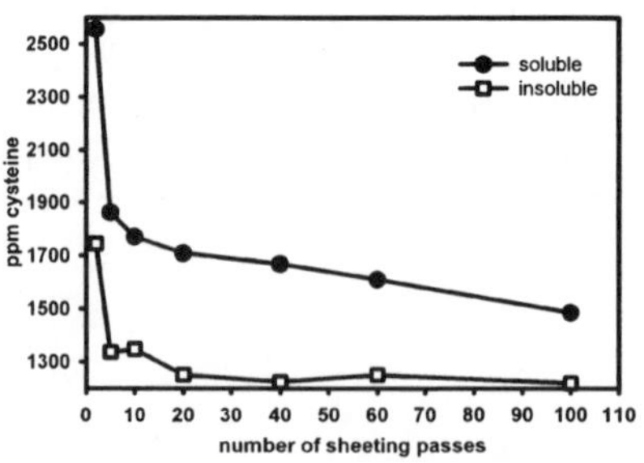

***Figure 3:** Effect of number of sheeting passes on the relative proportion of exposed thiols in the SDS-soluble and SDS-insoluble glutenin protein*

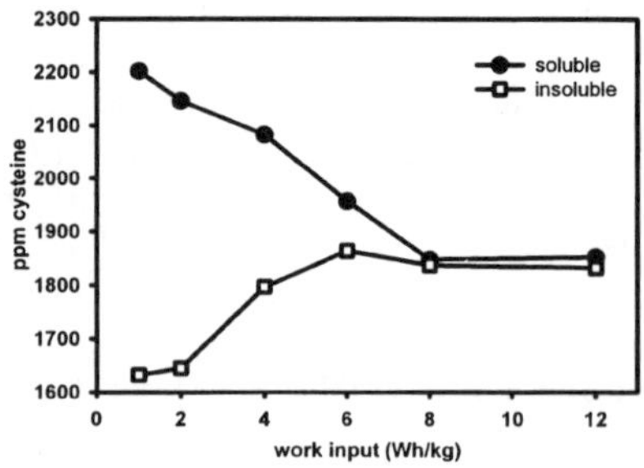

***Figure 4:** Effect of MDD mixing on the relative proportion of exposed thiols in the SDS-soluble and SDS-insoluble glutenin protein*

between the optimum baking performance and changes in the protein thiol exposure, although the rapid decrease in exposed thiol groups observed in Figure 3 appeared to coincide with the rapid increase observed in the rupture stress of the doughs.[2,3]

Figure 4 illustrates the effect that MDD mixing had on the level of exposed thiol groups in the two glutenin subclasses. The changes in the relative quantities of exposed thiol groups were different for MDD mixing compared to those observed for sheeting (Figure 3). Exposure of the protein thiols in the SDS-soluble glutenins decreased with mixing, indicating that SDS-soluble glutenins produced during MDD had a much lower exposed thiol content. For the larger SDS-insoluble proteins, thiol exposure increased with MDD mixing, indicating the rupture of S-S bonds to form exposed thiol groups. This observed increase in SDS-insoluble exposed protein thiol during MDD did not occur during the more gentle sheeting procedure, indicating that disulfide bond rupture may not be a required process in dough development. It may be that high stresses *per se* are not required to 'develop' doughs.

3.2 Molecular size of glutenins

The is a general concensus that the molecular weight (MW) of glutenin is in the multi-millions. There is also speculation that the sonication step used to extract SDS-insoluble glutenins reduces the MW of glutenin by shearing structural bonds. Glutenin proteins from a flour were run on SE-HPLC but with the addition of a Wyatt MALLS detector. Figures 5 and 6 illustrate the output for the two glutenin fractions.

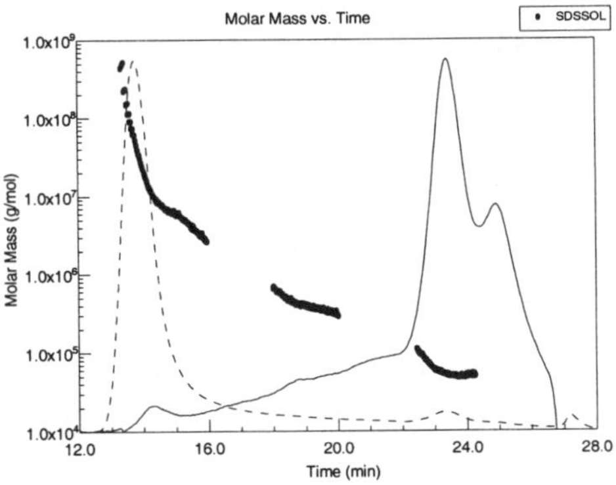

***Figure 5:** SE-HPLC-MALLS trace for SDS-soluble glutenin protein (solid line=RI, dotted line=MALS (90deg), heavy dots=calc. MW)*

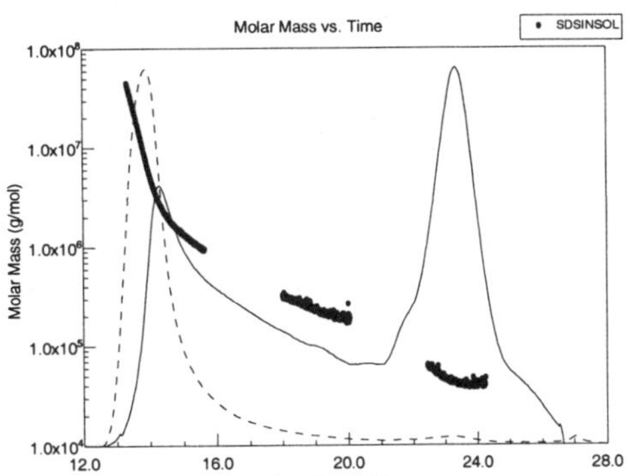

***Figure 6:** SE-HPLC-MALLS trace for SDS-insoluble glutenin protein (solid line=RI, dotted line=MALS (90deg), heavy dots=calc. MW)*

For the flour SDS-soluble fraction (Figure 5), protein present in the 12-16 min range had MW's ranging from ~3-500M. At elution times of 18-20 min, MW's were around 300k-1M. The "gliadin" peak at 23-25 min contained MW's around 40k-100k. Partial mixing studies revealed that the MW profile of the 12-16 min range varied with mixing energy input. Also, proteins extracted from doughs had higher MW's than those found in the flour. Although incremental mixing steadily increased the amount of SDS-soluble glutenin, the MW of the protein decreased. Overmixing decreased the amount of SDS-soluble glutenin but its MW increased.

For the flour SDS-insoluble fraction, protein present in the 12-16 min peak had MW's ranging from ~1-50M; notably smaller than that material found in the SDS-soluble sample. This result suggested that the sonication extraction step was breaking up large protein aggregates in order to render them soluble. However, sonication of the SDS-soluble proteins did not cause a reduction in the MW profile of the 12-16 min peak, indicating that the disruption of the large, SDS-insoluble aggregates only occurred when the residual pellet was sonicated.

4 CONCLUSIONS

Similar dough development mechanisms appeared to be operating during sheeting and MDD mixing but with a major difference being the amount of shear imparted to the dough components. Changes in the quantities of the two classes of polymeric glutenin protein were similar in both development processes, indicating that protein disaggregation was important in the process of dough development. However, changes in the exposure of protein thiols were not similar for the two processes. For the larger SDS-insoluble proteins thiol exposure increased with MDD mixing, indicating the rupture of S-S bonds to form exposed thiol groups. This increase in SDS-insoluble exposed protein thiol during MDD did not occur during the more gentle sheeting procedure, indicating that disulfide bond rupture may not be a required process in dough development and that high stresses *per se* may not be required to 'develop' doughs.

SE-HPLC-MALS indicated that protein aggregates in the SDS-soluble fraction ranged up to ~500 million in MW whilst those in the SDS-insoluble fraction ranged up to (only) 50 million, the smaller size appearing to be a result of the sonication extraction procedure. Although MDD mixing caused an increase in the quantity of SDS-soluble glutenin, the MW of the proteins decreased. Overmixing resulted in an increase in MW.

References

1. R.H. Kilborn, and K.H. Tipples, *Cereal Chem.,* 1974, **51**, 648-657

2. M.P. Morgenstern, H. Zheng, M. Ross and O.H. Campanella, *Int'l. J. Food Props.*, 1999, **2**, 265-275

3. K.H. Sutton, M.P. Morgenstern, M. Ross, L.D. Simmons, and A.J. Wilson, Proceedings 49th Royal Australian Chemical Institute, Cereal Chemistry Conference, Melbourne, September 12-17, 1999, in press

4. K. Kobrehel, J.H. Wong, A. Balogh, F. Kiss, B.C. Yee and R.B. Buchanan, *Plant Phys.*, 1992, **99**, 919-924

5. R.B. Gupta, K. Khan and F. MacRitchie, *J. Cereal Sci.*, 1993, **18**, 23-41

Acknowledgements

The authors would like to thanks the NZ Lotteries Science Grants Board for aid with equipment purchases. This research was funded by the NZ Foundation for Research, Science and Technology and contributes to research Programme 5.1.2 (Molecular interactions between flour components) of the Quality Wheat CRC.

WHEAT GLUTEN PROTEINS : HOW RHEOLOGICAL PROPERTIES CHANGE DURING FROZEN STORAGE

Yves Nicolas, Robert Smit and Wim Agterof

Unilever Research, Olivier van Noortlaan 120, 3133 AT Vlaardingen, The Netherlands
Yves.nicolas@unilever.com

ABSTRACT

In order to study frozen storage damage of wheat dough, a gluten is used as a model system to observe the effect of frozen storage on gluten rheology. Gluten samples are stored at various sub-zero temperatures. At different frozen storage times, the rheological properties are determined by planar and biaxial extension measurements. The results show that stiffness of the gluten increases in time. The fundamental cause of the increasing stiffness and implications for the frozen storage stability of dough are discussed.

1 INTRODUCTION

Wheat gluten proteins are the most important proteins in wheat flour. After hydration, a gluten network is formed in dough which is able to stabilise and maintain gas bubbles during the breadmaking process. The gas holding capacity of frozen-thawed dough, however, seems to be seriously affected by frozen storage conditions[1]. Especially, the bread volume decreases due to the freezing step and longer storage times. This weakness might be a consequence of damage development in gluten during frozen storage. To verify this hypothesis, the extensibility of gluten, stored at different temperature, were measured over time.

2 MATERIALS AND METHODS

Gluten (Sigma) was kneaded with demineralised water (40/60 v/v) containing sodium azide (0.03% w/w) for 11 min in Bradender Farinograph. After centrifugation (1h, 40000 × g), the gluten was squeezed between glass plates to a constant thickness of 5 mm, and rested for 14h in water. After removing the plates, the gluten was cut in dumbbell or rectangular shapes 40x5x5mm or 10x75x5mm respectively (Figure 1).

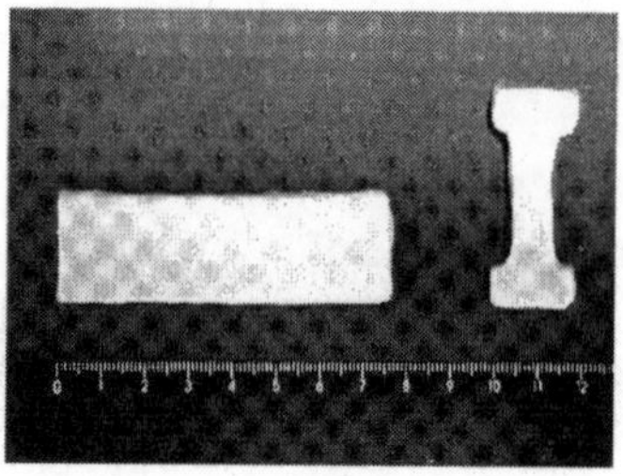

Figure 1 : *Rectangular and dumbbell shaped gluten samples*

The pieces were immersed in water, frozen and stored at different temperatures (-5, -10 and -28°C). After different storage times, the gluten was thawed and stretched in water up to a strain of 800 %, using an Instron tensile machine (speed rate 1000 mm/min) (Figure 2). Using a water bath prevents the drying of the sample but also reduces a lot the weight effect.

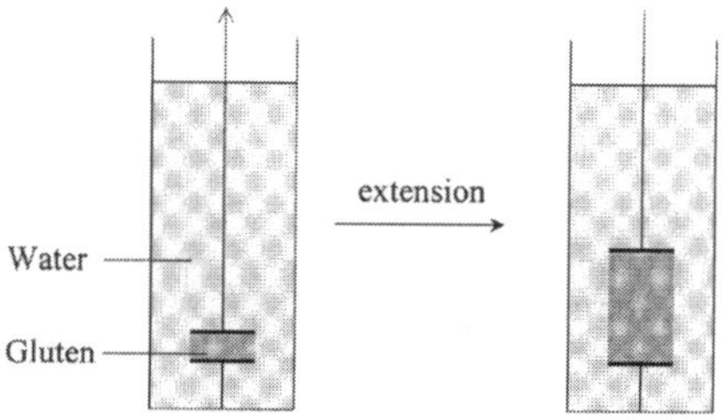

Figure 2 : *Set-up of the planar extension experiment for gluten in water.*

3 RESULTS

Figure 3 shows a typical stress-strain curve observed during planar extension of gluten. The centrifugation step removed some of the air bubbles included by the kneading and gave more reproducible stress results (CV ~ 10%)

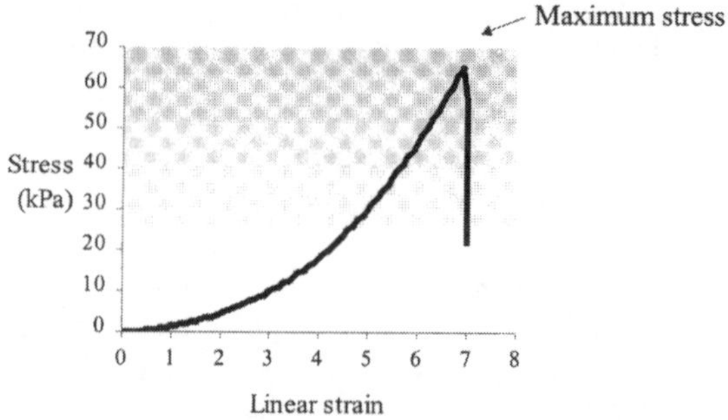

Figure 3 : *Stress-strain behaviour of gluten during planar extension.*

The effect of storage time on stress maximum for different storage temperatures is shown in Figure 4.

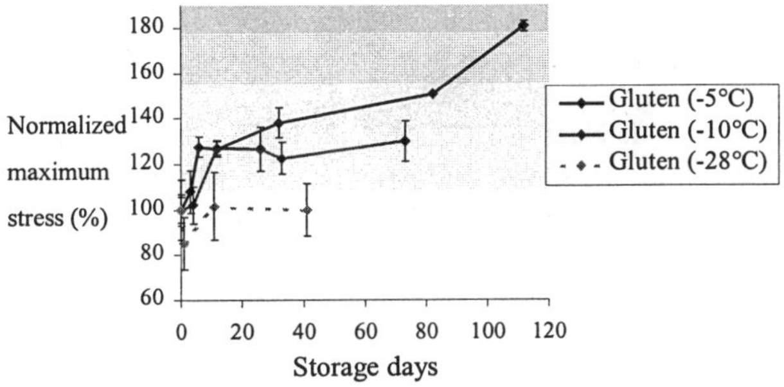

Figure 4 : *Maximum stress vs. storage time for gluten in planar extension (solid lines) and uniaxial extension (broken line) at different storage temperatures.*

The maximum stress increases for gluten stored at –5°C and –10°C. This implies that the gluten stiffens during frozen storage. The stiffening process is temperature and time dependent. At a low temperature (-25°C), the gluten stiffness seems to become constant in time.

4 DISCUSSION AND CONCLUSION

The increasing gluten stiffness suggests that the gluten concentrates during frozen storage. This concentration effect might be due to ice crystal formation. It should be remarked, however, that ice crystals do not seem to damage the gluten network by the formation of voids, because then we should expect a weakening instead of a strengthening.

The constant, and thus stable, behaviour of the –25°C sample might be caused by (i) its glassy state (the glass transition temperature is approximately –15°C[2], so the sample is glassy at –25°C and rubbery at –5 and –10°C), and (ii) a reduced growth rate of the ice crystals.

The consequence of the increasing stiffness might be a serious reduction of the proofing and baking performance of dough. The network resistance will be higher so a higher gas pressure will be needed to produce the bread volume. Furthermore, the increasing stiffness will be accompanied with a decreasing strain-to-break and thus a decreasing gas-holding capacity.

References

1. Inoue, Y., Bushuk, W. *Food Science and Technology* 1996 (New York) **72,** 367.
2. Kokini, J.L., *et al* E. 1994. *Trends in Food Science & Technology* **5,** 2181.

ANALYSIS BY DYNAMIC ASSAY AND CREEP AND RECOVERY TEST OF GLUTENS FROM NEAR-ISOGENIC AND TRANSGENIC LINES DIFFERING IN THEIR HIGH MOLECULAR WEIGHT GLUTENIN SUBUNIT COMPOSITIONS.

Y. Popineau[1], J. Lefebvre[1], G. Deshayes[1], R. Fido[2], P.R. Shewry[2] and A.S. Tatham[2].

1. INRA, Centre de recherche de Nantes, B.P. 71627, Rue de la Géraudiere, 44316 Nantes Cedex 03, France, 2 IACR-Long Ashton Research Station, Department of Agricultural Sciences, University of Bristol, Long Ashton, Bristol BS41 9AF, United-Kingdom

1 INTRODUCTION

The viscoelasticity of wheat gluten depends in a large part on the protein content and protein composition. The role of storage proteins (prolamins) that form gluten after hydration and mixing is prominent. In addition to the ratio of gliadin monomers to glutenin polymers, glutenin aggregation is an important parameter (Cornec et al. 1994). This is influenced by the subunit glutenin composition, i.e. relative proportions of low and high molecular weight subunits, but also by the type of HMW subunit (Popineau *et al.* 1994). In this work we studied the effect of differences in the HMW glutenin subunit compositions of glutens from near-isogenic and transgenic lines on glutenin aggregation and gluten rheological behaviour.

2 MATERIALS AND METHODS

Table 1. *HMW subunit compositions of the near-isogenic and transgenic wheat lines grown.*

	Line	HMW Subunit composition	Expressed trangene
Near isogenic lines Olympic xGabo	standard	1 /17+18/5+10	none
	1A, 1D null	-/17+18/-	none
	1A, 1B null	-/-/5+10	none
	1A, 1B, 1D null	-/-/-	none
Transgenic lines L 88-31	control	17+18	
	B72-8-11b	17+ 8	1Dx5
	B102-1-1	17+18	1Ax1
	B102-1-2	17+18	1Ax1
L 88-6	control	1, 17+18, 5+10	
	B73-6-1	1, 17+18, 5+10	1Dx 5

Studies were performed on four near-isogenic lines (NIL) of wheat obtained by crosses between the genotypes Olympic and Gabo provided in 1994 by R. Gupta and G. Lawrence (CSIRO, Australia) and on transgenic wheat lines grown in the field at IACR-Long Ashton Research Station in 1998 (Table 1).

The subunit compositions of glutens was analysed by SDS-PAGE and by capillary electrophoresis. Extractability, size distribution and aggregation of prolamins in flours were analysed by SE-HPLC on a Superose 6 column (1 x 30 cm; flow rate 0.3 ml / min; detection at 220 nm) equilibrated in 0.0125 M Na borate buffer 0.1% SDS. Samples were prepared by a three-step extraction: i) 0.0125 M Na borate buffer pH 8.5, 0.5 % SDS; ii) 0.0125 M Na borate buffer with 2% SDS and sonication (6W, 30 s); iii) 0.0125 M Na borate buffer, 2% SDS containing 1% DTT (dithiothreitol). All three protein extracts were analysed by SE-HPLC.

Rheological measurements were performed on glutens fully hydrated with water with a Carri-Med CSL 100 constant stress rheometer (cone-plate geometry; cone angle: 4°, diameter 2 cm). After loading into the measuring device, the sample was left to rest for one hour to allow dissipation of stress. The sample was then submitted to the following sequence of tests at 20° C : i) a first frequency scan, ii) after a few hours rest, a creep and recovery test (20 h creep, 60 h recovery, iii) a second frequency scan. Frequency scans covered the 0.001 - 36 Hz frequency window range.The amplitude of strain at all frequencies was kept close to 3%. The value of the stress applied during creep was the stress amplitude required for a 3% strain amplitude at 0.001 Hz upon the first frequency scan.

3 RESULTS AND DISCUSSION

3.1 Glutens from NIL

3.1.1 Biochemical characterization. The lines differed widely in polymer and aggregate compositions (Table 2). Glutenin contents were proportional to the number of HMW subunits expressed in the lines. The 1A,1B null line, however, had a slightly higher glutenin content than the 1A, 1D null line. The standard line had the highest content of unextractable glutenin polymers. In contrast, the triple null line was almost devoided of unextractable polymers. This means that the polymerisation of glutenin is very limited in the absence of HMW subunits. When compared to the 1A,AD null line, the 1A, 1B null line exhibited higher contents of unextractable polymers. Structural features must explain this difference in the ability of subunits 5+10 and 17+18 to form polymers of large size.

Table 2. *SE-HPLC analysis of prolamin compositions of glutens from NIL*

	gliadin %	extractable glutenin %	unextract. glutenin %	unextract. glut./total glutenin
standard	46	23	33	0.61
1A, 1D null	59	30	11	0.27
1A, 1B null	54	23	22	0.50
1A, 1B, 1D null	65	31	4	0.11

3.1.2 Viscoelaticity Determined from Combination of Dynamic and Creep-recovery Data. In the linear domain, any one viscoelastic function in shear can be calculated from any other provided the latter is known over a time or frequency range large enough, using the constitutive equation of linear viscoelasticity. This procedure allows the time-scale of observation of the viscoelastic behaviour to be extended. The results are shown expressed in terms of G'(ω) & G"(ω) in Figure 1.

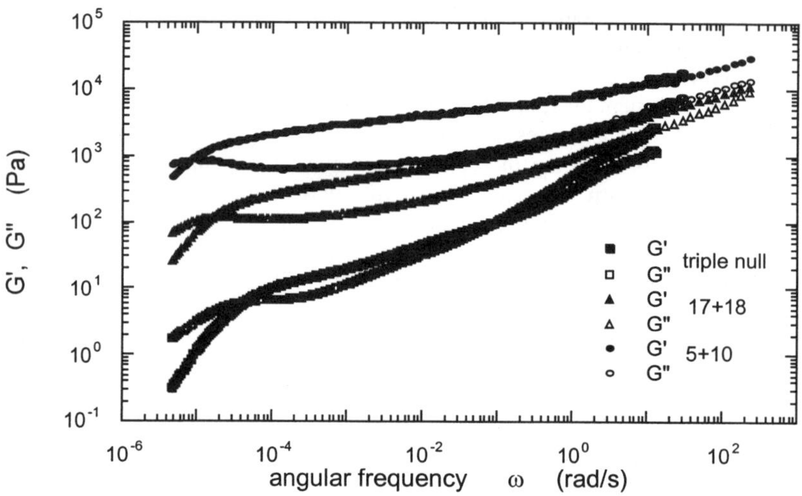

Figure 1. *Mechanical spectra of NIL obtained by combination of dynanmic and creep and recovery assays*

The nearly eight decades-wide frequency range covered encompasses most of the interesting regions of the viscoelastic behaviour, except for the softening transition from the glassy to the rubbery regions, which is not entered into at the higher frequency end of the window except in the case of triple null gluten. Two dissipative peaks are visible, corresponding to two distinct concentrations of dissipative mechanisms on the frequency (or time) scale. These flank the rubbery plateau, the limits of which can be taken as the intersection points of G'(ω) with G"(ω) curves. In the case of the standard, 5+10 and 17+18 lines, the first G'-G" crossover is seen, but not the second, although the frequency of the latter seems to be little higher than the high frequency limit of our experimental window. In the case of the triple null line, both crossovers are within the experimental window. The shape of the spectra is characteristic of transient network structures (Cornec *et. al.*, 1994). Table 3 shows that within the frequency range and at the temperature considered: i) the control line and the 5+10 line show viscoelastic properties which are very close to each other, ii)the main difference between the 17+18 line and the standard or 5+10 lines is in the height of the viscoelastic plateau, which is two times lower for the former than for the latter; iii) the triple null line differs from the control line and from the double deleted lines in all parameters, the height and length of the viscoelastic plateau being dramatically decreased, the position of this plateau shifted to lower frequencies by more than two logarithmic decades, and the width of the associated loss peak reduced.

Table 3. *Rheological characteristics of glutens from NIL*

NIL	G_N^0	f_0 Hz	η Pa s	ω1 rad/s	J_e^o (m^2/N)	ω2 rad/s
standard	3210	0.60	$1.49\ 10^8$	$9\ 10^{-6}$	$0.79\ 10^{-3}$	> 300
5+10	3130	0.36	$1.96\ 10^8$	1.10^{-5}	$0.79\ 10^{-3}$	> 300
17+18	1140	0.28	$1.69\ 10^7$	$2\ 10^{-5}$	$5.49\ 10^{-3}$	150
triple null	19	<0.001	$3.80\ 10^5$	$4\ 10^{-5}$	0.109	0.1

Thus, HMW glutenin subunits are indispensable to the expression of gluten viscoelasticity, their absence resulting in a breakdown of gluten rheology. However, they are not equivalent to each other with respect to their "viscoelastic potential". The double deletion 1A/1B does not seem to have appreciable consequences on gluten viscoelasticity in the plateau region. In contrast, the double deletion 1A/1D results in a considerable drop in the height of the viscoelastic plateau.

3.2 Glutens from Transgenic Lines

3.2.1 Biochemical characterization. The total glutenin content depended on the number of HMW subunits expressed in the lines (Table 4). The 1A, 1D null control line (L88-31) contained less glutenin than the L88-6 control line, which contained 1A, 1B and 1D subunits. As expected, insertion of trangenes increased the proportions of HMW subunits. However, the total protein contents were not modified. The 1Ax1 and 1Dx5 subunits encoded by the transgenes accounted for 50 and over 70% of HMW subunits, respectively, in the transformed lines.

Table 4. *Glutenin subunit composition of transgenic lines of wheat determined by capillary electrophoresis.*

Line	total glutenin % total protein	HMW GS % total glutenin subunits	Glu 1A % HMW GS	Glu 1B % HMW Subunits		Glu 1D % HMW Subunits	
			x	x	y	x	y
L88-6 Control	44	36	16	33	16	26	10
L88-6 + 1Dx5	53	44	4	8	4	73	4
L88-31 Control	33	18	0	75	25	0	0
L88-31 + 1Ax1	37	31	50	38	12	0	0
L88-31 + 1Dx5	44	28	0	21	8	71	0

Table 5. *SE-HPLC analysis of proteins extracted from control and transgenic lines.*

Genotype	Line	Expressed subunit	0.5 % SDS		2% SDS + US	2% SDS + DTT
			gliadin % total protein	extractable glutenin % total protein	% total. protein	% total protein
L88-31	Control		66.3	21.1	10.6	2.0
	B72-8-11b	1Dx5	57.5	15.9	8.3	18.4
	B102-1-1	1Ax1	65.1	17.7	15.9	2.2
	B102-1-2	1Ax1	63.3	23	11.8	2.0
L88-6	Control		53.4	12.9	29.9	3.1
	B73-6-1	1Dx5	47.0	11.5	12.1	29.4

3.2.2 Extractability and Aggregation of Prolamins. The effects of over-expression of subunits 1Ax1 or 1Dx5 in two different genotypes were compared in Table 5. Over-expression of subunit 1Ax1 notably increased the proportion of glutenin extractable with 2% SDS + U.S. However, the amount extractable with DTT did not change. Over-expression of 1Dx5 subunit modified glutenin aggregation more extensively. The proportion of DTT-extracted proteins increased considerably. This change was greater when subunit 5 was expressed in the L88-31 background which was probably due to the different amounts of subunit 5 in the glutenins from the two genotypes. This indicates the ability of subunit 5, which contains an additional cysteine residue, to promote the formation of highly crosslinked (by SS bonds) and aggregative glutenin polymers.

3.2.3 Rheological properties. The mechanical spectra of the three lines of genotype L88-31 are presented in Figure 2. Line L88-31, which contained only HMW glutenin subunits 17 and 18, showed the lowest values of G' and G". Furthermore, the high frequency limit of the elastic plateau could be observed within the frequency range. This spectrum is similar to that of the 17+18 NIL studied above. Expression of subunit 1Ax1 resulted in an moderate increase in viscoelasticity and the upper limit of the plateau was clearly shifted towards higher frequencies. However, storage and loss moduli were affected moderately. This can be related to the limited change observed in glutenin aggregation. On the other hand, expression of subunit 1Dx5 increased gluten viscoelasticity drastically. Both moduli were enhanced. Remarkably, the viscoelasticity of gluten from the L88-31 1Dx5 line was higher than that of gluten from the L88-6 control line, which contains 1A, 1B and 1D encoded HMW glutenin subunits (spectrum not shown). This is related to the lower gliadin content and the higher proportion of covalently cross-linked (or tightly aggregated) glutenin polymers in glutens where 1Dx5 subunit is over-expressed.

4 CONCLUSION

It is possible to characterise gluten rheology in shear over a very wide time-scale or frequency-scale by combining data from dynamic measurements with those of the creep and recovery test The composite spectra are similar to those of transient networks. Two distributions of relaxation or retardation times are visible in the case of the 5+10 and17+18 NIL, delimiting the viscoelastic plateau. In the case of the triple null NIL, the

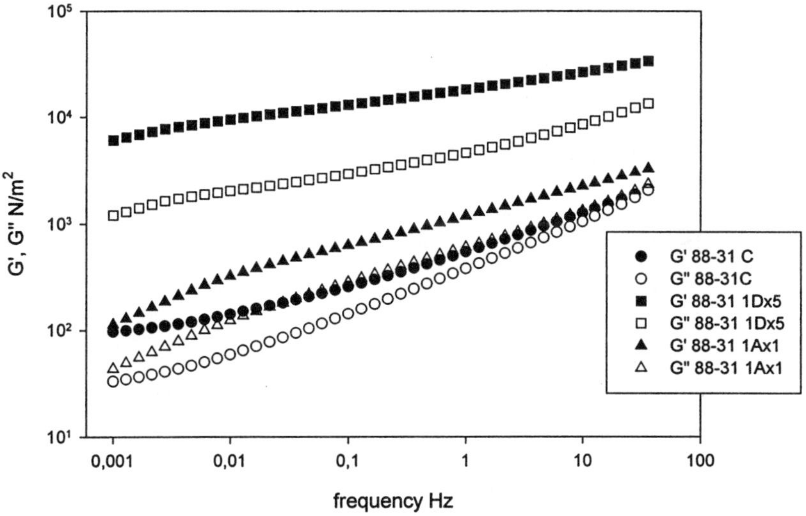

Figure 2. *Mechanical spectra of glutens extracted from transgenic lines (genotype 88-31)*

viscoelastic plateau is much shorter and lower. Globally, the rheological properties of the glutens from the near-isogenic lines were related to their glutenin properties and gliadin contents : i) the viscoelasticity collapse of triple null gluten corresponded to the highest gliadin concentration, the lowest content of large glutenin polymers and the quasi-absence of strongly aggregated glutenin, ii) the 1A 1B null line exhibited both a higher viscoelasticity and a higher content of strongly aggregated large glutenin polymers than the 1A 1D null line. The study of transgenic lines emphasized that small differences in subunit structure can influence their functionality and contribution to gluten structure and rheology. Expression of subunit 1Ax1 increased glutenin aggregation, but induced no change in crosslinking by SS bonds. This moderately increased gluten viscoelasticity. Expression of subunit 1Dx5 increased glutenin aggregation, possibly through covalent crosslinking between aggregates. This resulted in a drastic enhancement of viscoelastic properties. This was attributed to the presence of an additional cysteine residue available for intermolecular crosslinking in subunit 1Dx5.

References

1. M. Cornec, Y. Popineau and J. Lefebvre. *J. Cereal Sci.*, 1994, **19**, 131.
2. Y. Popineau, M. Cornec, J. Lefebvre and B. Marchylo, *J. Cereal Sci.*, 1994, **19**, 231.

Acknowledgements

Part of this research was supported by the European Community. FAIR CT96-1170 Eurowheat. IACR receives grant-aided support from the Biotechnology and Biological Sciences Research Council of the United Kingdom.

SIGNIFICANCE OF HIGH AND LOW MOLECULAR WEIGHT GLUTENIN SUBUNITS FOR DOUGH EXTENSIBILITY

I.M. Verbruggen, W.S. Veraverbeke and J.A. Delcour

Laboratory for Food Chemistry, Katholieke Universiteit Leuven, Kardinaal Mercierlaan 92, B-3001 Heverlee, Belgium

1 INTRODUCTION

Wheat gluten proteins (gliadin and glutenin) are important for breadmaking because they are responsible for the visco-elastic properties of dough. The elastic properties of dough are generally ascribed to glutenin. Glutenins are polymers consisting of high and low molecular weight glutenin subunits (HMW-GS and LMW-GS) linked through disulphide bonds. The significance of glutenin subunits can be investigated by increasing their concentration in the flour. Because glutenin subunits are a part of the glutenin network, incorporation of the added subunits into the glutenin network of the base flour is necessary to give a realistic indication of their functional role.

In the present study, the effects of LMW-GS and HMW-GS on the extension parameters maximum resistance (MR) and extensibility (EX) were evaluated with 'addition' and 'incorporation' protocols. The effect of alkylation of free sulphydryl groups was also evaluated.

2 MATERIALS AND METHODS

Total HMW-GS and LMW-GS were isolated from Soissons flour as previously described[1]. Alkylation of the glutenin subunits was performed with 4-vinylpyridine.

Flour of the wheat variety Minaret was used as the base flour. The method for 'incorporation' of glutenin subunits in the glutenin network of a base flour[2,3] was scaled-up for a 10-g mixograph (National Manufacturing, Licoln, NE, U.S.A.) and reducing agent and oxidant concentrations were adapted to obtain partial reduction and optimal restoration of the Minaret flour used. Flour (10.0 g, 14 % moisture base) with or without added glutenin subunits (50.0 mg) was mixed with 4.38 ml deionised water and 500 μl dithiothreitol (DTT) solution (0.6 mg/ml) for 45 s. After a resting period of 5 min, 500 μl potassium iodate (KIO_3) solution (0.35 mg/ml) was added to the partially reduced dough. Mixing was then continued for 30 s. To allow reoxidation, the dough was rested for 5 min before further mixing to peak dough resistance.

In what follows, the term 'addition' is used when glutenin subunits were added to the flour without a reduction/oxidation protocol. In this case, DTT and KIO_3 solutions were replaced by water.

Load extension curves were obtained with the TA-XT2 Texture Analyser using the SMS/Kieffer dough and gluten extensibility rig[4] (Stable Micro Systems Ltd, Godalming, U.K.). For each experimental sample, extension tests were performed on at least two doughs (with 10 measurements/dough). Extension parameters MR (mN) and EX (mm) of a sample were calculated as the average of the results of these measurements. The parameters of each sample (with addition of glutenin subunits) were compared to those of the controls (without addition of glutenin subunits) and the Tukey method of multiple comparisons was used to detect statistically significant ($\alpha = 0.05$) differences[5].

3 RESULTS

3.1 'Addition' and 'incorporation' of HMW-GS

The effects of 'addition' and 'incorporation' of unalkylated and alkylated HMW-GS on extension parameters MR and EX are shown in Table 1 and Figure 1a,b. 'Addition' of HMW-GS caused a significant increase in MR whereas EX was not significantly changed. Alkylated HMW-GS did not significantly change MR and EX. The effects of 'incorporation' of HMW-GS and alkylated HMW-GS were similar to those observed upon 'addition' of these subunits. HMW-GS caused a significant increase of MR and did not change EX. 'Incorporation' of alkylated HMW-GS did not significantly change either MR or EX.

Table 1 *Effects of 'addition' and 'incorporation' of HMW-GS and LMW-GS on extension parameters MR and EX*[a]

Treatment	Additive	MR[b] (mN)	EX[b] (mm)
'Addition'		248.0 ± 13.2	70.8 ± 3.7
'Addition'	HMW-GS	328.8 ± 19.2[c]	68.8 ± 3.4
'Addition'	HMW-GS-alk	264.1 ± 1.8	68.2 ± 2.3
'Addition'	LMW-GS	192.1 ± 4.4[c]	63.2 ± 3.0[c]
'Addition'	LMW-GS-alk	270.6 ± 1.9	56.5 ± 2.7[c]
'Incorporation'		233.7 ± 13.1	72.8 ± 4.1
'Incorporation'	HMW-GS	328.2 ± 12.4[c]	69.4 ± 3.5
'Incorporation'	HMW-GS-alk	248.0 ± 2.6	67.6 ± 0.2
'Incorporation'	LMW-GS	147.3 ± 10.0[c]	74.9 ± 4.1
'Incorporation'	LMW-GS-alk	278.4 ± 0.3[c]	58.3 ± 2.0[c]

[a] Abbreviations used : MR, maximum resistance; EX, extensibility; LMW-GS, low molecular weight glutenin subunits; HMW-GS, high molecular weight glutenin subunits; alk, alkylated.

[b] Mean values and standard deviation.

[c] MR or EX is significantly different (Tukey method, $\alpha = 0.05$) from MR or EX of the same sample without addition of glutenin subunits.

3.2 'Addition' and 'incorporation' of LMW-GS

As shown in Figure 1 and Table 1a,b, 'addition' of LMW-GS significantly decreased both MR and EX. Alkylated LMW-GS caused a slight increase in MR and clearly decreased EX. Much as 'addition', 'incorporation' of LMW-GS caused a significant decrease in MR. In contrast to what was observed with 'addition', EX was not significantly changed by 'incorporation' of LMW-GS. 'Incorporation' of alkylated LMW-GS resulted in effects on both MR and EX similar to those of 'addition' of these subunits: a small increase in MR and a clear decrease of EX.

a) b)

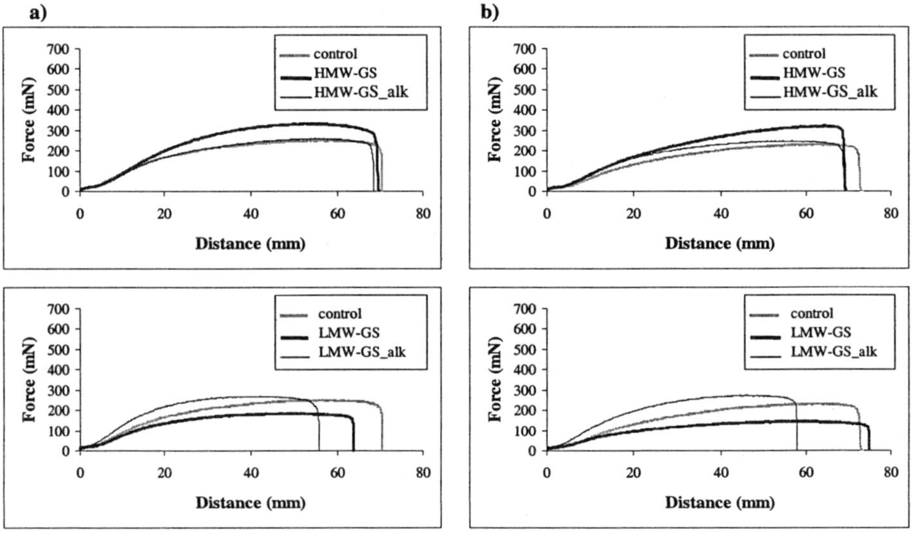

Figure 1 *Load extension curves of doughs prepared from Minaret flour without addition of glutenin subunits (control) and with addition of 50 mg glutenin subunits (LMW-GS, HMW-GS) or alkylated glutenin subunits (LMW-GS_alk, HMW-GS_alk). (a) 'addition', (b) 'incorporation'.*

4 DISCUSSION

Since alkylation of the sulphydryl (SH) groups prevented the incorporation of these subunits in the glutenin network, the effects observed with alkylated glutenin subunits can only be attributed to the presence of the monomers. As expected, no difference was observed between 'addition' and 'incorporation' of alkylated subunits (Table 1, Figure 1a,b). However, alkylated LMW-GS and HMW-GS affected the extension parameters in a different way. Alkylated LMW-GS caused a slight increase in MR and a significant decrease in EX whereas alkylated HMW-GS did not significantly change either MR or EX. The difference in behaviour of the alkylated glutenin subunits is probably caused by the different intrinsic properties of LMW-GS and HMW-GS and/or by the different effects induced by the introduction of the alkylated agent derived substituents. Alkylation may indeed affect the behaviour of LMW-GS more than that of HMW-GS because the former contain more SH groups.

'Addition' of unalkylated LMW-GS and HMW-GS had effects on MR and EX that differed from those of 'addition' of their alkylated forms (Table 1, Figure 1a). This difference is probably caused by the presence of free SH groups in unalkylated glutenin subunits and/ or the effect of substituents introduced by alkylation. These free SH groups may participate in SS/SH exchanges and/or offer the possibility of (partial) 'incorporation' of glutenin subunits in the presence of oxygen. 'Addition' of LMW-GS significantly decreased both MR and EX whereas HMW-GS caused a significant increase in MR but did not significantly change EX. It seems unlikely that SS/SH exchanges in the presence of free SH groups were responsible for the changes in MR because 'addition' of LMW-GS and HMW-GS had an opposite effect on MR. However, (partial) incorporation of glutenin subunits in the presence of oxygen may explain the observations. Changes in EX upon 'addition' of LMW-GS or HMW-GS were similar to those of 'addition' of their alkylated forms (Table 1, Figure 1a). Therefore, these effects can be ascribed to the intrinsic properties of the non-incorporated glutenin subunits.

'Incorporation' of LMW-GS caused a significant decrease in MR whereas 'incorporation' of HMW-GS caused a clear increase of MR. EX was not changed with 'incorporation' of either LMW-GS or HMW-GS (Table 1, Figure 1b). The effects of 'incorporation' of glutenin subunits were similar to those of 'addition', except for the effect of 'incorporation' of LMW-GS on EX. This observation may indicate that, as mentioned above, even with 'addition', glutenin subunits were (partially) incorporated into the glutenin network. However, it can reasonably be expected that the 'incorporation' protocol resulted in a more successful incorporation of glutenin subunits than the 'addition' protocol. The latter can also explain why 'incorporation' of LMW-GS did no longer decrease EX.

5 CONCLUSION

In this study, the effects of 'addition' and 'incorporation' of glutenin subunits on dough extension parameters MR and EX were investigated. Alkylated and unalkylated glutenin subunits had different effects on MR and EX. This was probably caused by the effect of free SH groups in unalkylated glutenin subunits and/or the effect of substituents introduced by alkylation. 'Incorporation' of LMW-GS and HMW-GS had totally different effects on MR and EX. Incorporated HMW-GS caused a clear increase in MR whereas LMW-GS decreased MR. EX was not significantly changed by either HMW-GS and LMW-GS. The similarity in effects obtained with 'addition' and 'incorporation' indicated that even with 'addition' glutenin subunits can be partial incorporated into the glutenin network in the presence of oxygen.

References

1. I.M.Verbruggen, W.S. Veraverbeke, A. Vandamme and J.A. Delcour. *J. Cereal Sci.*, 1998, **28**,32.
2. F. Békés and P.W. Gras, in *Gluten Proteins 1993*, Association of Cereal Research, Detmold, Germany, 1994, p. 170.
3. F. Békés, P.W. Gras, and R.B. Gupta. *Cereal Chem.*,1994, **71**, 44.
4. R. Kieffer, J.-J. Kim and H.-D. Belitz, *Z. Lebensm. Unters. Forsch.*, 1981, **172,** 190.
5. J. Neter, W. Wasserman, and M.H. Kutner, in *Applied linear statistical models*, eds. T. Richard, Jr. Hercher and E. Shiell, R.R. Donnelley and Sons company, USA,1990, p. 568.

WATER ACTIVITY IN GLUTEN ISSUES: AN INSIGHT

Luc De Bry

Golden Crescents' Technologies & Associates (G.C.T.A.), Waverstraat 52 - B-2860 Sint-Katelijne Waver, Belgium.
Email: luc.de.bry@skynet.be

1 WATER ACTIVITY OF WATER / ALCOHOL MIXTURES

As stated by a group of researchers of the United Kingdom, "within the wheat protein literature, there appears to be no more emotive topic than protein classification"[1]. Very

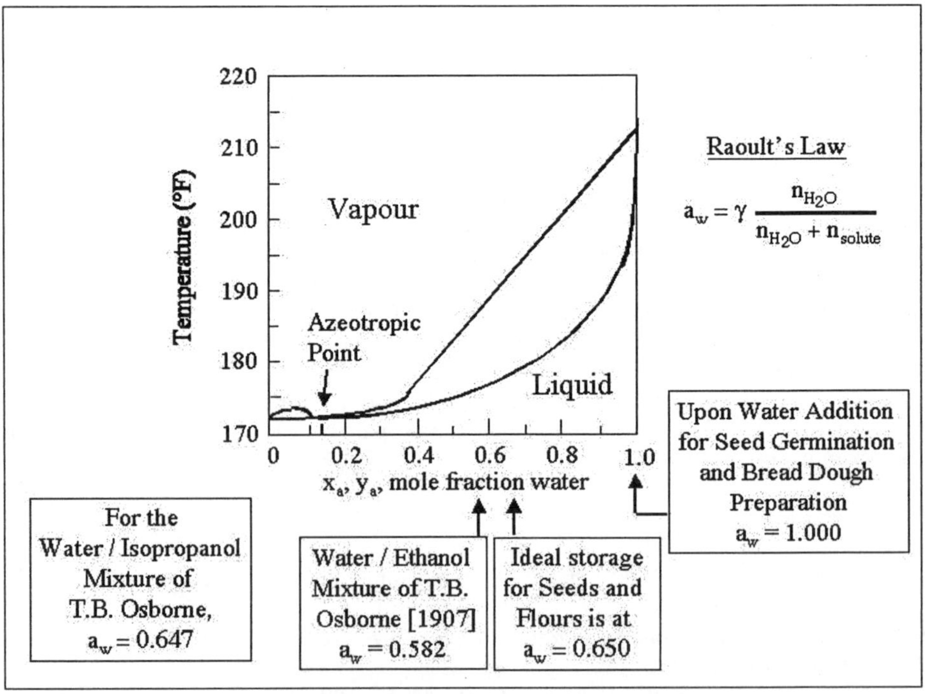

Figure 1. *Phase diagram of the water/ethanol azeotropic mixture. The graph in the figure is an adapted mirror image from a textbook[2] (Abbreviation: a_w = water activity).*

few studies have investigated the effectiveness of water/alcohol mixtures in wheat protein research[3]. The two liquids form hydrogen bonds together[4,5]. Water activity in various gluten issues has been calculated using Raoult's law, and mapped in Figure 1[5, 6]. Low water activity causes protein aggregation, hence the greater effectiveness of isopropanol for solubilising wheat proteins. Furthermore, as temperature increases water activity, hot alcohol solutions are more effective than cold.

2 WATER ACTIVITY CONTROLS WHEAT GENETIC ACTIVITY

The natural variability of wheat proteins is just as emotive. To understand what controls the expression of genes coding for wheat storage proteins should prove useful.

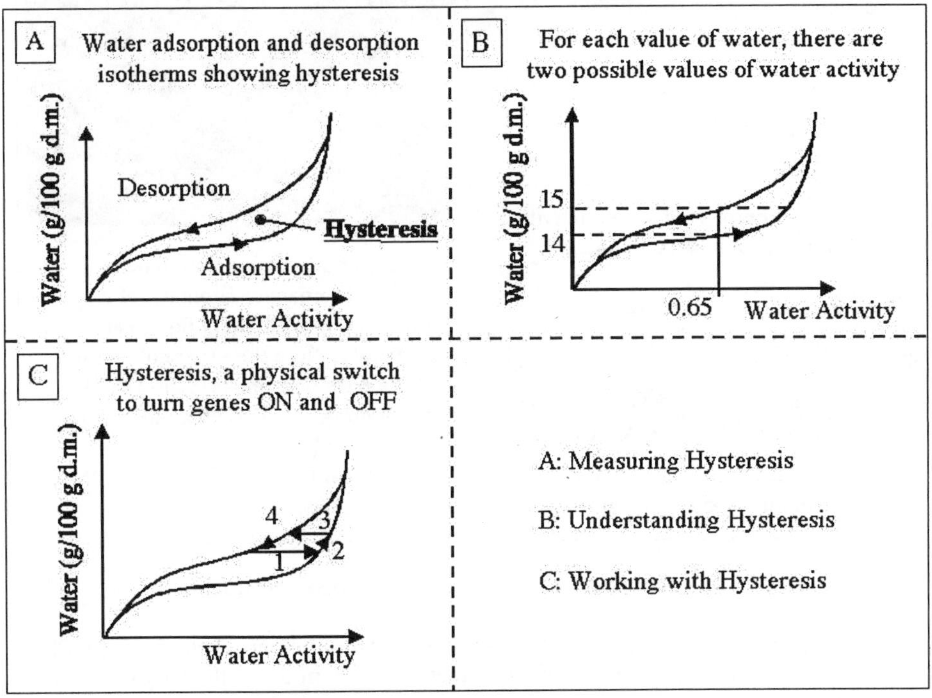

Figure 2. *Understanding and managing hysteresis in crop issues should prove as useful in wheat protein research as it is in microbiology and pharmacy.*

Evidence for water activity controlling genetic activity are reviewed in microbiology applications[6]. Figure 2 describes fundamentals of hysteresis in water sorption isotherms and the basic genetic response to water activity. For instance, when a rise in temperature causes an increase in water activity (1), it positions seed cells on their adsorption isotherm and switches on the genes coding for gliadins and heat-schock proteins (2). Upon cooling, water activity decreases, hence causing seed cells to move to their desorption isotherm (3) and switching the gene off (4). Thus, this tiny hysteresis system could be an elegant genetic feedback circuit, with a physical switch to turn genes ON and OFF.

3 WATER ACTIVITY CAUSES AND SOLVES GLUTEN RHEOLOGICAL TOXICITY

All seeds and tubers contain anti-nutritional factors. As they cannot run away from parasites and predators, these chemical defences form a fairly effective way to protect plant progenies from pathogens and phytophagous animals. Wheat anti-nutritional factors and activity are described in Figure 3.

Wheat Storage (?) Proteins	are	Anti-Nutritional Factors (!)
Albumins		Anti-Amylases Anti-Trypsins Anti-Lipases
Gliadins		Allergens
Glutenins		Digestion Blockers

Gluten isotherm with water activity averages of bakery doughs.

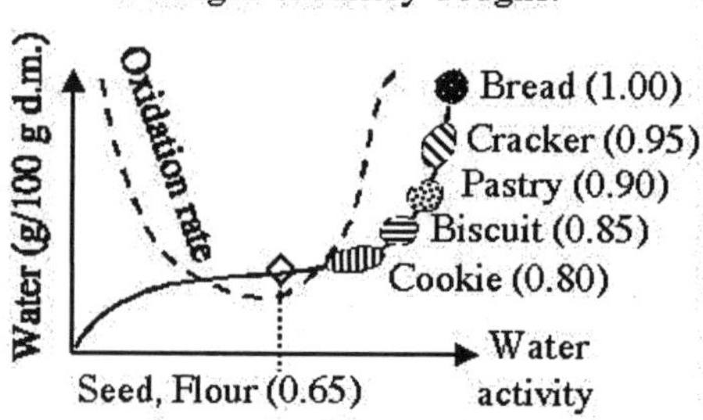

How Wheat Proteins kill Mammals

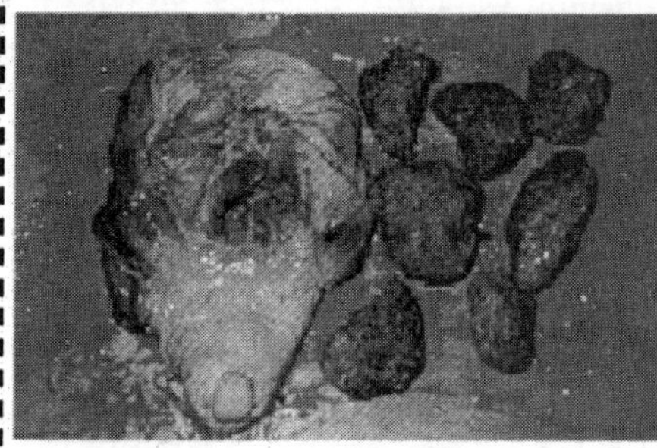

Wheat flour forming dough balls in pigs' stomachs. Stomach of an 85kg pig opened along the greater curvature. Note the severe ulceration of the pars oesophaga (centre) and seven dough-balls with a dimension up to 9 cm (right). (Scan from Penny R.H.C. et al., 1993, with permission)

Figure 3. *Raising water activity of flours causes gluten rheological toxicity*

A group of veterinarians recently raised the question about whether or not wheat can form dough balls in pigs' stomachs, hence starving them to death. That must be the first reported and clear-cut evidence of gluten mechanical toxicity[7]. During discussion with the researchers, it appeared that, due to price advantage, the pigs' diet was increased with wheat flour. Mastication with the jaws, plus saliva, increasing water activity and oxidation rate, contribute to develop gluten. Once the stomach is full of indigestible gluten balls, monogastric mammals can no longer eat. The additional anti-nutritional activity from the other wheat proteins may also have played a role in the loss of pigs. In Figure 3, mapping water activities of major dough types on a gluten isotherm gives an idea of gluten rheological toxicity. Fortunately, water activity and glass transition can solve the challenge of gluten anti-nutritional activity. Stretching gluten by bubble formation in bread, or by lamination in crackers and biscuits before protein denaturation occurs upon baking may be seen as part of the detoxification process. Figure 4 illustrates gluten detoxification for some doughs. In the end products, and in addition to Maillard aroma and colour, sensory softness or crispness can also serve as useful food safety indicators[8].

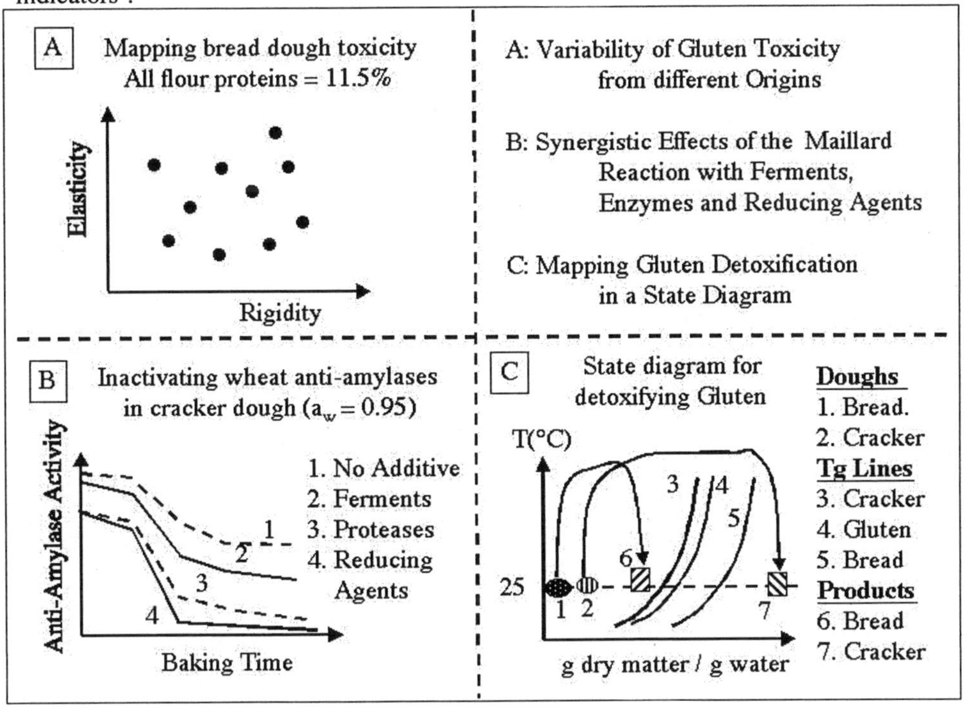

Figure 4. *Detoxifying wheat and gluten anti-nutritional activity. It has to be baked!*

References

1. M. Kreis, P.R. Shewry, B.G. Forde, J. Forde and B.J. Miflin. In *Oxford Surveys Of Plant Molecular & Cell Biology*, Vol. 2, ed. B.J. Miflin, Oxford University Press, 1985, 253
2. A.S. Foust, L.A. Wenzel, C.W. Curtis, L. Maus and L.B. Anderssen. *Principles Of Unit Operations*, J. Wiley & Sons, N.Y., 1960, pp. 450
3. J.A. Bietz, on AACCnet. 2000, http://www.scisoc.org/aacc/osborne/
4. F. Franks and J. E. Desnoyers, *Water Science Reviews*, 1985, **1**, 171
5. M. Shapero, D. Nelson and T.P. Labuza. J. Food Sci., 1978, **43**, 1467
6. Y.R. Roos, R.B. Leslie and P.J. Lillford. *Water Management in the Design and Distribution of Quality Foods* - ISOPOW 7, 1999, Technomic Publishing Co., Basel, pp.602
7. R.H.C. Penny, H.J. Guise, T.A. Abbott and D.H. Kerr. *Veterinary Record*, 1993, **133**, 297
8. L. De Bry. In *Maillard Reactions In Chemistry, Food, And Health*, eds. T.P. Labuza, G.A. Reineccius, V.M. Monnier, J. O'Brien and J.W. Baynes, Royal Society of Chemistry, Cambridge, 1994, 28

Gluten Protein Synthesis during Grain Development and Effects of Nutrition and Environment

ANALYSIS OF THE GLUTEN PROTEINS IN DEVELOPING SPRING WHEAT

R.J. Wright[1], O.R. Larroque[2], F. Békés[2], N. Wellner[3], A.S. Tatham[1] and P.R. Shewry[1]

1. Department of Agricultural Sciences, University of Bristol, Institute of Arable Crops Research, Long Ashton Research Station, Long Ashton, Bristol, BS41 9AF, U.K. 2. CSIRO, Division of Plant Industry, North Ryde, NSW, Australia. 3. Institute of Food Research, Norwich Laboratory, Norwich Research Park, Colney, Norwich, NR4 7UA, U.K.

1 INTRODUCTION

The functional properties of wheat in food products, such as bread and biscuits, are primarily determined by the amount, composition and properties of the gluten proteins. These proteins confer a combination of extensibility and elasticity. The balance between these two physical properties determines the end use of dough, with elastic doughs being preferred for bread making and more extensible doughs for making biscuits and cakes. Gluten is a complex mixture of proteins that are divided into two groups: gliadins and glutenins. The gliadins are monomeric proteins which interact by non-covalent forces whereas the glutenins form high M_r polymers (1×10^6 to 10×10^6) stabilised by inter-chain disulphide bonds. The glutenins are considered to be the major determinants of gluten visco-elasticity while the gliadins may act to plasticise the gluten mass.

In this investigation, the gluten proteins in developing caryopses have been studied using size-exclusion high-performance liquid chromatography (SE-HPLC), flow field-flow fractionation (F-FFF) and Fourier transform infrared spectroscopy (FT-IR). Sodium dodecyl sulfate-polyacrylamide gel electrophoresis (SDS-PAGE) was used to analyse the fractions collected from SE-HPLC. These techniques allow the characterisation of the gluten protein polymers and their accumulation during grain development.

2 MATERIALS AND METHODS

2.1 Materials

Spring wheat (cv. *Axona*) was grown in pots under controlled conditions (day length of 16 hours; day temperature 20 °C; night temperature 16 °C, and 100% moisture). Each head was tagged at the onset of anthesis of the central florets and the whole caryopses were harvested at 12, 14, 18, 22, 26, 30 and 40 days after anthesis (DAA).

2.2 Methods

2.2.1 SE-HPLC analysis. Total flour proteins were extracted from developing caryopses and samples were injected onto a Biosep sec-S4000 column[1]. Fractions were collected at 1 minute intervals and freeze dried.

2.2.2 SDS-PAGE of SE-HPLC fractions. SDS-PAGE of fractions was performed under reducing and non-reducing conditions[2].

2.2.3 F-FFF of SE-HPLC fractions. Freeze-dried fractions were dissolved in phosphate buffer and loaded onto an F-FFF column[3]. The average retention time for each fraction was recorded.

2.2.4 FT-IR of developing gluten. Protein bodies were isolated from fresh caryopses at specific developmental stages using sucrose density gradient ultracentrifugation[4]. The resulting gluten pellets were washed with water and analysed using FT-IR[5].

3 RESULTS AND DISCUSSION

3.1 SE-HPLC analysis

The chromatograms (Figure 1) from the different developmental stages had broadly similar profiles but some differences were observed, most noticeably being the areas under the fractions.

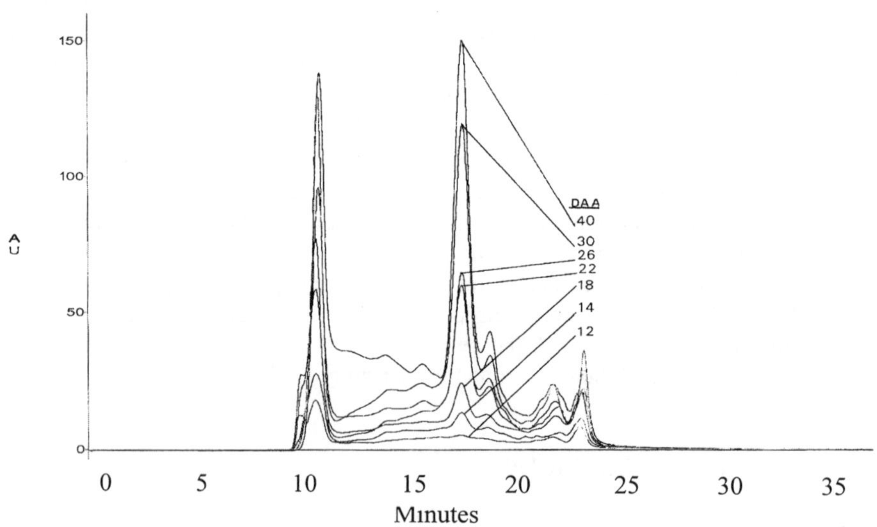

Figure1 *Overlaid SE-HPLC chromatograms of developing caryopses. Scaled to a per caryopsis weight basis.*

3.2 SDS-PAGE of SE-HPLC fractions

SDS-PAGE of reduced and non-reduced fractions from the SE-HPLC separations (Figure 2) shows that the first six fractions contain only HMW glutenin subunits, with

LMW glutenin subunits and gliadins being eluted in later fractions. When separated under non-reducing conditions fractions 1-6 showed smeared protein bands of increasing mobility, indicating that they contained polymeric protein with overlapping ranges of molecular mass.

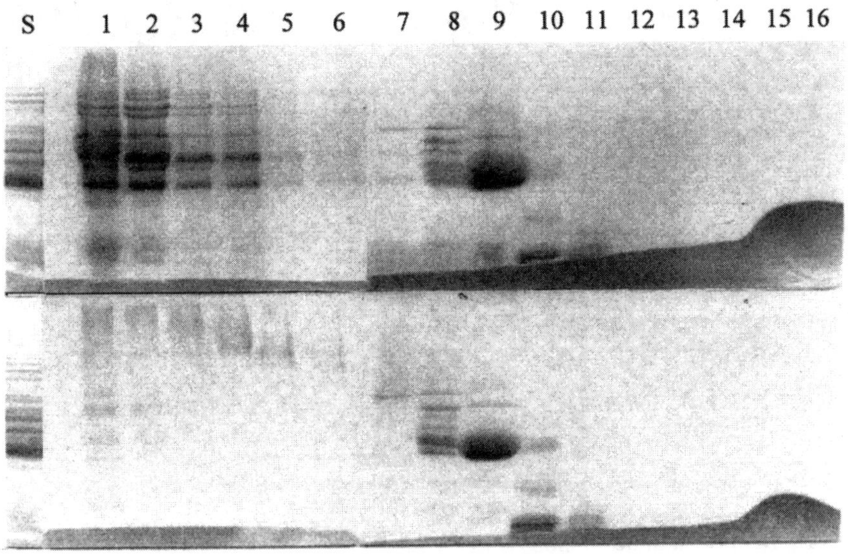

Figure 2 *SDS-PAGE of fractions from SE-HPLC reduced (top) and unreduced (bottom).*

3.3 F-FFF of SE-HPLC fractions

Analysis of fractions from SE-HPLC using F-FFF (Table 1) shows a decreasing average retention time, indicating the presence of proteins with a smaller average molecular weight. These values give a more accurate picture of the size of the fractions collected from SE-HPLC.

Table 1 *F-FFF of SE-HPLC fractions showing average retention times for each developmental stage.*

Fraction	12	14	18	22	26	30	40
1	10.782	10.967	11.000	11.200	10.840	10.671	11.047
2	------	10.502	10.739	10.940	11.000	11.159	10.736
3	10.496	10.628	10.880	10.716	10.687	10.708	10.750
4	10.482	10.235	10.439	10.314	10.265	10.722	10.316
5	9.135	9.561	9.897	9.616	9.317	9.326	9.609
6	7.428	7.460	7.717	7.749	7.554	7.285	7.550
7	5.235	5.841	5.893	5.959	5.657	5.388	5.443
8	3.803	3.317	3.311	3.492	3.752	3.320	3.310
9	2.957	2.330	2.315	2.413	2.456	------	2.595
10	2.405	2.165	2.101	------	2.465	2.842	2.448

3.4 FT-IR of developing gluten

The non-deconvoluted FT-IR spectra of isolated gluten shows an increase in the peak around 1660 cm^{-1} (the amide I peak) between 30 and 40 DAA, suggesting that changes in the pattern of protein hydrogen bonding/ conformation were occurring during dehydration of the developing caryopsis.

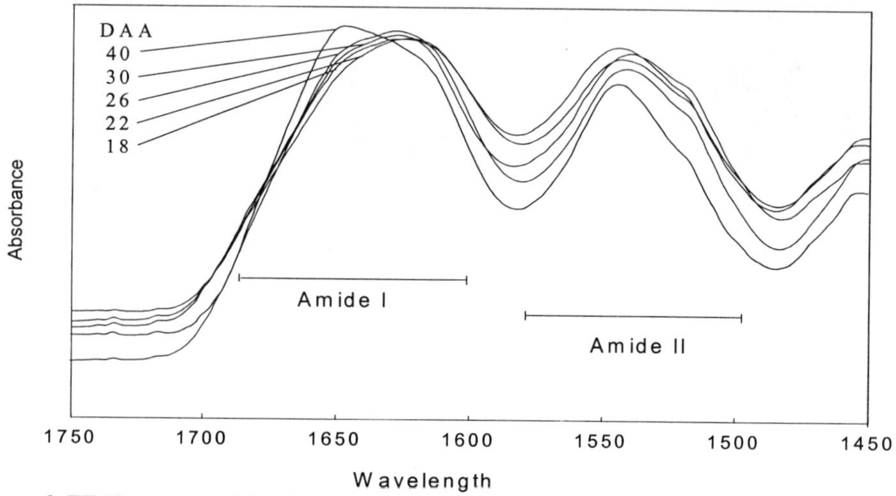

Figure 3 *FT-IR spectra of developing gluten*

4 CONCLUSIONS

Little difference was observed in the protein polymer size distribution during development using SE-HPLC and F-FFF. SE-HPLC is able to separate polymeric protein for further analysis in conjunction with other methods. The use of FT-IR has shown differences in protein hydrogen bonding patterns between 30 and 40 DAA, corresponding to dehydration, resulting in less extended chains and more β-sheet structures.

References

1. O.R. Larroque, M.C. Gianibelli, I.L. Batey and F. MacRitchie, *Electrophoresis,* 1997, **18**, 1064.

2. P.R. Shewry, A.S. Tatham and R.J. Fido, in *Plant gene transfer and expression Protocols,* ed. H. Jones, Humana Press, Totowa, New Jersey, 1995, 49, Chapter 34, p. 399.

3. O.R. Larroque, Daqiq, N. Islam and F. Békés, (in press).

4. B.J. Miflin, S.R. Burgess and P.R. Shewry, *J. exp. Bot.,* 1981, **32**, 199.

5. N. Wellner, P.S. Belton and A.S. Tatham, *Biochem. J.,* 1996, **319**, 741.

Acknowledgements

IACR receives grant-aided support from the Biotechnology and Biological Sciences Research Council of the United Kingdom.

SDS-UNEXTRACTABLE GLUTENIN POLYMER FORMATION IN WHEAT KERNELS

T. Aussenac and J.-L. Carceller

Department of Plant Physiology, Ecole Supérieure d'Agriculture de Purpan, 75 voie du T.O.E.C., 31076 Toulouse cedex 03, France. Email : aussenac@esa-purpan.fr.

1 INTRODUCTION

Differences in breadmaking quality among flours are to a large extent determined by differences in the polymeric protein fractions. The total glutenin polymer quantity is known to be correlated with various technological parameters[1,2]. Moreover, a certain amount of these polymers remains unextractable in various extraction systems (e.g. acetic acid solution or SDS phosphate buffer). Those unextractable polymeric proteins appear to be correlated positively with the baking performance. Furthermore, Gupta *et al.*[3] showed that the amount of unextractable polymer is more directly linked with certain technological parameters (especially those correlated with mixing) than the total glutenin quantity. The proportion of unextractable polymers among the glutenin polymers appears likewise to be an important parameter for the technological properties[3]. Regarding the importance of the diverse polymer protein parameters (quantity and quality) in breadmaking quality, it seems interesting to study the accumulation of those fractions in kernel during grain filling. Very few papers have reported information about the accumulation of those specific fractions during grain filling[4,5] and furthermore, no definite conclusion appears to have been reached about the formation of the unextractable polymers even though some authors[7] found a partial correspondence between the formation of SDS-insoluble polymers and the rapid loss of water (grain desiccation).

In this study, we followed the accumulation of polymeric proteins and the changes in molecular size distribution of these protein fractions (SDS-soluble and SDS-insoluble polymers) by SE-HPLC and RP-HPLC analysis with natural and premature desiccation. In the same experiment, the sulphydryl status of glutenins was investigated to determine redox changes during grain development.

2 MATERIALS AND METHODS

2.1 Materials

The two common wheat cultivars used in this study were Soissons and Thésée possessing *Glu-D1* subunits 5+10 and 2+12, respectively. These cultivars were grown at the experimental farm of ESAP Toulouse, France (1996-1997 and 1997-1998). At the first day of anthesis, each spike was tagged and dated. Subsequently, 30 spikes per replicate (four replicates) were collected at 2-day intervals from 5th day after anthesis (DAA) up to the 53rd DAA. For all the biochemical determinations, the fresh grains were freeze-dried, ground in a Janke A10 grinder and kept at -20°C. For applying artificial desiccation of the kernels, samples at 15, 18, 21, 24, 27, 31, 34 and 38 DAA were used. The fresh grains (30 spikes) were oven-dried at 30°C (a temperature normally recorded during the grain desiccation) for 3 days until the grains reached a constant humidity.

2.2 Methods

2.2.1 Molecular size distribution and glutenin subunit quantification. SE-HPLC was used to determine the relative size distribution of polymeric proteins and RP-HPLC was used to quantify glutenins subunits as described previously[6].

2.2.2 Sulphydryl-disulphide status. *In vitro* and *in vivo* mBBr labelling[7] of grain storage proteins prior to their extraction was applied to study the redox status of glutenins in developing grains.

3 RESULTS AND DISCUSSION

3.1 Changes in molecular size distribution of polymeric proteins under natural and premature desiccation

Although polymer accumulation was a continuous process commencing as early as 7 DAA, the accumulation of SDS-insoluble polymers began only during the late stages of the grain development (Figure 1). For both Soissons and Thésée, the period of most rapid accumulation of SDS-insoluble polymers coincided with the period of rapid water loss (Ph 3) (after 32 DAA).

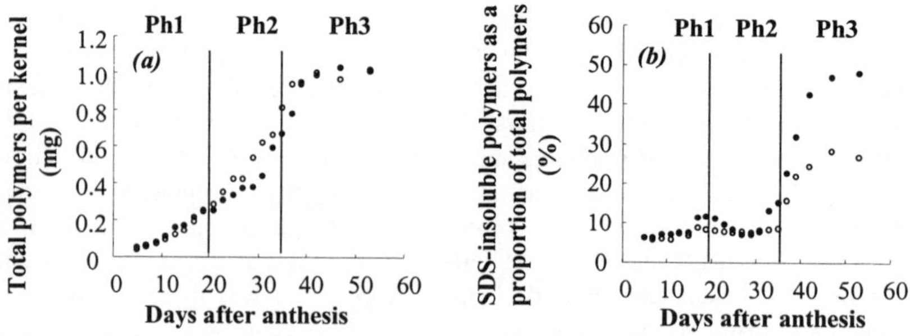

Figure 1 Accumulation of total polymers, and evolution of the percentage of SDS-insoluble polymers in total polymers as a function of the days after anthesis. (a) Total polymers per kernel for (●) Soissons, and (O) Thésée. (b) SDS-insoluble polymers as a proportion of total polymers (%). (●) Soissons, and (O) Thésée. (*harvest 1997).*

The question arises as to whether there is a relation between the appearance of SDS-insoluble polymers and the rapid water loss from the grains. The similarity in the timing of the events suggests this to be the case. In order to verify this hypothesis, we applied premature desiccation by ventilation at 30°C of grains during the cellular enlargement phase (Ph 2). The dehydration of kernels during development led to the deposition of SDS-insoluble polymers in the two genotypes (Figure 2). These observations are in total agreement with the results obtained under natural conditions. We can confirm that it is not necessary to reach a certain amount of total polymers accumulated for the massive accumulation of the unextractable polymeric fraction to occur, since we can observe insoluble polymer formation at any time. These results contradict the hypothesis of Gupta *et al.*[4] according to which SDS-insoluble polymer formation would be induced only when a certain amount of total polymers was accumulated, i.e. 60-75%. At the same time, the polymerisation index (SDS-insoluble polymers/total polymers) was not constant throughout the physiological stages. This index gradually increased during the cell enlargement phase and could be closely related to the modification of the HMW-GS/LMW-GS ratio. These results, which show the qualitative importance of HMW-GS for the formation of SDS-insoluble polymers, are in total agreement with the observations made by Gupta *et al.*[8] and Carceller and Aussenac[6] using isogenic lines, translocations or various genotypes.

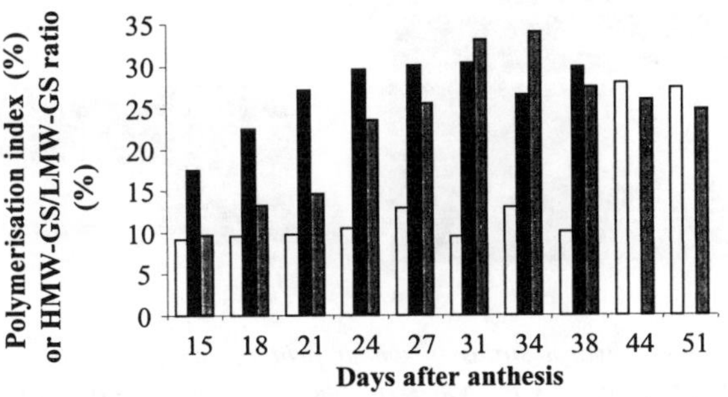

Figure 2 Evolution of the polymerisation index (SDS-insoluble polymers/total polymers) and the HMW-GS/LMW-GS ratio as a function of days after anthesis. (□) polymerisation index of controls (freeze-dried), (■) polymerisation index of desiccated kernels (oven-dried at 30°C), and (▩) HMW-GS/LMW-GS ratio. (*cv. Soissons - harvest 1998).*

3.2 Sulphydryl-disulphide changes in proteins during the grain development

A primary focus of this study was to elucidate the redox status of thiol (-SH or S-S) groups in protein during grain filling and maturation. To this end, available -SH groups in total protein extracted from grain harvested between anthesis and maturation were fluorescently labelled by incubation with a specific sulphydryl probe, monobromobimane (mBBr) according to Gobin *et al.*[20]. Analyses revealed that the sulphydryl status of endosperm proteins changed during grain development (Figure 3).

After the major protein fractions had been synthesised, at 33 DAA, they displayed a relatively high content of available -SH groups. Beyond that point - up to about 50th day - the abundance of -SH groups progressively declined so that at maturity most proteins, including glutenins, were largely oxidised. These results are in total agreement with the observations made by Gobin *et al.*[7]. However, contrary to these authors, a certain amount of -SH groups remain present at maturity irrespective of the genotype studied. The present study demonstrates that the major wheat storage proteins residing in the protein bodies undergo redox changes during the development of the grain. The proteins in general, and more particularly the glutenins, are synthesised in the sulphydryl (-SH) state and become oxidised during grain maturation. The question arises as to whether there is a relation between these changes in redox state and the appearance of SDS-unextractable polymers. The similarity in the timing suggests this to be the case. In order to verify this hypothesis, we applied artificial desiccation by ventilation at 30°C of grains during the cellular enlargement phase (Ph 2) with or without prior *in vivo* mBBr derivatisation of free protein -SH groups. As mentioned above, the dehydration of kernels during development led to the deposition of SDS-insoluble polymers (Figure 4).

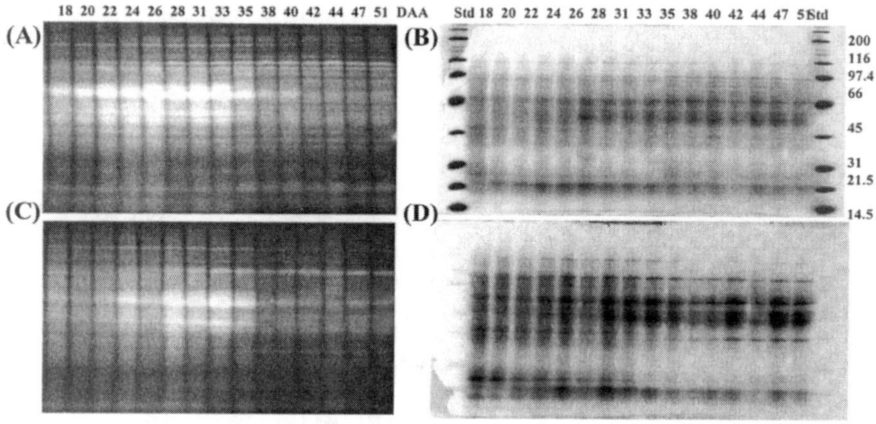

Figure 3 *Change in sulphydryl status of wheat proteins during grain development. (A,B) cv. Soissons, (C,D) cv. Thésée. (A,C) Coomassie blue staining, (C,D) mBBr-derivatised proteins* (harvest 1997).

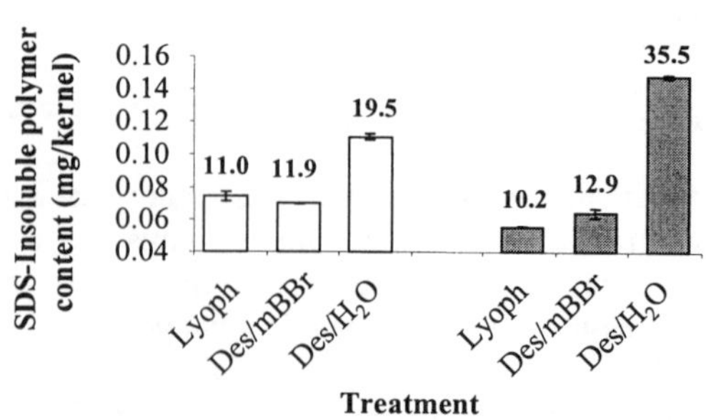

Figure 4 SDS-insoluble polymer content and polymerisation index (%) obtained (numbers) after artificial desiccation of wheat kernels (20 DAA) with or without prior *in vivo* mBBr derivatisation of free protein -SH groups. (Lyoph) freeze-dried controls, (Des/mBBr) desiccation after mBBr derivatisation, and (Des/H_2O) desiccation without mBBr derivatisation. (□) cv. Thésée (*Glu-D* 2+12), (■) cv. Soissons (*Glu-D* 5+10).

However, the *in vivo* derivatisation of the glutenin free -SH groups by mBBr prior to grain desiccation blocked the formation of SDS-insoluble polymers. Both the SDS-insoluble polymer content and the polymerisation index obtained for the Des/mBBr treatment and the control (freeze-dried kernels) were very close. Based on these results we can conclude that the oxidation of glutenin sulphydryl groups appears to be initiated by grain desiccation (rapid water loss from the grain) and led to the SDS-insoluble polymer formation. However, one key question remains unanswered : why are certain free -SH groups not affected by the oxidation phenomenon?

References

1. T. Dachkvitch and J.-C. Autran, *Cereal Chem.*, 1989, **66**, 448.
2. N. K. Singh, R. Donovan and F. MacRitchie, *Cereal Chem.*, 1990, **67**, 161.
3. R. B. Gupta, I. L. Batey and F. MacRitchie, *Cereal Chem.*, 1992, **69**, 125.
4. R. B. Gupta, S. Masci, D. Lafiandra, H. S. Bariana and F. MacRitchie, *J. Exp. Botany*, 1996, **47**, 1377.
5. P. J. Stone and M. E. Nicolas, *Aust. J. Plant Physiol.*, 1996, **23**, 727.
6. J. -L. Carceller and T. Aussenac, *Aust. J. Plant Physiol.*, 1999, **26**, 301.
7. P. Gobin, P. K. W. Ng, B. B. Buchanan and K. Kobrehel, *Plant Physiol. Biochem.*, 1997, **35**, 777.
8. R. B. Gupta, Y. Popineau, J. Lefbvre, M. Cornec, G. J. Lawrence and F. MacRitchie, *J. Cereal Sci.*, 1995, **21**, 318.

ENVIRONMENTAL EFFECTS ON WHEAT PROTEINS

E. Johansson

Department of Plant Breeding Research, The Swedish University of Agricultural Sciences, S-268 31 Svalöv, Sweden.

1 ABSTRACT

The composition, size distribution and relative amount of wheat storage proteins, protein subunits, protein groups and protein polymers have been investigated in wheat cultivars grown in Sweden during different years, in different locations, with different nitrogen fertilizer rates and with and without fungicide treatment. The results from these studies showed that;

- The amount of large polymers insoluble in SDS, the amount of albumins and globulins, and the percentage of large unextractable protein polymers in total large protein polymers, were influenced by as well cultivation year as cultivation location.
- Cultivation year and cultivation location giving rise to strong gluten were correlated with a higher amount of insoluble large protein polymers and a higher percentage of large unextractable protein polymers in total large protein polymers.
- The gluten strength of different cultivars can be explained by the composition of storage proteins and protein subunits, as well as by the percentage of large unextractable protein polymers in total protein polymers.
- The amounts of albumins and globulins were influenced by the fungicide treatment.
- The amounts of albumins and globulins were increased by wet growing seasons, with cultivation locations further to the north and with fungicide treatment.

The studies showed that the cultivar is important for determining gluten strength and cultivars can be choosed in order to get the right gluten strength. However, also the weather situation, in these studies determined by cultivation year and place, influences the gluten strength to a large extent by influencing the amount and size distribution of the protein polymers. Gluten strength can also be influenced by treatments during the cultivation. The fungicide used in this study increased the amounts of globulins and albumins.

2 INTRODUCTION

One problem in breeding for improved bread-making quality is to produce cultivars with not only good quality, but also with an even quality from year to year[1,2]. Dough properties and baking performance of wheat are strongly dependent on both genotype and environment[1-3]. Bread dough properties result from a balance between different components, starch, gluten, proteins, lipids, water, and so forth, and the interactions between these components[4]. However, the proteins are seen as the most important factor in determining the bread-making quality in a cultivar[5]. Several studies have shown the importance of the protein composition for the bread-making quality[6-13]. Protein composition is genetically determined[6,11]. Also the amount and size distribution of different protein components have been found to influence quality[14-18]. The amount and size distribution of different protein components have been found to vary both due to environmental conditions[19,20] and genetic determination[21].

The aim of the present investigation was to study the influence of environment on protein composition, and amount and size distribution of different protein components. A better understanding of how different protein components vary due to variation in environment and how this influences on the bread-making quality is assumed to lead to better possibilities to breed for higher stability in bread-making quality.

3 MATERIALS AND METHODS

The plant material comprised several spring and winter wheat cultivars grown in Sweden during different years, in different locations, with different nitrogen fertilizer rates and with and without fungicide treatment with the fungicide Amistar.

The cultivars were tested for quality at Svalöf Weibull AB, Svalöv, Sweden[12]. For investigating the protein composition, proteins were extracted from white flour and separated on polyacrylamide gels in the presence of SDS[13]. For investigating the amount and size distribution of protein groups and protein polymers and monomers, proteins were extracted from white flour and separated by RP-HPLC and SE-HPLC[16,19].

Statistical analysis, Spearman rank correlations, analyses of variance (ANOVA), principal component and principal factor analyses, were carried out[22].

4 RESULTS AND DISCUSSION

4.1 Cultivation year

The total amounts of different protein components (gliadins and glutenins and amounts of HMW glutenins, LMW glutenins and omega gliadins), were more influenced by cultivar and nitrogen application compared to cultivation year. The cultivation year was found to be of greater importance for the amount and size distribution of protein polymers (Table 1). In particular, the amount of SDS-insoluble large protein polymers (iLPP) and the percentage of large unextractable protein polymers in the total large protein polymers (LUPP) were found to be influenced by the cultivation year. Cultivation years giving rise to high gluten strength were correlated with a higher percentage of LUPP, compared to years giving rise to low gluten strength. Cultivation years with wet growing seasons were correlated with a high amount of albumins and globulins compared to cultivation years with drier growing seasons.

Table 1 *Mean squares from the combined analysis of variance of protein polymers. N=nitrogen application, Year =cultivation year, Df= Degrees of freedom, sLPP=SDS-soluble large polymeric protein, iLPP=SDS-insoluble large polymeric protein, LUPP=large unextractable polymeric protein in the total large polymeric protein. *, and ***=significant at P=0.05, and 0.005.*

	Df	sLPP (10^{11})	iLPP (10^{9})	LUPP (10^{-2})
Cultivar	1	4.1***	5.5	1.0*
Year	2	2.5***	664.4***	4.0***
N	10	3.6***	527.9***	1.1***

4.2 Cultivation location

Variation in cultivation location significantly influenced the amounts of albumins and globulins, the amount of iLPP and the percentage of LUPP. The amount of iLPP decreased with growing location further to the north while the amounts of albumins and globulins increased. This might be explained by variation in pre-harvest sprouting between the different cultivation locations.

4.3 Cultivar

Cultivar differences was found in the total amounts of different protein components, gliadins, glutenins, HMW glutenins, LMW glutenins and omega gliadins. However, the amounts of gliadins, glutenins, HMW glutenins, LMW glutenins and omega gliadins were correlated with differences in protein concentration between different cultivars. The higher the protein concentration, the higher the total amounts of gliadins, glutenins, HMW glutenins, LMW glutenins and omega gliadins. The cultivar also influenced the amounts of SDS-soluble and insoluble protein polymers.

Cultivar differences in gluten strength was found to be due to differences in storage protein composition, glutenin/gliadin ratio and percentage of LUPP. Cultivars with high gluten strength had a storage protein composition implying higher gluten strength, a higher total amount of HMW glutenin subunits, a higher glutenin/gliadin ratio and a shift of protein polymers from SDS-soluble towards more SDS-insoluble protein polymers compared to cultivars with a lower gluten strength.

4.4 Nitrogen application

The nitrogen application influenced all protein fractions containing gliadins and glutenins. Increased nitrogen application leads to an increase in all fractions containing gliadins and glutenins. The amounts of protein fractions containing mainly albumins and globulins were not significantly influenced by the nitrogen application. The relationships between different protein groups containing gliadins and glutenins were also unaffected.

4.5 Fungicide treatment

The fungicide treatment was found to correlate with a decrease in protein concentration and a decrease in falling number. Fungicide treatment also influenced the amounts of several of the protein polymers and monomers. Among others, the amounts of albumins and globulins were increased by the fungicide treatment.

References

1. E. Johansson and G. Svensson, *J. Sci. Food Agric.,*1998, **78,** 109.
2. E. Johansson and G. Svensson, *J. Agric. Sci.*, 1999, **132,** 13.
3. C.J. Peterson, R.A. Graybosch, P.S. Baenziger and A.W. Grombacher, *Crop Sci.*, 1992, **32**, 98.
4. D.D. Kasarda, in *Wheat is Unique*, ed. Y. Pomeranz, American Association of Cereal Chemists, St Paul, 1989, p. 277.
5. J.S. Wall, in *Recent Advances in the Biochemistry of Cereals*, eds. D. L. Laidman and R. G. Wyn Jones, Academy, London, New York, 1979, p. 275.
6. P.I. Payne, L.M. Holt, E.A. Jackson and C.N. Law, *Phil. Trans. R. Soc. Lond. B,* 1984, **304,** 359.
7. P.I. Payne, M.A. Nightingale, A.F. Krattiger and L.M. Holt, *J. Sci. Food Agric.*, 1987, **40,** 51.
8. T. Sontag, H. Salovaara and P.I. Payne, *J. Agric. Sci. Finland,* 1986, **58,** 151.
9. G.J. Lawrence, H.J. Moss, K.W. Shepherd and C.W. Wrigley, *J. Cereal Sci.*, 1987, **6,** 99.
10. A.K. Uhlen, *Norwegian J. Agric. Sci.*, 1990 **4**, 1.
11. E. Johansson, P. Henriksson, G. Svensson and W.K. Heneen, *J. Cereal Sci.*, 1993, **17,** 237.
12. E. Johansson and G. Svensson, *Cereal Chem.*, 1995, **72,** 287.
13. E. Johansson, *Plant Breed.*, 1996, **115**, 57.
14. J.M. Field, P.R. Shewry and B.J. Mifling, *J. Sci. Food Agric.*, 1983, **34,** 370.
15. K.H. Sutton, *J. Cereal Sci.* 1991, **14**, 25.
16. R.B. Gupta, K. Khan and F. MacRitchie, *J. Cereal Sci.*, 1993, **18,** 23.
17. J.L. Andrews, R.L. Hay, J.H. Skerritt and K.H. Sutton, *J. Cereal Sci.*, 1994, **20**, 203.
18. H. Wieser, W. Seilmeier and R. Kieffer, in *Gluten Proteins 1993*, Association of Cereal Research, Detmold, Germany, 1994, p. 141.
19. H. Wieser and W. Seilmeier, *J. Sci. Food Agric.*, 1998, **76,** 49..
20. E. Johansson, G. Svensson and S. Tsegaye, *Acta Agric. Scandinavica,* 2000, *(in press*).
21. F. MacRitchie, *Cereal Foods World,* 1999, **44,** 188.
22. SAS, *User's Guide: Statistics.* SAS Institute Inc, Cary, NC, USA, 1985.

Acknowledgements

This work was supported by The Swedish Farmer's Foundation for Agricultural Research, The Cerealia Foundation R&D, VL-stiftelsen, The Royal Physiographic Society, and Svalöf Weibull AB.

EFFECTS OF GENOTYPE, N-FERTILISATION, AND TEMPERATURE DURING GRAIN FILLING ON BAKING QUALITY OF HEARTH BREAD

A.K. Uhlen[1], E.M. Magnus[2], E.M. Færgestad[2], S. Sahlstrøm[2] and K. Ringlund[1].

[1] Agricultural University of Norway, Dept. of Horticulture and Crop Science, P. O. Box 5022, N-1432 Ås, Norway. [2] MATFORSK – The Norwegian Food Research Institute, Oslovn. 1, N-1430 Ås, Norway.

1 INTRODUCTION

Wheat quality for bread making is affected by the genotype and cultivation techniques, as well as by climatic conditions during grain filling. Genotypic variation in protein quality, affecting mixing requirements and the product quality of breads, is documented in many investigations. Besides the inherited differences in protein composition among genotypes, environmental factors are also found to affect gluten quality[1,2]. Several papers report effects of the temperature during grain filling on protein content and/or protein quality of the flours [3, 4, 5, 6]. Prolonged periods of high temperature stress are shown to affect dough strength negatively[7,8,2]. Effects of the temperature during grain filling at lower temperature ranges are less explored.

In most investigations of the relationships between flour quality characteristics and bread-making performance, test baking has been done in pans, and loaf volume has been the main parameter for evaluation of the products. However, hearth breads are produced in many European countries as well as pan breads. For hearth breads, the ability to retain a proper shape after proving and baking is an important character as well as the loaf volume[9].

The objective of the present study was to investigate the interrelationships between protein quality and quantity and the baking characteristics of hearth bread of wheat samples grown in two different years of contrasting temperatures during grain filling.

2 MATERIALS AND METHODS

Twenty spring wheat cultivars and breeding lines with broad variation in protein quality were selected for the study. Field trials, laid out as a split-plot design with two replicates, three levels of nitrogen fertilisation (80 kg N ha^{-1}, 120 kg N ha^{-1} and 160 kg N ha^{-1}) and the twenty genotypes, were conducted in 1997 and 1998 at Vollebekk experimental farm, Department of Horticulture and Crop Sciences, Ås, Norway.

Mean temperature and precipitation for May to September at the field location in 1997 and 1998 are given in Table 1. In 1997, the temperature was considerably warmer than normal in July and August, whereas the temperatures in July and August were cooler than normal in 1998.

Table 1. Mean temperature and precipitation at Ås from May to September in 1997 and 1998 compared to mean values for the period 1961-90.

	Temperature			Precipitation		
	1997	**1998**	**Mean 1961-90**	**1997**	**1998**	**Mean 1961-90**
May	9,1	10,9	10,3	57	21	60
June	15,8	12,7	14,8	61	124	68
July	18,4	14,9	16,1	69	61	81
August	19,9	13,7	14,9	63	96	83
September	12,1	11,9	10,6	82	96	90

The wheat samples from two of the three N fertiliser levels each year were used for baking experiments. To obtain flours with approximately equal protein levels from the two years, the low and intermediate N levels were chosen in 1997, and the intermediate and high N levels in 1998.

The wheat samples were analysed for protein content by NIT (Infratech 1255 FFA, Tekator AB, S-26 321 Höganes, Sweden) and SDS sedimentation volume (AACC Approved Method 56-70). The flour mixing properties were examined by using the Farinograph at mixing speed 63 rpm (ISO method 5530-1) and at mixing speed 126 rpm as described by Færgestad et al. (2000).

Baking quality was determined by a small-scale straight-dough hearth bread baking test using optimised mixing and fixed proving time [9]. Mixing was performed using the Farinograph bowl operated at 126 rpm, and mixing time was determined according to the Farinograph dough development time at high speed (126 rpm). The dough was fermented for 10 minutes at 27 ° C, and proved for 45 minutes at 37° C at 70% RH.

3 RESULTS AND DISCUSSION

Increased N fertilisation resulted in increased protein content of the wheat samples. On average for the 20 genotypes, the protein content increased from 12,6% to 15,5% in 1997, and from 13,0% to 13,9% in 1998, by increasing the N-fertiliser by 40kg N/ha. The SDS sedimentation volume and the Farinograph dough development time increased as the protein content increased. The loaf volume increased moderately with increased N fertilisation in 1997, whereas no significant difference in loaf volume was found between the two N-levels in 1998. The form ratios (height/width) of the loaves, however, showed a small, but significant decrease as the N fertiliser level, and thereby the protein content, increased in both years.

The genotypes showed broad variation in SDS sedimentation volume and flour mixing properties. High positive correlations were found in both years between SDS sedimentation volume and loaf volume (1997: R^2=0,69 P< 0,01 1998: R^2=0,69 P<0,01) as well as between SDS sedimentation volume and form ratio (1997: R^2=0,53 P= 0,04 1998: R^2=0,72 P<0,01).

The average SDS sedimentation volume and dough mixing properties of the wheat genotypes grown in 1997 and 1998 are shown in Table 2. The data are based on the first N level in 1997 and the second N level in 1998, having similar protein contents. Higher sedimentation volumes, longer mixing requirements and better dough stability were obtained in the samples grown in 1997 compared to 1998. These differences in protein quality are probably due to the temperature during grain filling, as this climatic parameter caused the most dominant variation in climate between the two years. The results showed that variation in climatic conditions during grain filling can give considerable variation in protein quality of the flours, giving inconsistency in wheat quality between different growing sites and year of harvest.

A few of the genotypes (7, 8, 10, 13) appeared to be less affected of the climatic conditions, and had about similar values of SDS sedimentation volumes (Figure 1) in both years. This indicates variation among the genotypes in environmental stability, which can possibly be exploited in breeding programs.

Table 2. Average protein content, gluten characteristics and specific loaf volume of the genotypes grown in 1997 and 1998. The data are based on the first N fertiliser level in 1997 and the second N fertiliser level in 1998, having similar levels of protein.

Year **Temperature July-Aug.**	**1997** **19,2 °C**	**1998** **14,1 °C**	**Significance level**
Protein content, %	12,6	13,0	*
SDS sedimentation volume, ml	60	47	***
Farinograph dough development time (126 rpm), min	2,5	2,2	*
Farinograph dough development time (63 rpm), min	4,8	3,7	*
Farinograph stability, min	6,89	4.10	*
Farinograph softening, FU	39,8	69,5	***

4 CONCLUSIONS

The weather conditions in the two years strongly affected the protein quality of the flour, giving higher SDS sedimentation volumes, longer mixing requirements and greater stability of samples grown at the higher temperature during grain filling in 1997.

The results indicated variation among the genotypes in environmental stablility. A few of the genotypes achieved similar values of SDS sedimentation volumes in the two years, whereas the other genotypes decreased significantly in protein quality in 1998 compared to 1997. Breeding cultivars with a better environmental stability can be important in areas with shifting climatic conditions between sites and years.

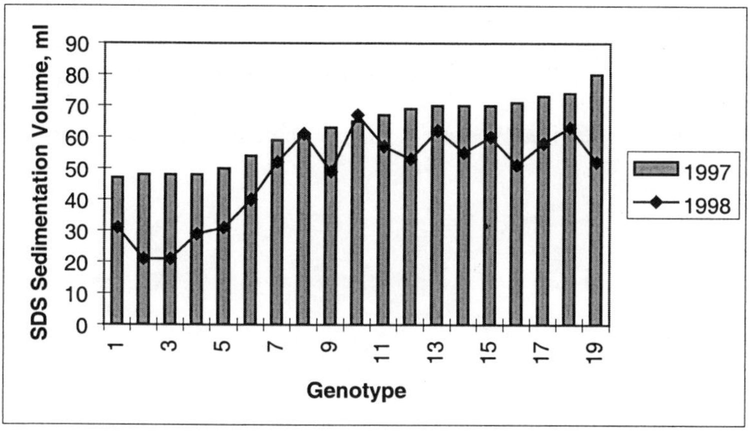

Figure 1. SDS sedimentation volumes of the genotypes grown in 1997 (bars) and in 1998 (line). The genotype order is according to increasing SDS sedimentation volume in 1997.

References

1. Petersen, C. J., Graybosh, R. A., Baenziger, P. S. and Grombacher, A.W. *Crop Sci.*, 1992, **32**, 98.
2. Graybosh, R. A., Petersen, C. I., Baenziger, P. S. and Shelton, D. R. *J. Cereal Sci.* 1995, **22**, 45.
3. Randall, P. J. and Moss, H. J. *Austr. J. Agric. Res.*, 1990, **41**, 603.
4. Scipper, A.. *Agribiological Res.* 1991, **44**, 114.
5. Peterson, C. J., Graybosh, D. R., Shelton, D. R. and Baenziger, P. S. *Euphytica* 1998, **100**, 157.
6. Uhlen, A. K., Hafskjold, R., Kalhovd, A.-H., Sahlstrøm, S., Longva, Å. and Magnus, E. M. *Cereal Chem.*, 1998, **75**, 460.
7. Blumenthal, C. S., Batey, I. L., Bekes, F., Wrigley, C. W. and Barlow, E. W. R. *Aust. J. Agr. Res.*, 1991, **42**, 21.
8. Blumenthal, C. S., Barlow, E. W. R. and Wrigley, C. W. *J. Cereal Sci.*, 1993, **18**,3.
9. Færgestad, E. M., Molteberg, E. L. and Magnus, E. M.. *J. Cereal Sci.*, 2000 (in press).

INTERACTIONS BETWEEN FERTILIZER, TEMPERATURE AND DROUGHT IN DETERMINING FLOUR COMPOSITION AND QUALITY FOR BREAD WHEAT

Frances M. DuPont, Susan B. Altenbach, Ronald Chan, Kerry Cronin, and Dao Lieu

USDA Agricultural Research Service, Western Regional Research Center, Albany, CA 94710, U.S.A.

1 INTRODUCTION

Heat, drought and fertilizer had separate but interacting effects on grain development and flour quality in greenhouse grown plants of a U.S. hard red spring wheat variety, 'Butte 86'.

2 MATERIALS AND METHODS

The effects of fertilizer on grain composition under different regimes of temperature and drought were evaluated. Until anthesis, wheat plants were grown in pots in climate-controlled greenhouses with a daily maximum temperature of 24°C . At anthesis, half of the pots were moved to a heated greenhouse that had a maximum temperature of 37°C for 5 hours each day. Soil moisture and fertilizer levels were regulated by a combination of hand watering and drip irrigation. The potting mix contained $CaSO_4$ (gypsum) but the fertilizer contained no additional sulfur. Addition of post-anthesis fertilizer (N:K:P = 20:20:20) by drip irrigation simulated field conditions in which soil nitrogen is available during grain development. Total RNA was prepared from grains that were collected during development[1]. At maturity, grain composition was determined and flour quality was assessed. Protein was determined by nitrogen analysis. Whole grain was ground in a UDY mill, a gliadin extract was prepared by extracting the flour with 50% propanol[2] and proteins were resolved by reverse phase HPLC (RP-HPLC) (DuPont et. al, submitted). SDS sedimentation volumes were determined using a final volume of 15 ml and 0.5 mg of UDY ground grain. The amount of insoluble glutenin was determined by LECO nitrogen analysis of the pellet remaining after removing soluble protein with 50% propanol[3].

3 RESULTS AND DISCUSSION

RP-HPLC of a 50% propanol extract of whole meal flour resolved the omega-gliadins (ω-gliadins) into four separate peaks (Figure 1). The first peaks to elute were identified as ω-gliadins coded on the 1B, 1D and 1A chromosomes (DuPont et. al, submitted). Some high mol weight and low mol weight glutenin subunits (HMW-GS and LMW-GS) also were present in the 50% propanol extract. The 1A ω-gliadins eluted along with the HMW-GS. Levels of ω-gliadin proteins increased with fertilizer, whereas little change was evident in regions of the chromatogram where other gliadins eluted. When protein percent increased in response to heat and drought, rather than in response to fertilizer, levels of ω-gliadins also increased (not shown). Preliminary experiments indicate that HMW-GS in the insoluble fraction also increased when grain protein increased.

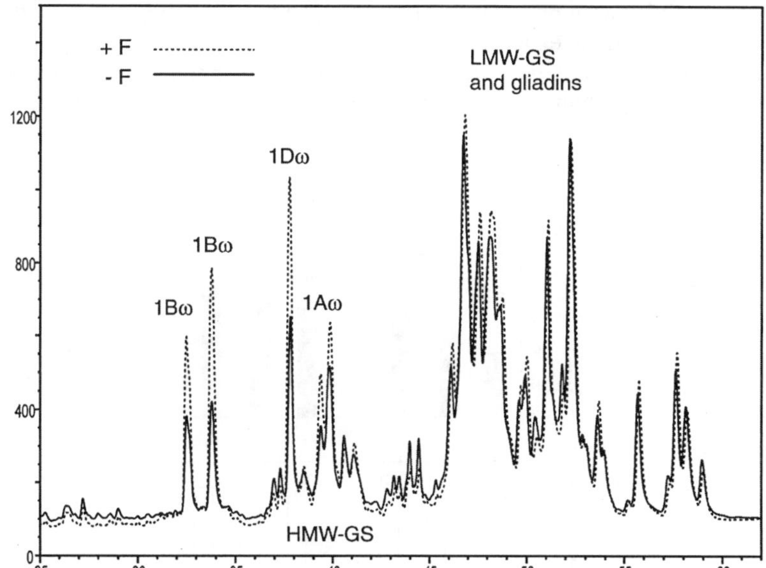

Figure 1 *RP-HPLC trace for proteins from grain of plants grown with (+F) or without (-F) post anthesis fertilizer at 24°C maximum daily temperature. Proteins were solubilized in 50% propanol and separated by RP-HPLC. Location of the chromosome 1A, 1B and 1D coded ω-gliadins, the HMW-GS, LMW-GS and the other gliadins are indicated.*

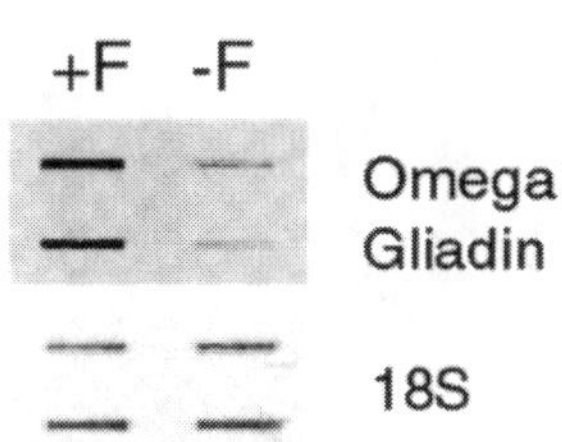

Figure 2 *Effect of fertilizer on mRNA transcript levels for ω-gliadins from grains of plants grown with (+F) or without (-F) post anthesis fertilizer and 24°C maximum daily temperature. The grains were collected at 20 days after anthesis.*

Levels of transcripts for the ω-gliadins decreased in the absence of post-anthesis fertilization (Figure 2). Although heat and drought increased protein percent and the amount of ω-gliadins in the absence of post-anthesis fertilizer, levels of transcript for the ω-gliadins did not change. However, temporal expression of gluten transcripts was sensitive to heat and drought (not shown).

Two measures of flour quality were performed. SDS sedimentation volumes had a high correlation with grain protein percent, regardless of the treatment (Figure 3). The amount of insoluble protein had a high correlation with grain protein percent (Figure 3), but there was little variation in the percent of insoluble protein, regardless of the treatment (not shown). A single experiment is depicted in Figure 3. In 29 samples from 6 experiments with various regimes of fertilization, temperature, and water, the average percent insoluble protein for flour from 'Butte 86' was 43.9 +/-2.3 percent of total protein.

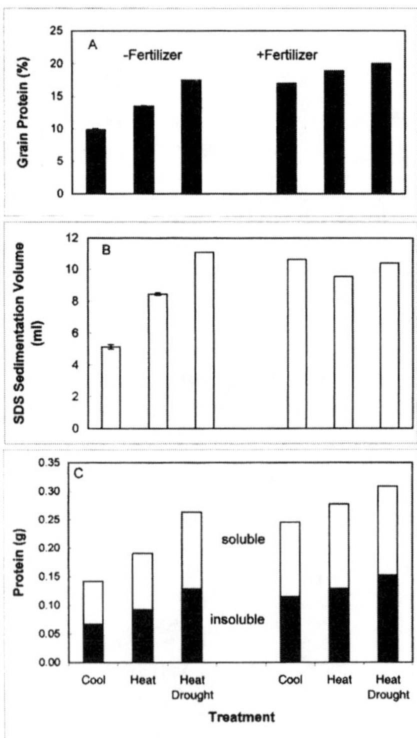

Figure 3 *Effect of treatment on SDS sedimentation volumes, grain protein percent and amount of grain protein in the soluble and insoluble fractions after extraction with 50% propanol.*

In the field, wheat plants send out an extensive root system to mine the soil for water and minerals. Thus the ability to retrieve mineral nutrients may be determined by interactions between genotype, soil composition, temperature, and moisture on root growth and ion absorption. Growing plants in pots in controlled environments with fertilizer applied by drip irrigation allows us to determine how grain development responds to defined conditions of temperature, fertilizer and water. Under these conditions, we found that temperature affected the rate and duration of grain fill, while fertilizer affected grain protein percent and protein composition.

When breeding wheat for improved flour quality and yield, it is important to define quality in molecular terms. It is also essential to understand the molecular basis for the interactions between genes and environment. As is well known, we found that the primary effects of temperature, water availability and fertilizer were on grain protein percent. We also found that variations in temperature, water availability and fertilizer combined to affect the amount of ω-gliadins and the SDS-sedimentation volumes in a similar manner to their effect on protein. However, there was little effect on the percent of insoluble protein. It is known that levels of ω-gliadins and high molecular weight glutenin subunits increase in response to sulfur deficiency in wheat[4]. In our experiments levels of ω-gliadins may have increased in response to an excess of nitrogen relative to sulfur, and this response may be regulated in part by regulation of transcript levels. Further research is needed to clarify this response.

References

1. S.B. Altenbach, *Theor.Appl.Genet.* (1998), **97**, 413
2. B.X. Fu and H.D. Sapirstein, *Cereal Chem.* (1996), **73**, 143
3. S.R. Bean, R.K. Lyne, K.A. K A Tilley, O.K. Chung and G.L. Lookhart, *Cereal Chem.* (1998), **75**, 374
4. C.W. Wrigley, D.L. Du Cros, J.G. Fullington and D.D. Kasarda, *J.Cereal Sci.* (1984), **2**, 15

INFLUENCE OF ENVIRONMENT AND PROTEIN COMPOSITION ON DURUM WHEAT TECHNOLOGICAL QUALITY

G. Galterio and M.G. D'Egidio

Istituto Sperimentale per la Cerealicoltura- via Cassia 176, 00191 Roma.

1 INTRODUCTION

It has long been known that the protein content of wheat influences the technological quality for products such as pasta, bread etc.[1-3]. The protein content of grains is influenced mainly by growing (soil fertility, water availability) and environmental conditions.

Over the last thirty years, breeding programs have produced new varieties of durum wheat characterised by higher ear fertility and reduced size. The increased wheat production (yield) has been associated with a decrease in protein level owing to the negative correlation between yield and grain protein percentage[4]. Previous findings showed that final products of different technological quality can be obtained from semolina or flour with similar protein content (i.e. pasta stickiness and cooking resistance, bread volume etc.). Also this can be explained by significant differences in the protein composition among wheat varieties and especially in the ratios between the protein fractions involved in the gluten network[5,6]. Several authors[7-12] proposed that the glutenin fraction is mainly responsible for dough rheological characteristics such as elasticity and extensibility, comprising protein polymers consisting glutenin subunits of low molecular weight (LMW-GS) and high molecular weight (HMW-GS).

The aim of this work is to evaluate the influence of environment and protein composition on the technological quality of durum wheat varieties.

2 MATERIAL AND METHODS

Durum wheat varieties (Colosseo, Creso, Duilio, Grazia, Ofanto, San Carlo and Simeto) which are representative of the cultivars widely grown in Italy, were grown in the year 1997-98 in different environments of central Italy, in experimental trials on 10 m^2 plots according to a randomised block design, at low rate of nitrogen fertilizer (80 kg/ha) and at a sowing density of 450 seeds/m^2.

Semolina samples were obtained by milling the grains in an experimental pilot mill (Bühler MLU 202); the technological analyses were performed using standard methods: gluten content (ICC 137), gluten quality expressed as Gluten Index (ICC 158) and Chopin alveograph (ICC 121). Protein fractionation was performed by SDS-PAGE and evaluated by scanning densitometer; densitometric data were obtained by scanning two gels for each sample and determining the average height readings[13].

3 RESULTS AND DISCUSSION

The data obtained for yield and protein content of the different samples showed a wide variability for these two parameters. Table 1 shows different quality traits for the seven durum wheat varieties grown in the four localities.

Table 1 *Yield and protein contents of different varieties in four environments*

	Yield t/ha				Protein % d.m.			
	Pisa	Macerata	Perugia	Roma	Pisa	Macerata	Perugia	Roma
Colosseo	6.9	7.0	5.6	4.6	15.9	12.1	10.3	10.3
Creso	7.6	6.2	5.2	3.2	15.9	12.8	10.6	11.6
Duilio	10.2	6.5	5.9	4.0	14.8	11.8	10.6	10.2
Grazia	7.0	6.2	5.6	4.1	16.7	12.0	10.6	10.1
Ofanto	6.6	6.3	5.2	3.8	15.3	12.1	10.8	10.6
Sancarlo	8.6	6.5	6.2	4.4	15.6	12.3	11.4	10.2
Simeto	7.0	5.7	5.6	3.1	16.3	9.3	11.0	11.0
Mean	7.7	6.3	5.6	3.9	15.8	11.8	10.8	10.6
Min-Max	6.6-10.2	5.7-7.0	5.2-6.2	3.1-4.6	14.8-16.7	9.3-12.8	10.3-11.4	10.1-11.6

The gluten characteristics were widely different showing good quality for Pisa, medium quality for Macerata and low quality for Roma and Perugia. A high positive correlation ($r = 0.763$ ***) was found between gluten content (%) and alveograph W. This means that the W, which is a measure of dough resistance, is linked to genotype, but is also influenced by total gluten content, a parameter strictly related to protein content and so highly influenced by the environment. It can be noted that samples obtained from the Roma and Perugia locations are characterised by similar protein contents, but by different gluten quality: the W values and P/L ratios are higher in Roma than in Perugia (Table 2). This could be explained by different protein compositions of the samples from the two environments, probably differences in the synthesis of protein subunits related to gluten quality.

Table 2 *Quality characteristics of different varieties in four environments*

	PERUGIA				ROMA			
	Dry Gluten % d.m.	Gluten Index	W	P/L	Dry Gluten % d.m.	Gluten Index	W	P/L
Colosseo	7.9	84	60	0.70	6.3	98	108	1.45
Creso	7.8	69	50	1.07	7.7	95	155	3.25
Duilio	6.8	80	60	1.29	5.7	98	105	3.79
Grazia	8.1	49	40	0.55	7.2	93	105	1.34
Ofanto	7.0	28	20	0.76	6.3	92	80	1.85
Sancarlo	8.0	88	75	1.44	5.6	99	144	5.96
Simeto	7.3	97	110	1.90	6.2	98	164	3.90
	PISA				MACERATA			
Colosseo	12.3	92	250	1.33	8.9	93	150	1.30
Creso	13.1	81	240	1.22	9.6	88	165	1.65
Duilio	11.0	86	250	1.94	8.0	90	155	2.80
Grazia	14.1	80	240	0.69	9.6	77	130	0.79
Ofanto	12.8	65	210	1.15	8.1	78	125	1.15
Sancarlo	12.7	95	350	1.51	8.9	96	220	3.88
Simeto	13.4	86	330	2.29	9.2	93	185	3.56

In bread wheat the findings of Payne *et al.*[14] have been a milestone in our understanding of the biochemical basis for wheat technological quality: the presence of some HMW-GS is highly correlated with technological quality. From analysis of the progeny of numerous crosses Payne *et al.*[15] developed a quality score taking in consideration the effects of each subunit or pair of subunits on the gluten quality evaluated by the SDS sedimentation test. An analogous quality score was developed by Pogna *et al.*[16] considering the alveograph W as a measure of the technological quality. It was demonstrated that these quality scores explain a high percentage (50-60 %) of the variability observed in the quality characteristics of bread wheat. In durum wheat it is not possible to directly apply these scores because only HMW-GS encoded by chromosome 1B are usually present. On the other hand, Boggini and Pogna[17] observed that some differences in the quality of durum wheat varieties are related to the presence of the 1B chromosome-encoded HMW-GS 7+8 > 20 > 6+8 > 13+16. In the present study the varieties considered are characterised by HMW-GS 7+8 (Duilio, San Carlo, Simeto), 6+8 (Creso), 13+16 (Colosseo) and 20 (Grazia and Ofanto). The densitometric values for the protein subunits separated by SDS-PAGE are reported in the Table 3.

Table 3 *Mean densitometric area of bands in six molecular weight ranges of SDS-PAGE patterns of total proteins from six varieties and two environments*

Variety		HMWG	IMWG	LMWG	LMW-2	30-20kDa	20-17 kDa	10-12 kDa
Colosseo PG		4,0	10,0	30,7	16,0	30,7	8,2	16,4
	RM	4,1	8,7	33,8	17,6	30,6	8,1	14,7
Creso PG		4,1	11,1	27,6	15,8	33,2	8,9	15,1
	RM	4,1	8,5	29,5	17,3	34,0	8,2	15,7
Duilio PG		4,0	9,1	30,0	15,7	34,3	7,6	15,0
	RM	4,5	8,9	31,2	16,7	33,9	7,4	14,1
Grazia PG		3,8	9,3	30,8	17,4	30,9	8,6	16,6
	RM	4,0	8,9	31,8	17,8	30,6	8,1	16,6
Ofanto PG		3,8	9,8	30,4	15,3	32,5	7,6	15,9
	RM	3,6	8,4	32,7	16,1	31,0	7,6	16,7
SanCarlo PG		4,0	9,5	30,8	15,7	32,8	7,6	15,3
	RM	4,3	9,1	32,3	17,1	31,2	7,9	15,2
Simeto PG		4,7	8,1	29,7	16,1	32,5	8,6	16,4
	RM	4,7	7,2	32,5	17,3	31,2	8,1	16,3

The protein subunits were grouped in six regions on the basis of their apparent molecular weights:

1) HMWG : protein subunits with molecular weight > 90 kDa
2) IMWG : protein subunits with molecular weight 78-50 kDa
3) LMWG : protein subunits with molecular weight 45-30 kDa
4) protein subunits with molecular weight 30-20 kDa
5) protein subunits with molecular weight 20-17 kDa
6) protein subunits with molecular weight 17-10 kDa.

Regarding the protein subunits of region 1 (HMWG) it can be noted that the Roma samples showed higher levels of these subunits when compared with the Perugia samples. Higher levels of HMWG seems to determine a higher gluten tenacity and the densitometric values of HMWG are related to the P/L ratio ($r = 0.568^*$). The semolina obtained from these two localities also showed differences in the amounts of IMWG subunits, LMW-2 subunits (apparent molecular weight of about 45 kDa) and of LMWG. Region 2 generally contained the IMWG, consisting of ω-gliadins, albumins, globulins and glutenins; the subunits of region 2 are characterised by a low level of active SH groups, while the LMW-2 and HMWG regions contain glutenins and gliadins, protein subunits rich in SH groups[18, 19].

The semolina from the Roma trials showed a lower percentage of IMWG and higher percentage of LMW-2 and LMWG. An explanation could be that the Roma soil is of vulcanic origin and more rich in sulphur than the Perugia soil which is of alluvial origin;

therefore in the Roma trial the synthesis of proteins requiring sulphur was increased. Correlations were found between the densitometric determinations of subunits with higher molecular weight (HMWG, IMWG, LMW-2 and LMWG) and alveograph W: 0.646*, -0.715** and 0.627* for HMWG, IMWG and LMW-2, respectively. It is important to note that the relative ratios between the different subunits seem to influence the technological results more than the absolute amounts of the subunits. The higher correlations between alveograph W and the ratios of the densitometric determinations of the different regions (HMWG/IMWG r=0.776***, (HMWG+LMW-2) / IMWG r= 0.816***, LMW-2/IMW r= 0.794***) confirm this hypothesis.

4 CONCLUSIONS

The results of this study confirm that the composition of subunits HMW-GS, LMW-2, LMW-GS and IMWG play an important role in determining the rheological characteristics of doughs in durum wheat; the ratios between protein subunits (HMWG+LMW-2)/IMWG seems to be very important for viscoelastic properties of gluten.

References

1. Finney K.F. and Barmore M.A., *Cereal Chem.,* 1948, 25, **5**, 291.
2. Dexter J.E. and Matsuo R.R., *Can. J. Plant Sci.,* 1977, **57**, 717.
3. D'Egidio M.G., Fortini S., Galterio G., Mariani B.M., Sgrulletta D., Volpi M., *Qualitas Plantarum*, 1979, 28, **4**, 333.
4. Mc Neal F.H., Berg M.A., Mc Guire C.F., Stewart V.R. and Baldridge D.E. *Crop Sci.*, 1972, **12** 599.
5. Wasik R.J. and Bushuk W., *Cereal Chem.*, 1975, **51**, 322.
6. Heubner F.R. and Wall J.S., *Cereal Chem.*, 1976, **53**, 62.
7. Payne P.I., Corfield K.G. and Blackman J.A., *Theor. Appl. Genet.,* 1979, **55**, 153.
8. Payne P.I., Jackson E.A .and Holt L., *J. Cereal Sci.*, 1984, **2**, 73.
9. Branlard G. and Dardevet M., *J. Cereal Sci.*, 1985, **3**, 345.
10. Pogna N.E., Autran J.C., Mellini F., Lafiandra D. and Feillet P., *J. Cereal Sci.*, 1990, **11**, 15.
11 Gupta R.B., Bekes F and Wringley C.W, *Cereal Chem.*, 1991, **68**, 328.
12. Halford N.G., Field J.M., Blair H., Urvin P., Moore K., Robert L., Thompson R., Flavell R.B., Tatham A. and Shewry P. R., *Theor. Appl. Genet.,* 1992, **83**, 373.
13. Autran J.C. and Galterio G., *J. Cereal Sci.*, 1989, **9**, 195.
14. Payne P.I., *Biotech. Crop Improv. Protection*, 1986, **34,** 69.
15. Payne P.I., Nightingale M., Krattinger A. and Holt L., *J. Sci. Food Agric.*, 1987, **40**, 51.
16. Pogna N.E., Mellini F. and Dal Belin Peruffo A., B. Borghi ed., Hard wheat: agronomic, technological, biochemical and genetic aspects, CEC, 1987, pp 53.
17. Boggini G. and Pogna N.E., *J. Cereal Sci.*, 1989, **9**, 131.
18. Galterio G., Desiderio E. and Pogna N.E. 'Gluten Proteins', Association of Cereal Research Detmold (Germany) 1993, 528:537
19. Zhao F.J., Salmont S.E. ,Withers J.A., Monaghan J.M., Evans E.J., Shewry P.R. and McGrath S.P., *J. Cereal Sci.*, 1999, **30**, 19.

Non-Gluten Components

INTERACTIONS OF STARCH WITH GLUTENS HAVING DIFFERENT GLUTENIN SUB-UNITS.

Ian L. Batey[1]

1. CSIRO Plant Industry, Grain Quality Research Laboratory, PO Box 7, North Ryde NSW 1670, Australia

1 INTRODUCTION

Gluten proteins have always been considered the prime determinant of wheat quality, and starch has been considered an "inert filler" in terms of quality. This view has been dispelled, at least in the case of Japanese Udon noodles, by recent work by a number of authors who have shown the importance of starch in noodle quality, and have linked noodle quality to a specific gene on chromosome 4A.[1] In pan breads, reconstitution studies have shown that there seems to be little contribution to quality by starch as one wheat starch can be replaced by another wheat starch without significant change in loaf quality.[2] In dough mixing, there usually appears to be little effect from the starch component of the doughs, although starches fractionated according to granule size do have different effects (E.A. Asp & P.W. Gras, personal communication). This latter observation is likely to arise from competition for water in a system which is deficient in water. Small starch granules have a higher surface area to volume ratio, and will be more easily hydrated per unit mass than larger starch granules. Competition for water with the gluten proteins will cause changes in the mixing properties under these conditions.

There must be, however, some interaction between starch and gluten in breads and similar products. While different wheat starches appear to have no effect on the bread quality, the replacement of wheat starch with starches from other botanical sources can have significant consequences.[3] It is clear, therefore, that there is some contribution to quality from the starch and it could well be that it results from some type of interaction between the starch component and the gluten.

The energies involved in the interactions between starch and protein are likely to be small, or they would be easier to measure. In measuring protein properties, energies tend to be relatively large. For example, the energy requirement for mixing is about 11 watt hours per kilogram of flour, or 39.6 joules per gram.[4] In contrast, the energy involved in gelatinisation of starch is an order of magnitude smaller, and the energies of interaction between starch and protein most likely would be smaller still. In the work described here, the gelatinisation and viscosity properties of wheat starch have been studied and the effect of different gluten proteins on these properties are reported. The glutens came from lines derived from a cross between parents with completely different

Glu-1 and *Glu-3* alleles. This allowed a study of the effect of different combinations of glutenin alleles.

2 MATERIALS AND METHODS

Flours were from a cross between the varieties Cranbrook and Halberd, the progeny of which were propagated using doubled haploid techniques. Starch was extracted from the flours using a Glutomatic Model 2200 as described previously,[5] and was washed several times with water before being freeze dried. Viscosity was measured on a Rapid Viscoanalyser (RVA) (Newport Scientific, Warriewood, NSW, Australia) using 3.00g of starch or 3.00g starch plus 0.50g gluten per 25.0 mL water. Thermal properties were measured in stainless steel pans on a Pyris 1 differential scanning calorimeter (DSC) (Perkin Elmer, Norwalk, CT, USA) using 1 part flour or starch to 3 parts water.

3 RESULTS AND DISCUSSION

3.1 Viscosity

The standard starch used was a commercial wheat starch of moderate viscosity. Its viscosity was measured alone and with the addition of different glutens to raise the protein level to about that of flour. Addition of the hand-washed gluten resulted in an increase in the pasting viscosity compared with the starch alone, with the increase ranging from 19 to 92 Rapid Viscoanalyser units (RVU). The viscosities obtained are shown in Table 1.

Table 1. *Changes in the viscosity of commercial starch after the addition of hand-washed wheat gluten.*

	Starch viscosity (RVU)	Viscosity + gluten (RVU)	Difference (RVU)
Peak viscosity	237	256-329	19-92
Holding strength	155	155-188	0-33
Final viscosity	306	312-358	6-52

The results indicated that the addition of gluten had a significant effect on the gluten, and that this effect was variable. It is unlikely that the increase in peak viscosity could be attributed to the competition between gluten and starch for water because of the excess water situation present in the RVA. Water to solids ratio is almost 8 to 1 so the effect is more likely to be a direct additive effect of gluten viscosity or an effect arising from interactions between gluten and starch. Probably, there is a combination of these affecting the viscosity. The size of the effects observed by Morris *et al.*[6] suggest that the contribution of the gluten *per se* to the viscosity is small and that the interaction is a major contributor.

When the glutenin sub-units of the different glutens were examined, it was observed that there were significant differences between the increase caused by glutens containing sub-units *e* and *i* from the *Glu-1B* locus, and between sub-units *a* and *d* from the *Glu-1D* locus (Table 2). The effects of high molecular weight sub-units from *Glu-1A*, and all of the low molecular weight sub-units appeared to be minimal. This was confirmed by analysis of variance.

Table 2. *Mean values of the paste viscosity of a commercial starch with the addition of glutens containing different glutenin sub-units.*

Locus	Allele	Peak viscosity	Final viscosity
Glu-1A	a	294	317 - 355
	b	294	312 - 358
Glu-1B	e	286	312 - 350
	i	301	331 - 358
Glu-1D	a	283	312 - 350
	d	303	331 - 358
Glu-3A	b	295	317 - 355
	e	293	312 - 358
Glu-3B	c	291	312 - 352
	d	299	332 - 358
Glu-3D	a	296	317 - 355
	b	293	312 - 358

3.2 Gelatinisation

Gelatinisation temperatures for wheat starch from Australian wheats are typically in the range 51-58°C (onset temperature, T_o) and 56-63°C (peak temperature, T_p) (I. Batey, unpublished results). When measured on flour, gelatinisation temperatures are usually higher. On the samples examined here, the onset gelatinisation temperature measured on flour ranged from 1.4-3.6°C higher than the onset temperature of starch isolated from that flour (Table 3). Likewise, the peak temperature for flour was 1.8-3.7°C higher than for starch.

Table 3. *Range of gelatinisation temperatures for flour and starch.*

	Starch	Flour	Difference
Onset temperature (°C)	54.0-56.7	55.6-58.8	1.4-3.6
Peak temperature (°C)	59.6-62.7	62.2-64.5	1.8-3.7
Final temperature (°C)	64.4-69.3	67.9-71.2	0.4-4.4

Initially. measurements were taken at a water to flour or starch ratio of 2:1. However, at this water content, it was thought that the competition for water between gluten and starch may have affected the gelatinisation temperatures of flour. On repeating the measurements at a water to solid ratio of 3:1, the same results were obtained, indicating that the increase in gelatinisation temperature in the flour was real. As in the case of viscosity, this effect was greater in flours containing certain glutenin sub-units. Statistically significant differences in the increases were observed for the alleles at *Glu-1B* and *Glu-1D* (Table 4).

Table 4. *Effect of different glutenin sub-units on the difference in gelatinisation temperatures of flour and starch.*

Locus	Allele	Onset temperature difference °C	Peak temperature difference °C
Glu-1A	a	2.6	2.9
	b	2.3	2.6
Glu-1B	e	2.3	2.6
	i	2.7	3.0
Glu-1D	a	2.8	3.0
	d	2.2	2.6
Glu-3A	b	2.7	2.9
	e	2.3	2.7
Glu-3B	c	2.4	2.7
	d	2.7	2.9
Glu-3D	a	2.5	2.8
	c	2.5	2.7

The use of techniques such as viscometry and differential scanning calorimetry is relevant in measurements of interactions between starch and gluten. It is the gelatinisation of starch during cooking that results in the textural properties of most wheat-based foods. Effects on this texture from the protein component are likely to occur during the gelatinisation process and measurement of this process could lead to an insight into the nature of the interactions between starch and gluten. While the findings presented here are not unequivocal proof of these interactions, they certainly do point in that direction. If the interactions are proven, the effect of certain alleles on the starch properties may be a partial explanation of why certain glutenin sub-units have a positive effect on bread quality.

References

1. X.C. Zhao, P.J Sharp, G. Crosbie, I. Barclay, R. Wilson, I.L. Batey, and R. Appels *J. Cereal Science,* 1998, **27**, 7
2. F. MacRitchie *J. Cereal Science,* 1987, **6**, 259
3. W.R. Morrison, in *Wheat is Unique: Structure, Composition, Processing, End-Use Properties, and Products*, ed. Y. Pomeranz, American Association of Cereal Chemists, St. Paul, MN, 1989 p. 193
4. S.P. Cauvain, in *Technology of Breadmaking*, eds. S.P. Cauvain and L.S. Young, Blackie Academic and Professional, London, 1998, p. 37
5. I.L. Batey, B.M. Curtin, and S.A. Moore *Cereal Chem.*, 1997, **74**, 497
6. C.F. Morris, G.E. King and G.L. Rubenthaler *Cereal Chem.*, 1997, **74**, 147

Acknowledgements

Samples used in this work were provided from the Grains Research and Development Corporation's National Wheat Molecular Marker Program.

INFLUENCE OF WHEAT POLYSACCHARIDES ON THE RHEOLOGICAL PROPERTIES OF GLUTEN AND DOUGHS

A.C. Gama, D.M.J. Santos and J.A. Lopes da Silva

Departamento de Química, Universidade de Aveiro - 3810-193 Aveiro, Portugal

1 INTRODUCTION

Gluten is considered to be the main component responsible for the viscoelastic properties of dough. However, the wheat polysaccharide fractions do not act as inert fillers, but influence the viscoelastic behaviour of the dough[1,2,3,4].

The importance of the mechanical input for dough development is well known. It affects protein conformation and, likely, the interactions among flour components. In this work, our main objective was to study how the wheat polysaccharide fractions affect the rheological properties of gluten and dough, following a different approach. We have studied gluten-starch-pentosan model systems at constant water content and hydration time, without any mechanical input except an initial gentle mixture of components in order to homogenise the samples.

2 MATERIALS AND METHODS

Flours from two Portuguese wheat varieties (*Amazonas* and *Sorraia*) were donated by ENMP (Elvas, Portugal). These flours were fractionated into three main components (starch, gluten and water-solubles) following a modified procedure based on that of Czuchajowska and Pomeranz[5]. Water-soluble pentosans (WSP) were isolated from the water-solubles fraction.

Reconstituted systems were prepared by making a blend of the isolated components and water. The samples were rested during 2 h, at 15°C, for hydration and moisture equilibration. Each sample was placed between the cone and plate geometry (angle 2°, diameter 2 cm) of the rheometer (AR 1000, TA Instruments), where it was rested for 30 min before testing, in order to relax any residual stresses. The sample edges were covered with paraffin oil to prevent drying. The study was conducted at 50% (w/w) water content for pentosan levels between 0 and 2%, on a constant gluten matrix basis, and for starch/gluten ratios between 6 and 19, at 1% pentosans. Rheological tests were done in shear, under small amplitude of deformation.

3 RESULTS AND DISCUSSION

The viscoelastic properties of the gluten-water samples are shown in Figure 1. The mechanical spectra were qualitatively, with the ratio tan δ=G"/G' being very similar for both samples, and the elastic character predominant over all frequency ranges analysed. Gluten isolated from Sorraia flour had higher moduli (both G' and G") than that from Amazonas. However, after heating, both gluten samples had similar moduli (Fig 1B). Thermal treatment led to a reinforcement of the network, with higher moduli and also lower values of tan δ, especially in the low frequency zone. However, the frequency dependence of the moduli did not vary appreciably. This effect may be due to the effect of heating on the gluten protein system, causing irreversible changes in the interactions between the protein molecules[6,7]. Gelatinisation of some residual starch may also contribute to the final viscoelastic profile of these systems.

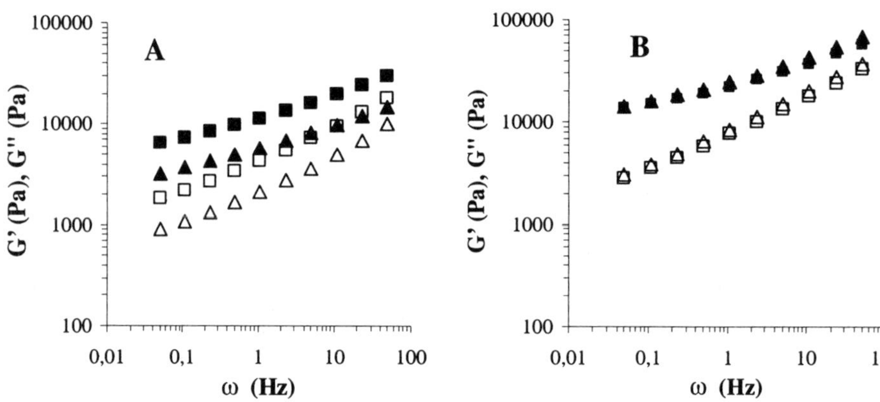

***Figure 1** - Mechanical spectra obtained at 20°C and 1% strain for Sorraia (squares) and Amazonas (triangles) gluten samples (50% (w/w) water content). Open symbols denote the loss modulus (G"), and filled symbols storage modulus (G'). **A** - Unheated samples; **B** - Spectra obtained after thermal treatment.*

Pentosans increase both the storage modulus and tan δ of the gluten networks (Fig. 2); however, these hydrophilic macromolecules do not appear to influence the thermal changes occurring in the gluten network around 60°C. In addition to any possible interactions between components, the different water affinities and sorption capacities of each might play a role in the viscoelastic behaviour of the unheated samples. The WSP exhibit the highest water sorption capacity and in spite of their low amount, may cause the gluten to act as if it contained less water. An increase in the G' of doughs caused by a decreasing water content has been previously reported[1,8]. In addition to increasing the storage modulus of the system, the presence of the WSP also increases the relative viscous character of the system, as shown by the higher tan δ values, especially in the lower frequency range.

In the presence of a large amount of starch, the effect of changing the amount of pentosans is not the same as observed for the gluten alone (Fig. 3A). In this case, and for the unheated samples, gluten association seems to be hindered by the presence of the hydrophilic polysaccharide chains. Pentosans increase the temperature at which the starch granules start to gelatinise (results not shown). A similar effect was observed for the reconstituted gluten+starch systems (Fig.3B).

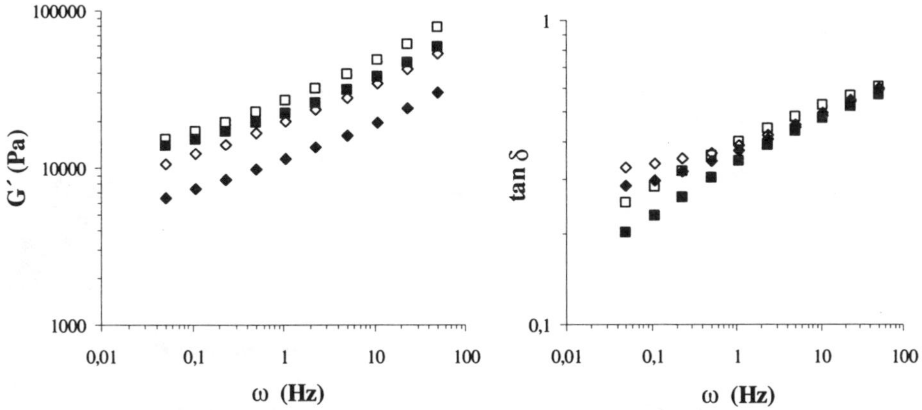

Figure 2 *- Effect of pentosans (1%) on Sorraia gluten viscoelasticity, for unheated (◊) and heated ( ) samples. Results obtained from frequency sweep tests at 20°C and 1% strain. Filled symbols denote samples without WSP, and open symbols samples with 1% WSP.*

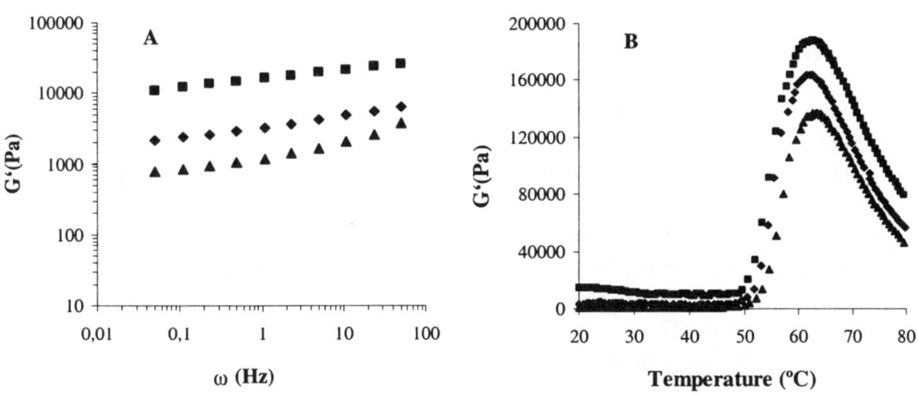

Figure 3 *- Effect of WSP content in reconstituted systems (gluten+starch+pentosans) at constant gluten contents (10%) for Sorraia flour fractions.* (■) *0.5% WSP,* (◆) *1% WSP,* (▲) *2% WSP.* ***A*** *- Mechanical spectra for the unheated samples;* ***B*** *- Changes in the storage modulus (G') during heating from 20°C to 80°C (2°C/min, ω=0.5 Hz, 1% γ).*

At constant WSP content, the relative amounts of starch and gluten clearly affect the rheology of these systems, especially the unheated samples. In this case, decreasing the gluten content clearly decreases the elastic character of the systems, an effect more pronounced for the Sorraia samples (Fig. 4A). After gelatinisation of the starch component (heated samples), changing the amount of gluten had a much smaller effect on the elasticity of the reconstituted networks. Changing the starch/gluten ratio had little effect on the temperature at which the starch fraction gelatinises (Fig. 4B).

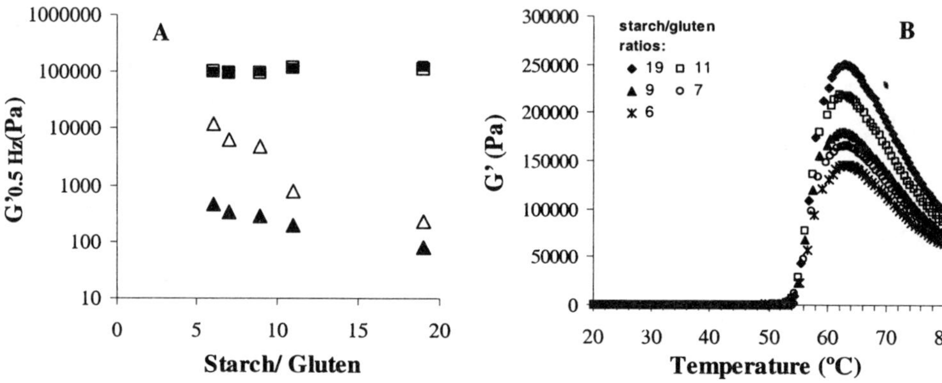

Figure 4 *- Effect of gluten/starch ratio in reconstituted systems at constant WSP contents (1%).* ***A*** *- Effect on G' of unheated (triangles) and heated samples (squares); filled symbols denote Amazonas, and open symbols Sorraia flour fractions.* ***B*** *- Thermal scans (2°C/min, ω=0.5 Hz, 1% γ) for the reconstituted systems. Results shown are for Amazonas flour fractions.*

References

1. P.C. Dreese, J.M. Faubion and R.C. Hoseney, *Cereal Chem.*, 1988, **65**, 348.
2. K.E. Petrofsky and R.C. Hoseney, *Cereal Chem.*, 1995, **72**, 53.
3. Y. Champenois, M.A. Rao and L.P. Walker, *J. Sci. Food Agric.*, 1998, **78**, 119.
4. K.A. Miller and R.C. Hoseney, *Cereal Chem.*, 1999, **76**, 105.
5. Z. Czuchajowska and Y. Pomeranz, *Cereal Chem.*, 1993, **70**, 701.
6. D. Schofield, J. Bottomley, M. Timms and M. Booth, *J. Cereal Sci.*, 1983, **1**, 241.
7. C. Larré, S. Denery-Papini, Y. Popineau, G. Deshayes, C. Desserme and J. Lefebvre, *Cereal Chem.*, 2000, **77**, 32.
8. G.E. Hibberd, *Rheol. Acta*, 1970, **9**, 497.

Acknowledgements

We thank FCT (Portugal) for financial support (PRAXIS XXI/PCNA/BIO/0703/96) and the National Station for Plant Improvement (ENMP) for providing the wheat flours.

EFFECT OF WATER UNEXTRACTABLE SOLIDS (WUS) ON GLUTEN FORMATION AND PROPERTIES. MECHANISTIC CONSIDERATIONS.

R.J. Hamer[1,2,3], M.-W. Wang[1,4], T. van Vliet[1], H.Gruppen[1], J.P. Marseille[3], P.L.Weegels[5]

1. Centre for Protein Technology, Wageningen University, Wageningen, the Netherlands. 2. Wageningen Centre for Food Sciences, Wageningen, the Netherlands. 3. TNO Voeding, Zeist, the Netherlands. 4. Department of Food Science, Wuhan Industry College, Wuhan, P.R.China. 5. Unilever Res. Lab, Vlaardingen, the Netherlands

1 INTRODUCTION

The gluten protein polymeric network plays a pivotal role in determining the end-use quality of wheat in many food products.[1] The properties of this polymeric network are strongly affected by wheat flour composition (protein, starch and pentosans etc.), ingredients (i.e. salt, fat), processing aids (i.e. enzymes) and process parameters (mixing time, mixing water, temperature). Changes in the quantity and properties of this polymeric fraction reflect changes in dough rheological properties and hence the quality of the original flour, the extracted gluten and the final product. Unravelling the underlying relationships and understanding gluten network formation is, therefore, of extreme importance. Most of the studies regarding gluten aggregation are related to proteins, reduction or oxidation conditions and processing. On the other hand, there is substantial evidence that water unextractable solids (WUS) also affect gluten formation and properties. WUS, consisting to a large extent of unextractable pentosans, are reported to have a strongly negative effect on bread-making quality[2]. It is for this reason that pentosan modifying enzymes are now widely used in bread-making. It is also known that the contents of WUS in wheat flour increase with increasing milling extraction. The wheat starch industry usually uses wheat flour of high extraction rate. Several theories aim to explain the effects of WUS, pentosans and pentosanase (such as water redistribution[3], covalent bonds between pentosans and proteins[4], interference with gluten aggregation[5]) based on either indirect or direct effects of pentosans. In all, the mechanistic action of WUS in relation to glutenin aggregation is far from clear and deserves further detailed study.

In this paper we report a miniaturised technique to study gluten aggregation. We used this technique to study the effects of WUS, a xylanase, and processing conditions on gluten yield, properties and composition of gluten. Based on these results, different mechanisms regarding the effect of WUS and xylanase on gluten aggregation and properties are discussed.

2 MATERIALS AND METHODS

2.1 Materials

Wheat flour was supplied by Meneba Meel BV. Xylanase I (batch ppj 4482) was a gift by NOVO Nordisk. The activities of xylanase, protease and amylase are 88.32±0.22U/mg enzyme, 5.86±0.02 U/mg enzyme and 4.72±U/g enzyme respectively.

2.2 Methods

2.2.1 Isolation of WUS preparations. WUS was isolated from wheat flour as described by Gruppen *et al*[6] . WUS(-) represents the starch containing isolate. WUS(+) is obtained from WUS(-) using amylase to remove starch. The AX content of WUS(-) and WUS(+) are 44% ± 0.4 and 73% ± 0.6, respectively. The water holding capacities of WUS(-) and WUS(+) are 11.9 g/g ± 0.3 and 12.g/g ± 0.2, respectively. WUS consists of various botanical components and its particle size ranges between 100 to 400 μm. A dispersion of WUS in water easily passes a 400 μm sieve.

2.2.2. Gluten yield experiments. Gluten formation was studied using a modified Glutomatic system, equipped with a custom-made pin mixer head. This allows the separation of 12 g of flour into starch and gluten mimicking the batter process on a miniature scale (see Figure 1). Gluten formation is critical in this system. Gluten is separated from the diluted batter with a 400μm sieve. This gluten yield is a measure of the rate of gluten formation. The system is also used to produce small quantities of gluten for rheological testing and chemical analysis.

2.2.3 Functional properties. Gluten samples were rheologically characterized using the miniature extensibility rig as developed by Kieffer[7].

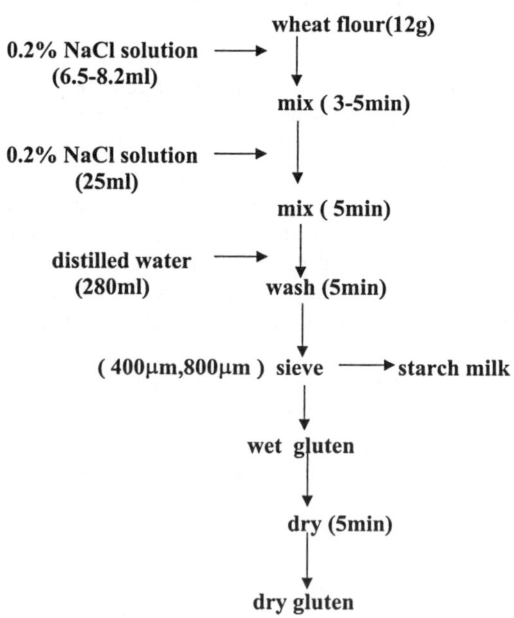

Figure 1. *Processing scheme of the modified Glutomatic system*

3 RESULTS AND DISCUSSION

3.1 Effect of WUS on gluten yield

3.1.1 Effect of amount and type of WUS on gluten yield. WUS(-) was added to flour at three levels, 1 %, 2 % and 4 %. In all cases a decrease in gluten yield was observed. A decrease of ca 20 % was obtained at 2% WUS(-) addition. Since no significant differences were observed between WUS (-) and WUS (+), 2%WUS(-) was mainly used in our experiments.

3.1.2 Compensation by increasing mixing time and mixing water. Addition of WUS can affect the availability of water for gluten development. Adding mixing water to correct for this also influences dough mixing conditions. We therefore studied both the effect of a higher water addition and an increased mixing time on gluten formation. The effect of WUS on gluten yield was maximal at 3 min mixing time and 7.7ml mixing water. This negative effect of WUS on gluten yield could be corrected to a large extent, but not completely, by increasing mixing time (2min) and mixing water (0.5ml) (see figure 2).

3.1.3 Correction by xylanase. Addition of xylanase to the control flour yields 8% more gluten. Xylanase also more than corrects for the loss of yield when 2 % WUS is added (see Table 1).

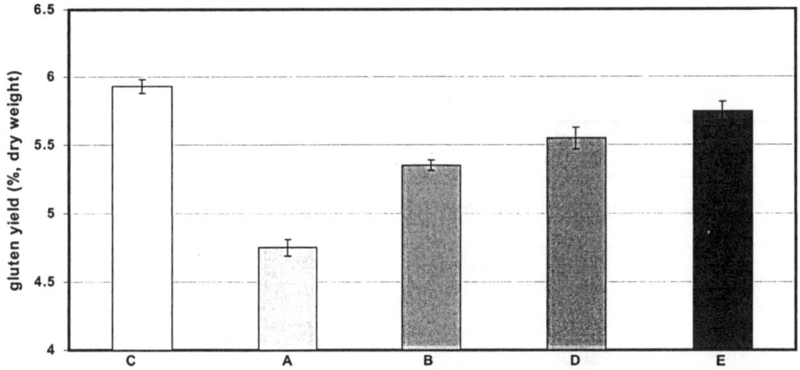

Figure 2. *Compensation for the effect of WUS by increasing mixing time and mixing water.*
A: 2% WUS(-), B: 2% WUS(-)+2min mix, C: control (3min mix, 7.7ml water), D: 2% WUS(-)+0.5ml water, E: 2% WUS(-)+2min mix+0.5ml water

3.2 Characterisation of gluten

3.2.1 Kieffer extensibility test. Addition of WUS typically produced a gluten with a higher maximum resistance to extension and a lower extensibility. Increasing mixing water and mixing time significantly increased maximum resistance to extension of gluten giving an extensibility comparable to the control. These results indicate that in addition to its water binding properties, WUS also has a direct effect on gluten properties. Addition of xylanase produced a more extensible gluten with a lower maximum resistance to extension. Adding both WUS and xylanase also resulted in a more extensible gluten with a lower maximum resistance to extension compared to gluten produced from flour with 2% of WUS (see Figure 3).

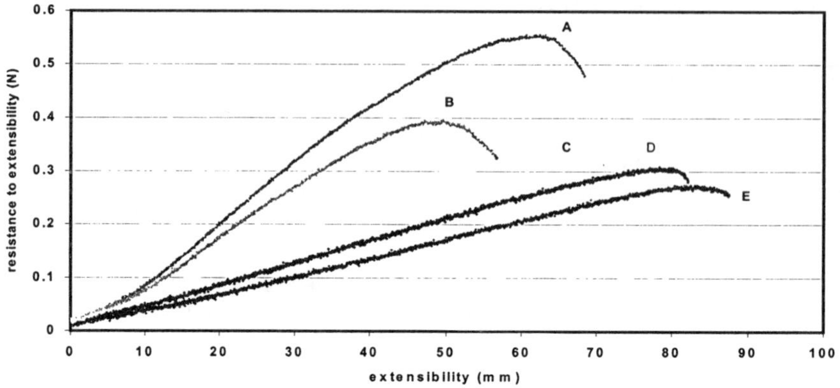

Figure 3. *The effect of WUS and xylanase on gluten rheological properties. Result of the Kieffer extensibility test.*
A: 2%WUS(-)+2min mix+0.5mlwater, B: 2%WUS(-), C: control (3min mix, 7.7mlwater), D: 2%WUS(-)+100ppm xylanase, E: 100ppm xylanase

3.2.2 Chemical composition. The chemical composition of different gluten samples is presented in Table 1.

Table 1. *The chemical composition and yield of different gluten samples*

Sample name	Content (D.M. ,%)			Yield (D.M. ,%)			
	Protein	Starch	AX	Gluten	Protein	Starch	AX
Control	86.2±0.36	2.7± 0.04	1.32±0.02	5.93±0.05	5.11	0.16	0.078
2%WUS(-)	86.9±0.22	2.9±0.04	1.67±0.06	4.73±0.07**	4.11**	0.14	0.079
2%WUS(-) +2minmix +0.5mlwater	86.7±0.24	3.5±0.07	1.45±0.01	5.72±0.04	4.96*	0.20*	0.083
100ppm xylanase	84.9±0.57	4.6±0.12	0.95±0.03	6.38±0.06*	5.42**	0.30**	0.060**
2%WUS(-)+ 100ppm xylanase	82.1± 0.63	8.9±0.20	1.01±0.02	6.67±0.05*	5.47**	0.60**	0.068*

Note: AX=Ara+xyl
Data are mean ± S.D. * means significant at $p< 0.01$. ** means significant at $p< 0.001$.

Addition of WUS led to a significant decrease in protein yield. Increasing mixing time and mixing water can improve protein yield, but cannot completely correct for the effect of adding WUS. Also, a longer mixing time led to a higher starch yield. Addition of xylanase resulted in a 6% increase in protein yield and a 23% decrease in AX yield, but was accompanied by an 88% increase in starch yield. The combination of WUS and xylanase improved protein yield, but resulted in an even higher starch yield. The effect of xylanase on protein yield is in agreement with earlier findings[9], but the effect of xylanase on starch yield has not been reported earlier.

4 CONCLUSIONS

WUS interferes with gluten formation in both a direct and an indirect way. WUS interferes indirectly by competing for water and changing the conditions for gluten development. This effect can be corrected for by the combination of adding more water and a longer mixing time. In addition, WUS has a direct effect by interacting with the protein particles forming the gluten. The particulate nature of WUS requires that the effect occurs through an interaction between WUS particles and gluten particles. This interaction interferes with the hyper-aggregation of these gluten particles to a continuous gluten protein mass and their subsequent covalent stabilisation. As a consequence, gluten yield is decreased and the resulting gluten sample has a lower extensibility. Both effects of WUS can be counteracted through the use of xylanase.

References

1. F. MacRitchie, *Cereal Foods World*, 1999, **44**, 188
2. X. Rouau, M-L. Ei-Hayek and D. Moreau, *J. Cereal Sci.*, 1994, **19**, 259
3. J. Maat, M. Roza, J.Verbakel, J.M. Santos da Silva, M. Bosse, M.R. Egmond, M.L.D. Hagemans, v. R.F.M. Gorcom, J.G.M. Hessing, v.d. C.A.M.J.J. Hondel,.and v. C. Rotterdam, in 'Xylans and Xylanases', Progress in Biotechnology Series Vol.7, (J.Visser, G.Beldman M.A.Kusters-van Someren and A.G.J. Voragen, eds.), Elsevier, Amsterdam,1992, p. 349
4. R.J. Hamer, 'Enzymes in the baking industry. In Enzymes in Food Processing', (G.A. Tucker and L.F.J. Woods, eds.) Blackie, Glasgow and London. 1991, p. 168
5. K.A. Tilley, G.L. Lookhart, R.C. Hoseney,and T.P. Mawhinney, *Cereal Chem.*, 1993, **70**, 602
6. H. Gruppen, J.P. Marseille, A.G.J. Voragen, R.J. Hamer, and W. Pilnik, *J. Cereal Sci.,*1990, **9**, 247
7. R. Kieffer, H. Wieser, M.H. Henderson and A. Graveland, *J. Cereal Sci.,*1998,**27**, 53
8. S.K.Patil, C.C. Tsen and D.R. Lineback, *Cereal Chem.*,1975, **52**,44
9. P.L. Weegels, J.P. Marseille and R.J. Hamer, *Starch /Staerke,* 1992, **4** ,44

THE IMPACT OF WATER-SOLUBLE PENTOSANS ON DOUGH PROPERTIES.

Wim J. Lichtendonk[1], Marcel Kelfkens[1], Roel Orsel[2], August C.A.P.A. Bekkers[3] and Johan J. Plijter[1]

1. TNO Nutrition and Food Research, Food Technology Department, P.O.Box 360, 3700 AJ Zeist, The Netherlands. 2. Present address: Quest International, P.O. Box 2, 1400 CA Bussum, The Netherlands. 3.Present address: Heineken, P.O. Box 530, 2380 BD Zoeterwoude, The Netherlands

1 INTRODUCTION

For wholemeal flours, the baking potential is usually lower than for white flours from the same wheat batch. Many studies have been performed trying to explain this phenomenon. Effects of fibres and water-soluble bran components such as pentosans and enzymes have all been reported. Also, the physical damaging of gas cells by bran components has been mentioned as a possible cause[1]. Thus, the lower baking potential of wholemeal flours is probably caused by a combination of factors.

In order to achieve quality control of wholemeal flours, e.g. by adjusting milling and blending procedures, a better understanding is required of the contribution of various factors is needed. These factors, especially the pentosans, have been topic of much recent research [2,3]. These studies, however, only scarcely describe the effect of the water-soluble pentosans on the visco-elastic characteristics of dough. Also, the interaction of the pentosans with the gluten network as present in dough is not well described. Nevertheless, the increased use of pentosan degrading enzymes in the baking industry underlines their technological importance. Thus, there is a gap in our knowledge regarding the functionality of pentosans in dough systems.

In this study, we set out to elucidate the effects of water-soluble pentosans on the visco-elastic properties of dough linked to their interaction with the gluten network.
As shown earlier in our lab. the rheology of Glutenin Macro Polymer (GMP) from flour gives important information about the quality of the flour. With the aid of this technique, we were able to determine the influence of pentosans and Xylanase on gluten functionality.

2 MATERIALS AND METHODS

All chemicals were purchased from Merck and were of analytical grade. Amyloglucosidase and pronase were purchased from Boehringer Mannheim and Merck respectively. Milli-Q (Millipore) purified water was used. Xylanases were obtained from Quest.

2.1 Preparation of water-soluble pentosans

Water-soluble pentosans were isolated from Hereward flour by extraction using trichloroacetic acid (TCA)[4], with modifications. Ten grams of flour were extracted using 40 ml of 18% TCA (w/v) for 16 hours at 4 C. After centrifugation, the supernatant was brought to a concentration of 75% ethanol (v/v) and the mixture was kept at 4 C for 16 hours in order to allow the pentosans to precipitate. The residue was isolated using centrifugation and washed with 70% ethanol at 4 C and dissolved in 50 ml of 50 mM Citrate buffer pH 4.6. The mixture was incubated using 14 units of amyloglucosidase for 16 hours at ambient temperature in order to remove traces of soluble starch. The pH of the mixture was brought to 7.5 using NaOH and 7 units of Pronase were added and allowed to react for two hours at 40°C. After the reaction, the mixture was dialyzed against water for 48 hours using a tubing with a cut-off of 3.5 kD and several changes of water. After dialysis, the sample was lyophilized.

2.2 Dough mixing and resting

Flour from the cultivar Camp Remy'92 was used. Dough was mixed in a 10 gram mixing bowl using a Brabender Plastograph at a constant temperature of 30° C. Water addition according to 500 BU after 5 minutes mixing at 30°C was used. Prior to the preparation of the dough, the flour, salt and Pentosans were premixed during 2 minutes, after which water was added. The dough were prepared at a fixed mixing time of 6 minutes and a mixing speed of 63 rpm. The dough were allowed to rest for 45 minutes, packed in plastic bags in a water bath at 30°C. After 45 minutes the dough were measured by Stress-Relaxation measurements in the Bohlin OR. For rheological studies on the 1.5% SDS insoluble glutenin fraction, doughs were frozen in liquid nitrogen, freeze-dried and milled by a Retch hammermill using a 0.25 mm sieve.

2.3 Flow-relaxation measurements

The Stress-Relaxation measurements were carried out using a Bohlin VOR strain controlled rheometer, with serrated parallel plates and a section of 30 mm. For Stress-Relaxation measurements, a piece of dough was brought into the rheometer between the serrated plates with a gap of 2 mm and covered with paraffin oil. To allow relaxation stress caused by insertion of dough into the rheometer, an equilibrium time of 10 minutes was used. The temperature was 30°C. The sample was deformed by a strain of 100% at a shear rate of 0.0208 1/s. After the deformation the peak stress was measured indicating the stiffness of a dough. The decrease of the stress in the dough caused by relaxation was measured during 1 minute. The time necessary for the dough to relax to a stress of 50 % of the peak stress was calculated.

2.4 GMP preparation and rheology

For the preparation of the Glutenin Macro Polymer (GMP), freeze-dried dough flour was suspended in 1.5% SDS and centrifuged at 80.000 g for 30 minutes in a Kontron Ultracentrifuge[5]. After centrifugation, the supernatant was decanted and the GMP isolated as a gel-like layer on top of the starch. For the rheological characterization of GMP, 1 gram of the material was carefully scraped off from the top of the gel and transferred into a Bohlin VOR rheometer between parallel plates with a gap of 1 mm. at 20°C. The behaviour of G' and delta was observed in a frequency sweep and a strain sweep. For comparison of GMP from different doughs, we used G' derived at 2% Strain and 0.15 Hz (linear area of measurement). The overshoot measurement was also carried out in the Bohlin VOR, but in a C8 concentric cylinder geometry at a temperature of 20°C and a shear rate of 0,146 1/s.

3 RESULTS AND DISCUSSION

In order to measure the effect of water-soluble pentosans on the dough rheological properties, we developed a novel testing procedure, the Stress-Relaxation test. In these measurements, the relationship between deformation (strain) and force (stress) as a function of time is studied. In case of a highly elastic dough, the stress built up after deformation of the sample will be stored for a longer time and relaxes only slowly. The relaxation half time, defined as the time needed for a dough to relax to a stress of half that of the peak stress, is therefore a measure of dough elasticity and was found to be very characteristic for the formulation of the dough. Dough is a visco-elastic material and the resultant stress after the relatively large deformation is a combination of viscous and elastic behaviour. It is generally accepted that the glutenin in the dough is responsible for the main part of the elastic behaviour. Thus, the Stress-Relaxation measurements yield information about the quality of the glutenin as present in dough. Different amounts of water in dough give a straight line in the Stress-Relaxation plot. Figure 1 shows that an increasing amount of water results not only in a decrease of stiffness, but also in a decrease of elasticity. The dough stiffness and elasticity as determined by the Stress-Relaxation measurements are shown for water additions ranging from 47.9 to 57.9 %. The Farinograph water absorption for the Camp Remy flour was 51.9 %. Water addition, as every baker knows, reduces the stiffness of a dough and allows correction for dough consistency. However, as shown in Figure 1, elasticity is also reduced. This may exert negative effects on the baking performance, as well as on the dough handling properties.

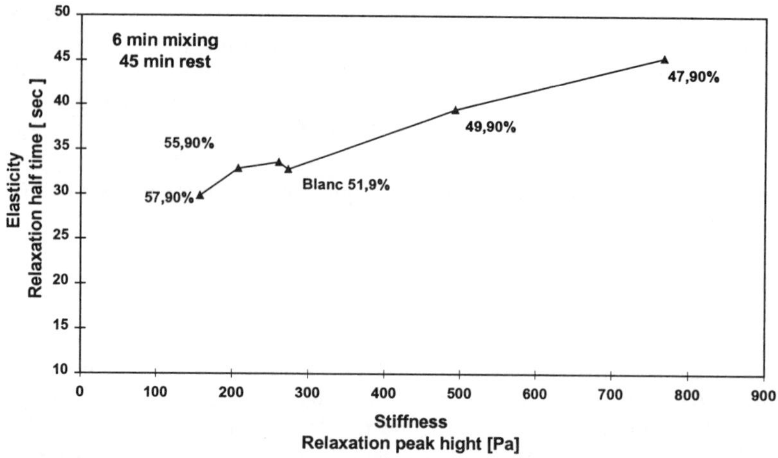

Figure 1 *The influence of water on the elasticity and stiffness of dough's*

In order to study the functionality of pentosans in dough, the rheological properties of dough supplemented with pentosans were studied. Upon adding 0.5 to 1.5% water-soluble pentosans, dough elasticity is reduced whereas the stiffness is increased (see Figure 2). The increase in dough stiffness is the result of water-binding by pentosans, similar to reduced water additions. As a control experiment, doughs were prepared with an extra 4% water added for every percent of pentosans. Clearly, the effect of pentosans on dough stiffness can be corrected for completely. However, the reduced dough elasticity cannot be explained based on the water binding capacity of the pentosans. As shown in Figure 1, a lower amount of water would increase the elasticity of the dough. Apparently, the pentosans interfere with the elastic components of a dough, i.e. the glutenin polymer. The experiments show that pentosans can markedly change dough properties.

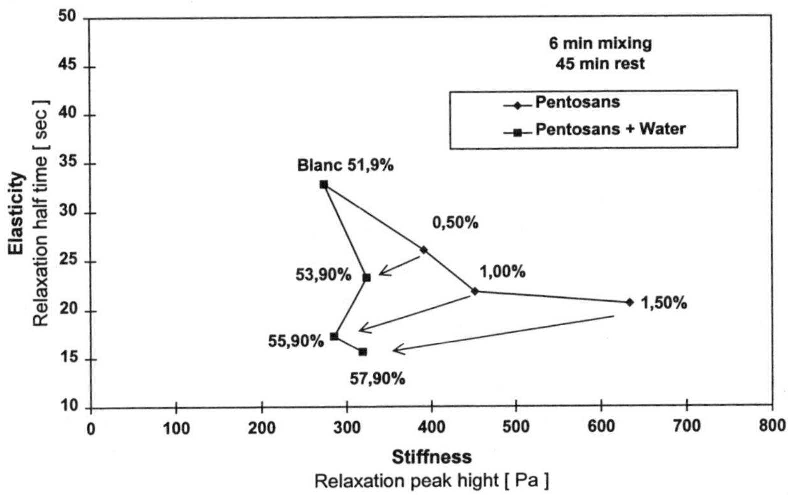

Figure 2 *The influence of pentosans and the addition of extra water on the elasticity and stiffness of a dough*

Pentosan hydrolyzing enzymes are increasingly being used in the modern bakery. In addition to their application potential, these enzymes can be used to study pentosan functionality in dough systems. In our study, we used Xylanase II. Addition of this enzyme up to 12 ppm increased elasticity and reduced stiffness (see Figure 3). A similar effect was found after adding xylanase to a dough supplemented with an additional 0.5% of added pentosans. Obviously, the xylanase employed is able to reduce the effect of added pentosans. The hydrolytic activity of the xylanase breaks down the pentosans which, as a result of this modification, have a lower water-binding capacity. Thus, xylanase activity in dough implies the release of water *in situ*, resulting in reduction of dough stiffness similar to the addition of extra water to the dough. However, xylanases can more than cancel out the negative effect of pentosans as doughs with added water-soluble pentosans and an increasing amount of xylanase are even more elastic than the blank. This indicates that xylanase activity eliminates the interference of the pentosans with the elastic components in dough. An explanation could be a reduction in the steric hindrance of pentosans in the gluten networks. The results shown in Figures 1 to 3 demonstrate the significant influence of water, pentosans and xylanase on dough quality properties such as elasticity and stiffness.

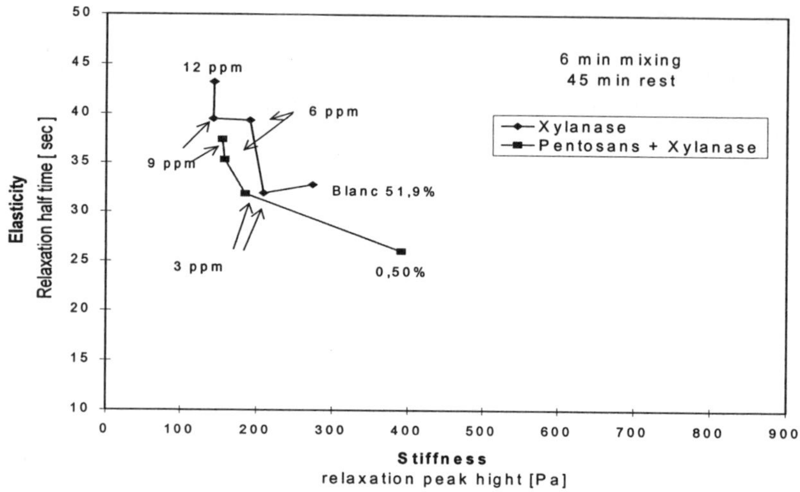

Figure 3 *The influence and dosage effect of xylanase on the elastic behaviour of a dough with or without additional added pentosans.*

The effect of water-soluble pentosans on the gluten network was further studied using GMP isolated from dough. During dough mixing the GMP is broken down into smaller polymer units. This has effects on the solubility of GMP in SDS. The mixing time needed to bring all GMP in solution depends on the variety, mixing speed, consistency of the dough and dough temperature. For Camp Remy we showed that the GMP gel layer is completely in solution after 10 minutes mixing. The doughs were divided in parts to rest for 30, 60 and 90 minutes. GMP was prepared by freezing the doughs in liquid nitrogen and freeze-drying. Subsequently, the dried doughs were milled in a hammermill. The dough powder was suspended in SDS 1.5% and centrifuged at 80.000 g. The result is a pellet of starch, with a gel layer of GMP on top. After pouring off the supernatant, the top of the gel can be scraped off carefully and put between the plates of the rheometer.

We see a reformation of the gel layer after dough rest. Figure 4 shows that the G' of GMP of dough is much lower than the G' of flour. The GMP is therefore, aggregated during resting of the dough. This aggregation is not only shown by an increasing amount of gel layer, but also by an increase in the G' of the gel layer during the resting time of the dough.

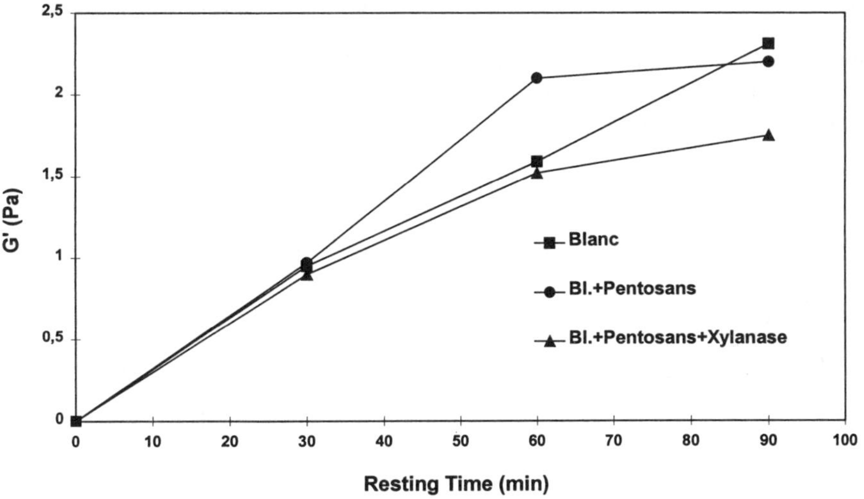

Figure 4 *Impact of pentosans and xylanase on G' of GMP*

After 30 minutes of rest we still did not find much difference between the dough samples, but at longer resting times more differences were found between xylanase treated dough and the other doughs such as the blank and the blank with extra pentosans added.

When xylanase is added, the GMP stays weaker. However, as can be seen in Figure 5, although the delta is lower, the GMP of the xylanase treated dough shows slightly more elastic behaviour.

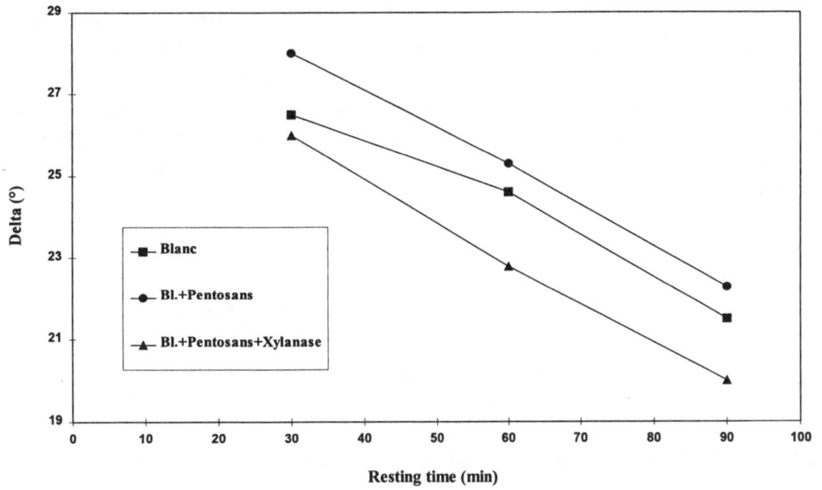

Figure 5 *Impact of pentosans and xylanase on delta of GMP*

Another method of determining the differences in the GMP samples is by overshoot measurement. An example is shown in Figure 6. This shows how pentosan hindrance occurs when GMP is continually under shear at a certain rate. On the other hand, it seems that xylanase promotes the orientation of the GMP polymers in the shear direction, so that the viscosity is decreasing faster.

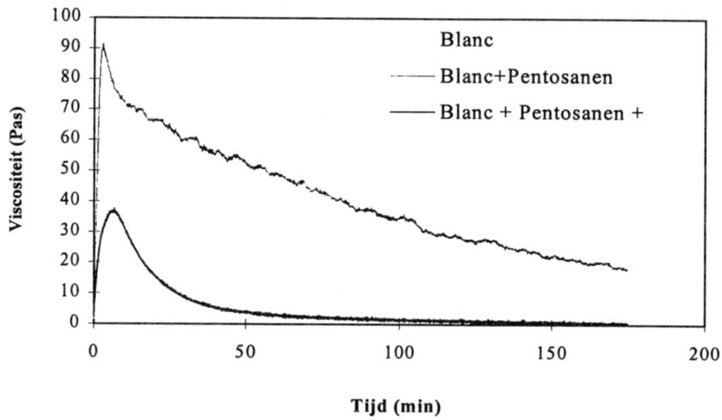

Figure 6 *Overshoot measurent on GMP from a dough which has rested for 60 min.*

4 CONCLUSION

Dough is a very concentrated system. The limited amount of water seems not to give the pentosans enough space to come fully in solution. An amount of 4 % water for every 1 % of pentosans is enough to give the same stiffness of the dough, but is probably not enough to dissolve the GMP. Therefore, it is reasonable to imagine that the pentosans will also form a polymeric gel. This could explain the higher G' of the gel layer of the dough with added pentosans. This gel, however, seems to have a more viscous and less elastic behaviour. At the same time it interferes with the GMP gel layer to give less elastic behaviour. It seems that xylanase changes the behaviour of GMP in a dough to that of a more linear polymer.

Literature

1 Z. Gan, T. Galliard, P.R. Ellis, R.E. Angold and J.G. Vaughan. *J. Cereal Sci.*, **15**, 1992, 151.

2 C.M. Courtin and J.A. Delcour. *J. Agric. Food Chem.*, **46**, 1998, 4066.

3 C.M. Courtin, A. Roelants and J.A. Delcour. *J. Agric. Food Chem.*, **47**, 1999, 1870.

4 A.H. Tran and P. Nordin. *Die Stärke*, **5**, 1977, 153.

5 A. Graveland, P. Bosveld, W.J. Lichtendonk and J.H.E. Moonen. *J.Sci. Food Agric.* **33**, 1982, 1117.

ISOLATION OF A NOVEL, SURFACE ACTIVE, M_r 50k WHEAT PROTEIN

J.E. van der Graaf[1], Z. Gan[1], J. Wykes[2] and J.D. Schofield[1].

1. The University of Reading, Department of Food Science and Technology, P.O. Box 226, Whiteknights, Reading, RG6 6AP, U.K. 2. RHM Technology Ltd, The Lord Rank Centre, Lincoln Road, High Wycombe, Bucks, HP12 3QR, U.K.

1 INTRODUCTION

The quality of leavened bread is judged by loaf volume and loaf volume depends upon the amount of gas held in the dough matrix during the proving and early baking stages. Wheat bread dough is a multiphase and multicomponent system composed of proteins, lipids, polysaccharides and other minor components and additives. During mixing and proving, dough develops a foam structure, the stability of which must be maintained until the later baking stages[1]. The stability of the foam structure depends upon several molecular species; the most important of those components, apart from gluten, are the non-starch polar lipids and surface active proteins.

The aqueous phase of dough, dough liquor, prepared by the ultra-centrifugation of dough, has long been known to have a beneficial effect on loaf volume[2]. This liquor contains most of the wheat flour solubles including proteins, which play an important role in gas retention during the proving and baking of bread[1, 3.] Using a foaming technique[4] three major size classes of these surface active proteins have been identified with approximate M_r of 66k, 50k and 37.5k (Figure 1). The proteins with approximate M_r 50k were observed to be particularly concentrated in the foam skeleton, despite having a limited presence in the total dough liquor, implying that these proteins are particularly surface active. Amino acid sequencing indicates that the N-terminal sequence of the protein has not been described previously. The aim of this project was to develop a purification procedure for this group of proteins that avoided the foaming step and that potentially could lead to surface denaturation.

2 MATERIALS AND METHODS

The flour used to develop an isolation procedure for the M_r 50k surface active protein was from the cultivar Hereward and supplied by RHM Technology Limited, The Lord Rank Centre, High Wycombe, Bucks, U.K.

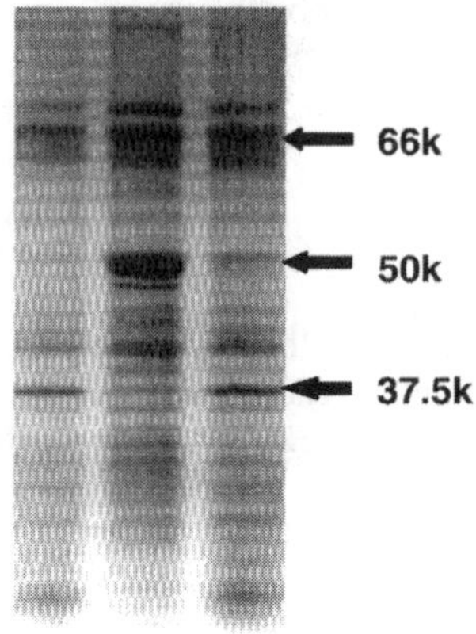

Figure 1

SDS-PAGE, under reduced condition, of proteins in (from left to right) the drained bulk phase, the foam skeleton and the total dough liquor[4].

3 RESULTS

After a saline extraction from flour a two-stage ammonium sulphate precipitation was carried out. SDS-PAGE showed the M_r 50k proteins to be concentrated by this procedure.

Trials were carried out on several combinations of buffer system to assess the degree of solubility of the protein in each. Several chromatographic techniques were then examined as the next step in the purification. These included anion and cation exchange, reversed-phase HPLC, chromatofocusing, size exclusion, hydrophobic interaction and gel-filtration. The degree of purification was monitored at each stage using SDS-PAGE.

The purification scheme finally adopted was ammonium sulphate precipitation followed by anion exchange chromatography on a PE Biosytems Poros HQ column with a buffer system comprising 1M sodium chloride, pH 6.3 (0.025M bis-Tris) with a salt gradient from 0M to 0.25M. The final hydrophobic interaction chromatography step used a Poros PE column with a buffer system comprising 1M ammonium sulphate, pH 7 (0.020M phosphate buffer) with a gradient from 1M to 0M. This resulted in a homogeneous protein as judged by SDS-PAGE.

Additional characterisation, together with baking trials will identify the potential functionality of this novel protein.

References

1. Z. Gan, P.R. Ellis and J.D. Schofield, *J. Cereal Sci.*, 1995, **21,** 215.
2. J.C. Baker, H.K. Parker and M.D. Mize, *Cereal Chem.*, 1946, **23,** 16.
3. F. MacRitchie, *Cereal Chem.*, 1976, **53,** 318.
4. Z. Gan and J.D. Schofield, in *Gluten '96,* ed. C.W. Wrigley, Royal Australian Chemical Institute, Cereal Chemistry Division, Melbourne, 1996, pp 379-382.

Acknowledgements

We acknowledge financial support from the Biotechnology and Biological Sciences Research Council and RHM Technology Ltd.

STARCH ASSOCIATED PROTEINS AND WHEAT ENDOSPERM TEXTURE

H.F. Darlington[1], H.A. Bloch[1,2], L.I. Tesci[1] and P.R. Shewry[1]

1. IACR-Long Ashton Research Station, Department of Agricultural Sciences, University of Bristol, Long Ashton, Bristol, BS41 9AF, UK. 2. Technical University of Denmark, Department of Biochemistry and Nutrition, Building 224, DK-2800, Lyngby,Denmark.

1 INTRODUCTION

Endosperm texture or hardness is an important quality characteristic in cereals. In wheat, grain hardness affects a range of characters including milling and end use properties. Hard wheats are preferred for bread making while soft wheats are used for cake and biscuit making. It has been shown that softness and hardness in wheat are associated with the presence or absence, respectively, of an M_r 15 000 protein (called friabilin) on the surface of water-washed starch granules[1]. Further work has shown that friabilin comprises a mixture of proteins, the major components being two hydrophobic proteins called puroindolines-a and -b [2] The molecular basis for the role of puroindolines in determinng grain texture is not understood in detail but they are thought to affect the strength of binding between the gluten (matrix) proteins and the starch granule surface. Thus, in soft wheats this binding is weak and the granules are readily released during milling. Starch granules are usually prepared by water washing which could potentially lead to redistribution of friabilin/puroindoline. We have therefore compared the amount, composition and distribution of friabilins on granules prepared using a dry sieving procedure. We have also studied the protein binding properties of water washed starch granules *invitro* in order to develop a routine method for exploring the molecular basis for starch:protein interactions.

2 MATERIALS AND METHODS

2.1 Materials

The uncoupled anti-friabilin monoclonal antibody (as used in the 'Durotest P' kit) was obtained from Rhone-diagnostics Technologies Ltd, Glasgow, UK. β-amylase (cat. no. A7130), bovine serum albumin (cat. no. A7096) and wheat grain α-amylase inhibitors (cat. no. A1520) were obtained from Sigma-Aldrich Co. Limited, Poole, UK. Total wheat albumins were prepared by mixing 40mg flour from cv. Mercia or Riband with 1ml water and suspending in a sonic bath for 30 minutes.

2.2 Methods

2.2.1 Non-aqueous Starch Separation. Milled flour samples were passed through a 38μm sieve and the <38 μm fraction retained as the non-aqueously extracted crude starch sample.

2.2.2 Friabilin Quantification. Friabilin was extracted from wheat flour and starch samples (50mg) using 1M sodium chloride (0.5ml) with suspension in a sonicating water bath for 30min. The supernatant was diluted 1:1000 and applied to a nitrocellulose membrane (200μl) using a Bio-Rad slot blot apparatus. The membrane was then probed using the anti-friabilin antibody and the density of the bands (OD) measured using a Bio-Rad 'Gel-Doc' system with 'Molecular Analyst' software.

2.2.3 Immnofluoresence Microscopy. Starch samples were attached to poly-L-lysine coated slides (BDH, UK) and probed with the anti-friabilin antibody and TRITC conjugate secondary antibody. Slides were viewed for immunofluoresence under a confocal epifluoresence microscope (Leica TCS, Germany).

2.2.4 Prime Starch Separation (Aqueous). Starch was extracted with water from wheat flour using the 'dough ball' method[3]. The resulting starch was then incubated with Proteinase K to remove any other residual proteins.

2.2.5 Puroindoline Preparation. Puroindolines were prepared from wheat flour using the method of Blochet *et al.* 1993[4].

2.2.6 Starch/Protein Binding. Starch (30mg) was preincubated with proteinase inhibitor (100mM phenyl methyl sulphonyl fluoride (PMSF) in 1ml 0.1M Tris/HCl buffer, pH 5.5) for 1 hour at room temperature to inhibit any residual Proteinase K activity. The washed pellet samples were then incubated with puroindolines or other proteins (0.5mg/ml in 0.1M Tris/HCl buffer, pH 5.5) overnight at 4 °C.

3 RESULTS AND DISCUSSION

3.1 Friabilin and starch

Friabilin levels were similar in flours from hard (Mercia) and soft (Riband) wheats, but higher amounts were present on the starch granules from soft wheat (Table 1). Greenwell and Schofield[1] demonstrated that friabilin was present on the surface of water washed starch granules from soft wheat but not hard wheat. The starch granules used in this study were extracted without water, but the same difference was observed. It was also shown by immunofluoresence confocal microscopy that the friabilin present in the starch preparations was associated with the starch granule surface (Figure 1).

3.2 Puroindoline and starch

To investigate the interactions between protein and starch, an *in-vitro* binding system was developed. Puroindoline preparations from hard (Mercia) and soft (Riband) wheat varieties were incubated together with prime (i.e. water-washed) starch granules from the

Table 1 *Total friabilin contents of Mercia and Riband wheat flour and non-aqueously prepared starch*

	Friabilin Content (OD%)	
	Whole Flour	Non-Aqueously Prepared Starch
Mercia (hard)	14.70	1.77
Riband (soft)	16.59	12.08

same wheat varieties. *N*-terminal amino acid sequencing of the puroindoline preparations showed that they contained puroindoline-a, 0.18/0.29 α-amylase inhibitor and purothionins. Following incubation, only puroindoline-a and purothionin bound to the starch. No differences were observed between the puroindoline or starch preparations from the hard or soft wheat varieties. Other proteins (BSA, wheat α-amylase inhibitor, barley β-amylase and wheat albumins) failed to bind to either starch granule preparation when incubated under the same conditions as the puroindoline preparations (Table 2). Similarly, no statistically significant differences were found between the two varieties in the concentration dependence of puroindoline binding, with a linear relationship being observed for concentrations between 0.125 and 2mg/ml and around half of the protein binding at each concentration.

Table 2 *Binding characteristics of prime starch from hard (Mercia) and soft (Riband) wheat varieties with puroindoline preparations and other proteins.* (✓ = binding ✕ = non-binding)

	Mercia starch	Riband starch
Proteins in Puroindoline Preparation		
Puroindoline-a	✓	✓
0.19/0.28 α-amylase inhibitor	✕	✕
purothionins	✓	✓
Other proteins		
BSA	✕	✕
α-amylase inhibitor	✕	✕
β-amylase	✕	✕
Wheat albumins	✕	✕

4 CONCLUSIONS

Friabilin levels differ in starches prepared from hard and soft wheats using a non-aqueous procedure. These results agree with the initial findings of Greenwell & Schofield[1] and show that the differences are not due to redistribution of friabilin during washing of aqueously prepared granules. Immunofluoresence confocal microscopy showed that the friabilin in these preparations was located on the starch granule surface. It was also shown that wheat starch granules bind puroindoline-a (a major component of friabilin) and purothionins *invitro*, but failed to bind a range of 'control proteins'. No differences were observed in the *invitro* binding of starch or puroindoline preparations from hard and soft wheat varieties, although clear differences in *invivo* binding were observed. This may indicate that the binding of puroindolines to starch requires a specific biological process or that other endosperm components present at the starch/storage protein interface may also play a role in determining binding, for example, lipids. The *invitro* binding method used in this work should facilitate the identification of such components and the determination of the molecular basis for starch:protein binding and hence grain texture.

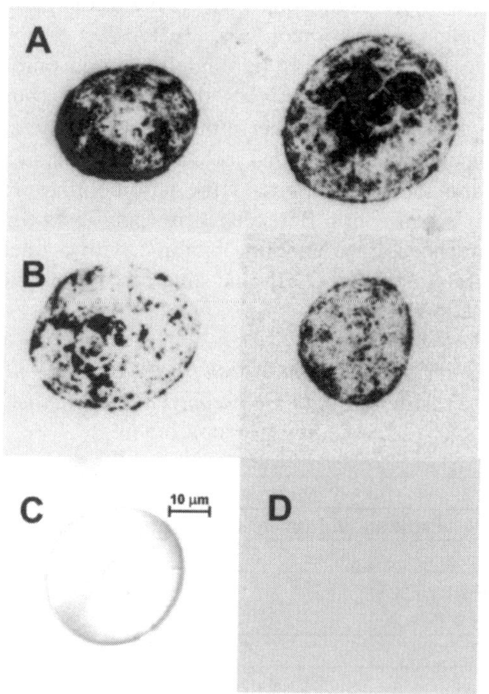

Figure 1 *Immunofluoresence confocal microscopy of non-aqueously prepared starch granules from A. Riband (soft wheat), B. Mercia (hard wheat) D. Ofanto (durum wheat). C. light microscopy of a starch granule from Riband. Friabilin is shown in black.*

References

1. P. Greenwell and J.D. Schofield. *Cereal Chemistry* 1986, **63**, 379.
2. M.J. Giroux, and C.F Morris. *Theoretical and Applied Genetics* 1997, **95**, 857.
3. M.J. Wolf. Wheat starch. In: Methods of Carbohydrate Chemistry vol 4. ed R.L. Whistler. Academic Press, New York, 1984, pp 6-9.

4. J.-E. Blochet, C Chevalier, E. Forest, E. Pebay-Peyroula, M.-F. Gautier, P Joudrier, M. Pézolet, and D.Marion. *FEBS Letters* 1993, **329**, 336.

Acknowledgements

IACR recieves grant-aided support from the Biotechnology and Biological Sciences Research Council of the United Kingdom.

INSECT AND FUNGAL ENZYME INHIBITORS IN STUDY OF VARIABILITY, EVOLUTION AND RESISTANCE OF WHEAT AND OTHER *TRITICEAE* DUM. CEREALS

Al.V. Konarev

All-Russian Institute for Plant Protection (VIZR), Podbelsky 3, St.Petersburg, 189620 Russia.

1 INTRODUCTION

Plant seeds contain various proteinaceous inhibitors of insect, fungal, mammalian and endogenous α-amylases and proteinases. Some 12 families of inhibitors can be recognized based on their amino acid sequences and target proteinases, including a group of cereal inhibitors that are related to prolamins[1]. Thc inhibitors may be involved in plant defense systems against harmful organisms[2,3] and may also play regulatory roles during plant development. Furthermore, plant inhibitors are of interest in relation to problems of host/parasite co-evolution[4], as markers in studies of plant diversity and evolution[4,5,6,7,8] and as potential drugs with antiviral and other activities. The biochemical properties of some hydrolase inhibitors are well studied in various plant families but the evolutionary variability of inhibitors in wheat and other cereals has not been investigated in detail.

In present work the polymorphism, biochemical properties, distribution, variability and genetic control of α-amylase and proteinase inhibitors were studied in lines, varieties and wild species of wheat and other representatives of tribe *Triticeae* Dum. using simple and effective methods. Special attention was given to inhibitors of insect α-amylases and cysteine proteinases, trypsin and chymotrypsin which are typical digestive enzymes of insects, mammals and fungi, and to inhibitors of subtilisin, a proteinase of microorganisms.

2 MATERIALS AND METHODS

More than 600 seed accessions of wheat (*Triticum* L.), *Aegilops* L., rye (*Secale* L.,), barley (*Hordeum* L.), *Elytrigia* Desv., and *Agropyron* Gaertn. were obtained from Vavilov Institute of Plant Industry, St.Petersburg. Insect amylases and serine and cysteine proteinases were isolated from 7 *Coleoptera* and 4 *Hemiptera* species. Fungal proteinases were represented by extracellular trypsin- and subtilisin-like enzymes of several *Aspergillus* species. Trypsin, chymotrypsin, subtilisin, papain, ficin, human saliva and pig pancreatic amylases were from Sigma and other suppliers.

Most of the study was carried out using simple and sensitive methods for detection of amylase and proteinase inhibitors among plant proteins, separated by isoelectric focusing

(IEF), electrophoresis or thin layer gel-filtration. These methods were also effective at all stages of purification of novel inhibitors by gel-filtration, affinity chromatography and other techniques. Amylase inhibitors were detected with polyacrylamide gel replicas from separating gels containing starch and α-amylase[6,9]. Proteinase inhibitors were detected with gelatin replicas from separating gels developed by proteinases[10, 11,9].

3 RESULTS AND DISCUSSION

Inhibitors (I) of various insect, fungal, mammalian and plant amylases (A) and proteinases including trypsin (T), chymotrypsin (C), subtilisin (S), cysteine proteinases (CP) and bifunctional inhibitors were identified in wheat and related cereals (Table 1). In wheat, the spectra of various inhibitors were found to be specific for individual species, genomes or even varieties and reflected evolutionary links between diploid and polyploid *Triticum* and *Aegilops* species (Figure 1). Monogenic control was characteristic for the majority of inhibitor components with genes on chromosomes 1B, 1D, 2B, 2D, 3B, 3D, 6B and 6D[4,6,11,12]. Only *Aegilops*-type inhibitors are present in the endosperm of *T. aestivum* with the A genome appearing to be silent for the majority of inhibitor genes. Three types of TI differing in molecular mass (M_r) and properties were found in the genus *Triticum*: *T. monococcum* and other diploid wheats with genome compositions A^bA^b and A^uA^u type (14 kDa), *Aegilops* section Sitopsis species (B^sB^s) type (19 kDa) and *Ae. squarrosa* (DD) type (12 kDa)[11,13,14]. The first type was also present also in the *T. timopheevii* (A^bA^bGG) group of species and *T. zhukovskii* ($A^bA^bA^bA^bGG$) and was similar in M_r and other properties to rye *Secale cereale* (RR) and barley *Hordeum vulgare* (HH)TI. The second TI type was characteristic of the *T. turgidum* (A^uA^uBB) group including *T. durum*. The two latter types were present in T. *aestivum* (A^uA^uBBDD) and showed high intraspecific variation in presence and composition. Inhibitors active against both chymotrypsin and microbial proteinases (C/SI) are the most heterogeneous and variable inhibitor system in wheat endosperm (Figure 2). Their polymorphism is comparable with that of prolamins and can be used in wheat variety identification. Some inhibitor systems of rye *S. cereale* were similar to those of *Ae. squarrosa.* Very high intrapopulational variation for insect AI, endogenous AI/SI and TI inhibitors was typical for cross-pollinated rye species. Mammalian AI (3R, 24 kDa) were genus specific. *Hordeum* species differed in insect AI composition.

The linkage between endosperm TI composition and the resistance of wheat varieties to grain pests with trypsin as their main gut proteinase was found [11].

The approaches used allowed some novel inhibitors to be found in wheat and other *Triticeae* cereals, including endogenous AI with M_r 19.5 kDa[15]; CPIs with M_r about 12 kDa controlled by chromosomes 2B and 2D[12], endosperm TIs in diploid and teraploid wheats[11]; bifunctional insect amylase/trypsin inhibitor in *H. bulbosum*[4]. Some of these have been studied since by other authors but others await characterisation in more detail.

Table 1 Distribution, genome /chromosomal control and variability of endosperm α-amylase and proteinase inhibitors in wheat and related cereals[4,6,10,11,12,13,14,15]

Type	Control	M_r (kDa) G-F	M_r (kDa) SDS	A^bA^b	A^uA^u	B^sB^s	DD	RR	HH	A^bA^b GG	A^uA^u BB	A^bA^b A^bA^b GG	A^uA^u BB DD
TI	A	14	14	+	+	-	-	M	M	+	-	+	-
TI	B	19	19	-	-	+	-	-	-	-	+	-	+
TI	1D	12	12	-	-	-	+	-	-	-	-	-	+/-
TI	3R	14	14	M	M	-	-	+	M	M	-	M	-
C/SI	A, B, D	12-24	6-12	+	+	+	+			+	+!	+	+!
C/SI	1B, 1D		6-12	-	-	-	-	-	-	-	-		+
C/SI								+	+				
CI	7DS	~26	~14	-	-	-	+	M, pI	-	-	-	-	+
CPI	A,2B,2D	12		+	+	+	+	+	+	+	+	+	+
EA/SI	A,B,D		19.5	+	+	+	+	M!	M	+	+	+	+
IAI	A^b	12	12	+r/-	-	-	-	-	-	-	-	-	-
IAI	B^s,B,6B	12	12	-	-	+!	-	-	-	+/-	+/-	-	+/-
IAI a&b	D, 6D	12	12	-	-	-	+	PI+/-	-	-	-	-	+
ImAI	A^b	24	12	+r/-	-	-	-	-	-	-	-	-	-
ImAI	A^u	24	12	-	+!	-	-	-	-	-	-	-	-
ImAI	B^s	24	12	-	-	+	-	-	-	PI	PI	pI	PI
ImAI	B, 3B	24	12	-	-	pI	-	-	-	PI	+	+	+
ImAI	D, 3D	24	12	-	-	-	+	-	-	-	-	-	+
ImAI	R, 3R	24	12	-	-	-	-	+	-	-	-	-	-
IA/TI	H^b	12	+	-	-	-	-	-	+	-	-		
TI (leaf and root)	A,B,D,3A, 1DS	6-18	6-18	+	+	+	+			+	+!		+!

"+", presence; "-", absence; +/-, intraspecific variability by presence or absence; +r, presence in rare accessions; !, high level of intraspecific polymorphism; M or pI, presence of inhibitor similar by M_r or isoelectric point; i, m & en, insect, mammalian or grain endogenous enzymes correspondingly.

4 CONCLUSIONS

1. The spectra of various inhibitors in *Triticeae* Dum. cereals can be specific for lines, single biotypes, varieties, species, groups of species, genomes, and genera.
2. The Inhibitor spectra reflect evolutionary links between diploid and polyploid *Triticum* and *Aegilops* species.
3. The inhibitors can be used both as markers and to confer plant resistance to pests.
4. Methods of detection of amylase and proteinase inhibitors developed for cereals[6,9,10,12] are applicable for representatives of *Solanaceae*, *Fabaceae*, *Compositae* and other plant families and can be used to identify novel inhibitor types[7,8,4,16]
5. Inhibitors can be effectively used in studies of plant diversity, evolution and plant-parasite co-evolution in combination with other protein and DNA markers.

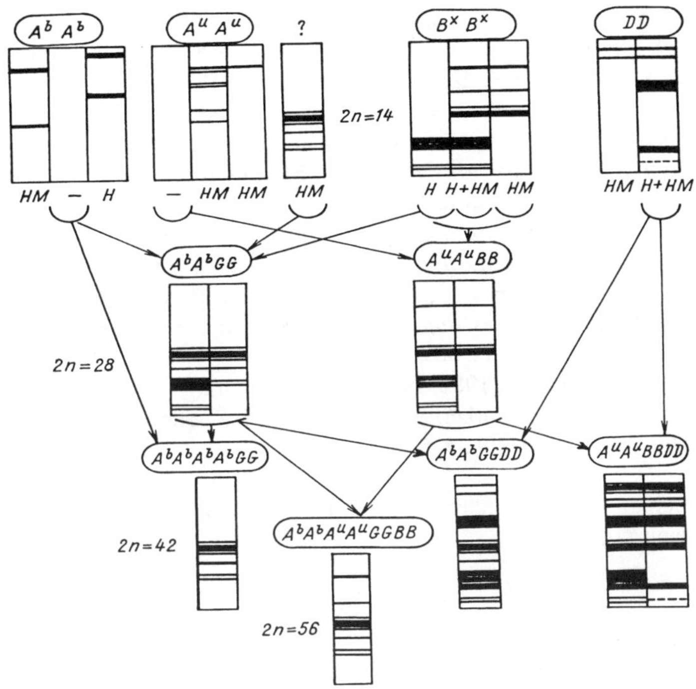

Figure 1 The evolution of insect(H) and mammalian (M) α-amylase inhibitor systems in genus *Triticum* L. (on results of IEF of seed proteins followed by detection of inhibitors).

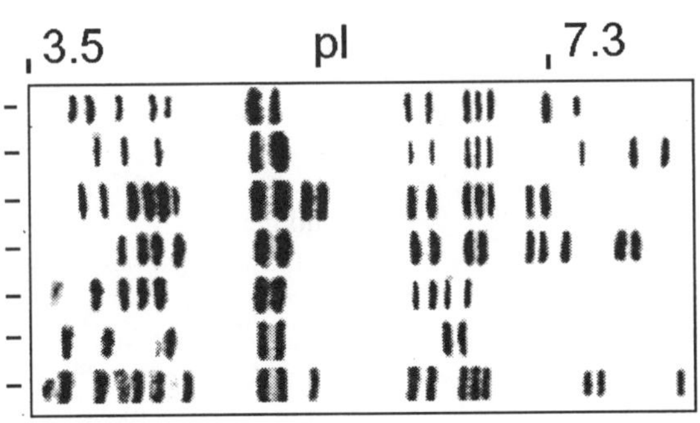

Figure 2 *The polymorphism of chymotrypsin/subtilisin inhibitors in several bread wheat varieties. Endosperm proteins were extracted with water and separated by isoelectric focusing in PAG. Protein bands were transferred from gel to the gelatine replica by diffusion and the latter was developed by chymotrypsin*

References

1. P.R. Shewry, In *Seed Proteins,* ed. P.R. Shewry and R. Casey, Kluwer Acad. Publishers, NL, 1999, p.587.
2. C.A. Ryan, *Annual. Rev. Phytopathol.*, 1990, **28**, 425.
3. P.R. Shewry, and J.A. Lucas, *Adv Bot.Res.*, 1997, **26**, 135.
4. Al. V. Konarev, *Euphytica,* 1996, **92**, 89.
5. V. Buonocore, T. Petrucci., V. Silano, *Phytochemistry,* V. 1977, **6**, 811.

6.Al. V. Konarev, *Soviet Agricultural Sciences*, 1982, **6**, 68.
7. Al. V. Konarev, I. N. Anisimova, V. A. Gavrilova, V.T. Rozhkova, R. Fido, A.S. Tatham and P.R. Shewry, *Theoretical and Applied Genetics*, 2000, **100**, 82.
8. Al. V. Konarev, N. Tomooka & D. A. *Euphytica*, 2000, (accepted, in press)
9. Al. V. Konarev and Yu.V. Fomicheva, *Biochemistry (Moscow)*, 1991, **56**, 628.
10. Al. V. Konarev, *Biochemistry (Moscow)*, 1986, **51**, 195-201.
11. Al. V. Konarev, *Soviet agricultural biology,*: Part 1 : *Plant biology,* 1987, **3,** 17.
12. Al. V. Konarev, *Soviet agricultural sciences* (*Doklady VASKhNIL*1984, **10**, 13).
13. Al. V. Konarev, *Soviet agricultural biology, Plant biology,* 1992, **5**, 10.
14. Al. V. Konarev, I. N. Anisimova, V. A. Gavrilova, and P. R. Shewry, In *Genetics and Breeding for Crop Quality and Resistance,* ed. G.T. S. Mugnozza, E. Porceddu and M.A. Pagnotta, Kluwer Acad. Publishers, NL, 1999, p. 135.
15. Al.V. Konarev, Bulletin VIR, 1982, **118,** 11 (In Russian).
16. S. Luckett, R.S. Garcia, J.J. Barker, Al. V. Konarev,, P.R. Shewry, A.R. Clarke and R.L. Brady, *J Mol. Biol.*, 1999, **290**, 525.

Acknowledgements: Author is grateful to Prof. P.R. Shewry for discussions and advice in preparing the manuscript.

PRODUCTION OF HEXAPLOID AND TETRAPLOID WAXY LINES

M.Urbano[1], B. Margiotta [1], G.Colaprico[1], D. Lafiandra [2]

1. Germplasm Institute, C.N.R., via Amendola 165/A, 70126 Bari, Italy. 2. Department of Agrobiology and Agrochemistry, via S.C. De Lellis, 01100, Viterbo, Italy.

1 INTRODUCTION

Waxy proteins are a group of proteins coded by genes located on chromosomes 7A, 7D and 4A (*Wx-A1*, *Wx-D1*, *Wx-B1* loci) of cultivated wheat which are responsible for amylose synthesis in the starch endosperm[1]. The identification of null genotypes at each of the three waxy loci has been reported by different research groups[2,3,4] and in the case of the bread wheat cultivars null at the *Wx-B1* locus better noodle making quality has been reported[5].

Electrophoretic separations carried out on different varieties, lines and accessions of bread and durum wheat, have permitted the detection of new null alleles at each of the *Wx-1* loci[6]. The amplification of waxy gene fragments by the polymerase chain reaction (PCR) technique has been performed on the same materials in an attempt to detect further polymorphism and to ascertain the cause of the lack expression of the null genotypes. Furthemore, the single null forms detected at each locus have been used to produce bread and durum wheat lines with different level of deficiency for these proteins in order to develop material to be used to understand the influence of different amylose content on the technological and qualitative properties of flour and semolina.

2 MATERIAL AND METHODS

2.1 Materials

Seeds of the bread wheat cv. Chinese Spring, corresponding nulli-tetrasomic lines together with four bread wheat nulls at the *Wx-B1* and two nulls at the *Wx-D1* loci have been used for PCR analyses.

Seeds of the durum wheat cv. Langdon, a durum wheat null at the *Wx-A1* locus and three polymorphic variants detected at the *Wx-B1* locus were also used.

Crosses were performed among the different null forms detected both in durum and bread wheat.

2.2 Methods

2.2.1 *Electrophoretic analyses.* Waxy proteins were extracted according to the method described by Zhao and Sharp[7]. Electrophoretic separation was performed on SDS-PAGE[8].

2.2.2. *Polymerase chain reaction analyses.* Genomic DNA was isolated from 150 mg of leaves as reported by Asemota[9] and PCR reactions have been performed using a pair of primers, prepared on the basis of published sequences of waxy genes of bread wheat cv. Chinese Spring[10], with the following sequences: a) 5' ACTTCCACTGCTACAAGCGCGGGGT 3'; b) 5' GCT GAC GTC CAT GCC GTT GAC GAT G 3'. Amplified product were analyzed on 8% acrylamide gel in TBE system.

3 RESULTS

3.1 Polymerase Chain Reaction (PCR) Analyses

The chromosomal location of the three waxy gene fragments, obtained through PCR amplification, in the bread wheat cv. Chinese Spring, is reported in Figure 1a. The absence of the fragments in the nulli-7D tetra-7A, nulli-7A tetra-7D and nulli-4A tetra-4D lines indicates that the fragment with slowest mobility (1017 bp) is controlled by the waxy gene located on the chromosome 7D, the fragment with intermediate mobility (953 bp) by the waxy gene present on chromosome 7A and the fragment with faster mobility (935 bp) by the gene on chromosome 4A.

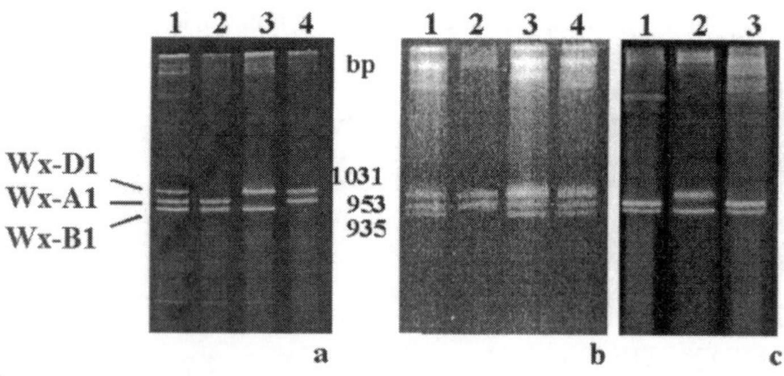

Figure 1: *electrophoretic separations on acrylamide gel (8%) of amplified waxy gene fragments in: a) bread wheat cv. Chinese Spring (1) and in the nulli-tetrasomic lines nulli-7D tetra-7A (2), nulli-7A tetra-7D (3), nulli-4A tetra-4D (4); b) cv. Chinese Spring (1) and bread wheat accessions lacking the waxy protein present at the Wx-B1 (2,3) and Wx-D1 (4) loci; c) durum wheat cv. Langdon (1) and durum wheat accessions lacking the waxy protein present at the Wx-A1 locus (2) and with polymorphic protein at the Wx-B1 locus (3).*

Three of the four bread wheat accessions characterized by the absence of the waxy protein at the *Wx-B1* locus and the two bread wheat accessions lacking the waxy protein encoded by the *Wx-D1* gene showed a PCR pattern similar to that present in Chinese Spring (Figure 1b: lanes 3 and 4). In the fourth genotype (Figure 1b: lane 2) the amplified fragment associated with the *Wx-B1* locus was absent.

In durum wheat a line characterized by the absence of the protein associated with the *Wx-A1* locus showed the amplified fragment related to the corresponding gene (Figure 1c: lane 2); but the size of the fragment was larger compared to the fragment normally present at this locus. Three accessions of durum wheat which exhibited polymorphism at the *Wx-B1* locus when analyzed by SDS-PAGE yielded an amplified fragment of normal size (Figure 1c: lane 3).

3.2 Production of waxy bread and durum wheats

A crossing program has allowed the combination of different null alleles with the production of lines characterized by the absence of two or three waxy proteins, both in durum and bread wheat.

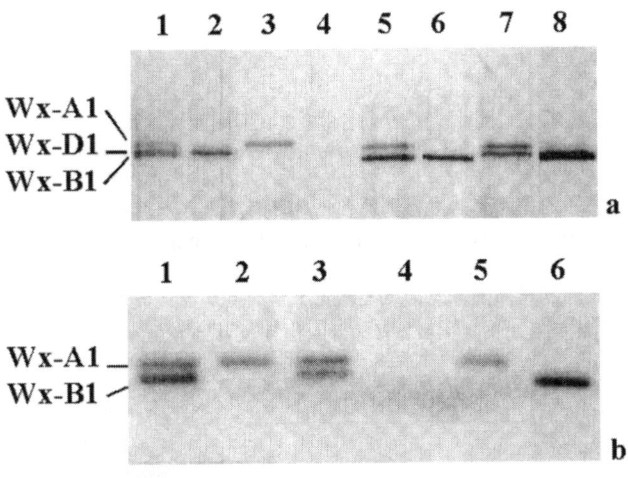

Figure 2: *SDS-PAGE separations of* ***a)*** *waxy proteins in bread wheat cv. Chinese Spring (1) and null combinations in a bread wheat segregating population* Wx-A1 *and* Wx-B1 *(2),* Wx-D1 *and* Wx-B1 *(3),* Wx-A1, Wx-B1 *and* Wx-D1 *(4),* Wx-D1 *(5),* Wx-A1 *and* Wx-D1 *(6);* Wx-B1 *(7),* Wx-A1 *(8);* ***b)*** *waxy proteins in durum wheat cv. Langdon (1) and null combinations in a durum wheat segregating population* Wx-B1 *(2,5),* Wx-A1 *and* Wx-B1 *(4),* Wx-A1 *(6).*

A segregating bread wheat population was analyzed by SDS-PAGE and all the eight possible null combinations for waxy proteins at *Wx-1* loci detected are reported in Figure 2a. Similarly, electrophoretic analyses of a durum wheat segregating population, allowed the identification of all the three possible null combinations at each of the *Wx-A1*, *Wx-B1* and *Wx-A1/Wx-B1* loci (Figure 2b).

4 CONCLUSION

It has been established that waxy protein polymorphism in wheat is not very high, especially when compared with other groups of proteins such as the storage proteins of wheat kernels. In fact, only one null type at the *Wx-D1* locus was detected in bread wheat[2] while no null form at the *Wx-A1* locus was found in durum wheat[3]. Screening of further material has permitted the detection of two bread wheat lines lacking the waxy protein at the *Wx-D1* locus and one durum wheat line carrying a null allele at the *Wx-A1* locus. These materials have been used as parents to produce waxy bread and durum wheat lines.
Production of isogenic lines possessing low amylose content in cultivated bread, but especially durum wheat varieties, will be useful to understand the relationship between amylose content and durum wheat processing properties.

References

1. J. Preiss, in *Oxford Surveys of Plant Molecular and Cell Biology*, ed. B. J. Miflin, Oxford University Press, Oxford , 1991, Chapter 7, p. 59.
2. M. Yamamori, T. Nakamura and T. R. Endo, T. Nagamine, *Theor. Appl.Genet*, 1994, **89**, 179.
3. M. Yamamori, T. Nakamura and T. Nagamine, *Plant Breed.*, 1995, **114**, 215.
4. R. A. Graybosh, C. J. Peterson, L. E. Hansen, S. Rahman, A. Hill and J. H. Skerrit, Cereal Chem., 1998, **75**, 162.
5. X. C. Zhao, I. L. Batey, P. J. Sharp, G. Crosbie, I. Barclay, R. Wilson, M.K. Morell and R. Appels, *J. Cereal Sci.*, 1998, **27**, 7.
6. M. Urbano, G. Colaprico and B. Margiotta, in *Proc. 6 th Int. Gluten Workshop*, ed. C. W. Wrigley, Cereal Chemistry Division, Royal Australian Chemical Institute, Melbourne, 1996, p. 66.
7. X. C. Zhao and P. J. Sharp, *J. Cereal Sci.*, 1996, **23**, 191.
8. P. I. Payne, L. M. Holt, E. A. Jackson and C. N. Law, *Theor. Appl. Genet.*, 1981, **60**: 229.
9. H. N. Asemota, *Plant Mol. Biology Reporter*, 1995, **13**, 214.
10. J. Murai, T. Taira and D. Ohta, *Gene*, 1999, **234** ,71.

OAT GLOBULINS IN REVERSED SDS-PAGE

Tuula Sontag-Strohm

University of Helsinki, Department of Food Technology, PO Box 27, 00014 University of Helsinki.

1 INTRODUCTION

The main storage protein in oats is globulin comprising about 75% of the total oat protein[1]. Prolamins, which are the main storage proteins in wheat, barley and rye, account for only 10% of the total oat protein[1]. Oat globulin is only partly soluble in salt solution and, therefore, mostly remains in the residue when globulins are extracted using the Osborne method. The residual globulin can be extracted by detergent-alkali or urea solutions combined with reducing agent.

Native oat globulin is a hexamer with molecular weight of 322 000. The M_r of globulin subunit is about 50,000, which can be reduced to give two subunit chains of M_r 20,000 and 30,000. The unreduced salt soluble globulin shows a subunit group at M_r 50,000 by SDS-PAGE, but the unreduced detergent-alkali globulin shows all the three groups of bands of M_r 20,000, 30,000 and 50,000[2]. The reduced globulin extracts show only the two smallest subunit groups on SDS-PAGE. The aim of this work was to characterise the largest oat globulin extracted by salt solution and detergent solution (without alkali) by reversed SDS-PAGE. In this method, the largest oat globulin could be separated after removal of smaller globulin subunit groups.

2 MATERIALS AND METHODS

Milled oat groats were defatted with 100% acetone and extracted sequentially with water, 1 M NaCl, 55% propan-2-ol and 1.5% SDS. The residue was extracted with 1.5% SDS containing 1% DTT.

The solutions of the samples were placed in a sample gel with 10.3% acrylamide and 1.3% crosslinker containing 0.4 M Tris-HCl, pH 6.8, and 0.6% SDS[3]. The oat globulin subunits migrate in this gel, but the largest globulins remain immobile until reduced. The polarity of electrodes was arranged so that the subunits migrated out (for 1 h) into the upper buffer reservoir. The electrode polarity was then reversed, reducing agent was added to the sample gel, and the run continued for 1.5 h with the normal electrode polarity for 12 % SDS-PAGE. The globulin sample extract was also separated in unreduced and reduced forms in 12% SDS-PAGE ready gels (Mini-PROTEAN II ready gels, Bio-Rad).

3 RESULTS AND DISCUSSION

SDS-PAGE of the unreduced salt and SDS extracts both showed an oat globulin subunit group at M_r 50,000 but the SDS extract showed also the smaller oat globulin subunit groups at M_r 20,000 and 30,000 (Figure 1). This is in agreement with the separation of the unreduced SDS-alkali extract[2]. In addition, the SDS-PAGE separation of the unreduced SDS extract (Figure 1) showed that the largest proteins remained in the bottom of the sample well. However, the reversed SDS-PAGE separation (Figure 2) showed that both globulin extracts had large globulin molecules that remained in the gel after removal of smaller globulin subunits. Reduction showed that these large globulins were composed of two subunit groups of M_r 20,000 and 30,000.

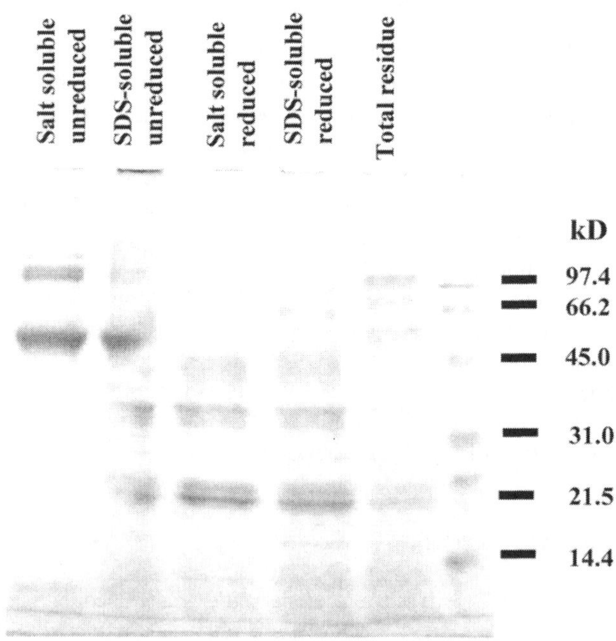

Figure 1 *SDS-PAGE of 1M NaCl soluble and 1.5% SDS-soluble oat globulins. The "SDS-soluble unreduced" track (second from left) shows partial reduction by diffusion of reducing agent from the adjacent "salt soluble reduced" track. As a result both reduced and unreduced forms of the proteins are observed.*

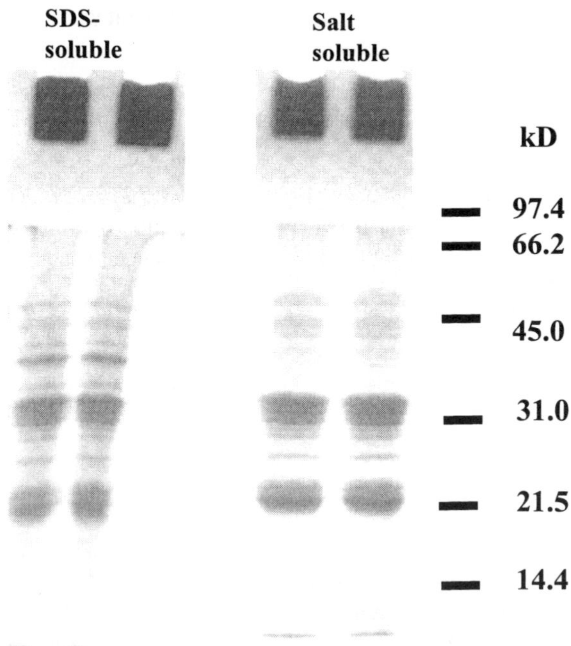

Figure 2 *Reversed SDS-PAGE of salt soluble and SDS-soluble oat globulin.*

The largest unreduced form of oat globulin could be extracted with 1 M NaCl solution as well as with 1.5% SDS solution. The reduced form consisted of two subunit groups of M_r 20,000 and 30,000. The smallest oat globulin subunit groups could be extracted with 1.5% SDS solution without reducing agent but not with 1 M NaCl. Oat groat contained three forms of oat globulin, the large unreduced protein, the medium M_r group (50,000) corresponding to the unreduced subunits and the smallest M_r groups (20,000 and 30,000) corresponding to the reduced subunit chains.

References

1. Peterson, D.M. & Brinegar, A.C. 1986 In oats: Chemistry and Technology, F. H. Webster, ed. AACC, St.Paul, MN, pp.153-203.
2. Lapveteläinen, A., Bietz, J.A. & Huebner, F.R. 1995. *Cereal Chem.* **72**: 259.
3. Sontag-Strohm, T. 1996 *J. Cereal Sci.* **24**: 87.

PUROINDOLINES: STRUCTURAL RELATIONSHIPS WITH TRYPTOPHANINS (AVEINDOLINES) FROM OAT (*AVENA SATIVA*).

M.A. Tanchak[1], and I. Altosaar[2]*

1. University College of Cape Breton, Dept. of Behavioural and Life Sciences, P.O. Box 300, Sydney, Nova Scotia, CANADA BlP 6L2. 2. University of Ottawa, Faculty of Medicine, Dept. of Biochemistry, Microbiology & Immunology, 40 Marie Curie Private, Ottawa, ON, CANADA KlN 6N5.
*Author to whom correspondence should be addressed. altosaar@science.uottawa.ca

1. INTRODUCTION

Puroindolines (PINs) are small, basic, cysteine-rich, wheat seed proteins possessing a distinctive tryptophan-rich domain[1]. PINs appear to be major components of the "grain softness protein" complex[2]. The amount of PINs associated with water-washed starch granules from wheat flour shows a strong correlation with the grain hardness of the wheat variety and with the baking properties of the flour. Generally speaking, flour from hard wheat varieties has no or small amounts of PIN associated with the starch granules and produces high quality bread dough. In contrast, flour from soft wheat varieties has larger amounts of starch-bound PIN and is used in the production of cookies, cakes and pastries.

PINs are divided into two different types, PIN-a and PIN-b, which differ with respect to the nature of the tryptophan-rich domain[3]. PIN-b has a shorter domain with fewer tryptophan residues and fewer positively charged amino acids. These differences are correlated with differences in the lipid-binding properties of the two types of PINs[4]. For example, PIN-a binds tightly to both phospholipids and glycolipids while PIN-b binds tightly only with negatively charged phospholipids and binds loosely with glycolipids.

In our study of oat (*Avena sativa* L.) seed proteins, we have identified the oat homologue to PIN, oat tryptophanin (OT) or, using the terminology of Gautier and co-workers, aveindoline[3]. Here we provide a description of OTs and discuss the similarities and differences between these proteins and PINs.

2. MATERIALS AND METHODS

2.1 Plant Material and Preparation of Clones

RNA and DNA were isolated from cultivated oats (*Avena sativa* L. cv. Hinoat). Etiolated seedlings, from surface-sterilized seeds germinated under sterile conditions, were used for the isolation of genomic DNA. Mid-maturation developing seeds were used for the isolation of RNA. RNA and DNA isolation were performed as previously described[5].

The cloning of 3B3-5, 3B3-7, 3B3T-3 and 3B3T-5 cDNAs has been described in detail[5]. Briefly, 3B3-5 and 3B3-7 were obtained by screening a λgt10 oat seed cDNA library with a probe specific for the 5' region of λ3B3[6]. 3B3T-3 and 3B3T-5 were products of a reverse transcriptase-polymerase chain reaction and were cloned into pGEM4Z[5,7].

The clone 3B3-1D was obtained from a polymerase chain reaction using oat genomic DNA and primers specific for the 5' and 3' ends of 3B3-5 coding sequence.

2.2 Sequence Alignments

Alignments were prepared using the MegAlign and Align programs from the DNASTAR software package. Default settings were used for all alignment parameters.

3. RESULTS AND DISCUSSION

3.1 Oat Tryptophanins

Four cDNA clones, 3B3-5, 3B3-7, 3B3T-3 and 3B3T-5, were isolated by screening of seed-derived cDNA libraries (3B3-5 and 3B3-7) and by the use of reverse transcriptase-polymerase chain reactions (3B3T-3 and 3B3T-5) (5). An additional genomic coding sequence (3B3-1D) has been amplified using conventional PCR. The derived amino acid sequences for these clones are shown in Fig. 1. 3B3-5, 3B3-7 and 3B3-1D are very similar sequences and probably represent allelic variants or very closely related members of a multi-gene family. 3B3T-3 and 3B3T-5 code for the identical amino acid sequence and differ only in the 5' and 3' untranslated regions of the cDNAs.

Alignments of the 3B3 and 3B3T derived amino acid sequences indicate that they are related proteins. For example, 79 of the 142 amino acids in the 3B3T clones are matched in an alignment with 3B3-5 and another 33 represent conservative amino acid substitutions[5]. Therefore, 3B3 and 3B3T clones most likely represent different members of a gene family.

Interestingly, in the alignment of 3B3 and 3B3T derived amino acid sequences, amino acid substitutions or changes occur throughout the length of the amino acid sequences except for one specific region where a 14 amino acid residue, tryptophan-rich sequence is conserved (Figure 1). This sequence is also encoded by the λ3B3 clone[6] and by the AV1 clone isolated from the wild oat, *Avena fatua*[8].

3.2 Puroindolines vs. Oat Tryptophanins

PINs are classified into two types, PIN-a and PIN-b. One characteristic that distinguishes the two types of PINs is the nature of the tryptophan-rich domain. Experimental evidence indicates that the structure and composition of this domain has a major influence on the lipid-binding properties of the PINs[9]. PIN-a has a longer domain (WRWWKWWK) with five tryptophans and three basic residues whereas PIN-b has a shorter domain (WPTKWWK) with only three tryptophans and two basic residues[3]. In all OT sequences cloned to date, including the AV1 clone from *A. fatua*, the corresponding sequence is WPWKWWK. This sequence is intermediate in composition relative to the two PINs with four tryptophans and two basic residues. At present, there is no information available about the lipid-binding properties of OTs.

3B3

```
MKIFFFLALLALVVSATFAQYVESDGSYEEVEGAHDRC   38
                                 S
                                 P

QQHQMKLDSCREYVADGCTTMRDFPITWPWKWWKGGCE   76
              ER

EVRNECCQLLGQMPSECRCDAIWRSIQHELGGFFGTQQ 114
             W
             W

GLIGKRLKIAKSLPTQCNMGPECNIPVTFGYYW      147
```

3B3T

```
MKALFLLAFLALAASAAFAQQYADTGVGGWDGCMPEKA   38

RLNSCKDYVVERCLTLKDIPITWPWKWWKGGCESEVRS   76

QCCMELNQIAPHCRCKAIWRAVQGELGGFLGFQQSEIM 114

KQVHVAQSLPSRCNMGPNCNFPTNLGYY           142
```

Figure 1 *Derived amino acid sequences for 3B3 and 3B3T clones. Amino acids are represented by their single letter codes. Sequence with single underline corresponds to the 3B3 N-terminal sequence obtained by Fabijanski and co-workers (6). Conserved 14 amino acid, tryptophan-rich sequence is indicated by double underlines. The complete sequence for 3B3-5 is shown. Other 3B3 sequences are shown only where they differ from 3B3-5. Letter in italic- amino acid found in Fabijanski's N-terminal sequence. Bold letters- amino acids found in 3B3-7 sequence. Underlined letters- amino acids found in 3B3-1D sequence.*

What is the relationship of the OTs to the PINs? Figure 2 shows the results of a multiple sequence alignment including two OT sequences, 3B3-7 and 3B3T, and one example for each of PIN-a and PIN-b. Table 1 summarizes the results of the alignment of different combinations of paired sequences. From these alignments, it is clear that all of these proteins share significant levels of amino acid sequence identity (Figure 2 and Table 1). In particular, the location of the ten cysteine residues is highly conserved.

```
    M K A L F L L A L L A L V A S A A F A Q  Majority
                      10                  20
1   M K I F F F L A L L A L V V S A T F A Q  3B3
1   M K A L F L L A F L A L A A S A A F A Q  3B3T
1   M K A L F L I G L L A L V A S T A F A Q  PIN-a
1   M K T L F L L A L L A L V A S T T F A Q  PIN-b

    - Y S E - - V G G S Y - E V X G G G G X  Majority
                      30                  40
21  - Y V E - - S D G S Y - E - E V E G P H  3B3
21  Q Y A D T G V G G W - - - - - - - - - -  3B3T
21  - Y S E - - V V G S Y - D V A G G G G A  PIN-a
21  - Y S E - - V G G W Y N E V G G G G G S  PIN-b

    D Q C P Q E X X K L N S C K D Y V X D R  Majority
                      50                  60
36  D R C Q Q H Q M K L D S C R E Y V A D G  3B3
31  D G C M P E K A R L N S C K D Y V V E R  3B3T
37  Q Q C P V - E T K L N S C R N Y L L D R  PIN-a
38  Q Q C P Q E R P K L S S C K D Y V M E R  PIN-b

    C X T M K D F P V T W P W - K W W K G G  Majority
                      70                  80
56  C T T M R D F P I T W P W - K W W K G G  3B3
51  C L T L K D I P I T W P W - K W W K G G  3B3T
56  C S T M K D F P V T W R W W K W W K G G  PIN-a
58  C F T M K D F P V T W P T - K W W K G G  PIN-b

    C E - E V R X E C C X X L G Q I A P Q C  Majority
                      90                 100
75  C E - E V R N E C C Q L L G Q M P W E C  3B3
70  C E S E V R S Q C C M E L N Q I A P H C  3B3T
76  C Q - E L L G E C C S R L G Q M P P Q C  PIN-a
77  C E H E V R E K C C K Q L S Q I A P Q C  PIN-b

    R C D A I W R S I Q G E L G G F L G F Q  Majority
                     110                 120
94  R C D A I W R S I Q H E L G G F F G T Q  3B3
90  R C K A I W R A V Q G E L G G F L G F Q  3B3T
95  R C N I I Q G S I Q G D L G G I F G F Q  PIN-a
97  R C D S I R R V I Q G R L G G F L G I W  PIN-b

    Q G E I X K Q L Q X A Q S L P S R C N M  Majority
                     130                 140
114 Q G L I G K R L K I A K S L P T Q C N M  3B3
110 Q S E I M K Q V H V A Q S L P S R C N M  3B3T
115 R D R A S K V I Q E A K N L P P R C N Q  PIN-a
117 R G E V F K Q L Q R A Q S L P S K C N M  PIN-b

    G P X C N I P X T X G Y Y W              Majority
                     150
134 G P E C N I P V T F G Y Y W              3B3
130 G P N C N F P T N L G Y Y                3B3T
135 G P P C N I P G T I G Y Y W              PIN-a
137 G A D C K F P S - - G Y Y W              PIN-b
```

Decoration 'Decoration #1': Box residues that match the Consensus exactly.

Figure 2 *Alignment of Ots and PINs. 3B3 is represented by 3B3-7. PIN-a: accession number CAA49538. PIN-b: accession number CAA49537*

Table 1: *Summary of Sequence Alignment Results.*

Clones or Accession No. (Protein Type)	Similarity Index (%)	Gaps	Gap Length (nucleotides)
3B3-7 vs 3B3T (OT) (OT)	52.3	3	8
3B3-7 vs CAA49538 (OT) (PIN-a)	53.9	3	3
3B3-7 vs CAA49537 (OT) (PIN-b)	54.9	3	5
3B3-7 vs CAB65472 (OT) (PIN-b)	53.6	3	5
3B3T vs CAA49538 (OT) (PIN-a)	49.0	4	9
3B3T vs CAA49537 (OT) (PIN-b)	57.9	3	9
3B3T vs CAB65472 (OT) (PIN-b)	55.9	3	9
CAA49538 vs CAA49537 (PIN-a) (PIN-b)	56.4	5	6
CAA49538 vs CAB65472 (PIN-a) (PIN-b)	58.3	5	6

Based on the alignment results shown in Table 1, it is not possible to make a definitive statement about the relationship between the OTs and the PINs. For example, when comparing the 3B3-7 OT to the two PINs, the calculated similarity indices are 53.9 with PIN-a and, depending on which PIN-b is used, 54.9 or 53.6 with PIN-b. Clearly, there is no basis for claiming that the 3B3-7 OT is more "PIN-a like" than "PIN-b like" or vice-versa. In the case of the 3B3T OT sequence, the similarity indices appear higher with PIN-b (57.9 and 55.9) than with PIN-a (49.0). The significance of this observation is unclear as the alignments of PIN-a with PIN-b also yield relatively high similarity indices of 56.4 and 58.3. Therefore, it would seem premature to consider 3B3T as the oat homologue of PIN-b. Copy number reconstruction experiments in our lab[5] indicate that the OTs belong to a potentially large gene family. Perhaps, this gene family has evolved in a divergent manner from the PIN gene family in wheat. If this is the case, there may not be a direct or precise correlation between members of the oat and wheat gene families. Alternatively, the OTs may represent genes whose homologues have not yet been identified in wheat.

4. SUMMARY

(1) A number of clones for OTs have been isolated and sequenced.

(2) OTs share significant levels of sequence identity and similarity with the PINs of wheat. OTs are the oat homologues of the PINs.
(3) OTs have a distinctive tryptophan-rich domain that is intermediate in composition to the corresponding domains of PIN-a and PIN-b.
(4) Based on amino acid sequence alignments, it is not possible to assign the 3B3 and 3B3T OTs to the PIN-a or PIN-b groupings in the PIN gene family.
(5) The OTs most likely represent members of the PIN/OT gene family that either do not exist in wheat or have not yet been identified in wheat.

References

1 J.-E. Blochet, C. Chevalier, E. Forest, E. Pebay-Peyroula, M.-F. Gautier, P. Joudrier, M. Pézolet and D. Marion, *FEBS Lett.*, 1993, **329**, 336
2 M.J. Giroux and C.F. Morris, *Proc. Natl. Acad. Sci. USA*, 1998, **95**, 6262
3 M.-F. Gautier, M.-E. Aleman, A. Guirao, D. Marion and P. Joudrier, Plant *Mol. Biol.*, 1994, **25**, 43
4 L. Dubreil, J.-P. Compoint and D. Marion, *J. Agric. Food Chem.*, 1997, **45**, 108
5 M.A. Tanchak, J.P. Schernthaner, M. Giband and I. Altosaar, *Plant Science*, 1998, **137**, 173
6 S. Fabijanski, S.-C. Chang, S. Dukiandjiev, M.B. Bahramian, P. Ferrara and I. Altosaar, *Biochem. Physiol. Pflanzen*, 1988, **183**, 143
7 M.A. Tanchak, M. Giband, B. Potier, J.P. Schernthaner, S. Dukiandjiev and I. Altosaar, *Genome*, 1995, **38**, 627
8 R.R. Johnson, M.E. Chaverra, H.J. Cranston, T. Pleban and W.E. Dyer, *Plant Mol. Biol.*, 1999, **39**, 823
9 M. Kooijman, R. Orsel, M. Hessing, R.J. Hamer, and A.C.A.P.A. Bekkers, *J. Cereal Sci.*, 1997, **26**, 145

Acknowledgements

This work was supported by a research grant from the Natural Sciences and Engineering Research Council of Canada. M.A.T. gratefully acknowledges financial support from the Research Evaluation Committee of the University College of Cape Breton.

Subject Index